THE HOUSE OF SAUD

THE HOUSE OF SAUD

The Rise and Rule of the Most Powerful Dynasty in the Arab World

by

David Holden

Richard Johns

Holt, Rinehart and Winston
New York

Copyright © 1981 by Richard Johns and the
Estate of David Holden
Chapter 25, 'The Return of the Ikhwan', copyright
© 1981 by James Buchan
Picture research by James Buchan and Anne Horton
First published in the United States in 1982 by Holt, Rinehart
and Winston, 383 Madison Avenue, New York, New York 10017.

Library of Congress Cataloging in Publication Data
Holden, David.
The house of Saud.
1. Saudi Arabia—History.   I. Johns, Richard.
II. Buchan, James.   III. Title.
DS244.52.H64 1981   953'.8   81-47474
ISBN 0-03-043731-8   AACR2

Printed in the United States of America
10 9 8 7 6 5 4 3 2

Foreword
by Lord Trevelyan

WE shall probably never know why David Holden died. Certainly, there is no reason why he should have earned the hostility of any party in the Middle East. Nor is there any apparent reason why any group should think it to its political advantage to murder him. That he was murdered is certain, and it was in the period immediately after he left Cairo airport on that fatal evening in December 1977.

All that we must now forget, remembering David's services to his country, to the British press, and to current history and literature, and build on what he left behind. That was, principally, this book, which was unfinished when he died. It has been completed by David's friend and colleague Richard Johns of the *Financial Times*. I feel sure that this is what David would have wished, and its completion is the best memorial we could give him.

What a story it is, the emergence of the unknown Saudi prince from the vast sands of the Empty Quarter to become such a power in a world built on oil. What fascination in the inter-twining of the histories of Britain and Saudi Arabia, in the stories of Captain Shakespear, killed whilst on duty with Ibn Saud, of Abdullah Philby, for many years friend and servant of the Saudi princes, whom I recollect on the last day of his life.

There is much here to ruminate on and many questions affecting our future, which I used to discusss with David. He was one of the finest foreign correspondents. I saw him at work in the Middle East, always calm, always so understanding of the people, whom he knew well. Let this book be dedicated to his memory.

London 1981

Contents

Illustrations

Author's Preface and Acknowledgements

DAVID HOLDEN started this book towards the end of 1976. He devoted a large part of the following year to it, but his work was interrupted in November 1977, when he could not avoid, as chief foreign correspondent of the *Sunday Times,* travelling to the Middle East to cover the immediate aftermath of Sadat's visit to Jerusalem. He went first to Egypt and then to Israel, spending two or three days on the occupied West Bank, before proceeding to Amman, the capital of Jordan, from where he took a flight back to Cairo on 7 December. Concerned that he had not made contact, the newspaper initiated inquiries. David's body was discovered in a city morgue on 10 December. He had been found lying in a roadside ditch on the outskirts of Cairo, shot in the back and stripped of all means of identification. Three and a half years later his death remains a mystery. An investigation by the *Sunday Times* 'Insight' team produced no evidence that pointed with any conclusiveness to who his murderers might have been and was never published. All that is known is that he was met at Cairo airport by persons whom he may or may not have expected to find waiting for him and that the fatal bullet was a 9mm low velocity cartridge fired at point-blank range.

The reporters assigned to the investigation gave some credence to the theory of mistaken identity. The distinguished journalist involved in speculation along these lines has the same first name and initials as David Holden. Earlier in 1977 he had been evicted from Egypt, having gravely offended Sadat. The man in question discounts the possibility of mistaken identity, however. And in fact the killing was not in the style of the Egyptians, not the least because it was perpetrated on their home ground and carried out with a ruthless efficiency. The same two aspects of the murder would tend to rule out any of the Palestinian groups. At no point, I should hasten to add, have I ever felt that

David's death had anything to do with his researches into the House of Saud. I believe that he died merely because he aroused, unjustifiably, the suspicions of some persons in the paranoid world of intelligence and subterfuge of which, I trust and hope, he was no part. The probability is that he was a victim of his indefatigable and forthright curiosity, which did not preclude the bluntest, as well as the most subtle, of questions.

David was amongst those journalists branded, even smeared, with being 'pro-Arab'. Certainly, he understood and to a great extent sympathized with the grievances and preoccupations of the Arabs, but despite his great patience he could be irritated by their time-consuming manners, their oblique evasions and downright lies. Essentially, David was a professional. His speciality and expertise lay in the Middle East, which he had reported and interpreted for two decades. But he was anxious not to become embroiled in and identified with the region at the expense of other international affairs. The pity is that he did not progress further with this book. He was cutting through the clichés covering the West's lack of comprehension of the House of Saud. David had great empathy and the ability to think his way into the minds of others, and as Don McCarthy, a distinguished Arabist and former British diplomat, who greatly admired David, says: 'He built a sound foundation in his earlier years as a correspondent, when he got to know in their relative quiescence many places which were later to hit the headlines. He viewed them and their peoples without prejudice, and when they became news he returned to their affairs with a notably balanced perspective. His writing combined clarity with excellence and was marked by a talent for scene painting which stopped healthily short of the purple passage.'

To turn to details, some points about this book's composition and presentation should be explained. The basic genealogical reference for the Al Saud dynasty, in particular the enumeration of King Abdul Aziz's sons, has been *Burke's Royal Families of the World*, Volume II. From Chapter 11 onwards the basic currency denomination is the dollar, not the least because that is the unit of account for Saudi Arabia's oil revenues. From this point on money figures have been expressed in sterling only when they were officially cited as such. Conversions have been calculated according to the central rate recorded for the end of each year by the International Monetary Fund. As the transliteration of Arabic names has no standard form, spellings adopted in this book are those that the average reader will, hopefully be most accustomed to. Academic presentation, including diacritical marks, has been avoided. In practice, therefore, the usage is on the whole more Levantine than Arabian.

David wrote his text in the first person, and this relates to him up to Chapter 10. Thereafter the first person is myself. David had written some 75,000 words when he died. They covered Chapters 1 to 10 in fairly complete form. He also left unfinished material that has been incorporated into Chapters 11 to 13. The text relating to the 1945 – 53 period (Chapters 11 and 12) was somewhat fragmentary and did not include an account of the last days and death of King Abdul Aziz. But he provided the basic structure and half the content for Chapter 13, which tells the story of King Saud's early years as ruler, an era of decadence and disintegration that seemed to fascinate him as much as it did myself. I had the use of his notebooks, packed with information gleaned during a six-week visit to Saudi Arabia made early in 1977, on which I have drawn heavily for the early part of Chapter 24.

My own labours began seriously in the autumn of 1978. I co-opted James Buchan, a colleague and friend, who lived in Saudi Arabia from 1978 to 1980, to write Chapter 25, 'The Return of the Ikhwan', which recounts the story of the seizure of the Grand Mosque in Mecca in November of that year. A classical scholar of Eton and student of oriental languages at Magdalen College, Oxford, he also revised Chapters 1 to 10, helped to cut chapters 16 to 23, and made a number of constructive suggestions, as well as impertinent remarks, concerning the latter. In addition, he enthusiastically hunted for photographs, some of which are included in this book.

The main responsibility for this writer's present fraught appearance and demented condition rests upon Jane Heller, the book's editrix. Publication does, however, owe much to her whiplash and staying power. Quite apart from being one of the best and most ruthless of her species, she has a heart of some precious substance, possibly gold, has become a trusted friend, and is very adept at Yemenite conversations.

The subject of this book is such that it is neither desirable nor possible to name all those who have assisted in compiling it. Most of the prime non-Saudi informants still have a vested or continuing interest in the Kingdom and could not risk offending its rulers. Similarly, I have had to exercise discretion about the attribution of

information to the House of Saud's subjects, who are very much beholden to their lords and masters. A dozen or so friends have been exceptionally helpful. Amongst those I can name is Don McCarthy, who read and commented on Chapters 1 to 21. He gave much needed comment and advice, as well as reproving me on several occasions. He will not agree with all the interpretations I have given in this book. Several others amongst those I have managed to contact have given me permission to mention their names. John Christie and Ian Seymour have been a constant source of stimulation and encouragement. Patrick Bannerman, a formidable researcher, gave great assistance in checking facts. Thanks also to my own lords and masters at the *Financial Times* for the exceptional indulgence and patience they have shown me over my various absences from duty. I am particularly grateful to J.D.F. Jones, who initially encouraged me to undertake the book and then refused to let me give it up. Ian Davidson also took a friendly interest in the project and spurred me on. My apologies are due to the extra burden thrust, as a result of my distraction, upon those honest troopers of the *F.T.* 'Kamel Korps' – James Buxton, Anthony McDermott, Patrick Cockburn, Roger Matthews, and the newspaper's man in Tel Aviv David Lennon.
    Ahlan ya Mamlakia!

RICHARD JOHNS
London, June 1981

Maps

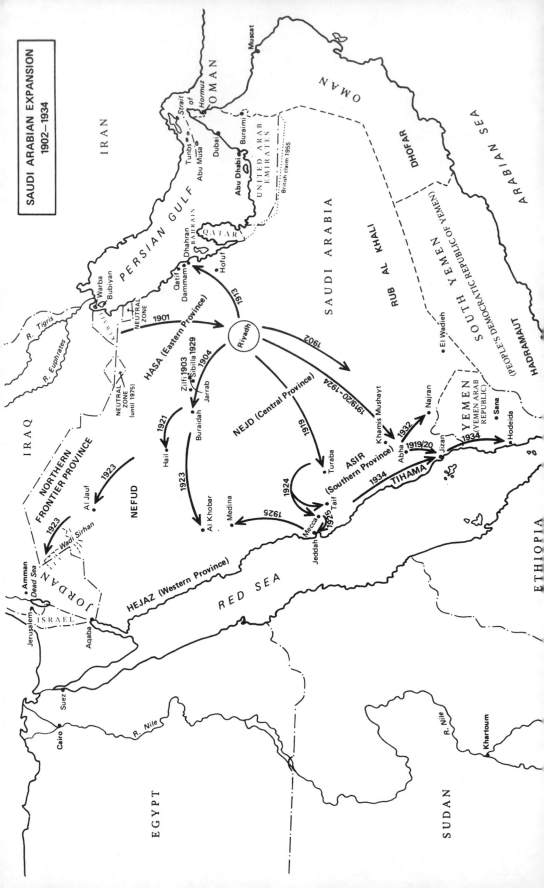

SAUDI ARABIAN EXPANSION
1902–1934

IRAN

OMAN

Strait of Hormuz

Muscat

Tunbs
Abu Musa
Dubai
Buraimi

PERSIAN GULF

Dhahran
Qatif
Dammam
Abu Dhabi
BAHRAIN
QATAR
Hofuf

UNITED ARAB EMIRATES

British claim 1955

OMAN

DHOFAR

ARABIAN SEA

Warba
Bubiyan
NEUTRAL ZONE

R. Tigris

R. Euphrates

1901

HASA (Eastern Province)

1913

Riyadh

SAUDI ARABIA

RUB AL KHALI

El Wadieh

SOUTH YEMEN
(PEOPLE'S DEMOCRATIC REPUBLIC OF YEMEN)

HADRAMAUT

NEUTRAL ZONE (until 1975)

Zilfi 1903
Sibilla 1929
1904
Jarrab

1921

Hail

1923

Buraidah

NEJD (Central Province)

1919

1902

1919/20–1924

Kharmis Mushayt

Turaba

ASIR
(Southern Province)

1932

Najran

Abha

1919/20

YEMEN
(YEMEN ARAB REPUBLIC)

Sana

IRAQ

NORTHERN FRONTIER PROVINCE

1923

NEFUD

Al Jauf

1923

Wadi Sirhan

1923

JORDAN

Amman
Dead Sea

ISRAEL
Jerusalem

Aqaba

1925

Al Khobar
Medina

1924

Mecca
Jeddah

1925
Taif

1934

TIHAMA

Jizan

1934

Hodeida

ETHIOPIA

HEJAZ (Western Province)

RED SEA

Suez

Cairo

R. Nile

EGYPT

R. Nile

Khartoum

SUDAN

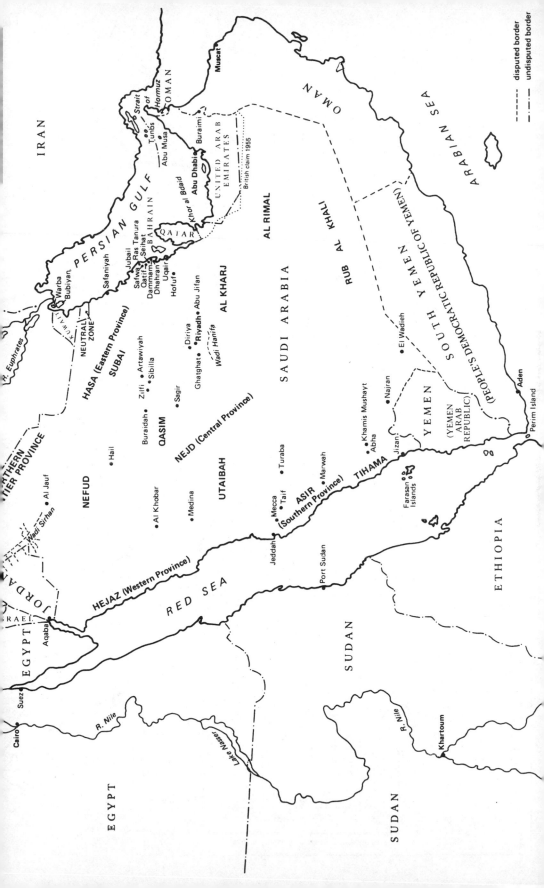

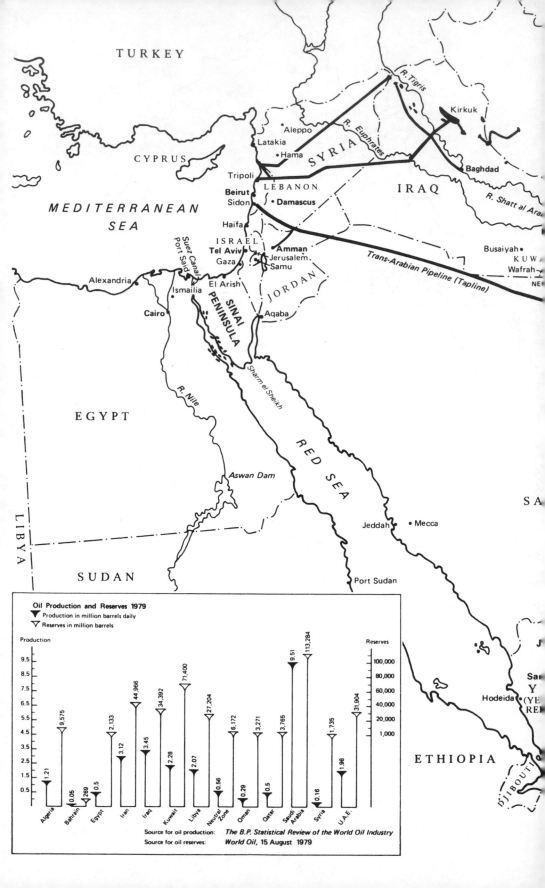

TURKEY

CYPRUS

MEDITERRANEAN
SEA

SYRIA

• Aleppo
Latakia
• Hama

R. Tigris

Kirkuk

R. Euphrates

Baghdad

IRAQ

R. Shatt al Ara

Tripoli

Beirut
Sidon

LEBANON
• Damascus

Haifa

ISRAEL
Tel Aviv
Gaza

• Amman
Jerusalem
Samu

Busaiyah •

KUWA
Wafrah •

NE

Alexandria

Suez Canal
Port Said

Ismailia

El Arish

JORDAN

Trans-Arabian Pipeline (Tapline)

Cairo

SINAI
PENINSULA

Aqaba

EGYPT

R. Nile

Sharm el Sheikh

RED SEA

SA

LIBYA

Aswan Dam

SUDAN

Jeddah

• Mecca

Port Sudan

Sa
Y
RE

Hodeida

(YE
RE

ETHIOPIA

DJIBOUTI

Oil Production and Reserves 1979

▼ Production in million barrels daily
▽ Reserves in million barrels

Production

Country	Production	Reserves
Algeria	1.21	9,575
Bahrain	0.05	269
Egypt	0.5	2,133
Iran	3.12	44,966
Iraq	3.45	34,392
Kuwait	2.28	71,400
Libya	2.07	27,204
Neutral Zone	0.56	6,172
Oman	0.29	3,271
Qatar	0.5	3,765
Saudi Arabia	9.51	113,284
Syria	0.16	1,735
U.A.E.	1.96	31,904

Reserves

100,000
80,000
60,000
40,000
20,000
1,000

9.5
8.5
7.5
6.5
5.5
4.5
3.5
2.5
1.5
0.5

Source for oil production: *The B.P. Statistical Review of the World Oil Industry*
Source for oil reserves: *World Oil, 15 August 1979*

CASPIAN SEA

Tehran

IRAN

es Suleiman

Bushire

PERSIAN GULF

Qatif
Dammam
Dhahran
BAHRAIN

Bahrain

QATAR

Ghawar

ARABIA

Saudi Arabia and U.A.E. 1974

Strait
of
Hormuz

Ras al Khaimah
Sharjah
Dubai

OMAN

GULF OF
OMAN

Muscat

Buraimi

Abu Dhabi

UNITED
ARAB
EMIRATES

Liwa   Zarrara

Shaibah

British claim 1955

AFGHANISTAN

PAKISTAN

OMAN

DHOFAR

SOUTH YEMEN
(PEOPLE'S DEMOCRATIC
REPUBLIC OF YEMEN)

HADRAMAUT

ARABIAN SEA

GULF OF ADEN

SOMALIA

- - - - - disputed border
- · - · - undisputed border
——— pipeline
🌢 oilfield

THE MIDDLE EAST

Political Development

IRAQ

Under British occupation, then mandate,
First World War until 1932; treaty
relations with U.K. 1932–55. Monarchy
under Hashemite dynasty 1921–58; republic
since military revolution of that year.
Intermittent war with Kurdish minority.

KUWAIT

Principality under Al Sabah dynasty.
British protection, formalized in 1899,
until 1961.

EGYPT

Monarchy until 1952, republic since military
revolution of that year. British control
until 1936; treaty relations with U.K.
1936–54. United with Syria
in United Arab Republic 1958–61.

LEBANON

Under French occupation, then mandate,
1918–46; then independent republic.
Intermittent civil war since 1975.

ISRAEL

Created out of Palestine, which was under
British mandate 1920–48.
Independent Jewish State since 1948.
Intermittent war with surrounding
Arab states since 1967.

SYRIA

Under French occupation, then mandate,
1920–46; independent republic from 1946.
United with Egypt in United Arab Republic
1958–61; later took name Syrian Arab Republic.

JORDAN

Kingdom under Hashemite dynasty, formed
by union of Transjordan and part of
Palestine 1948. Treaty relations
with U.K. until 1957.

LIBYA

Under Italian control until 1945, then
under U.N. mandate. Granted independence
1951. Monarchy until 1969; republic
since military revolution of that year.

YEMEN (Yemen Arab Republic)

Independent Imamate since 1918. Republic
proclaimed after revolution of 1962.
Civil war 1962–9.

SOUTH YEMEN (PDRY)

Independent republic formed out of British
colony of Aden and surrounding protectorates
after British withdrawal 1967.

IRAN

Independent kingdom under Pahlevi dynasty
from 1925 until revolution in 1979.

BAHRAIN

Sheikhdom under Al Khalifa dynasty.
Treaty relations with U.K. from 1820
until British withdrawal from Gulf 1971.

QATAR

Sheikhdom under British protection 1868
until British withdrawal from Gulf 1971.

UNITED ARAB EMIRATES

Formerly known as 'Trucial States' or
'Trucial Oman', seven sheikhdoms under
British protection from 1820s.
Independent federation formed after
British withdrawal from Gulf 1971, with
ruler of Abu Dhabi as first president.

OMAN

Under British protection until British
withdrawal from Gulf 1971.

1 Rebirth
1902

How youngly he served his country, how long continued. – CORIOLANUS

BURIED in the heart of modern Riyadh, no more than a stone's throw from the minarets of the city's grand mosque, is a small mud fortress called the Musmak. It was once the chief landmark of a little mud-walled town. Desert travellers could glimpse its stumpy towers from many miles away, jutting above the low, flat roofs of the surrounding houses. Now office blocks and car-choked highways have overwhelmed the modest desert town and visitors seeking the Musmak often retire in frustration after a morning's search believing that the bulldozers must have claimed it.

It is still there, though. Deep among the concrete and the hooting jams of shimmering metal, it stands just off the main perfume bazaar of Thumair Street. On most days the doors are closed and padlocked; but now and again, a small wicket gate used to be left ajar and a stool was placed outside where sat an old man. He may be dead by the time these words appear in print, and if he is not he cannot have much longer to go upon this earth, for he was born well before the turn of the present century and was already approaching manhood when the event occurred that was to make the Musmak famous.

The date can be placed precisely. It was 15 January 1902 when a young scion of the House of Saud forced the gates of the Musmak by a mixture of strength and cunning, put its occupants to the sword and proclaimed the revival of his clan as the lords of central Arabia.

The young man was the Emir Abdul Aziz bin Abdul Rahman bin Feisal al Saud – Prince Abdul Aziz, son of Abdul Rahman, son of Feisal of the House of Saud. In later years he was to become well known to the world at large as plain Ibn Saud, meaning literally the son of Saud; but that was a name adapted chiefly to non-Arab tongues and memories. Among his own people, even in his greatest days, he was better known by his given name, Abdul Aziz, the Servant of the Beloved or, as Christians might say, the Servant of God, for 'Beloved' is only one of the ninety-nine synonyms for the sacred name of Allah.

At the time Abdul Aziz enters this story he had only just turned twenty-one; to the best of a somewhat uncertain family recollection he was born in November 1880. But he was already recognized among his peers as a capable desert warrior and a natural leader of men. Physically, he was a giant among his wiry but generally slight companions, standing six feet four inches tall with a massive chest and shoulders in proportion. In energy and endurance he could outlast any of his fellows. His temper was quick and his rage fearsome. Like many men of action, he was not fond of books and read little beyond the Koran, but of that he was an eager and pious student and in prayer and religious duty he never failed. As a leader he was impulsive, loving the swift, bold gesture whether in war or peace. In victory he was generous, in defeat he was not dismayed. He was, indeed, the kind of man around whom legends are created.

For 150 years the Al Saud had been the principal chieftains of the central Arabian province of Nejd, and twice in that time they had done what no other Arabian leaders had accomplished since the days of the Prophet Muhammad: they had extended their domains by force from one coast of the peninsula to the other. But in Abdul Aziz's childhood they had fallen on hard times. Early in 1891, when his father, Abdul Rahman, was head of the family, they had been forced to flee from their home in Riyadh under threat of capture by their rivals, the Rashid clan, rulers of the Shammar tribe, whose capital was 350 miles away in the north Arabian town of Hail. Abdul Aziz, aged ten, was one of the refugees – a handsome, well-built boy, so it is said, but still too young to take his full share of the hardships of desert travel, especially in time of war. So he journeyed, as desert children did then, in a brightly-woven bag slung from the saddle of his father's camels. Alongside him was his elder sister Nura, in another bag, each of them swaying and jolting to the camel's pace.

They were on the road for two years, sheltering for a time in the little sheikhdom of Qatar on the shore of the Persian Gulf and living for many months with the bedouin of the Al Murrah tribe on the edge of the vast, southern sand desert called by southerners the Rub al Khali, the Empty Quarter, but by the Murrah simply Al Rimal, the Sands. Eventually their exiles' trail led them to Kuwait, at that time still nominally a part of the Ottoman Empire of Turkey. There the local ruling family of Al Sabah agreed to give them shelter, and the Turkish Government in

Constantinople – traditional allies of the Al Rashid against the Al Saud – undertook to provide a small pension as long as they kept quiet and accepted their fate with decent grace.

Abdul Rahman was an honest and pious man, even more strict in his devotions than Abdul Aziz, but he was not a gifted warrior and he was beset by family squabbles. He led raiding parties against the Rashid from Kuwait territory, but it began to seem as if Saudi power had gone forever.

In 1897 two events changed the local balance of power once more. One was the death of Muhammad bin Rashid, the Shammar ruler, whose personal dynamism had been largely responsible for the recent successes of his clan. The other was the murder of the Kuwaiti ruler by his brother, Sheikh Mubarak, a man of equivalent energy. The Rashid fell to fierce intestine strife, to the advantage of Mubarak and the Saudis.

Abdul Aziz was growing towards manhood. At fifteen he had married for the first time, only to see his young wife die a mere six months later. At eighteen he married again and in due course welcomed as his firstborn a son, called Turki after one of the great rulers of the Saudi past. But Abdul Aziz was chafing at the restrictions of an exile's life. Family legend describes how he would often go off alone into the desert beyond the walls of Kuwait Town to sit on the top of a favourite little knoll, gazing in the direction of Riyadh and Nejd, 500 miles away to the south, drawing patterns in the sand with the camel stick that he always carried with him.

By the turn of the century he was leading occasional raids against the Al Rashid and, although his enthusiasm at first outran his skill and the strength of his small forces, so that he was several times forced to beat a swift retreat, he began to build a reputation as a daring young man whom others were glad to follow. Yet it was more in desperation than in hope that one day in the autumn of 1901 he sought Mubarak's sanction for a deeper probe into Rashid territory that would take him all the way through the eastern desert of Arabia to seek out his childhood friends and hosts, the Murrah, on the edge of the Rub al Khali. His purpose was largely opportunist. At best, perhaps, he would give the shaky Rashid empire another jolt and encourage the Murrah to join the House of Saud in a war of reconquest. At worst, he might pick up a few camels.

Mubarak let him go with a warning to do nothing rash and, gathering forty of his closest relatives and friends, together with a few family slaves, he set off southwards by camel. Some bedouin joined them, attracted as they always were by the prospect of raids and booty, but Abdul Aziz and his companions had nothing to show for their efforts after two months in the saddle. By then they were far south, on the marches of the Rub al Khali. Autumn was turning into winter, the desert nights were cold and the Muslim fasting month of Ramadan was at hand when followers of the faith should take neither food nor water between sunrise and sunset and sensible men seek the company of their wives and families. The bedouin began to leave the party, and even the Murrah, whose aid and friendship seemed secure, were scattering with their camel herds still deeper into the southern desert in search of grazing raised by the winter rains.

By December the group was much reduced and needed a refuge in which to pass the long fast. Abdul Aziz chose the wells of Haradh, later the centre of a costly bedouin settlement programme, a station on the railway line from Riyadh to the Gulf, and the southernmost extremity of one of the biggest single oilfields in the world, but then no more than a few small springs in a cluster of date gardens midway between Riyadh and the coast. There, in the days of fast and enforced idleness, he conceived a bold attempt to recapture Riyadh.

Even to the battle-hardened warriors with him it must have seemed a madcap notion. Desert raiding was one thing, mobile and occasionally rewarding, with few people hurt and a discreet retreat always possible in the face of superior forces. To take over and hold a fortified town deep in enemy territory with no immediate reserves of men to call on was something different. Even if the attempt succeeded, which seemed unlikely, it would surely tempt Rashid revenge; and the small Saudi party looked ill equipped to withstand that without early reinforcement. The confidence of Abdul Aziz won over his doubting companions. The House of Rashid was losing its grip, he argued, and the people of Riyadh would welcome the return of their old Saudi rulers. With the townsfolk on their side, the desert tribes around them would rally to the Saudi flag as well. A quick victory would not only throw the enemy into confusion but bring to the Saudi force all the recruits it would need.

Towards the end of Ramadan, Abdul Aziz and his men set off towards Riyadh. Keeping to the unfrequented higher ground out of Haradh, north of the well-travelled Wadi Hanifa, where long lines of date gardens and vegetable plots now flank the railway from Riyadh to the Gulf, they halted for a day or two at the wells of Abu Jifan, eighty miles from Riyadh, to celebrate the feast of *Id al Fitr* – the Breaking of the Fast. As soon as the feast was over they moved on again and twenty-four hours later they were in sight of Riyadh itself. At this point, so most versions of the story tell us, a rearguard of ten men was detached from the forty companions with instructions to secure the baggage train, and, if nothing was heard from Abdul Aziz and the rest of the party by the following noon, to return at once to Kuwait and announce their capture. The remaining thirty went on to take cover in the thick palm groves that then lay near the city walls. Here Abdul Aziz posted another group in reserve under his brother Muhammad, while he went forward with the rest under cover of darkness to find a way into the town.

Exactly who was with him in this final assault party is unclear. The legends that now surround their deeds are none too precise on matters of fact. A few of them, it is said with good authority, were actually slaves, for slaves in old Arabia were often raised to positions of trust and any who fought with valour on such an occasion as this would surely win their freedom if they lived or a place in Paradise if they fell in battle. Certainly, there were at least two prominent members of the House of Saud besides Abdul Aziz himself – his first cousins Abdullah bin Jiluwi and Abdul Aziz bin Musaid, half-brothers of the Jiluwi branch of the family which had forked from the ruling line only two generations previously. Both were to spend much of the next half century as trusted lieutenants of Abdul Aziz in the expansion and consolidation of the Saudi Kingdom. Still another who was almost certainly one of the assault party was Muhammad al Sudairi, another first cousin of Abdul Aziz, this time from his mother's family. He and his descendants, too, were to play important roles in the development of the Saudi state. Beyond these we meet the Arab love of story-telling in which prosaic details of substance usually run a poor second to heroic style. Some accounts have it that fifteen men accompanied Abdul Aziz on the first vital foray into the town. Others say nine, still others say only six, and at least one, which quotes Abdul Aziz himself, has twenty-three.

No matter: a small and daring group crept through the palm groves in inky darkness – for the end of Ramadan is signalled by the appearance of the thinnest of thin new moons and two nights after the *Id al Fitr* there cannot have been more than the barest glimmer in the sky. Their arms were plentiful but primitive. Each man had a short-shafted spear clutched in one hand, a sword hung to one side from a leather belt, a big curved dagger and a pistol thrust into the waistband, and an ancient, long-nosed rifle and bandolier strapped across the back and chest. Thus burdened, they approached the city wall – and found themselves in luck. The Rashid Governor, Ibn Ajlan by name, had grown complacent and allowed part of the wall to fall away. It was a simple matter to lever an old palm trunk into position against it, run up to the top and jump down on the other side.

The town was in darkness, the populace asleep. Abdul Aziz and his men made their way towards the Musmak where, as they had learned in advance, Ibn Ajlan always slept for safety's sake. Across the street from the fortress gate was the private house where he kept his family, and next to that was another house where old servants of the Al Saud were living. A soft knock at the door brought a girl to enquire who was there. 'It is Abdul Aziz, son of Abdul Rahman,' was the whispered answer. 'Let him in!' The young prince was right about the welcome he would receive. Astonished and delighted, the family ushered the men inside and showed them the way across the roof to the Governor's house next door. There the sleeping women and servants were quickly overpowered, and Abdul Aziz and his men settled down at a lattice window, with coffee and quiet murmuring of verses from the Koran, to keep watch over the Musmak gate until dawn.

The muezzin in the nearby mosque – a simple, insignificant building then, with a square minaret of mud little taller than the roofs around it – signalled daybreak with the first call to prayer. As the sun rose, there were sounds from within the fortress to indicate that the Rashid garrison was stirring. With their prayers complete, Abdul Aziz and his men crept downstairs to await their moment to rush to the Musmak gate. It came just half an hour after sunrise when, from one end of the street, there sounded the clatter of hooves as Ibn Ajlan's horse was brought round from the stables while simultaneously the little wicket gate of the fort was opened from inside and Ibn Ajlan himself emerged, stooping low, with several of his men behind him.

With a shout of 'Al Saud!' and 'God is Great!' Abdul Aziz charged from the doorway of the house. Firing his rifle on the run he missed his shot at Ibn Ajlan and saw the Governor turn to duck back through the wicket. Behind him his cousin, Abdullah bin Jiluwi, hurled his spear in the same direction but that, too, missed and stuck quivering in the main door to one side of the opening.

In another moment Ibn Ajlan might have been safe, for although Abdul Aziz's momentum had carried him clear across the street, he was momentarily felled by a blow from the rifle butt of one of the Rashid guards. But Ibn Jiluwi's charge was equally fast and he was just in time to grab the disappearing Ibn Ajlan's legs and haul him unceremoniously back into the street. Then, as the Governor turned with his sword upraised to strike at Ibn Jiluwi a bullet caught him in the chest. Courtesy attributes this deadly shot to Abdul Aziz, who is said to have drawn his pistol as he leaped up from the ground to defend his cousin. But as the doorway must by that time have been a seething mass of struggling bodies, with swords, daggers and rifle butts flailing everywhere, a certain scepticism seems justified about who fired the fatal shot – and, indeed, about whether Ibn Ajlan died that way at all. Again, no matter: the mêlée was real and Ibn Ajlan certainly died, as did many of his companions. Most of them, it seems, fell inside the fort where, once the Saudi attackers had fought their way through the gate, the remaining guards were pursued through the dark, ground-floor chambers and up the narrow staircase to the flat mud roof. One man was cut down as he rushed into the little mosque of the fort in a vain attempt to claim sanctuary on holy ground. Another fell on the stairs. A third is supposed to have been badly burned in an indelicate place when he fell into the embers of the coffee hearth, still brightly glowing from the first brew of the morning.

But the fight was brief. With their Governor dead the guards were demoralized, and within an hour of sun-up on that January morning the young Abdul Aziz bin Saud appeared on the Musmak battlements bearing Ibn Ajlan's severed head and threw it to the gathering crowd in the street below. At twenty-one he was the master of his father's lost mansions.[1]

'Thanks be to God!' said the old man at the Musmak gate when I last saw him. 'I have lived to see all this.' He peered out through his dim old eyes at the sweltering traffic jam a few feet away and then turned back to the little wicket gate beside his stool. 'Do you see there?' he asked me, squinting closely at the wood beside the opening. 'Do you see?' He took my hand and placed it gently on the door itself. 'Do you see?' he asked again. I took my hand away and saw: a flash of metal half an inch long embedded in the wood. 'It is the spear point of Abdul Aziz himself,' said the old man, 'where he ran it through Ibn Ajlan, the Rashid Governor! By God, it is the truth I tell you!'

2 | Out of the Desert

As whirlwinds in the south pass through; so it cometh from the desert, from a terrible land. – ISAIAH 21 i

THE Arabia in which the House of Saud began its third ascent to power in 1902 was little different from that in which the family had twice flourished and twice been laid low. Indeed, life had changed very little since the Prophet Muhammad had proclaimed himself the Messenger of God nearly 1,300 years before. Religious faith, clan loyalties and personal honour were its moral inspiration, the desert its frame.

Many Arabian merchant families had a long tradition of international travel, at least within the Muslim world, for they traded regularly as far afield as Africa and Indonesia, Egypt and the Levant. Many bedouin tribesmen, also, knew the cities of Baghdad and Damascus because they often drove their camel herds there for sale in the markets and bought their few household needs from the city shopkeepers. Yet the seclusion of central Arabia was still stark. In the West, much of the peninsula was unknown territory and substantial parts were to remain so until after the Second World War.

In the whole of the nineteenth century, when Europe's new empires were expanding rapidly elsewhere around the globe and exploration mingled imperial service, romantic mania and scientific pursuit, only three Europeans, all English, succeeded in reaching Riyadh, and their accounts of their journeys were so full of hardship and danger that few sensible travellers wanted to follow them.

Commercially, central Arabia had nothing to offer even to an industrial world that was increasingly ready to tear itself apart for mercantile advantage. Its natural resources seemed minimal and its people were few and scattered; in 1900 they probably did not number more than three or four million from end to end of the peninsula, and one of Arabia's periodical droughts was soon to reduce the numbers of Nejdis. It was nearly 2,000 years since the collapse of the ancient land route for south Arabian frankincense. The kingdoms and caravan centres which had grown fat in those days from supplying the incense to the temples of the ancient world, from Babylon and Jerusalem to Athens and Egyptian Thebes, persisted only in scattered references in the Koran and bedouin song. Apart from the cities of Mecca and Medina, the Muslim Holy Places near the Red Sea coast, where no non-Muslim had ever openly set foot, only the port town of Jeddah was of any importance. A few tens of thousands of the faithful still made their pilgrimage each year, but the electrifying charge of the Prophet's revelation had subsided in a welter of lassitude. The great period of Muslim conquest after Muhammad's death was already more than 1,000 years distant, and the empire of the Ottoman Turks, which had been the chief political embodiment of Islam for five centuries, was dying in a long fall. Although the faith was still maintained in some parts of the Arabian interior with boundless zeal and much xenophobia – especially by the House of Saud and its allies – that seemed only further proof to the world beyond Arabia that the peninsula, on the whole, was a place to be avoided.

The hostility of the desert was notorious among foreigners. Richard Burton, who made his famous visit to the holy cities in disguise in 1853, described Arabia briefly but memorably as 'a haggard land'. So it was – and, in spite of everything, to foreign eyes it remains so. Jagged cliffs and coral reefs or treacherous sandbars and salt flats protect most of its coast. Dreadful heat and the menacing emptiness of the desert lie beyond. Only in a few places is nature relatively kind. The mountains of Yemen and Asir in the south-west and of Oman in the south-east feel the seasonal touch of the monsoon rains off the Indian Ocean and both have maintained a settled and occasionally even prosperous people for many thousands of years. North of Asir, the valleys of the Hejaz region in the west have also traditionally supported town and village life, like that of Mecca and Medina; and in the far north, arching through the fringes of Jordan, Iraq, and Syria, there is a thin zone of semi settlement where the desert shades gradually into the so-called Fertile Crescent between the Mediterranean and the Gulf. But within those extremities lies a huge quadrilateral as big as Western Europe in which only a pitiful scatter of oases has, until lately, contained any settled life. As Charles

Doughty described it in the Biblical and Arabized prose of his classic work *Arabia Deserta*: 'Here is a dead land, whence, if he die not, [the traveller] shall bring home nothing but a perpetual weariness in his bones...In a parcel of desert earth great as an house floor, you shall find not many blades and hardly some one of the desert bushes, of which the two-third parts are no cattle-meat but quite waste and naught.'[1]

Before the present century this desiccated, bitter land defeated most foreigners. Since the time of the Prophet no non-Muslim power had ever ruled more than a tiny fraction of the peninsula and no external force of any kind had done more than scratch its burning, stony surface. Augustus Caesar once sent an expedition to Yemen and the men christened its green, rain-tipped mountains *Arabia Felix*; but distance, hardship, and the resistance of the mountain tribes soon forced their withdrawal. The Persians from the east and the Dutch and the Portuguese from the west all maintained at various times and places a few military or trading posts along Arabia's shores, but none left anything more permanent than a fort or a string of cisterns; and in 1900 the two empires that then maintained a special interest in Arabia – those of Britain and Ottoman Turkey – were both, in different ways, excluded from its innermost affairs. The Turks were handicapped by long lines of communication and administrative decay. Although they claimed historical sovereignty over most of the peninsula, including the ancestral territory of the Al Saud, they had never chosen to colonize and their grip was spasmodic. Even in their heyday they had rarely done more in the interior of the peninsula than to play off the major tribes against each other – as in their support of the Rashid of Shammar against the Saud of Nejd – and by 1900 their empire was moribund. The Ottoman Sultan in Constantinople still enjoyed the title of Caliph – literally successor, or *Khalifa*, to Muhammad – and was formally recognized as the spiritual as well as the secular leader of the Muslim world, but his domains were shrivelling fast. Two centuries of steady retreat before the rising powers of Western Europe had squeezed Turkey out of most of its possessions in the Mediterranean and the Balkans. A foothold remained on the coast of Libya, but Greece was independent and Cyprus and Egypt had both passed into British hands. In the rest of the Middle East the Turkish provinces still stretched in name through a vast, unwieldy arc northwards from Yemen through the Hejaz to Palestine and Syria, then south again through Mesopotamia and the cities of Baghdad and Basra to Kuwait and the western shore of the Gulf. But many of them were semi-autonomous and, beneath an apparently placid surface, none of them was secure.

For a number of reasons, most notably parsimony and the disastrous lessons of the Second Afghan War, the British excluded themselves from new adventures. In 1900 they were at the pinnacle of their global power but they were determined not to go beyond Arabia's coasts. Having replaced the Turks as the dominant foreign power in Egypt and won control of the Suez Canal after its opening in 1869, their over-riding interest was to maintain the sea route to India, suppressing Arab piracy on the one hand and resisting rival imperial challenges on the other. Throughout the nineteenth century they followed a dual policy in Arabia to achieve those ends, propping up the Turks wherever their control seemed reasonably effective – especially along the Red Sea coast – and compelling a multitude of petty coastal rulers elsewhere to pay due regard to Britain's wishes on pain of deposition or a brisk naval bombardment. They nourished Aden but neglected its hinterland. By 1900 they had evolved a system of treaties with Arabia's coastal potentates that ensured their virtually unbroken control from Aden to the Gulf; but they remained utterly indifferent to the affairs of Arabia beyond the screen. As long as no rival power could interfere there, the British themselves were not merely content but eager to stay out of unrewarding desert entanglements.

The Al Saud were not, in the strict sense, desert dwellers. Like many other powerful Arabian families, they were born and bred to the settled life of the oasis. Riyadh itself was an oasis town. But as the centre of the region called Nejd, itself at the heart of desert Arabia, Riyadh's life was inescapably entwined with that of the wilderness around it. So, therefore, was that of the Al Saud. Their lineage officially derived from the desert tribes and their whole outlook was shaped by the traditional interplay between the nomadic bedouin and the people of the oasis – the conflict and alliance of desert and town.

To understand the Al Saud and their kingdom today it is essential to understand the bedouin, whose fiery enthusiasms have so often been imposed upon people apparently stronger or more numerous than themselves. The bedouin have never been great in numbers. Although some estimates suggest there may be as many as one million of them in Saudi Arabia today, a truer figure would probably be nearer half a million, no more than one-eighth of the present settled population. In the past, their numbers were proportionately greater, before a booming wage economy lured them out of the desert and modern weapons in the hands of an organized state crushed their fighting power. Yet in Arabia as a whole they have always been outnumbered by the settled folk and even in the desert interior they have probably never exceeded the total population of the oases. What they lacked in numbers, however, they made up for with a fitful but effective dynamism that enabled them to set the characteristic tone of much of Arabian life, in both pre- and post-Islamic times, through a constantly renewed process of conquest, intermarriage and absorption in the settled communities.

The first element in bedouin life was a natural instability and fragmentation of loyalties that arose from the demands of their wandering existence. Depending on the camel for their traditional livelihood, they roved in search of grazing for up to nine months of the year, sometimes covering thousands of miles; and their allegiances were apt to be as light as their baggage had to be. The family was their anchor, for in the desert each family was often alone, with its little group of black tents and its camel herd, separated by perhaps a day's ride or more from its nearest rivals or companions in the search for food. After the family came the tribe, united by wider and more complex links of blood. Within these two small circles the bedouin were often resolute, bound by codes of honour that were virtually absolute. But their loyalties otherwise were fickle. Left to themselves, they made and abandoned alliances with equal alacrity according to the needs and opportunities of the moment; and if one tribe would usually unite against a second in adversity, it might just as often divide against itself if that seemed advantageous to one group or another. Even families could be riven by need or opportunism. Fratricide, especially among half-brothers whose mothers might have come from rival families, was as common among bedouin as among settled folk. Few ruling families in Arabia, not excluding the Al Saud, are free from that taint and some have records of family murder to rival Jacobean tragedy.[2]

In all matters of power and politics bedouin values were intensely personal. By choice they recognized neither settled administrations nor established frontiers. Nor did they accept any social distinctions among themselves saving that of lineage. With so little in the way of material possessions to distinguish one man from another and nothing between them and eternity but faith, luck, and individual ability, the bedouin were egalitarian. In the desert their rulers lived like their people, with all hands dipping into the same, usually meagre, pot;

at the sheikh's *majlis*, or council, the ruler would listen to other voices on issues of tribal or family importance. This is a tradition to which even kings must still pay heed. From Abdul Aziz down to King Khaled today, all modern Saudi rulers have continued to hold open court for their subjects, just as their ancestors did, for they are perfectly aware that their credit among the tribesfolk would vanish overnight if they tried to suspend the custom. It was said of the great rebel leader, Feisal al Duwish, that he could execute a man without consulting the elders; with Abdul Aziz it was never true.

On the other hand, the bedouin firmly believed that all who were not as they were must be inferior beings. That is a not uncommon belief, of course, among the peoples of this world. Among the bedouin, however, the belief had two important practical results that still affect Arabian society and that were even more significant in the untouched life of Arabia a few decades ago. One was a deep reluctance to undertake any labour not fitting to their image of themselves as warriors and freemen, so that without the conquest, enslavement or co-operation of other people the bedouin were never able to form themselves into a lasting power. The other was a profound, almost obsessional, respect for the pride of family or tribal blood, especially as it concerned their women. Like the medieval troubadours of Europe, their poetry – the most characteristic, weightless bedouin art, cultivated around a million camp fires – sings forever of the perfection of womanhood because the tribe's honour resides in its daughters, through whom the blood line is continued.

Because of the isolation of the bedouin communities and the importance attached to the purity of the blood line, marriages between first cousins, or equivalent relations, were preferred, or else with carefully selected partners of equal status and purity in another tribe. As Islam permits each man to keep four wives at any one time – provided, says the Koran, that he can offer them all equal treatment – and as divorce is made easy for males under Koranic law, so that the magic number of four may be multiplied many times over in one man's life, this custom begot not only large numbers of children by a single father but also an immense ramification of family and tribal inter-relationships through several generations. Nephews married aunts, uncles were wedded to nieces and their children married each other to form a close-knit and, to the outsider, impenetrable mesh. This created a simple patriarchal order which at best enabled an astute and ardent patriarch to extend his influence to tents far beyond his own and at worst helped to sustain the tribe in adversity. Not surprisingly, genealogy ranks next to poetry as a bedouin art, for every man needs to carry in his head a tribal *Burke's Peerage* or *Almanach de Gotha* to guide him through the web of his own relations.

The Prophet himself set a notable example as a tribal patriarch, marrying often to unite many previously warring Arabian groups in what was, at the time, a deeply divided society. But Abdul Aziz and his descendants are the undisputed masters of such judicious wedlock in modern times.

With about 500 princes descending from Abdul Aziz, together with wives, daughters and collateral branches of the family, the House of Saud today cannot number less than 20,000 people. Abdul Aziz himself was by no means the only progenitor of this vast family. Some of his brothers and cousins were industrious patriarchs, too, and a few of his older sons carried on the tradition with zeal until only recently. But for the sheer range and influence of his marital alliances Abdul Aziz was unrivalled. The number of his wives has never been officially computed – characteristically, the family still declares that it is a private matter – but he fathered a total of forty-five recorded sons by at least twenty-two different mothers representing most of the major Arabian tribes and families with whom it seemed prudent to form some alliance. In addition, there were at least as many daughters from an even wider range of women including, no doubt, some unacknowledged mothers among the various concubines, slave girls, and 'wives of the night' whom it was customary for Arabian patriarchs to enjoy whenever desire recommended and opportunity arose. Although such multiple marriages are now far less common among the younger generation of Saudis, even the royal family, they were the accepted style among tribal sheikhs at the turn of the century – and for many decades afterwards – and they literally bound together the current Saudi Kingdom by the bonds of blood.

Much as the bedouin respected ancestry and a patriarchate of this kind, however, they never granted their leaders, whether titled sheikhs or kings, any prescriptive right to rule. Traditionally, the bedouin accepted certain families as natural leaders of the tribe, established through custom and performance; but if the performance faltered in one, individual custom demanded that another soon took his place. Because of the size of the patriarchal, sheikhly families, there was usually someone standing in the wings ready to assume power by consent or seize it by force – even if that meant, as it often did, killing half a dozen other rivals. Ruthlessness of this kind was a recognized attribute of authority, for weakness in the desert was no man's friend.

Whoever filled the role of leader, however, could only maintain his place through worldly success, especially through skill in war and politics, and generosity in his tent. Survival in a naturally hostile environment was, after all, the first and necessary aim of bedouin life, so it was essential to have a leader who seemed to have God, or luck, on his side. The sheikh must prove himself literally the 'father of his people'. He must know the family troubles of every man for settlement. Above all, he must not be miserly, and must keep open house. No name has a more unworthy meaning, or leaves a nastier taste in the mouth of the bedouin, than the epithet *bakhil*, or 'stingy one'.

Permanently on the edge of hunger otherwise, the bedouin saw other people's possessions as legitimate loot, especially if they were the camels of another tribe, and the armed raid was both the greatest of their pleasures and the most serious of their concerns. The natural precariousness of desert existence, where so much may depend on the chance rain shower or a dried-up well, was echoed throughout traditional bedouin life. Nothing in that life was certain. If God, or luck, favoured you there was every reason to take swift advantage of the fact, whatever the cost to other people; but if they deserted you, there was little to be done. Life, therefore, was lived between moral as well as physical extremes, with fatalism and improvidence at one pole and a deeply practical, even cynical, instinct for self-preservation at the other. As Charles Doughty described them, the bedouin were 'light-timbered ships' – often faint with hunger, always predatory and alive for the main chance of today, but altogether without thought for tomorrow; and in their ever-shifting courses they kept most of Arabia in a state of perennial and sometimes bloody flux until well into the twentieth century.

Against this natural instability of bedouin life the people of the oases opposed an equally natural stability. They owned the water and supplies without which even the bedouin could not survive, they offered markets for the nomads to trade in, and they built walled towns and citadels that bedouin raids could not penetrate. Under a powerful leader they could, and often did, impose themselves upon the desert flux. Yet they, too, needed the milk, meat and hides that the bedouin camel herders provided and – perhaps even more important – they needed bedouin protection to travel across the desert themselves. So the two sides lived in uneasy equilibrium, each maintaining a wary dependence on the other and, in the absence of any outstanding personality on either side, often falling into sporadic hostility.

Out of this relationship of opposites evolved a more or less common moral outlook whose chief characteristic was a profound ambivalence about most aspects of social life. Its basis remained the desert code, for in spite of the material power of the sedentary Arabs, it was the bedouin, with their image of a free, warrior people, who were accepted as the Arab 'aristocracy'. Most oasis dwellers liked to claim bedouin descent (and still do) even when they simultaneously decried the alleged deficiency of bedouin morals. Equally, though, whilst most bedouin affected to despise the effete customs of the settlers, they cast envious eyes – and, when possible, grasping hands as well – upon their superior worldly goods. The result was a society that seemed to teeter permanently on the edge of multiple schism.

Unity in anything but transitory association required not just an individual of stature but a cause as well. The cause, at least since Muhammad's day, was invariably religion, for nothing else would sufficiently subdue the disordered bedouin mind. Although their theology was usually indifferent, the bedouin saw and felt the work of omnipotent nature all about them and the man who could interpret their resulting triumphs and trials in terms of a Supreme Being responsible for them all could usually win their loyalty.

That was the immense attraction of the Prophet's message of Islam. The word, Islam, means submission – that is, to the will of God – and to the sceptic it might seem a translation of the manifest uncertainties of desert life into the expression of God's will. The implicit fatalism of this approach finds an echo in the practical inability of the bedouin to do very much about their circumstances – and incidentally explains the lassitude that commonly afflicts their life even now in between swift bouts of action. To justify their love of raiding, bedouin could easily consider their victims unbelievers whose goods were forfeit and their own divine reward. As many desert travellers found, loot and the love of God were never too far apart in the bedouin mind.

This was mere bedouin casuistry. Yet Islam suits well the bedouin cast of mind and society, not least because of its simple hierarchy. Islam has learned men called *ulema*, who specialize in the study and interpretation of the Koran, but it has no formal priesthood. Every man is required to communicate directly with God for himself, for the

Almighty is not conceived – as in many Christian minds – as some sort of Holy Roman Emperor, with hosts of flunkeys to intercede between ordinary mortals and Himself, but rather as an abstract being ever accessible to his community or, to the bedouin mind, as a sheikh in touch with his tribe.

Islam also tends to be exclusive. In principle, Islam is an inclusive, tolerant faith, and Jews and Christians are specifically accepted as 'People of the Book' – that is, those who are said in the Koran to have enjoyed earlier, but incomplete, revelations of God. But at the same time the superiority of Islam's final revelation is always emphasized and, as the first recipients of the Prophet's words, the Arabs have a special role. As God often reminds His audience through the Prophet's mouth, this final revelation is granted in the Arabic language 'so that ye may understand'. The language itself, therefore, has acquired a holy status and the Arabs, as its possessors, have seemed to gain, by extension, a special place in God's affections and in mankind's affairs. This attitude is all too often reflected in contemporary Saudi life where, alongside a sometimes surprising readiness to accept foreigners and foreign ways in material matters, a deep vein of religious chauvinism is constantly at work comparable to the tribal chauvinism of the bedouin.

A third parallel between desert life and the Koran's message can be seen in the picture of paradise offered to true believers. Repeatedly, heaven is envisaged as a place of rich gardens and fruit trees, where the faithful will recline on couches of silk brocade to be attended by virgins untouched by man, while running water – above all, running water – sounds sweetly all about them. Clearly, this is the promise of the oasis, so important to the desert dweller. On the other hand, the promise can only be redeemed by the acceptance of certain clear laws and restrictions, especially those known as the five pillars of Islam. These require the faithful to abandon all gods but God, to pray five times a day (starting, as the bedouin naturally do, at crack of dawn and ending just after sunset), to give alms to the poor, observe the fast of Ramadan (which celebrates the first moment of Muhammad's encounter with God), and to make the Pilgrimage to Mecca, called the *Hajj*, at least once in their lifetime if possible.

But the most important aspect of Islam is the call to unity. The Muslim faith admits no saints, no Holy Trinity, no Son of God – though Jesus, Mary, Moses and many others are given their due in the Koran as prophets or important figures in the revelation of God to mankind. The essential point is that God is one and indivisible: His divinity cannot be shared. Even Muhammad is only His Messenger, retailing His words exactly as they were conveyed to him, either through God Himself (who does seem, incidentally, to have acquired some human attributes in the process) or through the Angel Gabriel.

Here the Koran differs radically from the Bible. Although it draws heavily on stories from both the Bible and the Jewish Torah, sometimes almost word for word, it purports to be not merely a human recounting or interpretation of the experience of God but the actual and final revelation of God in His own words. This gives it an absolute authority for Muslims that only the most fundamentalist Christians can find in the Bible. Together with the *Hadith* – the recorded traditions and sayings of Muhammad himself – the Koran provides a total and explicit moral code; and although, inevitably, many of its provisions sit uneasily with the demands of twentieth-century life and much re-interpretation of its meaning has been attempted since the Prophet's day, the *sharia*, or 'path', that the Holy Book supplies remains the basis of all true Muslim life. In Saudi Arabia it is even more than that: it remains there the only recognized and enforceable code of law, so that the country is held in a 1,300-year-old corset of town and desert morality that is deemed to be universally and eternally applicable – a fate that leads to a great many conflicts and expediencies.

The insistence on unity as well as submission, however, was crucial to Muhammad, probably because it was the reverse of the actual disunity that existed in Arabia in his time – as, indeed, on many occasions since. Arab history is full of examples of the dramatic interplay between the two and the bedouin of Arabia have been among the foremost exponents of both. The Muslim conquests in the century after the Prophet's death saw them carry all before them in a vast *jihad*, or holy war, in which their particular notions of unity, equality and devotion were impressed upon millions of people from Spain to India.

But bedouin staying power was short and their conquests soon disintegrated or were taken over by the conquered, while the bedouin themselves were absorbed or returned to their shifting, nomadic ways. So characteristic was this rapid rise and fall of bedouin zeal and power that the fourteenth-century Arab historian Ibn Khaldun attempted to interpret the whole of Arab history up to his own time in terms of a constant cyclical relationship between the desert and the town. In the first stage, he suggested, the nomads of the desert would place their military strength behind some outstanding leader, usually of a religious bent, and impose themselves through conquest upon a certain area. In the second stage the superior administrative skills and learning of the sedentary folk would enable them to take over the military-religious state of the nomads, so that by the third stage it was transformed into something altogether more secular and sophisticated – and, therefore, inimical to the free, bedouin ways. In the fourth stage the bedouin would typically lose interest in the whole enterprise and, drifting back to their nomadism, they would leave the state to decay, morally and physically, until a new leader arose to start the cycle once more.

Like much social theory this is probably too neat, yet it does correspond to many of the facts of traditional Arabian life in which, from the time of Muhammad onwards, most successful rulers were closely associated with some form of religious zealotry and bedouin power, and few of the states they created outlasted their founder's death by more than a generation or two. It certainly illuminates the story of the House of Saud, whose origins and achievements, both in this century and the past, have been intimately associated with this characteristically Arabian exchange between the nomad and the settler.

*   *   *

Like most distinguished Arabian families, the Al Saud's beginnings are lost in the mists of Arab mythology; but as members of a small sub-section, called the Masalikh, of a great north Arabian tribal confederation, called the Anaizah, they belong to the Adnan line, one of the two parent stocks, the other being the Qahtan, distinguished by the bedouin as fathers of all the Arab tribes.

Like most other big tribal groups in Arabia, the Anaizah – who may number altogether as many as one million people – are a mixture of nomads and settlers, and the House of Saud itself has been among the settled folk for at least 400 years. Its first traceable ancestor is usually identified as Mani al Muraidi, a sheikh from the nomadic Durra tribe of Oman who were – and to some extent still are – famous for their herd of riding camels known after the tribal name as the Daraiya, or Diriya. Invited by the people of the Wadi Hanifa to settle among them, possibly in order to improve their own camel stock, Mani took his family there in the sixteenth century of the Christian era and founded a little colony that also took the name of Diriya. One of

Mani's descendants bore the name of Saud and from him stems the line and title of Al Saud that now rules most of the peninsula.

The aristocracy of this descent in Arab terms is the first element in the House of Saud's claim to power. Its members are not and never were – as many people outside Arabia often imagine – just 'unknown desert sheikhs'. To the world beyond Arabia they were unknown, of course, like practically everything else about that forbidding place. But among their own people the purity of their Arab blood and the extent of their tribal connections always entitled them to respect. Yet although this noble descent was indispensable as a basis for leadership, it could never have established Saudi supremacy on its own. Plenty of other families in Arabia enjoyed an equivalent status and the House of Saud was in no way distinguished from the rest until the middle of the eighteenth century. Then, in what must now be regarded as the most fateful meeting of minds in Arabia since the time of Muhammad, Sheikh Muhammad bin Saud, the ruler of Diriya, and great-great-great-grandson of Mani, the first identifiable Saudi ancestor, gave shelter to an itinerant preacher of Nejd named Sheikh Muhammad bin Abdul Wahhab.

The preacher was a Muslim revivalist of the sternest hue – a John Calvin, so to speak, of Arabia come as if in fulfilment of Ibn Khaldun's desert cycle to re-create and re-purify the God-fearing, military-religious state of the Prophet. The world of Islam by then was much in need of rejuvenation and reform. Centuries of schism between the orthodox followers of the Prophet, known as Sunni Muslims, and many smaller, revisionist sects had sapped its strength and will. Under Ottoman leadership the faith had become a bureaucracy and the rising power of the Christian West seemed a constant rebuke to its division and degeneracy. Peninsular Arabia, especially, was once more sunk in the sort of corruption, idolatry and lawlessness that Muhammad himself had encountered. Instead of worshipping God alone, men were offering prayers to a multitude of saints. Stones, trees and statues were deemed sacred, the once simple mosques had become elaborate temples, and the laws of the Koran were treated with contempt. In addition, the country was in a state of political chaos and more or less continuous intestine warfare. In Nejd, a host of petty princelings contended with each other for power, and townspeople and bedouin were constantly at each other's throats.

Abdul Wahhab was a true zealot, come to cleanse the stinking stables of Arabia once more with the Word of God. But the Word alone proved insufficient for the task. Like the Prophet, Abdul Wahhab needed a Sword as well – and, to his eternal joy, he found one in Muhammad bin Saud and his family.

Although Muhammad bin Saud was only one of the numerous quarrelling Nejdi sheikhs at the time, little more important than the rest, he evidently grasped (with the aid of some persuasion from his wife who was impressed by Abdul Wahhab's exceptional piety) that an alliance with a man who had a message would give him the edge over all his rivals, enabling him to unite bedouin and townsfolk in a new *jihad* to extend his personal dominion. Accordingly, in 1744, he married off his son, Abdul Aziz, to a daughter of the preacher, and thus sealed a compact between the two families that has been continued unbroken by their descendants ever since.

Contemporary Saudi Arabia, for all its money and the new corruption and idolatry that wealth has encouraged, remains in theory and to a surprising extent in practice a Wahhabist state, officially dedicated to the preservation and propagation of pure Islam as propounded by Muhammad bin Abdul Wahhab. While the Saudi family continues to direct its temporal fortunes, in alliance now with most of the great tribes of Arabia, the descendants of Abdul Wahhab are still its chief spiritual advisers. Today they are known simply as the family of the Sheikh (Al al Sheikh in Arabic) out of respect for the original preacher; and no Saudi government is complete without one or more of them close to the seat of supreme power. Their frequent marriage links with the Al Saud have also continued the original alliance – King Feisal's mother, for example, was one of the Al al Sheikh – and their influence is directly traceable even in the message inscribed in Arabic upon the green background of the Saudi national flag: 'There is no god but God and Muhammad is His Prophet.' That is the creed of Islam, officially accepted by Muslims everywhere, but Saudi Arabia has made it the emblem of the state.

The return to this concept of the absolute one-ness of God, on which Muhammad originally insisted, is so much the essence of Wahhabism that inside Saudi Arabia the term Wahhabism is not admitted. To all true followers of Abdul Wahhab it is a coinage of the movement's critics only, implying that the preacher merely founded another sect of Islam instead of restoring the true and only faith. Although the rest of the world may go on calling them Wahhabis, they call themselves unitarians (*muwahaddun*). Moreover, true to the message of Muhammad as they see it, they insist not only on the unity of God but on all the other unities that flow from that – the unity of things temporal and spiritual and of all true believers as well. In logic, therefore, the Wahhabist, or unitarian, state can admit no distinction between religious belief and political action, nor should it recognize any divisions of class, kind, or even of nationality among those who truly believe. Unbelievers, on the other hand, are beyond the pale by definition and deserve all the punishment the Koran lays in store for them.

The leadership of Wahhabism since the eighteenth century has been vested by right as well as custom in whichever member of the House of Saud is named as temporal ruler, for Wahhabism, above all, denies any division between secular and spiritual authority. Although the Al al Sheikh continue to provide influential religious advisers, the Saudi King is necessarily also the Imam (religious leader) of his people; and many of the country's current problems stem from the growing difficulty of reconciling the two responsibilities in the modern, materialist world.

Historically, even more than today, what distinguished the Saudis from the rest of the Muslim world was simply the intensity with which they pursued Islam's unitarian vision under Wahhabi tutelage. The strength of its conviction was reflected in its early success. Within sixty years of the first alliance of the House of Saud and the House of the Sheikh, the bedouin forces of the Wahhabis had won control of most of Arabia in a *jihad* that echoed the achievements of the first Muslim conquerors. They captured the holy cities of Mecca and Medina in the west and controlled the Arab shore of the Gulf in the east. Their raiders harassed the interior of distant Oman in the south and ranged as far afield as the outskirts of Damascus and Baghdad in the north.

The success of the Wahhab-Saud alliance, however, was also its downfall. Partly, its conquests fell apart because the original religious impulse began to fade as usual, and was replaced by sheer force, xenophobia and self-aggrandizement. But equally important was the fact that the alliance challenged powers far greater than itself. In the east, the British were consolidating their position in the Gulf and when they found Wahhabi pirates raiding their stations on the Persian shore and attacking their ships at sea they soon turned the guns of the Royal Navy upon them. By degrees they imposed a formal maritime truce and an effective hegemony upon the coastal tribes that was to last in treaty form right down to 1970. In the west and north the

Ottoman Empire was prodded into ponderous action. As Caliph of the entire Muslim world, the Ottoman Sultan was, at least in his own eyes, the rightful guardian of Islam's Holy Places, and the maintenance of Turkish rule in Mecca and Medina was central to his spiritual authority. Yet the Wahhabis were destroying all traces of Ottoman occupation there and indulging their puritan zeal by molesting pilgrims, smashing shrines and mosques, and burning hubble-bubble pipes, musical instruments and any other aids to innocent, worldly enjoyment. Such presumption was a threat to the very existence of the Empire and in 1811 the Sultan called upon his powerful new Viceroy in Egypt, Muhammad Ali, to help drive the desert upstarts out.

It took Muhammad Ali seven years to complete the task, but by 1818 his son, Ibrahim Pasha, had led his armies not just to Mecca but 500 miles further into the Arabian wilderness, to the Nejdi heartland of Wahhabism itself. There he entered the little town of Diriya, the Saudi capital, and laid it waste. If the British had had their way, he would have gone even further and marched on to the shore of the Gulf to help them in suppressing the Wahhabi pirates. But Ibrahim by that time had tired of Arabia. Dispatching the Saudi ruler to Constantinople to have his head chopped off, and distributing all the food stores of Diriya and Riyadh to his weary soldiers, leaving the inhabitants to starve, he retreated westwards again for home. By so doing he prompted unconsciously what must rank as one of the most bizarre episodes of early Western travel in Arabia, for he was chased unavailingly clear across the peninsula by a dogged young British army captain who had been sent from the Gulf with a presentation sword for the Egyptian commander and a message requesting his aid against the pirates.

George Forster Sadleir was the captain's name, and his journey of more than 1,000 miles in eighty-four days was the first recorded crossing of Arabia by a European. On the way he also became the first European ever to reach Riyadh – then only one of Diriya's tributary villages – and he inspected the ruins of what had been the first Saudi capital. He found the inhabitants of both places in a wretched state, with not a grain of wheat or barley or a horse left among them, and the date gardens of Diriya destroyed and utterly deserted. The town was never fully reoccupied and the remains are still there, much as Sadleir found them. They are a favourite picnic spot for the people of Riyadh now.

The House of Saud did not recover from this disaster for twenty years, but then, first under Muhammad bin Saud's grandson Turki (after whom Abdul Aziz bin Saud's firstborn was named in 1900) and later under his son, Feisal, the family established itself with a new capital in Riyadh and began a fresh campaign of Wahhabist conquest and conversion. Feisal was the father of Abdul Rahman and the grandfather of Abdul Aziz, and his is the name that is now accepted by family tradition as the first in line of the present-day House of Saud, from whom all true princes of the full blood royal should be able to establish their descent. He was also a highly successful desert warrior and the statesman who once again extended the Al Saud's Wahhabist rule over much of central and eastern Arabia. But he, too, ran foul of the great powers, and of the desert intrigues and family jealousies released by his success.

In 1862 Napoleon III of France sent a spy to report on Feisal's activities, apparently hoping to use him in some way as part of France's long imperial campaign against the British in the Middle East. The expedition was a political failure but the spy became the second European traveller to reach Riyadh. His name was William Gifford Palgrave, and just where his real loyalties lay is still hard to discover, for although he was a Jesuit priest in French pay he was also an English Jew who had disguised himself for the trip as an Arab doctor. As might be expected of a man who chose to entangle himself in such complex layers of identity, his account of his journey is rich in intrigue, threat and drama, culminating in a graphic picture of one of Feisal's sons demanding strychnine from him to poison his elder brother and then menacing Palgrave himself with poisoned coffee when the traveller refused.

All unaware of this curious and abortive mission, however, the British sent in their own man, Colonel Lewis Pelly, from the Gulf three years later. Arriving in 1865 to warn Feisal against any fresh Wahhabist adventures in the territories now controlled by Britain around the Gulf, he became the third and last of the nineteenth-century European travellers to reach the Saudi capital – and the first actually to fix its geographical position with any accuracy. His visit was less dramatic than Palgrave's, perhaps because Pelly was a more straightforward man, but he, too, found the atmosphere at the Saudi court a good deal less than welcoming. Feisal himself was polite, but blind and frail with age and surrounded by stern Wahhabist zealots who deeply resented the presence in their midst of the representative of an infidel power. In face of their disapproval Pelly retreated, determined to maintain Britain's guard in the Gulf against further renewal of Saudi incursions there.

Yet it was the combination of a Muslim power and the folly of Feisal's own sons that actually wrecked this second Saudi empire. When Feisal died, only a few months after Pelly had seen him, his two eldest sons, Saud and Abdullah, began a running battle for the spoils of the Kingdom that lasted for the next twenty years. In the process they gave the Turks a second chance to reduce Saudi power. In 1871, seizing the opportunity of an invitation from Abdullah to help him against his brother, Turkish soldiers occupied the whole of the eastern province of Hasa, on the shore of the Gulf, as far as the little sheikhdom of Qatar. At the same time, the Turkish Government found another, rival leader to support in the dynamic Muhammad bin Rashid of Hail who was just then emerging from a bloodbath of his closest relations as the undisputed leader of the Shammar tribe. This time the Turks did not need to emulate Ibrahim Pasha's feat of half a century earlier and march on the Saudi capital. Cut off from the sea in the east, attacked by the Rashid from the north and mortally wounded by its own family hostilities, the proud Saudi state crumbled again. In another ten years the Al Rashid had virtually taken over the Saudi domains, leaving Feisal's third son, the pious but ineffective Abdul Rahman, in Riyadh as little more than a licensed governor.

The eventual flight from Riyadh, the exile and the return in 1902 which heralded the third – and, so far, final – rise of the House of Saud did not seem of much moment to the world of 1902. Those few people who knew of the events in Riyadh saw them as no more than another, not especially significant, change of fortune in Arabia's perennial tribal flux. They could not know that in Abdul Aziz bin Saud there had come a young man who would combine in himself the traditions of an old world and the opportunities of an infant new one to create a unity in his homeland such as it had not known even in the wake of the Prophet himself.

3 The Consul and the Oilman

1903 – 8

I say it without hesitation, we should regard the establishment of a naval base, or of a fortified port, in the Persian Gulf by any other Power as a very grave menace to British interests, and we should certainly resist it with all means at our disposal.[1] – LORD LANSDOWNE

FOR the first twenty-two years of Abdul Aziz bin Saud's life, throughout his early childhood in the ruins of the second Saudi Kingdom in Riyadh and his teenage exile in Kuwait, there is no record of his ever having met a European. No doubt in Kuwait he encountered some of the Indian merchants who traded around the Gulf, and the occasional Turkish official probably crossed his path – for, after all, his father was receiving a Turkish pension and Kuwait was still nominally part of the Turkish Empire. But no representative of the great Western powers of the day had enjoyed any contact with the House of Saud since Colonel Pelly's visit to Riyadh on behalf of Britain in 1865, and it was nearly forty years after that notable adventure and more than a year after Abdul Aziz's recapture of the Saudi capital before the next recorded meeting took place between a European and any member of the House of Saud. This time it was, of all people, a Russian who made the contact.

The first news of the encounter reached the British Foreign Office in a telegram dated 11 March 1903 from Colonel Kemball, British Consul in Muscat, the steamy little Arabian port and capital of the old Sultanate of that name (now the Sultanate of Oman) where a Royal Navy vessel had just arrived after shadowing a group of French and Russian warships around the Gulf. 'H.M.S. "Sphinx", which arrived from Kuwait yesterday, reports having met there the Russian and

French cruisers which were there from 4th to 8th inst. Abdul Aziz bin Feysul, who had just arrived at Kuwait, was visited by the Russian Consul at Bushire [then the chief port of the Persian shore], and as he subsequently informed the Commander of the "Sphinx", was promised by the Russian Consul assistance in the shape of rifles and money.'[2]

Here was, indeed, a little bombshell, for 'Abdul Aziz bin Feysul' was no other than the young conqueror of Riyadh himself (characteristically, neither Colonel Kemball nor the Commander of the *Sphinx* seems to have been quite sure of his full name) to whom the Russian Government was now offering support against his rivals. Few things could have been better calculated to stir British representatives in the Gulf to protest and the British Government itself into imperial action.

By what would turn out eventually to have been a stroke of luck, the revival of the Saudi Kingdom had coincided with the growth of a new international interest in the fate of Arabia and its surroundings. In the last two decades of the nineteenth century Europe's imperial struggles had multiplied. In Africa, France, Germany and Britain had just carved up a continent between them in a pell-mell rush to see that no one of them totally pre-empted the advantage of the others. In Asia, Russia was steadily expanding her power and influence eastwards – to Vladivostok and the China Sea in one direction, to the interior of Afghanistan and Persia in the other – while both France and Germany were wooing the Turkish Government for a privileged place in its Arab lands. In Persia, British City interests had been attempting to wrest monopolies and concessions from timber to tobacco out of the Shah's beleaguered and all but bankrupt Government.

The chief target of the rival thrusts, or so it seemed to many British at the time, was British rule in India, incomparably the richest of all Britain's imperial possessions. Any hint of European competition there induced something like hysteria in the Viceregal Government in Delhi and Simla and in its far-flung minions in the provinces and dependencies that were traditionally attached to British India. Among these were the scattered Consuls and Political Agents (the latter a specifically Indian Government title) who supervised Britain's interest in Arabia's coastal sheikhdoms and the Gulf. Most of these officials answered in the first place to a superior officer called the Political Resident who was based at Bushire, on the Persian shore of the Gulf. The Resident in turn reported to the Viceroy in Delhi. Only then was the collated news of Arabian affairs transmitted to the British Government in London, refracted, as it were, through the prism of Indian interests.

For some time before the twentieth century opened, Delhi and its agents had been forwarding indignant accounts of Russian activity in Persia and Afghanistan and among the tribes of India's North-West Frontier, where the names of Kabul, Kandahar and the Khyber Pass had already become to a newly jingoist British public both familiar warning signals of the 'meddling of the Russian bear' and stirring reminders of past feats of British arms in defence of India. So when France allied herself with Russia after the British occupation of Egypt, in 1882, and began to interest herself in the naval affairs of the Gulf as well, the Government of India's reaction was one of suspicion and alarm.

The first French effort to probe British India's defences in the Gulf was made in Muscat in the 1890s when a new French Consul there tried to lure the Sultan away from his British protectors by political intrigue and offers of trade. Unhappily for the Consul, and for French policy, however, he over-reached himself through clandestine involvement in the Sultan's illicit slave trade; and in the embarrassment of discovery the French cause went into swift decline. The Russians worked with more discretion. They already shared with Britain what amounted to an informal protectorate over the Shah of Persia and they had recently installed their Consul at Bushire alongside the British Resident. At the end of the nineteenth century they were also deep in negotiations with Abdul Aziz's friend, Sheikh Mubarak of Kuwait, for the right to build a new railway line from Kuwait to Tripoli on the Mediterranean coast of Syria, to circumvent the British controlled Suez Canal in the passage from Europe to the Orient. But they, too, were forestalled when the Viceroy in Delhi, Lord Curzon, got wind of their plans and promptly sent the Resident across from Bushire to attempt to frustrate them.

Blithely ignoring Turkey's old claim to sovereignty in Kuwait, the Resident negotiated a secret treaty with Sheikh Mubarak in 1899 that, in essence, gathered him into the British fold in the same way as many lesser rulers of the Gulf had already agreed to Britain's overlordship on pain of naval retribution. 'Mubarak . . . of his own free will and desire,' the document said, with something less than the strictest regard for truth, 'does hereby pledge and bind himself, his heirs and successors, not to receive the Agent or Representative of any Power or Government at Koweit, or at any other place within the limits of his territory, without the previous sanction of Her Majesty's Government.' He bound 'himself, his heirs and successors not to cede, sell, lease, mortgage, or give for occupation or for any other purpose any portion of his territory to the Government or subjects of any other Power without the previous consent of Her Majesty's Government . . ."³ The treaty was kept secret in order to avoid offence to the Turks. The world remained unapprised, therefore, of the vehemence of British feelings in the matter and the Russians continued with their railway talks.

To make matters worse from the British point of view, Germany soon entered the stage with a rival railway scheme. Sensing in the decay of the Turkish Empire a golden opportunity for commercial expansion, providing new markets for the growing industries of the Ruhr, the German Government had launched a *Drang nach Osten* (drive to the East) which was a conscious effort to turn the Ottoman provinces into a German economic preserve. At its core was the plan for a Berlin–Baghdad railway that would tie half the Middle East to Germany by bands of steel and open the possibility of a formal political alliance with Turkey that might threaten or even demolish the British position in the Middle East. With the Russian example before them, it was natural that the Germans should extend their ambitions to the Gulf; so now they, too, dangled before Sheikh Mubarak the attractions of a railway line that might make his little sheikhdom one of the great new commercial ports of the Eastern world. When, on top of this, the Russians started a heavily subsidized shipping line from Kuwait to Odessa and then, in the spring of 1903, joined the French in mounting a series of naval visits to Gulf ports, Delhi's anxiety was almost uncontainable. The offer of Russian arms and money to 'Abdul Aziz bin Feysul' was the last straw.

'I say it without hesitation,' the Foreign Secretary, Lord Lansdowne, told the House of Commons on 5 May 1903, 'we should regard the establishment of a naval base, or of a fortified port, in the Persian Gulf by any other Power as a very grave menace to British interests, and we should certainly resist it with all means at our disposal.'⁴ By implication that went for any rival links with local Arab rulers too, and – to press that point home for all to see – the British Government fired a second, more theatrical shot across the bows of its imperial competitors in the region. It dispatched Lord Curzon himself on a personal tour of the Gulf to let the Arab chieftains know in what high regard His Majesty's Government held them.

Curzon was the first Viceroy of India ever to visit these outlying provinces of his empire, and as he was by any odds one of the most remarkable embodiments of the British Raj, his tour made a profound impression, more than equalling that of the French and Russian gunboats. It is true that there were a few carping British critics at the time who called his excursion 'Curzon's prancings in the Persian puddle',[5] but he soon proved that in his person ceremonial flummery and diplomatic effectiveness were handsomely united.

He was a tall and angular man with a beak of a nose, a high-domed forehead, and an aristocratic temperament that brooked no cross. Nobody could have been better fitted to impose a sense of British superiority upon the Gulf. From Muscat to Kuwait he went, striking respect and awe into all who saw and heard him. In the frequent absence of adequate harbours and jetties at his ports of call, he was often carried ashore through the shallows in makeshift sedan chairs supported on the shoulders of British sailors or Arab fishermen. Decanted thus upon Arabia's beaches, in white-plumed hat and gold-braided Viceregal uniform, he was like some fantastical emperor made flesh. In Kuwait Sheikh Mubarak ordered a special victoria carriage from Bombay to carry him from the landing place into town, and his suite was mounted on Arab horses with several thousand Arab escorts firing their rifles wildly into the air in welcome. According to Curzon's own account, Sir Arthur Hardinge, the British Minister to Persia who was riding alongside him, was thrown over his horse's head at one point 'and deposited with no small violence upon the ground. Nothing daunted, he courageously resumed his seat and, amid a hail of bullets, continued the uneven tenor of his way.'[6]

It is hardly surprising that Curzon's tour is still recalled to this day in the Gulf as a famous occasion. In the British Embassy in Bahrain they still treasure the home-made carrying chair in which the great Viceroy was, or at least might have been, brought ashore. More to the point, Curzon's visit worked. Between his panache and Lansdowne's solemn warning, the Arab rulers were given to know on which side their bread was buttered, and all three of Britain's imperial rivals learned the limits of their influence. In 1905, Russian prestige in the Gulf dimmed further when its fleet sailed from the Baltic half way round the world in support of the Tsar's policy of Far East expansion, only to be destroyed, amid international consternation, by the brand new navy of Japan. After that, Russian influence faded from the Middle Eastern scene until a communist government resumed Moscow's Arab probings, several decades later.

But the Russian Consul had helped to give Arabia – and especially the new power of the House of Saud – a fresh significance in British eyes. Henceforth successive British and Indian governments were to be increasingly pestered by Arabian problems and concerned about keeping out rival powers.

Abdul Aziz himself was not displeased by the failure of the Russian overture. Indeed, in promptly repeating it to the Commander of the *Sphinx* he seemed deliberately to be inviting British intervention, showing a shrewd appreciation both of diplomatic tactics and of the reach of British power. Over the next few years he frequently tempted the British to go still further by hinting to them through various intermediaries that he would like nothing better than an 'understanding', or even a treaty, to place him on the same footing as those numerous Gulf rulers, including his friend Sheikh Mubarak, who were already secure under Britain's wing. Several of Britain's representatives in the area, and the Viceregal Government, also tended to this course. Accustomed to dealing with petty princelings of all sorts and still obsessed with an ever-present threat to India, they argued that a resourceful young sheikh like Ibn Saud, as they now began to call him, was probably worth some judicious support, if only to prevent him from turning elsewhere.

In local terms, they had some justice on their side. Abdul Aziz's exploit in regaining Riyadh had soon rallied the people of Nejd to the Saudi cause. His father, Abdul Rahman, agreed to surrender the family titles to him, so that he became known henceforth as Imam – religious leader – of the Wahhabis and Emir (Prince) of Nejd as well. Within three or four years, with the occasional aid of Sheikh Mubarak, he was able to eliminate all Rashid influence from Nejd in a series of desert skirmishes, and in 1906 his forces actually killed the new Rashid ruler, Abdul Aziz bin Rashid, in the field. When the Turkish Government sent several battalions of their own men against him in support of the Rashid, hoping to emulate the feat of Ibrahim Pasha at Diriya a century before, they were soundly thrashed by the Saudi cavalry, leaving numerous artillery pieces behind them in the desert. The prince was a man to reckon with.

But in London the Foreign Office saw it otherwise. Ibn Saud, they argued, was still largely an unknown quantity; his territory might be expanding but it was still small and of no intrinsic importance. Its material resources at the time would scarcely have sufficed for a small parish council in rural England. Nejd had barely recovered from a terrible drought when, it was said, men had eaten dogs in the streets of Riyadh.

As Prince of Nejd, Ibn Saud enjoyed a share of the date crop in the oases around Riyadh and profited, like other desert sheikhs, from the camel breeding of the desert tribesmen under his control. But his entire income probably did not exceed £50,000 a year and for many years he was able to transport the entire contents of his treasury in a couple of camel bags – just a few boxes of gold sovereigns, Indian rupees, and the big Maria Theresa silver thaler, or dollar, that was once the standard currency of the Austro-Hungarian Empire and that remained until only a few years ago a favoured coin in many parts of Arabia.[7]

Moreover, this sketchy and impoverished territory was still cut off from any direct contact with the British in the Gulf by Turkish control of Hasa, the province of the eastern Arabian shore, and the Foreign Office sceptics felt, on the whole, that it was better left that way. For was there not a risk, they asked, that encouragement of a Saudi revival would lead to a repetition of those Wahhabi excesses that had so troubled the British in the Gulf in the previous century? More important, was it not obvious that by helping Ibn Saud Britain must foster Turkish suspicions – already mounting over Britain's arbitrary actions in Kuwait – and thereby possibly hasten a Turco-German pact which could upset the balance of power throughout the Mediterranean and Europe? Certainly, it was plain within a year or two of Ibn Saud's return to Riyadh that he was intent upon smashing altogether the power of Turkey's allies the Al Rashid, and once that was done there would be nothing to stop him from reviving old Wahhabi greed for the spiritual benison and temporal riches of Mecca and Medina. If those cities were threatened, the Turks would have to fight to protect the last shade of the Sultan's imperial authority. And if the British so much as hinted that they were supporting Ibn Saud, they might then be sucked into a war they did not want for a cause they could not favour against their strongest traditional ally in the Middle East.

Caution won the day in London. Ibn Saud was left to make his own way in the void of Arabia's interior. Yet Europe was beginning its slide towards the first of its great twentieth-century wars. By 1908, the first tremors could be felt very clearly in Arabia when the old Ottoman imperial system was finally challenged from within by the revolt of the Young Turks.

The Sultan's government in Constantinople had been a byword for lassitude and corruption for over a century, and the young Turkish army officers who led the revolt were the spiritual forerunners of many a later soldiers' rebellion against decadent regimes in the Arab world. Although they respected the Sultan's person and maintained his titles as Emperor and Caliph, they reduced him to the role of a figurehead while they pursued their earnest goals of national reform and regeneration. In doing so, they released a bitter European scramble to tear the remaining flesh off the bones of the Turkish Empire.

In a series of wars, invasions and rebellions over the next five years Turkey lost nearly all her surviving Balkan provinces, had her foothold in Libya usurped by Italy, and her sovereignty in Crete denounced in favour of union with Greece. The repercussions in Turkey's Arab provinces were less dramatic but no less significant, for the secular nationalism of the Turkish officers provoked a corresponding feeling among educated young Arabs. Suddenly, these territories, too, stirred with the breath of release; and although the nationalism of the intellectuals and army officers in Damascus and Baghdad found no echo in peninsular Arabia, the mere hint of an Arab revival gave new impetus to international rivalry throughout the Middle East. By 1913, when Turkey's débâcle in Europe was virtually complete, the future of Turkey's Asian provinces had become one of the decisive questions of international relations – and Ibn Saud's role as a leading prince of Arabia inevitably assumed a greater significance in proportion to this shift.

Ironically, in the same year as the Young Turk revolt against the decadence of the Sultan's regime, Turkey's imperial government at last inaugurated the peninsula's railway age. The Berlin – Baghdad railway was still little more than a glint in German eyes, and after 1903 nothing had been heard of the proposals to build a port and railway terminus at Kuwait. But in the west the Turks had been building a narrow-gauge track for several years to link Damascus with the holy city of Medina through the province of the Hejaz. In 1908 it was completed and, while the Damascus caravan welcomed a faster, cheaper and safer way of reaching the Holy Places, cutting out the weeks of patient and dangerous travel by camel, Turkish military men could send soldiers southwards into Arabia without having to make the passage of the British-controlled Suez Canal. Here was the first twentieth-century breach in Arabia's physical defences, and one that was destined to play a minor role as the military focus of the Arab revolt against Turkish rule during the First World War.

But it was still another event of 1908 that cast the longest shadow over Arabia and the future of the House of Saud – the discovery of oil in southern Persia near the little town of Masjid es Suleiman. Although no one could have grasped its repercussions at the time, the strike was to transform the balance of economic power not only in Persia but throughout the world. The story behind it went back seven years to just about the time that Ibn Saud was laying plans for his desert foray to Riyadh, when a rich, walrus-moustached English speculator, who had made his fortune in the Australian goldfields, persuaded the Shah of Persia to put his name to a remarkable document. It was signed on 8 May 1901, and for the sum of £20,000 in cash it gave the Englishman William Knox d'Arcy an exclusive right to look for, produce and sell natural gas, petroleum, asphalt, and ozerite throughout the Persian Empire for a term of sixty years. It was the first oil concession of the twentieth century and the first ever to be signed with any Middle Eastern ruler. Although its terms were destined to be revised out of recognition long before its sixty years were up, it set the pattern for the early growth of the Middle Eastern oil industry.

At the moment of signature the industrial world was still coal-fired and the oil age was in its cradle. Only in the United States, where John D. Rockefeller had created his Standard Oil monopoly, had any of the great modern oil companies made their mark. Elsewhere, the discovery and exploitation of oil was more often a struggling, hit or miss business in which individual entrepreneurs either made a spectacular personal killing or went dramatically bust. Two such individuals had already achieved some prominence in competition with Rockefeller: the Englishman, Sir Marcus Samuel, whose company traded under the name of Shell, shipping oil from a few small fields in the Far East and the Caucasus, and the Dutchman, Henri Deterding, who had made a promising start in the Dutch East Indies.

The Middle East was virgin territory when the century opened. Only a young and still unknown Armenian called Calouste Gulbenkian had sniffed the possibilities of scientific exploration in northern Mesopotamia, where he had personally investigated some ancient naphtha seepages. The Turkish Government, however, had mulishly sat on all his proposals. Further south, around the Arab shore of the Gulf, there had been no reports of oil, and even if there had been, few speculators would have been willing to risk a penny among the lawless tribes.

D'Arcy's party endured dysentery, plagues of locusts and an outbreak of smallpox. They wrestled with obstructive Turkish customs officials in Basra to release their equipment for use in Persia and they organized mule and camel trains across the sweating, malarial plains of the Mesopotamian delta and up to the fringes of the Zagros mountains. But they found no oil; and within three years even d'Arcy's purse was beginning to feel the strain of failure.

Yet in those three years, engineers and manufacturers on both sides of the Atlantic had learned new ways to exploit oil's huge potential. Among them was a small group of British naval officers led by Admiral Sir John (later Lord) Fisher, a friend of Sir Marcus Samuel of Shell. Fisher – perhaps not entirely coincidentally – was convinced that to sustain the Navy's supremacy against the German threat the coal-fired boilers in its ships must be replaced by the new oil-fired kind. This in turn required that Britain must find a source of the new fuel that would compare in security, if not in access, to the coalfields of Lanark and South Wales.

Apart from a small field on the island of Borneo, and a bigger one just discovered in Burma, Britain's imperial territories looked unpromising. Nor was Sir Marcus Samuel's company any longer of much help, for in 1905 he was on the brink of entering a forced merger with Henri Deterding, on unfavourable terms, to form the Royal Dutch Shell Company – and the British Government was suspicious of the foreign entanglement. D'Arcy's lone venture in Persia therefore acquired fresh significance. Politically, southern Persia was under British control, in effect if not in name, and Persian oil would be nearer to Britain than that from the Far East. Prodded eagerly by Fisher the Government agreed to find d'Arcy a partner to help him stay in business. They lit upon the new Burmah Oil Company, a consortium of safe Scotsmen whose happy discovery in Burma, then under the equally safe Indian Government, had given them resources to invest elsewhere.[8]

With Burmah's aid d'Arcy carried on; but it was another three years before the gamble paid off. Just before dawn on 26 May 1908 the rig at Masjid es Suleiman, some 130 miles north of the Gulf coast, trembled and with a thundering roar a jet of black oil shot into the air through the latticework of the derrick, drenching the delirious drillers.

When the news reached Britain there was high excitement. A new group called the Anglo-Persian Oil Company was formed, with

£1,000,000 from Burmah as the basic stake, and preference shares were sold to an enthusiastic public queuing ten deep at bank counters. D'Arcy got his money back and £900,000 worth of Burmah shares on top.

Oil was clearly destined to become the biggest thing since steam. Already there were half a million cars on the roads of the United States and in London the first double-decker motor buses were on the streets. In France, Blériot was building the string-and-cloth monoplane that was to make the first crossing of the Channel the next year.

Anglo-Persian began at once to build the first regional oil pipeline from Masjid es Suleiman to the Gulf, where it created the port and company town of Abadan, and the Middle East's first oil refinery. At the same time the company and the British Government entered a long and complex series of negotiations in which the company sought a Government subsidy on the grounds that it was now the safest source for the Royal Navy's fuel oil while the Government sought a guaranteed price advantage in return. The result, triumphantly announced by Winston Churchill, as First Lord of the Admiralty, just three months before the outbreak of the First World War, was that the Government bought 51 per cent of Anglo-Persian's shares in exchange for control of the company's prices to the Navy, thus putting itself squarely into the international oil business. Thanks to that arrangement, succeeding British governments have remained in the oil business ever since, albeit often with growing embarrassment, as Anglo-Persian has been transformed progressively first into Anglo-Iranian in 1935 and then, in 1954, into the giant British Petroleum.

British companies also took up Gulbenkian's old reports on the Mesopotamian naphtha seepages. The Turkish Government, with the new army officers in charge, proved more receptive now to that idea and within five years a new consortium had been formed to explore and exploit the area. Originally three-quarters British and one-quarter German, with the famous 5 per cent for Gulbenkian himself, it was called the Turkish Petroleum Company; but in due course it, too, like Anglo-Persian, changed its name (and some of its shareholding) and re-emerged as the modern Iraq Petroleum Company.

Around the shores of Arabia, the British Government in India was moving swiftly to protect Britain's future oil interests in another way. In 1911 the ruler of Bahrain was persuaded to sign an exclusive treaty promising to grant no oil concessions without the prior agreement of Delhi. In 1913 the ruler of Kuwait obligingly followed suit.

These were only the first of many such agreements to be signed over the next twelve or fifteen years between Britain and Arab rulers in the Gulf and they confirmed in an unmistakable way the shift in Britain's Arabian interests that had been foreshadowed by the Russian Consul's interview with 'Abdul Aziz bin Feysul' in 1903. For over one hundred years those interests had been exclusively maritime, directed to the preservation of the Indian Empire and its trade; but now it was the land of Arabia, just as much as its surrounding seas, that was beginning to hold global attention. In future, if oil should ever be found on Arabian territory, Britain – and the other foreign powers, including the United States – would be compelled to look to that land and ensure its security. Thus, between the Consul and the oilman, the world was catching up with Arabia at last and Ibn Saud's years of isolation were nearing their end.

4 Shakespear Rides Out

1910 – 15

Meanwhile, as of yore, the small states and nomads war together from Kuweit to Aden and from Muscat to Sinai, wildly and bloodily, in the ingenuous belief that they act of their own free will to avenge real or imagined insults and to feather their own nests. Little do they know that in all their seeming exercise of freedom they are really subserving a great political game.[1] – BARCLAY RAUNKIAER

DEEP in the eastern desert of Arabia one night in February 1910 a young Englishman was encamped with a small group of Arab guides and companions listening eagerly to the gossip of the day. Tall, well built and handsome, and smartly dressed in the khaki uniform of a British Army officer, he could have passed at a glance for one of the imperial heroes of the *Boy's Own Paper*. He was quite unlike most of the Europeans who had preceded him in the Arabian interior. They had generally assumed Arab robes to protect themselves from Wahhabi fanaticism or plain bedouin hostility; none had attempted, let alone achieved, more than brief expeditions of geographical or political exploration. The new man in his military uniform was deliberately announcing himself as an official representative of His Majesty's Government, seeking a more permanent relationship with the desert tribes; for he had recently been appointed Britain's Political Agent in Kuwait and he was determined to use that job to promote, if he could, a deeper British interest in the affairs of central Arabia.

He was William Henry Irvine Shakespear. If any one man embodied the arrival of the twentieth century in Arabia it was he. In a way this was paradoxical, for Shakespear not only looked like a *Boy's Own Paper* hero but tended to behave like one too. Born into the heyday of imperialism, thirty years earlier, he had become a captain in that most splendid of all Indian regiments, the Bengal Lancers. His insistence on wearing Army uniform in the desert sprang, at least in part, from his conviction that to adopt a disguise was beneath the dignity of any Englishman, most of all that of a servant of His Majesty. But Shakespear also broke the mould of his kind. He had an unusual flair for oriental languages and a restless, driving temperament. He was a passionate motorist and a talented photographer and had a keen, but unfulfilled, ambition to become an air pilot long before such pursuits were taken seriously. He was never afraid to make up his own mind on controversial issues and challenge his superiors if he thought it necessary. He combined discipline with resolution and genuine independence and now, in the job to which his talents had brought him, he was characteristically eager to assert himself on behalf of Britain in the Arabian interior.

He had already formed the impression that the man he must seek out was the rising new Prince of Nejd, Abdul Aziz bin Saud. He knew of the overtures that Ibn Saud had made to the British over the previous few years and of London's deep reluctance to entertain them. He was aware of the current of sympathy among his colleagues in the Gulf, on the other hand, for closer contact with the Saudi ruler; and it was his dearest wish that he might become the first foreigner to meet Ibn Saud since those curious encounters with the Russian Consul and the commander of the *Sphinx* in Kuwait in 1903. As long as London maintained its fear of offending Turkey and encouraging a rebirth of militant Wahhabism, he was hamstrung. So he had adopted a compromise of pursuing innocent geographical exploration in the desert hinterland of Kuwait, as far as the fringes of Saudi territory, while trying simultaneously to befriend the tribes there and pick up enough favourable intelligence about Ibn Saud's activities to support his plea for closer links with the reviving House of Saud.

What actually befell him on this February night, however, was pure grist to Whitehall's mill, as if to emphasize yet again in the eyes of fearful officials there that central Arabia was still a place to stay away from. He was in the middle of his first major desert journey and his guides had just come back from a chance meeting with some fellow tribesmen full of a colourful bedouin story about Ibn Saud who, so they claimed, had just had a miraculous and at the same time tragic escape from death at the hands of his own family. Two of his nephews, they declared excitedly, had secretly put poison in the coffee that he always drank immediately after his early morning prayers. But before he could touch it, the wives of the two men – who were Ibn Saud's own sisters – had tasted the coffee to make sure it was properly prepared. Within seconds they had fallen dead, so powerful was the poison, and their husbands had fled immediately to escape Ibn Saud's vengeance by claiming asylum in one of the Turkish garrisons on the eastern coast.

Here was rich stuff indeed, and we may be sure the storytellers made the most of it, with every embellishment of language they could muster and no nuance of ancestry or family rivalry left unexplored. So absorbed were they and their little audience that they failed to notice that their tents had been surrounded in the darkness until, of a sudden, they heard the click of rifle bolts being drawn back and, in the abrupt silence that followed, the whispers of unseen men. Grabbing their own guns they scrambled away from the dangerous light of their camp fire, but in the confusion of shots and shouting that filled the next few minutes one of the guides was hit and fell, bleeding from the mouth. It was only then that the shouts of hostility gave way to cries of recognition as the combatants realized they knew each other. The attackers turned out to be old acquaintances of the storytellers and soon both sides were embracing each other and sitting down to the fire again with more coffee, and more stories to tell.

Shakespear was understandably enraged by this calamitous display of desert caprice. His best guide was badly wounded, his camp was a shambles, and he had only just escaped death himself. Seizing one of the intruders he beat him almost senseless, then picked up the injured guide and carried him to his tent. But he could do nothing for him. The bullet had pierced the man's lung and within an hour or two he was dead. Plainly, Arabia was still a dangerous place and Shakespear could have been forgiven if he had concluded, between the story of Ibn Saud's poisoned coffee and the fatal attack on his own camp, that London was right, after all, in wanting as little to do with it as possible.

Yet when he reached Kuwait again at the end of his ill-fated trip, he was immediately given the opportunity of meeting Ibn Saud, and any doubts he might have felt about Arabia's future under him were promptly silenced. The Nejdi prince had once more come to Kuwait, accompanied by some of his brothers, to visit his old friend and adviser Sheikh Mubarak, and on the very night of Shakespear's return, there was a banquet in Ibn Saud's honour to which the Political Agent was invited. It was, so Shakespear noted, a more than usually lavish affair, denoting the high esteem in which Ibn Saud was held. 'At a very rough guess entertainment and presents for these princes cannot have cost less than £20,000,'[2] he wrote. Next day the Agent was able to offer his own entertainment to the visitor and although neither custom nor the resources of His Majesty's Government permitted him to rival Sheikh Mubarak's munificence, he did his best with an English menu of roast lamb with mint sauce, roast potatoes and tinned asparagus – the first time, surely, that Ibn Saud and his retinue had ever tasted such curious fare.

The result was a cordial meeting of minds. Henceforth Shakespear never wavered in respect, affection and support for Ibn Saud, and his feelings seemed to be reciprocated. Although at this first meeting the two men discussed no politics, they evidently enjoyed each other's company and each seems to have recognized in the other from the start a potentially valuable partner as well as a kindred spirit. As Ibn Saud left, he invited Shakespear to pay him a visit in Riyadh.

It was an invitation the Englishman would have loved to accept at once but, in spite of his highly favourable description of the meeting, London remained as wary as ever. In later years many critics were to attack the Whitehall officials and their ministers for 'missing the boat' with Ibn Saud by not acting on Shakespear's advice. But Turkey had been Britain's Middle Eastern buffer against European rivals for at least a century, and the more dangerous those rivals grew in the last few years before the First World War the more important it seemed to maintain Turkey's dominions in Asia as the cardinal factor in Britain's entire Eastern policy. And as those dominions still included, by Turkish reckoning, the Arabian heartland of Nejd, the British Government felt compelled, categorically and repeatedly, to reject all Captain Shakespear's notions about independent relations with, or friendly visits to, the man who was now the actual ruler of Nejd. All the accumulated wisdom of past experience suggested that Arabia's shifting tribal loyalties were never likely to be permanently settled and that if the House of Saud could rise yet again under Ibn Saud it could as easily fall once more if he were to disappear from the scene.

The affair of the nephews and the poisoned coffee seemed to prove the point. For behind that bit of desert gossip gaped a genuine dynastic schism that might easily have stopped the revival of Saudi fortunes and is still, even today, avoided by some Saudis as a topic too delicate for frank discussion. The dispute grew out of the fratricidal rivalries begun more than forty years before when Ibn Saud's uncles – the elder brothers of his father Abdul Rahman – had fought for the succession to the Saudi throne. The ambitions then released were not forgotten. They help explain Ibn Saud's bold attack on the Musmak. In this dynastic squabble, possession was everything. But even after Ibn Saud had captured Riyadh he was challenged by jealous cousins and nephews who claimed seniority. Rather than accept the supremacy of the junior branch of the family they threw in their lot with the Rashid, hoping thereby to hasten the defeat of Ibn Saud in the field and claim Riyadh once more for themselves.

They lost their first battle, however, in 1906 when, in a desert skirmish with the Rashid forces, Ibn Saud captured three of the renegades in the enemy camp. In other Arab hands their fate would almost certainly have been instant death, but Ibn Saud proved a merciful opponent and forgave them, offering them a home and a respected place in family affairs by his side. They accepted his offer – and in doing so earned the family nickname of the *Araif*, a term used by the bedouin for camels lost in one raid and recovered in another. But the *Araif*'s gratitude was merely dissembled. In 1910 the attempt by the two nephews to poison Abdul Aziz proved to be only the herald of another general *Araif* rebellion against his leadership in which parts of several tribes, including the powerful but notoriously fickle Ajman of the eastern region, took the field against Ibn Saud's forces. For another six years the *Araif* and their allies were to offer a persistent threat to Ibn Saud's life and authority until most of them were crushed in battle and the survivors decided to accept his rule, the last and most powerful of them, Saud Kebir, surrendering in 1916. He was generously pardoned and given the ruler's favourite sister, Nura, as his wife – she who had fled Riyadh, with Abdul Aziz, in a camel bag.

A second incident of 1910 was just as threatening and had far wider repercussions. It was the first direct clash between Ibn Saud and a new, rival ruler in the Hejaz, Hussein bin Ali, appointed Sherif of Mecca by the Turks in 1908. The title of Sherif was an ancient and honourable one, especially in the Hejaz, indicating direct descent from the Prophet; and as a member of the noble House of Beni Hashem, of the Prophet's own tribe of Quraish, Hussein bin Ali claimed descent in the male line from Muhammad's daughter, Fatima. In Arab eyes, therefore, he was a man of incomparably aristocratic lineage, purer of blood and holier by tradition than any other Arab ruler. By temperament and appearance, too, he commanded authority. Already nearly sixty at the time of his appointment, he had been obliged in a very Turkish way to spend much of his adult life as the Sultan's 'honoured guest' in Constantinople in order to keep him out of mischief among the Arabs at home. He found the Hejaz decidedly provincial after that, for all its significance as the Muslim Holy Land; and he soon sought to extend his authority there into an arena better suited to his own assessment of his merits. Anxious not to offend his Turkish patrons by seizing too much power, yet eager to impress them by his command, he saw a promising opportunity in the still unsettled affairs of central Arabia.

To the Sherif, so conscious of his own worldly experience, age and breeding, the young Ibn Saud was no better than a tribal upstart, for all that he, too, was of pure Arab blood. Yet Ibn Saud had proved unpleasantly successful so far and, if left unchecked, he might be tempted to emulate his ancestors and try to take over again the Sherif's Hashemite homeland in the Hejaz. Constantinople would be happy if the Sherif prevented that and happier still, perhaps, if he could just take the Saudi leader down a peg or two. So it came about that in the summer of 1910 the British Consul in Jeddah reported to his chiefs in London how he had learned from the Hashemite court that Sherif Hussein had recently demanded an annual tribute in cash from Ibn Saud and was secretly exhorting some of the Saudi subjects to switch their allegiance to himself and his Turkish masters.

Hussein can hardly have expected that he would get anything from Ibn Saud but a short, sharp answer; but he knew of the embarrassment of the *Araif* uprising and, supremely confident of Turkish support and his own power, he sent a military expedition into Ibn Saud's western territory to show the desert chieftain that he meant business. There he had a stroke of luck, for his men captured Ibn Saud's favourite brother, Saad, who was in the same area raising tribal recruits for the campaign against the *Araif*. In return for his release Hussein extracted a promise from Ibn Saud that he would accept Turkish overlordship and pay the Sherif a small annual tribute.

It was a triumph for Hussein and a substantial blow on behalf of Turkey. But it was a humiliation for Ibn Saud, and he was not a man ever to forget it. As soon as his brother was returned to him he repudiated the promise he had made to the Sherif, protesting that it had been extracted under duress, and he determined that when the next confrontation arrived he, Ibn Saud, would be the master. Thus began a contest that set the House of Saud against the House of

Hashem for the next four generations, to be ended only when the present King Hussein of Jordan, great-grandson of Sherif Hussein of Mecca and the last reigning survivor of the Hashemite dynasty, buried the hatchet at last with Ibn Saud's son, King Feisal, in a mutually protective alliance of monarchs signed in the 1960s.

With the preservation of the Turkish alliance ever in mind, the British Government told Shakespear firmly it could not countenance any further encouragement of Ibn Saud. 'The objections to a policy of adventure,' the India Office wrote loftily, had not diminished.[3] He was instructed instead to confine his desert expeditions to the immediate surroundings of Kuwait and do nothing to let Ibn Saud believe that the British regarded him as anything more than just another Arabian tribal chief.

Ibn Saud's success in the eight years since he had made himself master of Riyadh was by no means total, yet his territory by now stretched north into the region called Qasim between Nejd and the rump of the Rashid domains around Hail. Westwards his influence reached almost to the mountains that mark the borders of the Hejaz, and southwards he controlled the desert up to the Empty Quarter. In the east the Turks still commanded Hasa, along the Gulf coast, but Turkish soldiers had not ventured into the desert against Ibn Saud after their rout alongside Rashid forces a few years before.

Ibn Saud was respected and feared also for his temper and moral strictness, and he was beginning to make a name for himself in statesmanship and tribal politics. He had a reputation for ardour in the marriage bed which stood him in especially good stead. By 1910 he had fathered four living sons and several daughters by a variety of wives, and tribal sheikhs all over central Arabia were happy to offer their daughters to a man who had so effectively proved his potency in more ways than one.

News of Ibn Saud spread on the Turkish grapevine to Constantinople and so to Europe, where ambitious explorers as well as statesmen and officials became anxious to know more of the powerful new chieftain who was imposing his will on Arabia's unruly tribes. But the man to break Riyadh's isolation was not to be Captain Shakespear. Many years later, after Shakespear was long dead, when Ibn Saud was asked to name the greatest of the Europeans he had met in his life, he replied unhesitatingly, 'Captain Shakespear!'[4] This was accepted by all who knew the Saudi leader as genuine testimony to the mutual respect between the two men. In view of this friendship it was all the more galling for Shakespear that Riyadh remained, by order, out of his reach while other Europeans who had never met Ibn Saud were able to precede him there. One of them was a British colleague, Captain Gerald Leachman. He set off from Damascus and reached the Saudi capital at the end of 1912 before the British authorities could stop him. Completing his unofficial trip a few weeks later by continuing eastwards to the Gulf, he became the first European to cross Arabia from coast to coast since Captain Sadleir, nearly a century before.

Leachman was beaten to Riyadh by a determined young Danish traveller named Barclay Raunkiaer who was sent off by the Royal Danish Geographical Society some nine months earlier to reconnoitre Arabia's eastern and central deserts in preparation for a bigger expedition the Society hoped to mount two years later. Raunkiaer picked up a virulent fever in Kuwait that proved to be tuberculosis and he was held up by the ill-tempered Emir of the little town of Buraidah – a crabbed and dirty place where Charles Doughty had also been much abused and where, even to this day, only a fool would smoke a cigarette in public. When the Emir finally released him he was almost too sick to travel, but he pressed on somehow into Riyadh, only to find that the ruler had left his capital two days previously to attack some tribes, men of the *Araif*, to the west of his territory who had been plundering his caravans.

Raunkiaer was the first European to see Riyadh for half a century and the first to leave any record of the place under the rule of Ibn Saud. As far as he could see it had changed not at all since Palgrave and Pelly had been there in the 1860s. His illness compelled him to leave Riyadh sooner than he would have liked in search of medical treatment from the British in Bahrain (he never recovered from his infection and died in Denmark two years later). He did, however, meet Ibn Saud's father, Abdul Rahman, who – as usual – was deputizing as ruler in his son's absence. Abdul Rahman soon turned their talk into an eager review of international politics. 'While we drink coffee and tea alternately, we engage in lively conversation about Kuweit, about [Ibn Saud's] expedition, about the interests of England and Turkey in Arabia, about the Turco-Italian war [in which Italy had just won possession of Tripoli] and finally about the relative power of the European States. On the last subject, especially in the matter of Africo-Asiatic politics,' Raunkiaer noted wryly, 'I could do no more than confirm the chieftain's deep-rooted belief in the hegemony of the British Empire.'⁵

Shakespear still could not move the British Government from its view that the Turks were its best friends in Arabia. By the spring of 1913, not long after Raunkiaer had finished his trip, the British and Turkish Governments had so far agreed on their common interest that they were on the point of signing a new convention specifically recognizing Turkish suzerainty over Ibn Saud, and the wires between London and the Gulf were hot with rebukes for Shakespear for continuing to extol the qualities of a man the Foreign Office now preferred to regard as a rebel against Turkey's rightful authority. If Ibn Saud was a rebel, however, he was about to prove himself an effective one, for in May he suddenly decided to assert his growing power in a new direction by throwing the Turks out of Hasa.

At the time, Hasa consisted of the long oasis of Hofuf, with its abundant natural springs, the smaller oases of Qatif and Seihat, largely populated by farmers of the Shia persuasion, and a string of tiny fishing villages. Although the Seihatis, at least, were not enamoured with the prospect of Saudi rule and actually sought British protection, the decay of Turkish power and the growing lawlessness had alienated the leading Sunni merchants in Hofuf. According to tradition in Hofuf, the leader of this 'fifth column' was one Ibrahim Qusaibi, from a family that was later to distinguish itself in the service of Ibn Saud. Whatever this family's actual role, the Turkish garrison of Hofuf surrendered when Ibn Saud easily secured the walls. Within a week or two, the whole of Hasa was in Saudi hands and Ibn Saud now had an outlet to the sea. His exiguous income was doubled as he gathered in the taxes from the date gardens and the customs dues from the small harbours. He now faced the British, a mere twenty miles away on the island of Bahrain.

As Europe moved rapidly towards war and Germany's long-standing policy of Eastern penetration was pursued with ever more determination and success, the threat of provoking a Turco-German pact had begun to haunt every waking hour at the Foreign Office. Adamantly, its officials insisted that no second thoughts about Ibn Saud could be entertained and the Anglo-Turkish convention was signed a few weeks later, formally consigning Ibn Saud to the lowly status of Mutasarrif, or administrative officer, of the Ottoman Sanjak (province) of Nejd. In London the Foreign Secretary, Sir Edward Grey, noted sternly that from now on Abdul Aziz bin Saud 'must be dealt with as a Turkish official or not at all'.

It was a fiction and could not last. The Turks had never controlled Nejd – except briefly through their proxy Ibrahim Pasha – and they could not now control Hasa either. The only substance to their claims was that lent by British strength; and in quiet deference to that Ibn Saud swallowed his disappointment again, made a few polite noises in the direction of Constantinople, and was graciously pleased to accept a small Turkish salary for his nominal Governorship.

To Shakespear, Albion was not merely perfidious but obtuse. However, he was mollified by at last receiving London's consent to his cherished project of visiting Riyadh. His tour of duty in Kuwait was almost up and the visit was to be a consolation – part of a great final flourish to his brief Arabian career in which he proposed to cross the entire peninsula from coast to coast. To allay Foreign Office suspicions about his purpose, and Turkish reaction to the trip, he offered to avoid all political involvement or conversation on the way and confine himself purely to geographical exploration. Reluctantly, under great pressure from Sir Percy Cox, the Resident in the Gulf, and the Viceroy of India himself, Whitehall sent its agreement. Clearly, Ibn Saud's new position removed the problem of Turkish sensibilities. The Viceroy's chief argument was that it would be a pity if the field were closed for political reasons to Englishmen while foreigners (including a young Russian lady called Countess Molitor) were planning crossings.[7] After five weeks on camel-back from Kuwait, Shakespear reached the Saudi capital in March 1914.

He had a much more rewarding time than Raunkiaer, for he was greeted warmly by his friend Ibn Saud and was able to stay long enough to indulge his enthusiasm for photography in a series of family portraits and local scenes that comprise the first record in pictures of the heart of Arabia. From it we can see Riyadh was still little more than a mud-walled village, even though, as Shakespear recorded, 'Quite a third of the town is taken up by the homes of the Saud family.'[8] In spite of his promise to London, politics dominated his talks with Ibn Saud who reaffirmed yet again his faith in what he touchingly described as the 'Great Government' of the British Empire[9] and urged Shakespear for the last time to negotiate an alliance for him. Shakespear was all sympathy but could promise nothing; and his last sight of the Saudi leader before he returned to England was of him mounted among his men as he led them out of Riyadh to yet another raid against the *Araif* – a noble and romantic figure but still apparently an outcast.

That should have been the end of the Shakespear story. After the completion of his desert crossing, which he accomplished without mishap, becoming only the third European after Sadleir and Leachman to make such a trip, he made his way back to England. But no sooner had he reached that country than all the calculations of Britain's Eastern policy for a century were overthrown by the Serbian nationalist Gavrilo Princip when he jumped in front of the coach carrying the Archduke Franz Ferdinand in Sarajevo and shot him.

As the great powers promptly split into two warring camps it soon became clear that their struggle was bound to spill over into the strategic heartland of Turkey's empire in the Middle East. But with Britain now allied to France and Russia on one side, Turkey swayed in natural reaction towards the opposing camp of the Central Powers, Germany and Austria-Hungary. The German *Drang nach Osten* had at last succeeded and by November 1914 Britain and Turkey were formally at war.

For Ibn Saud the repercussions were immediate and gratifying. From being ostracized by Britain in the greater interest of the Turkish alliance he was now sought as an ally himself against the Turkish threat to Britain's Eastern empire, and his friend, Captain Shakespear, was recalled at once for a fresh tour of duty in Arabia with instructions directly contrary to those he had previously found so irksome. This time he was to join Ibn Saud in Riyadh and ensure at all costs that he was kept out of the Turkish camp.

When Shakespear reached the Saudi capital for the second time in January 1915 he found his old friend already under Turkish pressure and eager to renew his demand for a treaty with Britain. Shakespear responded gladly. Within a week he had drafted a document that won Ibn Saud's approval, recognizing him as the independent ruler of Nejd under British protection. Then, sending off the document by messenger to his old Agency in Kuwait in the confident expectation of acceptance and action in higher quarters, he joined Ibn Saud on the warpath. Riding together out of Riyadh with 6,000 men, they went to meet the forces of the old enemy, Ibn Rashid, near the little town of Zilfi, some 250 miles to the north, on the edge of the great arc of sand desert called the Nefud that sweeps down from Saudi Arabia's northern borders into the heart of Nejd.

The battle that followed, at a spot known as Jarrab, was at one level a proxy war between Britain and Turkey, with Ibn Saud and Shakespear standing side by side against the Rashid forces who, as usual, enjoyed Turkish help. But it was also a completely traditional desert affray of bedouin horsemen, camel cavalry and thousands of barefoot tribesmen battling hand to hand among the sand dunes. Shakespear, so his Arab companions related afterwards, was hugely excited by the scene, exulting that no European had ever before witnessed anything like it. He took up station behind the infantry in the centre of the Saudi line, alongside a young gunner called Hussein. In front was the green Wahhabist banner with its inscription of the Muslim creed. To left and right the cavalry were spread upon the flanks, with the slender Arab horses tied, unmounted, to the fast war camels so as to save their energies until the last moment for the first and swiftest charge. The infantry were a wild-looking lot, with their long, dirty shirts hitched up about their calves to free their movements and their checkered head scarves discarded to show their long black curls. A few wore old woollens and sheepskins against the winter chill, but most crouched unblinking in the cold in nothing but their cotton shirts. Bandoliers were stretched across every chest, great curved daggers were stuck into every belt, and rifles of a multitude of makes and ages bristled at all angles from their massed ranks where they waited behind the dunes for the Rashid forces to attack.

Discipline was rarely the strong point of any bedouin army; only an eye for the main chance and the hope of loot held them together, and the battle of Jarrab proved no exception. When the Rashid cavalry appeared, charging over the sand and salt flats with their war banners streaming in a great cloud of dust, shrieking their tribal war cries and shouting *Allahu Akhbar!* (God is Great!), the Saudi army at first held firm. The cavalry counter-charge from the flanks went in to divert the Rashid thrust and soon there was a mêlée of hand-to-hand fighting with dagger, sword and rifle butt across the little plain in front of Shakespear's position. Shakespear himself was in full view, standing on top of a dune in his British khaki uniform and solar topee, alternately taking photographs of the scene and trying to direct the fire of Hussein, the gunner. Presently, however, out of the dust part of the Rashid cavalry emerged, charging at full speed upon the Saudi infantry in front of Shakespear. For a few moments there was ragged fire from the wild men behind the sand dunes, but as the mixed cohorts of camels and horses thundered down upon them the bedouin ranks broke. Within minutes they were fleeing in hundreds past Shakespear's position, and Hussein, seeing his own gun emplacement left unprotected, paused only long enough to jam the breech before he, too, fled calling to Shakespear to do the same. But the Englishman stood his ground. When last seen, his solar topee was gone but he had drawn his service revolver and, bareheaded, was shooting at the oncoming cavalry at almost point-blank range while the Saudi forces melted from the field.

Ibn Saud afterwards blamed one of his tribal contingents, the Ajman, for changing sides in the middle of the fight; and with their record of treachery and sporadic friendship with his family rivals, the *Araif,* that is probably as good an explanation as any. Certainly, the Ajman later helped the Rashid forces to clean up the booty on the field, which was probably all they were concerned about anyway.[10] But whatever the explanation, Shakespear was dead, and with him had gone Ibn Saud's most devoted and eloquent of foreign admirers. Yet, by a tragic irony, his death helped to accomplish what his life had left undone.

For an official representative of His Majesty's Government to be killed in an inter-Arab battle of this kind was a unique event; and in the changed circumstances of the First World War it reinforced in the most dramatic way possible the message that Britain could not stand aloof from the internal affairs of Arabia any longer. At the very least, any further successes by the pro-Turkish forces of Ibn Rashid had to be redressed in Britain's favour. As a result, Ibn Saud became a British protégé at last, assured of a reliable source of arms and money for the first time.

With Turkish aid now flowing to Ibn Rashid more freely than before, Ibn Saud's needs were actually more pressing then ever. In July 1915 he wrote to Sir Percy Cox explaining that the war had closed many of the normal avenues of trade for his people who were unable to sell their date crop to the Hejaz, or to send their camels to market in towns of northern Arabia and Mesopotamia, because of Turkish interference. Also, he added ingenuously, he had been put to very heavy expense owing to the necessity for keeping large forces under arms over a long period. Cox sent him 300 captured Turkish rifles and 10,000 rupees in temporary help. But it was not until June 1916 that His Majesty's Government approved the supply of another 1,000 Mausers, 200,000 rounds of ammunition, and a £20,000 loan.

By the end of 1915, the bureaucratic mills of the British Government, grinding with accustomed slowness between London and Delhi, managed at last to turn Shakespear's draft treaty into its final form. Signed on 26 December 1915, it formally recognized Ibn

Saud as the independent ruler of Nejd and its Dependencies under British protection. In return, Ibn Saud undertook to follow British advice and to refrain from any interference in the affairs of neighbouring sheikhs who were also in treaty relations with Britain. In effect, he had joined the sheikhs of Bahrain, Kuwait, Abu Dhabi and all the rest. But whereas the other chieftains ruled only coastal territories, Ibn Saud commanded the innermost reaches of Arabia.

Although the Saudi treaty was to be outmoded within ten years by a further rapid extension of Saudi power, it marked Britain's decisive abandonment of its old policy of avoiding Arabian entanglements and a fundamental change in the relations between the warring sheikhs of the peninsula and the outside world. Henceforth, Arabia and the world would be increasingly enmeshed, in defining frontiers, settling disputes, pacifying rebellious tribes, awarding oil concessions, and in general imposing an altogether novel permanence and rigidity upon the hitherto timeless flux. It was the end of Arabian innocence.

5 The Arab Revolt
1916

Some Englishmen, of whom Kitchener was chief, believed that a rebellion of Arabs against Turks would enable England, while fighting Germany, simultaneously to defeat her ally Turkey...So they allowed it to begin, having obtained for it formal assurances of help from the British Government. Yet none the less the rebellion of the Sherif of Mecca came to most as a surprise, and found the Allies unready. It aroused mixed feelings and made strong friends and strong enemies, amid whose clashing jealousies its affairs began to miscarry.[1]

T. E. LAWRENCE

THE changes wrought by the 1914–18 war upon the Arab lands of the Middle East, like many of the repercussions of that great conflict, were revolutionary. The war changed the map of the Middle East by destroying the Turkish Empire and releasing in its place all those forces of modern nationalism which are now so important throughout the Arab world. But it also left a legacy of Arab mistrust and confusion that has not yet been overcome. In these few years of profound and costly transformation, the Middle East as we now know it was born.

The House of Saud played little part. The negative advantage of Saudi neutrality, or passive friendship, was as much as Britain needed, or was prepared, to promote. Yet once Turkey had declared itself on the side of Germany it was natural that Britain should seek more active Arab allies to protect its Middle Eastern interests and harass the Turkish armies. The choice fell upon the House of Hashem, led by Ibn Saud's new enemy, the diminutive, vainglorious but wily Sherif Hussein of Mecca, whose name was inscribed, at Britain's instigation, as leader of what came to be called the Arab Revolt.

This was then, and has continued to be, a source of controversy. It was suggested at the time, especially by some British officials in the Gulf – where, after Shakespear, Ibn Saud's virtues were increasingly admired – that Hussein's elevation to the role of wartime Arab leader and Ibn Saud's virtual exclusion from the battle were glaring examples of British misjudgement, comparable to the cavalier treatment of the Saudi leader before the war.[2] As the Saudi state has grown in wealth and stature over the years since then, while the Hashemite legacy has contracted to a single, unsteady kingdom on the left bank of the Jordan, that argument has come to seem all the more persuasive. But what has helped perhaps more than anything else to keep the dispute alive is that it has become entangled with a more personal and flamboyant debate about the role and reputation of that strange, ambivalent figure who made his name out of the Arab Revolt – T. E. Lawrence. The truth is that whatever the long-term effects may have been of Lawrence's lingering spell, his practical influence on policies in his own time was less than might appear and certainly had little if anything to do with the British choice of Sherif Hussein as their chief wartime Arab ally in preference to Ibn Saud. That was governed by far more serious considerations.

Behind all the romance that Lawrence and his friends eventually concocted, the Arab Revolt was largely a marriage of convenience between Britain and the Hashemites, cemented by about £1,000,000 in British gold.[3] The Arabs, who can be devastating realists when they choose, have usually been more candid than the British about this. Once, about twenty years ago, I encountered the paramount sheikh of a powerful bedouin tribe in southern Jordan whose family had dealt with Lawrence in the First World War. He was old then, so I asked him if he remembered the man most Arabs knew as 'Laurens'. 'Yes,' he said, sucking his gums reflectively. 'And what did you think of him?' I asked. There was the merest suspicion of a sniff. 'Laurens? He was the man with the gold.'

The basis of the Anglo-Hashemite marriage was the notion that there might exist in the Arab lands of Asia a growing resentment of Turkish rule that could be employed to Britain's benefit. That was certainly true, but the resentment was unevenly spread, had different origins or emphases and was therefore difficult to harness as a coherent force against the Turks. In cities like Damascus, Jerusalem and Baghdad, for example, there was a small, new middle class of professional people, as well as some Arab officers in the Turkish army, who were ideological nationalists of a modern stamp, seeking freedom for an Arab 'nation'. But they were unarmed and too few in numbers to mount an effective challenge to the Turks, although they could –

and eventually did – muster the power of the word to become the Revolt's chief Arab propagandists.

In the deserts of Syria, Jordan and Arabia, on the other hand, there were plenty of rebellious bedouin tribes, like those around Ibn Saud, who resisted the Turks because they resented the interference of any organized government in their lives, but who had no concept of modern nationhood let alone any desire to fight for it. They were far more numerous than the modern nationalists and more naturally warlike, but they had almost as little in common with the ideologists as they had with the British and they seemed unlikely to accept the leadership of either. Moreover, Turkey still enjoyed the advantage with most Arabs of being a Muslim power whose Sultan had claimed the title of Caliph for centuries. The Turks could therefore employ religion as their unequivocal ally against the Christian British. The British, on the other hand, found it hard to decide how far they could dangle the bait of national freedom before the Arabs without losing all control of the political outcome.

These last difficulties were felt especially acutely in British India. If, said Delhi, the Turks were provoked by British meddling with the Arabs to declare a Muslim *jihad* there would be a grave risk of disorder among the millions of Muslims in India, too. Equally, if the British promoted Arab national freedom as a legitimate war aim they would find it all the harder to stifle the demands of Indian nationalism. The Indian Government dragged its feet. But in London, anything seemed welcome that might relieve, if only in terms of morale, the appalling pressure of trench warfare in Flanders. So the old disputes between Whitehall and India were reopened in a new way, with the former looking for any means to whip up Arab support and the latter hoping nervously that it would not be necessary.

It was from Cairo that the first proposals for an Arab revolt were circulated. The Egyptian capital had been Britain's main Middle Eastern headquarters since the 1880s. Unlike the minuscule British posts in the Gulf, which were controlled from India, it was answerable directly to London, and British representatives there inevitably had London's interests most at heart. At the start of the war the chief among them was no other than Field Marshal Lord Kitchener, Britain's most famous military commander of the day, then serving in his penultimate post as British High Commissioner. (He was shortly to be appointed Minister of War in London, a job he held until his death in 1916 when his ship, H.M.S. *Hampshire*, was torpedoed by a German submarine.)

Even before the war began, Kitchener had obtained a hint that an Arab rebellion could be raised when he met, in a casual way, the young Emir Abdullah, Sherif Hussein's second son. Delicately, Abdullah had asked what the attitude of His Majesty's Government might be if a conflict broke out between the Hejaz and the Turks. Politely, and with the old Anglo-Turkish alliance in mind, Lord Kitchener had answered non-committally. Abdullah would not give up. Two months later, in a meeting with Kitchener's Oriental Secretary, Ronald Storrs, he wondered whether Britain might be so kind as to send his father a present of some of those nice new machine-guns that the British made so well, so that he might more readily assert his authority among the benighted folk of Arabia. Storrs was no more forthcoming. But when the old alliance with Turkey was overtaken by the war in August, Kitchener saw to it that these conversations were dusted off and others of more purpose embarked upon.

Responsibility for those talks fell upon a typically small, brilliant but hastily assembled wartime circle of diplomats, scholars, linguists and military men called the Arab Bureau. Founded in February 1916 by the head of the Department of Military Intelligence in Cairo, Sir Gilbert Clayton, the Bureau was never composed of more than fifteen officers, of whom Lawrence, though relatively junior, was later to become the most celebrated. It was their dream of Arab uprising and Turkish collapse that finally got the Arab Revolt started, and it was mere tactical common sense that suggested to them that the Sherif of Mecca should be its leader.

For a start, Hussein's age, experience and breeding entitled him to respect. As a former resident of the Turkish capital he knew enough about current political movements to talk some of the language of the modern nationalists, while as a descendant of the Prophet and a true son of Arabia he could reasonably expect to win the ear of some of the bedouin. His home base in the Hejaz was strategically more significant than that of Ibn Saud in Nejd. Commanding much of the Red Sea coast of Arabia, it complemented the already established British position in Egypt and the Sudan on the western shore and was nicely interposed between the main Turkish provinces to the north and their southernmost Arabian extension in Yemen, where Turkish forces threatened the British port and naval base at Aden. If Sherif Hussein came over to the British side, therefore, both the security of

Aden at the southern gate of the Red Sea and the vital passage of the
Suez Canal in the north might be assured.

The Hejaz was also the Muslim Holy Land and its Sherif was a man
of consequence throughout the Muslim world. If he joined the British
against his Turkish co-religionists many millions of Muslims elsewhere
might be persuaded to follow his example. If, instead, he accepted
Turkey's leadership and, perhaps, lent his name to a *jihad* against the
Christian powers, then Muslims from North Africa to India might
also revolt against the allies.

Finally, Sherif Hussein's men were in direct contact with the
Turkish army. The completion of the Hejaz railway from Damascus
to Medina had enabled the Turks to strengthen their garrisons in the
Muslim Holy Land, and although that was in one way a disadvantage
to the Sherif it also offered him both reason and opportunity to assert
himself against the Turks by attacking their soldiers and their railway
line in the interests of enlarging his own power.

In the sort of bargaining that inevitably preceded any Anglo-
Arabian alliance, these were all highly marketable Hashemite assets
which Ibn Saud could not hope to match. It was true that he had
extended his territory just before the war from land-locked Nejd to the
coastal province of Hasa, so that he could claim some influence on the
shore of the Persian Gulf. But the Gulf did not then come near to the
Red Sea in importance. Even the new oilfield in Persia at the head of
the Gulf did not seriously change this. Although there was a natural
concern for its safety, the practical Turkish threat to its security was as
yet small.

Within a week of the outbreak of war with Turkey, a division of
the Indian Army was landed at the mouth of the Shatt al Arab and
within three weeks it had captured the Turkish local capital of Basra,
fifty miles upriver. Although its first attempt after that to march on
Baghdad, a further 500 miles to the north, foundered in one of the
most ignominious British failures of the war, there was never much
serious prospect after the fall of Basra of the Turks menacing the brand
new installations of the Anglo-Persian Oil Company and the small but
precious flow of oil to the Royal Navy. So Ibn Saud's aid on the
British flank in the Gulf was of far less consequence than that of
Hussein on the Red Sea coast.

Nor could Ibn Saud compete with the Sherif as an Arab or Muslim
leader. Outside his tribal territories he had as yet no following and few
contacts. The Wahhabism he professed was unpopular elsewhere, his

bedouin supporters were notoriously wayward and, to the urban
intellectuals of Damascus, Jerusalem and Baghdad, their primitive
Arab tribalism offered an unwelcome contrast to the worldly
sophistication of Sherif Hussein. The bedouin might be effective in the
field against the similar forces of the pro-Turkish Ibn Rashid
(although, as the battle of Jarrab and Shakespear's death had
demonstrated, even that could not be relied upon), but they were not
in contact with the main body of the Turkish enemy and it was hard
to see, at that juncture, how they could be any better employed.
For – and this was the crucial argument deployed against the Saudi
connection in Cairo and London – even if the British had put most of
their arms and money on Ibn Saud instead of on Sherif Hussein, the
most likely outcome, given the jealousy that existed between the two,
would have been a pro-Turkish alliance between the Sherif and Ibn
Rashid that would probably have been more than Ibn Saud could have
handled and that would certainly have risked a serious setback for the
British position all round Arabia. So, in spite of his new treaty with
Britain and the arms and money he obtained, Ibn Saud was consigned
yet again to the periphery of British interests while his rival in Mecca
was thrust into the leader's role.

   The Sherif would have been less than sensible, however, if he had
not exploited his advantage to extract promises from Britain of a rosy
political future for himself; and in the months of secret bargaining
with the Arab Bureau that preceded his final switch from Turkish
vassal to British ally, he steadily raised his terms for co-operation. It
was 1916 before a final agreement was reached in a tortuous exchange
of letters between the Sherif and Kitchener's successor in Cairo, the
run of the mill civil servant at the end of his career named Sir Henry
McMahon. In general terms the letters agreed that in return for
Hussein's aid against the Turks, the British would send him arms and
money and support independence for the Arab people after the war
throughout most of the Turkish provinces in Asia. The only specific
exceptions to this great promise were identified by McMahon as the
territories of Arab chiefs and princelings in treaty relationships with
Britain – which adequately protected the position of Ibn Saud along
with many other rulers in peninsular Arabia – plus what were
described as 'portions of Syria lying to the west of the districts of
Damascus, Homs, Hama and Aleppo'. A more general reservation
was added, however, to the effect that the agreement would only

apply to those areas where Britain was 'free to act without detriment to the interests of her ally France'.⁴

Obviously there was a good deal of diplomatic ambiguity here which the Sherif, perhaps, ought to have cleared up before he finally went into action. (This was much complicated later on when the Arab Bureau lost the only Arabic copy that Ruhi, Sir Ronald Storrs' Bahai translator, had made of important sections of the correspondence. Years later it was found jammed behind a drawer in Storrs' desk.) But the main burden of Britain's promise seemed clear enough: Arab independence throughout most of the Turkish provinces. If Hussein could become the agent of that apotheosis he would emerge from the war as the most powerful man in the Middle East. It was too good an opportunity for an ambitious ruler to miss and, leaving aside the French reservation and the Syrian exclusions for further consideration when the war was won, the Sherif proclaimed his trust in Britain's good faith and set the wheels of revolt in motion.

It was a decision he lived to regret for, unknown to him, the British had been negotiating meanwhile with their allies, the French, for a second and in some senses contradictory agreement about the post-war fate of the Arab lands. Known as the Sykes-Picot Agreement, after its two authors, Sir Mark Sykes for Britain and M. Georges Picot for France, this divided most of the Arab lands north of peninsular Arabia into two spheres of influence or control. France, it was established, would acquire supreme rights in Lebanon and Syria, and Britain would win similar status in Iraq (then still known to the British as Mesopotamia) and what is now Jordan. Palestine, however, would have an international administration as befitted its status as a Holy Land for all three great Middle Eastern religions. In short, contrary to what appeared to have been promised in the Hussein-McMahon letters, only some unspecified parts of peninsular Arabia would be left after the war to enjoy any genuine independence.

As if this were not enough to sow eventual confusion and mistrust, Britain then concluded a third and still more contradictory agreement to dispose of Palestine in another way, through the Balfour Declaration of November 1917. This took the form of a letter of intent, written by Britain's Foreign Secretary, Sir Arthur Balfour, to Lord Rothschild, an eminent representative of the Zionist Federation: 'His Majesty's Government view with favour the establishment in Palestine of a national home for the Jewish people and will use their best endeavours to facilitate the achievement of this object, it being clearly understood that nothing shall be done which may prejudice the civil and religious rights of existing non-Jewish communities in Palestine, or the rights and political status enjoyed by Jews in any other country.'[5]

Besides being in conflict with the two previous undertakings, to the French and to Sherif Hussein, this third agreement was even more disastrously in conflict with itself. For at that time Palestine was an almost wholly Arab country, with 600,000 Arabs living there and only 80,000 Jews. To swell the numbers of the latter to any significant extent without simultaneously infringing the rights of the former would have required the British Government to square the circle.

As succeeding decades were to demonstrate with clarity and bitterness, it could not achieve that miracle. What was worse, it gradually emerged that the authors of the Declaration, both British and Jewish, never really intended that the circle should be squared. The Secretary of the Zionist Federation, Dr Chaim Weizmann, asserted more than once at the time that the true Zionist aim was to make Palestine as 'Jewish as America is American and England is English'.[6] Recently disclosed documentary evidence leaves no more room for doubt on that point. Balfour himself confessed in a confidential minute written within a year or two of the Declaration itself that 'so far as Palestine is concerned, the Powers have made no statement of fact which is not admittedly wrong, and no declaration of policy which, at least in the letter, they have not always intended to violate'.[7]

Many attempts have been made over the intervening decades to defend or explain away these British deceptions, but the best that can truly be said is that they were by no means out of keeping with the standards of imperial diplomacy in general sixty years ago and that Britain, anyway, was fighting for its life and could not afford to be too scrupulous about the methods it employed to win new allies and defeat its enemies.

The years of 1916 and early 1917, before the decisive American entry into the European war, were certainly the nadir of that struggle for Britain and her allies. In the Middle East the collapse of the Indian Army's first march on Baghdad, in 1916, after the bloody failure of an Allied invasion force to defeat the Turkish at Gallipoli in the previous year, had left the whole Arab front in stalemate. In Europe the toll of the trenches seemed increasingly insupportable, and in the Atlantic the German submarine blockade was proving effective. In such desperate straits the susceptibilities of the Arabs were not likely to be

uppermost in British minds. If they had to be sacrificed on the altar of
realpolitik, so be it; it would not be the first or the last time a Great
Power behaved like that. And if it laid up trouble for the future, that
was surely preferable to the prospect of having no future at all. So,
amid much duplicity, the Anglo-Hashemite alliance was forged and
the Arab Revolt began.

It opened in the Hejaz in the summer of 1916, in the month of
Ramadan, and at first its momentum was fitful. After a rousing start,
when Mecca was captured by the Sherif's troops, the Medina railway
line was cut and Jeddah fell to a rabble of tribesmen bent on loot, it
skidded swiftly to a halt in a characteristic Arab mixture of langour,
boastful talk and demands for more money. But fresh infusions of
British gold, British arms and British leadership got it going again.
From August 1916, Sherif Hussein was to receive around £100,000 a
month – twenty times what the British were paying Ibn Saud to keep
him quiet – and regular supplies of weapons were flowing in from the
Sudan. Then, with Lawrence himself in the field, the attempted
Turkish counter-attacks were frustrated and, in what appears in
Lawrence's own account as a dashing series of romantic camel rides,
punctuated by thirst, pain and ambush, the Sherif's raggle-taggle army
advanced northwards towards the valley of the River Jordan. By July
1917, one year after the revolt had started, the little port of Akaba at
the head of the Red Sea had fallen and virtually the whole of the Hejaz
was free. In retrospect this seems to have been a moment when Sherif
Hussein might reasonably have argued that he had now fulfilled his
part of the bargain and that other Arabs should henceforth take up the
cause alongside the British, while he bent his energies to consolidating
his rule at home. But his ambition had long outstripped his own
territory and, growing now by the success it fed on, it began to lead
him into paths of temptation from which there would be no return.

Hussein was, in fact, obsessed not so much with opening the door
to widespread Arab independence as with the chance of carving out a
new Arab empire for himself and his family. Within a few months of
the start of the revolt he had persuaded the nobility of Mecca to
proclaim him 'King of the Arabs' – a title entirely of his own
invention – and he had thereupon delivered imperious demands for
taxes and fealty to Ibn Saud and several other Arabian princelings. As
his men advanced under British leadership he grew still more
overweening, advancing fresh claims to Ibn Saud's western desert
territories and thrusting his sons into positions of power and

prominence through the good offices of his British patrons. After the fall of Akaba there seemed no limits to Hashemite ambitions, for by that time regular British armies were on the march as well, driving the Turks back on all fronts and, apparently, creating vast new opportunities for the House of Hashem to establish its authority in their wake. In the east, Baghdad had fallen to the Indian Army at its second attempt and most of Iraq was already free. In the west, a powerful mixed force under General Allenby was advancing northwards from Egypt towards Jerusalem and Damascus. This was no moment, in Hussein's eyes, to eschew the succulent fruits of victory; rather, it was the time to pursue them with even greater zeal.

Naturally, the British agreed. The contribution of the Arab Revolt so far had been far from decisive in the Middle Eastern struggle, but it had certainly been useful; and now, with Damascus and a final crushing victory over the Turkish Army in their sights, the British welcomed Hashemite support all the more warmly, pouring still more arms and money into the revolt to raise the tribes against the Turks all the way to Damascus. Armoured cars and artillery came to join the camels and the scurrying riflemen. A flight of Royal Air Force biplanes was sent to aid the Arab scouting parties with aerial reconnaissance and occasional bombing raids. This was Lawrence's finest hour. Harrying and skirmishing along the line of the railway, blowing up a bridge here and a culvert there, the strengthened Arab forces that he fought with earned their money, enjoyed the heat of battle and learned a few new tricks of desert war, so that by the time Damascus fell at last to the Allied armies in September 1918, their revolt had staked a genuine claim to its promised reward.

In mingled excitement and complacency, therefore, Sherif Hussein and his sons prepared to claim their due as the true begetters of Arab freedom. While General Allenby drove up to Damascus from Jerusalem in a grey Rolls-Royce to receive the cheers of an expectant Syrian populace, the Sherif's eldest son, the Emir Feisal, simultaneously stepped down from a special train that had puffed its way over the last few intact miles of the Hejaz railway into the Damascus station. There, amid more cheers, he was proclaimed in all solemnity as Viceroy to his father, the aspiring King of all the Arabs. In that moment the Hashemite cause seemed supreme, as if the Prophet's empire was about to be reborn in the hands of his descendants. But behind the glory there was deceit, in the achievement there was rottenness. Very soon the Hashemite triumph would give way to humiliation – and one of the chief agents of that change would be the neglected and increasingly resentful Ibn Saud.

6 Sowing the Wind
1917 – 21

On the frontier of Iraq, Jordan or Kuwait the mere mention of the word Ikhwan *was once enough to spread terror and send the Bedouin at full speed across the desert to seek protection within the walled towns. Who then were these harbingers of terror in Arabia?*[1] – HAFIZ WAHBA

IF LUCK, success and gold are what matter in a desert leader, Abdul Aziz bin Saud now possessed little of all three. He was plagued by persistent rebellion on a scale he had not known since his recapture of Riyadh. He would have liked to gain his revenge on the Rashid by launching a full-blooded assault on their capital at Hail; but, although a few local British officials supported him in that desire, the governments in Delhi and in London were mostly at one in continuing to argue that such an operation would require more arms than they could easily spare for a political and strategic return of only marginal proportions.

The British had been throwing him scraps from the imperial table accompanied by floriferous expressions of their undying goodwill. In November 1916 he was invited to a great imperial durbar in Kuwait and invested as a Knight Commander of the Indian Empire. He had his hand X-rayed at a British military hospital, saw his first aeroplanes at a special military demonstration in Basra, and was given a ceremonial sword by the British Commander in Chief in Mesopotamia. That month he even got another small consignment of arms and a temporary stipend from the British Treasury of £5,000 a month.

And, to his well-bred consternation, he was introduced for the first time to an administrator of the 'Great Government' of Britain who also happened to be a woman. Gertrude Bell was already well known in Britain as a writer and traveller in the Arab lands. She had visited Hail and been entertained by the Rashid in typically suspicious fashion just before the war and had acquired a reputation as one of those formidably competent and determined ladies whose adventures enlivened so many parts of the British Empire in its later years. A tall and imposing figure in her ankle-length dresses and floppy hats, she was more than the equal of most of the men she met or worked with; and when war came to the Middle East she was quick to press her services upon the Government and was appointed to the little circle of the Arab Bureau. From there she joined the staff of Sir Percy Cox, now in Iraq, and was introduced to the Saudi prince during his visit to Basra. Ibn Saud's disapproval was plain: an unveiled woman taking part in public ceremonies as the equal of the men around her was something totally outside his experience and deeply repugnant to his beliefs. Yet he did not fail to impress Miss Bell, as much as he had impressed all his other European contacts, for in reporting their meeting afterwards to the Arab Bureau she offered a portrait of him that has never been bettered:

> Ibn Saud is now barely forty, though he looks some years older. He is a man of splendid physique, standing well over six feet, and carrying himself with the air of one accustomed to command. Though he is more massively built than the typical nomad sheik, he has the characteristics of the well-bred Arab, the strongly marked aquiline profile, full-fleshed nostrils, prominent lips and long, narrow chin, accentuated by a pointed beard. His hands are fine, with slender fingers, a trait almost universal among the tribes of pure Arab blood, and, in spite of his great height and breadth of shoulder, he conveys the impression, common enough in the desert, of an indefinable lassitude, not individual but racial, the secular weariness of an ancient and self-contained people, which has made heavy drafts on its vital forces, and borrowed little from beyond its own forbidding frontiers. His deliberate movements, his slow, sweet smile, and the contemplative glance of his heavy-lidded eyes, though they add to his dignity and charm, do not accord with the western conception of a vigorous personality. Nevertheless, report credits him with powers of physical endurance rare even in hard-bitten Arabia. Among men bred in the camel saddle, he is said to have few rivals as a tireless rider. As a leader of irregular forces he is proved daring, and he combines with his qualities as a soldier that grasp of statecraft which is yet more highly prized by the tribesmen.[2]

But by 1917 Ibn Saud was slipping steadily further out of Britain's reckoning as a major force in Arab affairs. As the Anglo-Hashemite revolt spread northwards into Jordan, freeing the Red Sea from the old Turkish threat, his bargaining position was deteriorating month by month. By the end of the year, when Jerusalem was in Allenby's hands and Baghdad had fallen to the British, at the second attempt, peninsular Arabia had reverted to its old position as a strategic backwater wherein Ibn Saud languished in obscurity as nothing more than the Prince of Nejd and its Dependencies, while Sherif Hussein and his family prepared to claim their due as the liberators of the Arab world.

It was not a prospect that Ibn Saud could be expected to accept without demur, and the more Sherif Hussein saw the light of victory and aggrandizement before him, the more the Saudi leader chafed and fretted and the more he resolved to assert himself before history altogether passed him by. Aware of these feelings of frustration in their lesser protégé, the British tried harder to placate him, sending in November 1917 a new mission to Riyadh to explain their faith in the Sherif and to explore a little further Ibn Saud's own demand for a final settlement with the Rashid. The man they chose for this high purpose was Harry St John Bridger Philby, a 32-year-old finance officer in Iraq under Sir Percy Cox, and the man who was to become within a few years the most assiduous chronicler of Ibn Saud's life and times as well as Arabia's greatest explorer.

In his way, Philby – a stocky, long-jawed, deep-eyed figure, with the slightest suggestion of a bantam strut – was almost as complex a character as Lawrence but, while Lawrence kept all his puzzles hidden behind a dazzling screen of charm, Philby laid his bare for all to see. Where Lawrence wove spells of fantasy and imagination, Philby was all for bluntness. Where Lawrence wrought his wartime work into an exquisite, evocative and, it seems, partly fanciful tale of desert loves, betrayals and derring-do, Philby set about his great work of exploration with meticulous method and scrupulous regard for fact and produced from it, alas, a series of the dullest books that have ever been written about Arabia. Yet he was a man with a demon, whose abilities could have taken him to the highest rank of government or administration but whose nature forced him to spend nearly all his life in a state of fierce and usually frustrated contrariness.

Both a loyal servant of Britain's empire and an imperious critic of it, he spent nine years of his career trying unsuccessfully from the inside to shape British policies to his ideas in the Middle East and nearly forty more trying, equally unsuccessfully, to tear them apart again from exile. He was driven by a frankly expressed desire for fame and influence, and dreamed for many years of being elected to Parliament; yet he sacrificed nearly all his opportunities for advancement through his perversity, and doomed many of his great talents to frustration in the Arabian wilderness, far from the friends who would and could have helped him. It was typical of him that when he eventually stood for election it was on an individual anti-war platform of his own in 1939, just when Britain had geared itself to an irrevocable clash with Nazi Germany. Not surprisingly, he was resoundingly defeated.

On his infrequent visits to London he was, nevertheless, the complete and apparently established Englishman, bowler-hatted and sober-suited, who divided his time between the officials of Whitehall, the mandarins of his club, the Athenaeum, and the cricket field at Lord's. He was deeply loyal to his English wife and family and passionately wished that his son, Kim, should do what he never did and climb the ladder of the establishment, rung by rung, to fame. There was something tragically appropriate in the fact that, having selected the innermost core of that establishment upon which to make his mark, Kim should indeed have climbed to a fame that finally exceeded his father's – but only because he chose to betray through his secret Soviet service the Britain that, in spite of everything, his father loved.

St John – known as Jack to his English friends and Abdullah to his Arab comrades – won respect through his obstinate honesty even when his wrong-headedness was most infuriating; and for many years he was one of the few men whose fierce strictures and contrary arguments Ibn Saud himself would accept with grace and even humility.

At the time of his first mission to Ibn Saud both Philby's vices and his virtues were already clearly marked and they became clearer still as the mission proceeded. He had no sooner arrived in Riyadh than he fell out with one of his senior British colleagues on the issue of who should be in overall command of the mission. Having won that battle, he at once took up Ibn Saud's cause with partisan fervour, arguing that he should be armed and paid immediately to launch the final attack on his enemies in Hail. Then, without waiting for an answer to his messages – which would have meant a stay of several weeks in Riyadh while the camel post trod slowly to the coast and back and the mills of British bureaucracy ground out a reply – he set off impetuously to fulfil one of his heart's desires: a full crossing of central Arabia from coast to coast. Shakespear had crossed the peninsula by the northern route less than four years previously, but no European had traversed the whole of the desert centre for a century, since Captain George Forster Sadleir had unavailingly pursued Ibrahim Pasha in the hope of presenting his ceremonial sword. Philby, in Riyadh, was already one-third of the way and, with the blessing of his new Saudi friends, he was determined to complete the journey.

To be fair to him, he had excuse beyond his own ambition. While he was conferring with Ibn Saud, other British representatives from the Arab Bureau in Cairo were on their way to Jeddah in the hope of persuading Sherif Hussein to be less haughty with his fellow Arab princes. They had proposed that one of them should journey afterwards to Riyadh to acquaint Ibn Saud with the outcome of their meeting and so, perhaps, bring the two men to a better understanding. But Sherif Hussein was in no mood to compromise. Contemptuous as ever of the desert chief beyond his eastern horizon, he refused permission for the onward journey, saying he could not guarantee the safety of the mission once it left his territory for the rebellious and anarchic lands of the Wahhabi tribes. Philby, therefore, announced that he would do the journey in reverse and bring Ibn Saud's views to the Sherif – thus proving incidentally, but not without malice aforethought, that Hussein's claims about the lawlessness of Saudi territory were without foundation. Ibn Saud was equally eager to speed him on his way for, as Philby discovered in the first of his many private meetings with him, his rancour towards the Sherif had grown by now to hatred.

So Philby set out with a light heart having, to his own satisfaction, reconciled the requirements of diplomacy with the promptings of his demon. He arrived in Jeddah on the last day of the year after a remarkably swift and uneventful passage that triumphantly upheld Ibn Saud's claim to central Arabian authority and established beyond question Philby's own nerve and skill as a desert traveller. Some time later the Royal Geographical Society awarded him their Gold Medal for his feat; but he won nothing but black marks for his diplomacy. The Sherif was enraged by his unsanctioned arrival and, far from being impressed by Ibn Saud's apparent control of the interior, became all the more scornful of his crude pretensions and all the more obstinate about enforcing his own claims to supremacy. Nor was he in the least amused by Philby's candid advocacy of the Saudi cause when they met in Jeddah. With his characteristic relish for an argument, the Englishman thought it was all rather fun to be taking down such a pompous little potentate, writing complacently to his wife Dora afterwards that, 'It is not often one gets the chance that I have had of making a fool of a king.'³ But kings do not like being made to look foolish and Philby did both himself and Ibn Saud little good by his cocksure adventurism.

His intervention was especially unfortunate in its timing for it came just after the publication of the Balfour Declaration and Hussein's almost simultaneous discovery of the existence of the Sykes-Picot Agreement. The latter was particularly wounding to the Sherif, for it had been kept secret until then and was only sprung upon the world through the unlikely medium of the Bolsheviks who, newly come to power in their October Revolution in Russia, had found the text of the agreement among some recent Tsarist files and hastened to publish it as an advertisement of the untrustworthiness of Hussein's erstwhile allies. With Hussein's suspicion of British intentions thus inflamed, Philby's insensitivity on behalf of Ibn Saud was all the more painful.

Sherif Hussein at this time was better prepared for war than Ibn Saud. In return for his leadership of the Arab Revolt, his soldiers had received British arms and training and his treasury was comfortably stocked with British gold. Ibn Saud and his followers had but one fearful asset that Hussein could not command: the old Saudi menace of militant Wahhabism now revived in a new and especially virulent form. If foreign aid and gold would not recover his patrimony, he had yet one recourse: to raise the flag of holy war, like his forefathers, and unite Arabia in the name of God.

By 1918 there were many among his followers who were waiting impatiently for that call. They had seen their leader shackled or rebuffed repeatedly by the infidel power of Britain, while others, less deserving, had won gold and promises beyond all reckoning. In the Holy Land of the Hejaz they saw presumption incarnate in the person of Hussein – a self-styled Guardian of the Holy Places who had actually gone to war on behalf of a Christian power against the Turks who, whatever their other vices, were at least of Muslim faith. They saw, too, idolatry and corruption once more usurping the sacred places of pilgrimage and they itched to smash the idols as their ancestors had done before them.

As early as 1912, some of these desert zealots had started a

Harbingers of terror: Ibn Saud's army on a desert march, photographed by
the ill-fated Captain Shakespear on 8 March 1911

The first venture by a member of the House of Saud outside Arabia.
Aboard H.M.S. *Kigoma* on his way to London in September 1919, the
fourteen-year-old Feisal is sketched by an Englishwoman

The Uqair Conference, November 1922. Ibn Saud poses with Sir Percy Cox. Standing behind them, with the elaborate nonchalance of the gatecrasher, is the country's first oil concessionaire Major Frank Holmes

Aboard the sloop *Lupin* in the Gulf, February 1930: Ibn Saud and the Hashemite King Feisal of Iraq cement the more amicable relationship that followed the collapse of the Ikhwan revolt

Ibn Saud in 1935 with the eldest of his grandsons

A new toy: Saud and camera at the *Hajj* camp at Mina in 1936, photographed by St John Philby. Khaled, aged about twenty-four, is third from left

Ibn Saud, with stick, and Saud
lead the way to the Friday
prayer along the walls of the
Murabba Palace, Riyadh,
November 1935

The pleasure-loving Saud and
his sterner half-brother
Muhammad set off for a day's
shopping in London, August
1938

movement of evangelism known as the Brotherhood[4] – the Ikhwan – and at a little village called Artawiyah in the homelands of the Mutair tribe, some 300 miles north of Riyadh, they had established the first of many religious settlements, or *Hujar*,[5] where the pure doctrine of Wahhabist unitarianism was preached with varying degrees of coercion to the surrounding bedouin tribes. The results were immediately impressive. Artawiyah rapidly grew to a town of 10,000 souls. Men who had been traditionally the scourge of any and every government, unpredictable and fickle as mayflies, settled down to a life of fierce dedication. They wore turbans and short robes rather than the trailing *thobe* of the townsman. They abandoned their tents and their nomad life, sold their sheep and camels and gave themselves up to farming, to prayer and to the Koran. Sixty-seven years later, rebels from Saudi society were to look back at their example for inspiration.

But for Ibn Saud this movement was a practical as well as a religious blessing, for at one stroke it turned the bedouin poachers into gamekeepers; instead of creating trouble, they became the guardians of his new Islamic state. If Ibn Saud did not actually found the first settlement, he soon appointed himself the movement's leader. Indeed, he went so far as to claim to outsiders that it had no identity separate from himself, reassuring the British Envoy, Harold Dickson, then based in Bahrain, with the words, 'Oh Dickson, don't worry. I am the Ikhwan, no one else.'[6] Deliberately he began to expand its power among his people. In 1916 he ordered that all the bedouin tribes owing allegiance to him must also give up herding and join the Ikhwan, and their sheikhs were brought to Riyadh in relays for special religious instruction. They were to receive subsidies from the treasury and, in return, respect the King as their Imam and swear to uphold Wahhabist orthodoxy.

That this was an incitement of religious fanaticism seemed a small price to pay. Indeed, in time of war it made them a reliable source of fighting men to supplement Ibn Saud's meagre sources of townsmen. The Ikhwan would spare neither themselves nor anyone else in their role as the soldiers of God. 'I have seen them hurl themselves on their enemies,' wrote one Arab eye-witness of their exploits, 'utterly fearless of death, not caring how many fall, advancing rank after rank with only one desire – the defeat and annihilation of the enemy. They normally give no quarter, sparing neither boys nor old men, veritable messengers of death from whose grasp no one escapes.'[7]

Like all fanatics, however, they were also a menace to any government, including Ibn Saud's. They assumed power without responsibility and practised techniques of persuasion and enforced conversion that were stark indeed. They whipped the faithful who missed attendance at the mosque, forcibly prevented all enjoyment of music or tobacco, and threatened death to any Christian who did not enjoy the ruler's personal protection. Other Muslims, even bedouin outside the *Hujar*, also fell victim to their wrath. They might as easily execute one man for failing to recite the Koran properly as they would kill another for being holy and well read and yet so benighted as to refuse to join their ranks. If they met a man with a moustache longer than the Prophet's – who is said to have worn his close-clipped – they would stop him and bite it off to show him his offence. The chief of the Mutair tribe, Feisal al Duwish, once took a pair of scissors to Ibn Saud's trailing robe. In short, they were a dangerous and sour body of men who, as they grew in numbers, became as much of a challenge as a comfort to their ruler – and would, in due time, emerge as the greatest threat of all to the security of the new Saudi state.

By 1918 they were already strong enough to be troublesome and were seething at Ibn Saud's continued acquiescence in the face of infidel provocations from Britain and the apostate Sherif Hussein. Eager to show their own chief the true path, as well as to puncture the Sherif's pretensions, they were pressing their conversions upon the Utaibah and Subai villages in that disputed territory between Nejd and the Hejaz where the Sherif had earlier, in 1910, outwitted Ibn Saud by the capture of his brother and extracted recognition from Riyadh of his sovereignty.

The Sherif naturally objected and appealed to his British allies for protection. The Ikhwan, equally naturally, redoubled their efforts to gather the villages into their God-fearing grasp. Ibn Saud, caught between his desire to retain Britain's support and his need to keep the Ikhwan happy, pursued a policy of calculated ambivalence, murmuring soothing diplomatic niceties at one moment and muttering that he could no longer restrain his people at the next. The British themselves were equally ambivalent. The pro-Hussein lobby in Cairo and London conscious of the problems that must follow the Turkish retreat and the need to retain Hashemite loyalty in what was bound to be a difficult peace settlement in the Middle East, urged full support for the Sherif. The pro-Saudi faction in Arabia and the Gulf, led by an increasingly impassioned and scornful Philby, argued that with the success of the Arab Revolt Hussein had now outlived whatever usefulness he might have had and his obsession with pan-Arab leadership would drag Britain to disaster. Ibn Saud, they insisted, was now Arabia's premier ruler and Britain's most promising friend.

For months the diplomatic notes went their tortuous courses from Jeddah to Cairo and on to London, then east again to Baghdad, Bushire, Kuwait, and Riyadh, then back again to London, accumulating fresh glosses and qualifications at every turn. There were threats that British subsidies might be cut off from one ruler or the other in order to bring them to their senses. A brave attempt was even made to arrange a personal meeting between them on board a British warship in Aden harbour. But in the end there could be no reconciliation, no meeting, no compromise. The two men were set on an implacable collision course and, after a series of minor clashes in the disputed territory, their forces met in earnest for the first time on the night of 25 May 1919, near the little village of Turaba, only sixty miles from Hussein's holy capital of Mecca. Hussein's second son Abdullah was in charge of the Sherifian forces, with 5,000 well-trained men and a full complement of modern weapons. Ibn Saud himself set off from Riyadh while the Utaibah leader, Sultan bin Bijad bin Humaid, hurried to Turaba with 1,100 camel-borne troops. But the pitched battle that both sides were preparing for was never joined. Instead, the Utaibah Ikhwan, armed in the traditional way with swords, spears and an assortment of ancient rifles, went scouting ahead on their camels by night and fell upon Abdullah's camp in the darkness. Screaming that the sinners should prepare themselves for Paradise, they swept through the lines like a terrible, expunging flame. By morning there was nothing left of Abdullah's professional army. Its guns were captured, its men were dead or scattered, and Abdullah himself had fled in his nightshirt. The sand was wet with the blood of the apostates. The soldiers of God and the Holy Book raised their dripping swords in triumph and fell to prayer. As late as the early 1970s, the bones of unburied Hejazi dead could still be found on the field.

The victory at Turaba, however, proved to be only a ranging shot in the Ikhwan's whirlwind campaign of conquest. It left Hussein humiliated – although not yet chastened – and brought the British promptly to his rescue. Fearful that Ibn Saud must now descend on a defenceless Mecca and his religious warriors put the inhabitants to the sword, they informed him immediately that he must restrain his men or face dire consequences. A flight of aeroplanes was sent from Egypt to the Hejaz to show that London was not to be trifled with, and there were anxious exchanges with officials in the Gulf about what targets on the Saudi coast the Royal Navy might bombard if Ibn Saud disregarded the British warning. None, came the truthful but discouraging reply, for alas, there were no targets worth the shelling; but happily the British were spared any further anguish over that when the Saudi leader answered their demands in tones of the sweetest reason. He would be happy to restrain his men if Britain's High Government would only restrain Hussein; and in the name of Allah, the Merciful, the Compassionate, he would gladly submit the whole dispute to arbitration if his brother, Hussein, would do the same.

It was a shrewd reply, well calculated to evoke new sympathy among British officials who found Hussein, by contrast, recalcitrant and emotional, alternately threatening war against the upstart Saud and abdication if the British did not regain his territory for him. Ibn Saud's subsidy was discontinued for only a month; for the truth was that now, with the end of the war in Europe, the balance of power in Arabia was moving steadily against Hussein.

As the peace conference ground through 1919 at Versailles, Hussein's personal ambitions seemed ever more remote from reality. Blissfully unaware as yet that the war had fractured the foundations of their own power, the victorious European allies were once more carving up the world between them and it was immeasurably more important to the governments in London and Paris to achieve a division that might guarantee Europe's own stability than to attempt to satisfy the cravings of remote Arab princes. Indeed, few people at Versailles even knew who – or what – these Arabs were. The Middle Eastern war had always been a bit of a sideshow and the Arab Revolt, as Lawrence had realized, was only 'a sideshow of a sideshow'. When the old lions of Europe like David Lloyd George and Georges Clemenceau were busily sniffing around each other's musk marks, looking for a chance to scratch out some new advantage, there was little point in the Grand Sherif of Mecca protesting that their minions had already promised him his share of the loot. If they heard him at all in those rarefied circles, it was only to dismiss him as of no account. For the minions who had helped him were all gone. The Arab Bureau's influence had vanished with victory and its officers were soon dispersed. Lawrence did his best to hold the British Government to its promises, shepherding Hussein's eldest son, Feisal, from Damascus through the corridors of Versailles and using his reputation and his well-placed friends to salvage what he could from the wreckage of Arab aspirations. But against French opposition, British indifference and the commitment to a Zionist Palestine, even his wizardry was ineffective.

In the end, Feisal was driven out of Syria permanently by the French to be installed, against much local opposition, as Britain's puppet King of Iraq in Baghdad, while his brother, Abdullah, was given nominal charge of what has been aptly called the 'vacant lot' of Transjordan – an arbitrary and impoverished chunk of desert between Palestine, Syria and Iraq, whose sudden elevation to a dubious statehood was the fruit of a hasty night's conclave between Lawrence and Winston Churchill in his capacity as His Majesty's Secretary of State for the Colonies. Lawrence for a time professed that these presents had got Britain out of the affair with honour, after all.[8] But set beside the vague visions of Arab freedom inspired by Hussein's ambition, wartime success and British ambiguity, they were poor stuff and they left the Grand Sherif himself with nothing more than the Hejaz to his name and his new title as King. Boiling with resentment and a conviction of betrayal, it was not surprising that he seemed emotional and stubborn. But it did him no good in British eyes. Philby and the others were cruelly correct in arguing that he had outlived his usefulness; for his strident – albeit unrecognized – claim to be King of the Arabs only jeopardized British relations with other Arab leaders, while his awkward insistence that Britain should honour the spirit of the McMahon correspondence filled British officials with a growing impatience that was all the sharper for the tinge of guilt.

Confined to an even more remote province of Arabia than his rival, Ibn Saud found the intrigues of European peace-making no less puzzling and frustrating. But, in spite of the Ikhwan's fearsome triumph at Turaba, he sensed that the moment had not yet come for a confrontation with Hussein. The Rashid were still active on his northern flank, even though the defeat of the Turks had removed their main support. To the south-west, the Idrisi rulers of Asir, between Yemen and the Hejaz, were also hostile and ready to ally themselves with Hussein against him. Beyond them lay Yemen, under the Imam Yahya – strengthened now by the collapse of Turkey and the withdrawal of their last troops from his stern mountains. And all around the south and east of the peninsula from Aden to Kuwait lay the multitude of territories under British protection where he could not move without meeting British power or putting his subsidy at risk.

His statesmanship won its due reward. Over the next year or two, the British upgraded both his subsidy and his title, lavishly donating an extra £20,000 on top of his existing £5,000 a month, and in 1921 accepting his own suggestion, supported by the *ulema*, that he should be described more respectfully henceforth as the Sultan of Nejd and its Dependencies instead of merely as its prince or ruler. It may seem ludicrous now, but these titular distinctions were treated with the utmost seriousness by the British, probably because they reflected their experience in India, where princely titles not merely abounded but also mattered. In Arabia they were far less important, but the British Government maintained an elaborate code of protocol for them, nevertheless, specifying among other things how many guns should be fired in salute to each title. An emir of the Lower Gulf could rate as little as one gun. A sultan got the full treatment of twenty-one.

The British also escorted Ibn Saud's second living son, the young Prince Feisal, on a lengthy tour of Britain and Europe in 1919. It was the first time any of the family had ventured beyond Arabia and, whether by chance or design, it bestowed on Feisal at the tender age of fourteen the first hint of that special authority in international affairs that he was to display in his later years on the throne.

It was not, superficially at least, the most auspicious of visits, and the small, slight, pale, large-eyed prince, wearing on full-dress occasions a dagger about as big as himself, had as much to endure as to enjoy. For a start the Government unwisely chose the months of October and November for the trip and young Feisal and his party were depressed by the damp and gloomy inconstancy of the end of the British year. Something also went wrong with the hospitality arrangements, for after their first night in a small London hotel the Saudis found themselves unexpectedly on the street, and only the resourcefulness of Philby, appointed as their guide, got them a bed with friends for the next few nights. Even so they were forced to go out to eat at the Grosvenor Hotel, and tramping to and fro in the rain, in their long desert robes, they made a sorry, sodden spectacle. 'A Government Bungle', said a headline in the *Daily Graphic*, 'Wandering Arabs and the Hospitality Committee'.

Their misfortunes continued, through a Welsh snowstorm on Snowdon – where someone had rashly invited Prince Feisal to try a spot of mountain walking – an unhappy meeting with Lord Curzon in the Foreign Office who treated them like children (or perhaps as a Viceroy should), and an even less happy meeting in Paris with the other Prince Feisal, Hussein's son. The Saudi prince stayed away from this last encounter and his older namesake took the opportunity to be rather grand about his wartime leadership on the battlefield and decidedly insulting about the Ikhwan. The remaining Saudis walked out in protest, fuming with rage.

In between these frequent contretemps, the visit followed the usual ceremonial pattern. The incongruities of history and culture between hosts and guests were many and obvious: the young prince from the desert inspecting a captured German submarine, appearing on Pathé newsreels, riding a donkey cart in Ireland, gazing in astonishment through modern telescopes at Greenwich, watching with bland incomprehension the ritualized antics of the D'Oyly Carte Company in Gilbert and Sullivan's *Mikado* or of the naked ladies in *Chu Chin Chow*, and listening, no doubt with equal incomprehension, to whatever orotund welcomes a whole procession of dignitaries might deliver to him, from the Lord Mayor of Cardiff to the Secretary of the Royal Metal Exchange. But there were moments, too, when what seemed bizarre became more poignant, as on a tour of the battlefields of the Western Front where the torn and blasted landscape was yet unhealed by time and the stench of death still hung in the winter air. And there was, of course, the obligatory meeting with the King. George V, the bluff old sailor, received the Saudis in the Throne Room of Buckingham Palace, with Queen Mary and her sister, the Princess Mary, to help him. Such an exposure of the women of the Royal Household must have surprised Feisal almost as much as his father had been shocked by his interview with Gertrude Bell on his trip to Basra in 1917. But the prince took it with perfect calm and politeness, exhibiting throughout the cool self-control of a born diplomatist.

Back in Arabia, however, there were murkier affairs afoot. In the cool uplands of south-western Asir, where the high Arabian plateau suddenly heaves still higher into a jagged wilderness of crags and mountain peaks, terraced gorges and fertile valleys, pro- and anti-Turkish factions among the tribes had been locked in desultory siege and battle for several years, and when the last supporting Turkish troops withdrew in 1919 King Hussein of the Hejaz sought to strengthen his southern flank by new alliances among the warring groups. The Idrisi family, who settled at Mecca in the early nineteenth century, had thrown in their lot with the British against the Turks and been recognized as rulers of all Asir by the Allies. From their base on the coast, they were now attempting to bring the tribes to heel. The Aidh chiefs of the main Asiri town of Abha appealed to Ibn Saud for help and a force of Ikhwan was dispatched to their aid. In command, at least in name, was the young Feisal, just back from his genteel London theatres and his jaunty Irish donkey carts in time to be blooded in a full-blown tribal war.

After the scientific carnage of the Flanders battlefields, the first Asir campaign of 1919 – 20 must have seemed to him a comparatively joyful excursion. Under the direction of one of Ibn Saud's cousins and companion at the assault on Riyadh, Abdul Aziz bin Musaid, together with the Ikhwan who had already routed Abdullah at Turaba, the column easily pushed the Idrisis back to their capital of Jizan on the Red Sea coast. Not long afterwards the Aidh became troublesome and the column occupied Abha and highland Asir. The tribe revolted again in 1924 but were quickly crushed by another expedition led by Feisal. Meanwhile in 1920 an Ikhwan band under Feisal al Duwish routed a Kuwaiti force that had attempted to dislodge a Mutair settlement on the grounds that it was the territory of Sheikh Salim, Mubarak's successor. The Kuwaitis were so alarmed they fortified Kuwait Town.

In the north, the Turkish defeat had left a void and their old allies, the Rashid of Hail, were once more exhibiting that distressing tendency of all desert dynasties to collapse in fratricidal strife and murder. As a succession of intrigues and assassinations ate away the family's authority, so did Ibn Saud's influence among their people grow. Ibn Saud moved smoothly and swiftly into position for an almost bloodless takeover.

It came in November 1921. The Ikhwan defeated the last of the Rashid forces in the field and encamped outside Hail, planning a final assault upon the mud walls of the city. At their head was the fanatical and skilful commander of the Dushan Mutair, Feisal al Duwish, whose reputation for merciless warfare boded ill for the townspeople and the surviving members of the Rashid family. Ibn Saud, fortunately, had other ideas. Hastening to take command of the Ikhwan in person, and bringing with him non-Ikhwan units, he sent a secret message to the town's Deputy Governor encouraging him to surrender peacefully. He was a sensible man – and, no doubt, a frightened one too. But he had first to outwit the Governor, a Rashid whose pride would not countenance surrender. So he answered with another secret message, smuggled into Ibn Saud's camp in darkness, that he would arrange to hand over the city on condition that the lives of the remaining members of the Rashid family were protected and that other people who had taken refuge in Hail from Ibn Saud's justice would be permitted to go free.

Feisal al Duwish protested at these proposals but he was overruled. The next night, the vanguard of Saudi troops slipped through the gates of Hail, left open on secret orders from the Deputy Governor, and surrounded the last palace of the Rashid. The coup was complete – and the House of Saud was revenged at last upon its oldest enemies. But once again in the moment of triumph Ibn Saud showed his stature by preferring statesmanship to vengeance. The conditions of surrender were strictly enforced. There was no bloodshed and even the looting was firmly controlled. The surviving representatives of the Rashid clan were taken to Riyadh to live as Ibn Saud's guests and the widow of one of those who had been but lately murdered became his wife and in due course bore him yet another of his many sons. He was named Abdullah and a little over forty years later he was to become the Commander of the National Guard, the force that eventually replaced the Ikhwan which had helped to capture his mother's birthplace for the House of Saud. Ibn Saud's triumph, however, was of far more than merely dynastic consequence. In taking over the responsibilities and domains of the House of Rashid, he advanced Saudi power another 300 miles to the north where it met with the new authority of Britain and its Hashemite protégés, Abdullah in Transjordan and Feisal in Iraq.

As Gertrude Bell, now Oriental Secretary in Baghdad, was quick to remark, this changed the whole balance of Arabian power. 'It will bring Ibn Saud into the theatre of trans-Jordanian politics,' she wrote to her father in the following month, 'and probably into the Franco-Syrian vista also.'[9] Indeed, it did even more than that. From the head of the Gulf to more than halfway down the Red Sea coast Ibn Saud was confronted now by a solid crescent of Hashemite territory with no saving buffer of land between them, no agreed frontiers to define their respective limits, and only the conciliatory influence of Britain to keep the two families from war. The Ikhwan were aflame with ambition and these marchlands of Arabia were now to be their prey.

7 Reaping the Whirlwind

1922 – 30

I have left my desert and my sheikhship and sacrificed my wealth in seeking the gifts of God and fighting the infidels.[1] – FEISAL AL DUWISH

SIR PERCY Zachariah Cox was not one of those Englishmen of the East who made a name for himself with romantic exploits. Unlike Lawrence, he led no desert raids and dreamed no princely dreams. Unlike Shakespear or Philby, he was no great traveller or explorer. He rarely, if ever, transgressed against superior orders, he kept no diaries and produced no memoirs.

For most of his career he was known to Ibn Saud more by letter and repute than in person. He never went to Riyadh and only rarely visited even the fringes of Saudi territory. But as a lifelong servant of the British Government in a succession of posts in and around the Gulf, from Muscat to South Persia and Bushire to Baghdad, he gradually acquired a knowledge of the area and a standing among its people that was, in his day, probably unrivalled. He was the first Political Resident in Bushire; he held Lord Curzon's hand in Muscat – if such a liberty, even in metaphor, is not beyond all imagining – on that grand Viceroy's famous tour of the Gulf. He was successively the boss of Shakespear and Philby as well as of Gertrude Bell, and even from such a trio of fierce individualists he won nothing but respect and admiration.

So perhaps his influence on Ibn Saud is not so surprising. Although exercised at a distance as a rule, and through many intermediaries, and only occasionally boosted by personal meetings in Kuwait and elsewhere, it reflected the ascendancy of a calm and orderly mind and a determined personality over a man who, for all his native talents, knew he needed an interpreter to the world. It is arguable that Cox's influence on Ibn Saud and, therefore, on the development of the Saudi Kingdom, was greater than that of any other foreigner – at least until the years of oil wealth and nationalism arrived to change so much both of Arabia and of the imperial world he knew. Certainly Gertrude Bell thought so: 'It's really amazing,' she wrote in 1922, 'that anyone should exercise influence such as his...I don't think that any European in history has made a deeper impression on the Oriental mind.'[2]

In 1922 Cox was in his last post as High Commissioner in Baghdad, where he had just helped to create a throne for King Hussein's son, Feisal, to compensate him for the loss to the French of his hoped-for kingdom in Syria. That was not an action likely to endear him to Ibn Saud, and with the collapse of the Rashid and the capture of Hail, Cox's old protégé was now glaring at his new one across an uncharted desert borderland that was claimed by both sides and inhabited by wandering bedouin whose allegiance was uncertain. It was an explosive situation, made worse by the rising terror of the Ikhwan, whose persistent raids on the tribes to the north of them in Transjordan and Iraq brought appeals for protection to the new government in Baghdad and provoked counter-raids upon Ibn Saud's recently acquired territories in turn. To avert still wider conflict that might destroy Britain's position as a friend of both sides, Cox resolved that sovereignty in the disputed frontier area must be defined and charted, and he persuaded Ibn Saud to journey down from Riyadh to meet him on the Gulf coast to discuss the matter. The place he chose for their talk was the little port of Uqair, or Ujair, just across from the island of Bahrain.

The Uqair Conference in November 1922 was almost as much of a landmark in Saudi affairs, in its way, as Ibn Saud's recapture of Riyadh, for it placed specific and guaranteed limits upon Saudi expansion for the first time, and imposed upon the bedouin notions of tribal grazing lands and the rigid concept of fixed national frontiers. It was, in short, the moment when Ibn Saud's new kingdom began to assume the outline of a contemporary nation-state in place of the amorphous structure of a traditional desert fiefdom; and the contrast between the two was never more eloquently expressed than in the appearance and character of the two men who met to bridge the gap of centuries between them.

We can see them now in a photograph most happily taken at the time, seated on a pair of hard, upright chairs in the sand outside the tent of the Saudi ruler. He sits there in his enveloping desert robes as if the chair were scarcely big enough to contain him. Beside him, every inch a civil servant, sits the prim Sir Percy in black jacket, pin-striped trousers, a grey trilby and a spotted bow tie with his starched white cuffs neatly shot about his wrists and his high-laced boots incongruously scuffed with sand. But for a hint of worldly rakishness in the tilt of his trilby, he is the very image of Whitehall in the wilderness.

There was no doubt, however, about who won the encounter. Cox had brought with him a representative of Feisal's government in Iraq to establish the Iraqi claim and at first he left the two Arabs to argue out their case without more than a word or two from him. After five days of histrionic nonsense Cox took Ibn Saud aside and reprimanded him like a headmaster with an errant schoolboy. If he could do no better in the way of negotiation, Cox would take over and fix the boundary himself. 'This ended the impasse,' wrote the only eye-witness of this meeting afterwards. 'Ibn Saud almost broke down, and pathetically remarked that Sir Percy was his father and mother, who had made him and raised him from nothing to the position he held, and that he would surrender half his kingdom, nay the whole, if Sir Percy ordered.'³

And so it was agreed. At a general meeting a little later, Cox took a red pencil and carefully drew a line on the map from the head of the Gulf to the Transjordan frontier. It gave to Ibn Saud a large chunk of territory claimed by Kuwait until Feisal al Duwish's feat of arms in 1920, but transferred another large slice of Ibn Saud's territory into Iraqi hands. Then, to meet the needs of the bedouin tribes along the border who traditionally passed to and fro to find grazing for their flocks, Cox traced out two enclaves. These, he said, would be neutral zones, where sovereignty would be shared by the respective neighbours. It was a true judgement of Solomon and, with a final histrionic burst of tears and emotion all round, in which even Sir Percy joined, it was accepted.

Apart from leaving ill-feeling among the Kuwaitis, whose pretensions had been so summarily rebuffed, the arrangement was to lead to persistent trouble over the next few years as the tribes – and the Ikhwan – fought against the unnatural corset it set on their shifting movements; but as a frontier it has remained unchanged ever since, a relic of British imperialism in its last confident years, a monument to an Englishman of remarkable influence and, above all perhaps, a herald and a symbol of the new world that was now striving consciously to tame the wilder shores of Arabia.

We cannot, however, leave the Uqair Conference quite there, for if we look again at the photograph of Ibn Saud and Sir Percy, we will see standing between and behind them another and even more potent augury of the changes that would soon begin to transform Arabia and the fortunes of the House of Saud. It is a puffy-looking figure in a high-crowned solar topee and stiff collar, posed with the elaborate nonchalance of a nervous gatecrasher at a party. It is, in fact, a speculative New Zealand mining engineer and businessman called Major Frank Holmes who had arrived in Bahrain a few weeks before with an obviously phoney account of himself as a butterfly collector and a commission from an obscure company in London called the Eastern General Syndicate Ltd. He carried a letter of invitation from Ibn Saud, a large white umbrella lined with green, a green gauze veil to go over his helmet and keep the flies off his face, and more than fifty boxes, bags and crates containing presents for the Saudi ruler. He was, of course, a natural successor to William Knox D'Arcy – a man in search of a Middle Eastern concession, but this time in Arabia itself.

In the fourteen years since D'Arcy's team had first struck oil at Masjid es Suleiman in southern Persia, the Middle Eastern oil industry had expanded only slowly. The war had provided vast new incentives for increased production through its forced development of all the new oil-consuming industries from transport to petro-chemicals, but it had also seriously inhibited exploration. Most of the industry's growth, therefore, had been concentrated in America and East Asia where there was no fighting to threaten oil company lives or investment. Now that the war was over, however, development of the Middle Eastern fields was accelerating. The world price had tripled in the course of the war. The French Government was insisting on a share of the Turkish Petroleum Company in Iraq, and an American consortium, formed by an industry and Government anxious about declining U.S. reserves, was also pressing for entry on the grounds of international equity. The wholly British-owned Anglo-Persian Oil Company (formed from D'Arcy's original concession) was struggling, with British Government support, to keep out any further interloper from Persia or the lands around the Gulf.

In most of the Gulf territories, from Kuwait south to Muscat and westwards from there to Aden, British political control was well established and the series of exclusive oil agreements that had begun with Bahrain in 1911 was completed in Muscat by 1923, giving Britain a legal claim to a monopoly of concessions throughout the sheikhdoms. Only Saudi Arabia remained outside this net – for although Ibn Saud had accepted the British trucial system during the war he had not signed any accompanying oil concession agreement. As the British Government's representative, Sir Percy Cox would have preferred him to do so at Uqair, but he was forestalled by Major Holmes who had shrewdly seen the opening and realized that if he could win a concession from Ibn Saud, he might be able to sell it again at a vast profit to an international oil company, once the presence of oil was confirmed. Sir Percy, having pressed his luck already over the matter of the disputed frontier, and knowing that Anglo-Persian was highly sceptical of the chances of discovering oil in Saudi territory anyway, decided to let the matter ride.

A few months later Ibn Saud signed his first oil concession agreement with the paunchy Major. It is hard to believe that he had much grasp of what he was doing. What he understood – like most Arabian rulers at that time – was not oil but water, and he rather hoped that the foreigner's promised experts would discover some for him while searching for the other stuff. Meanwhile, his ever-empty treasury got a small but welcome boost, for in return for the right to prospect throughout Ibn Saud's eastern provinces, from the new boundary with Kuwait to the fringes of Qatar, Major Holmes and the Eastern General Syndicate undertook to pay him the magnificent sum of £2,000 a year.

*   *   *

With most of his northern frontier apparently settled at Uqair, Ibn Saud's attention inevitably switched back again to the west and the running sore of his relations with King Hussein of the Hejaz. In spite of Britain's intervention after the battle of Turaba and Ibn Saud's promise to restrain the Ikhwan, the hostility between the two men remained unabated. Now that Abdullah was established in Transjordan and Feisal was on his new throne in Iraq, Ibn Saud was genuinely fearful of Hashemite encirclement and more than ever determined to break Hussein's remaining power. He was also envious of the riches of the Hejaz. Hussein, on the other hand, was furious at the Saudi success at Turaba and more than ever cantankerous about Britain's betrayal of her wartime promises. Influenced, no doubt, by his mounting sense of ill use – and perhaps also by his advancing years, for he was now nearly seventy-five – his personal rule in the Hejaz was growing more venal, autocratic and eccentric. Pilgrims from Nejd were banned altogether from the Holy Places as a reprisal against Ibn Saud and pilgrims from elsewhere faced extortion from his agents on the one hand and robbery by uncontrolled freebooters of the Harb tribe on the other. In the 1923 season, the mother of the Emir of Afghanistan was alone mulcted of £700.

Yet as Hussein's support and reputation sank, his ambitions continued to rise, until in 1924 he took a step that finally severed him from all but family sympathy. When the new republican government in Turkey announced the abolition of the Ottoman Caliphate, he pronounced himself the natural successor to the office of Caliph of the entire Muslim world. He was not, in fact, the only pretender to this august title. King Fuad of Egypt, the Sultan of Morocco and several obscure Malayan princes also laid claim to it with varying degrees of conviction, and the last of the Ottoman Caliphs, Sultan Muhammad VI Vahideddin, who had only recently been exiled by the Turks, still enjoyed recognition as the leader of Islam in some quarters. But King Hussein's gesture was in a different category. It was, so he said, his right and duty as a descendant of the Prophet and the Guardian of the Holy Places to take up the sacred burden which the Turks had laid down, and he promptly organized announcements in Mecca to that effect and junketings among his friends and relations in Damascus and Amman to celebrate his new status.

It appears, too, that he banned the Ikhwan from making the *Hajj* on the grounds that they were a risk to foreign pilgrims. In June 1924, a great council was held in Riyadh of *ulema*, tribal notables, and Ikhwan under Abdul Rahman, Ibn Saud's father. The redoubtable Sultan bin Bijad made a speech in which he said the patience of his men was exhausted. They would make the *Hajj* by force if necessary. The *ulema* ruled that this was valid ground for war. The *Hajj* was then at its height. But by September, most of the pilgrims had departed from Mecca and it was becoming clear that the Muslim world – and particularly the powerful Khilafat, or Caliphate, Committee representing 70 million Indian Muslims – was out of patience with Hussein.

That month some 3,000 Ikhwan under Sultan bin Bijad and Khaled bin Luwai, the Ikhwan leader at Khurma, swept down on the little hill town of Taif, the summer capital of the Hejaz, high on the escarpment overlooking the road to Mecca and the Red Sea coast, where Hussein's youngest son, Prince Ali, commanded the garrison of Hejazi troops. In more determined hands Taif might have resisted the Saudi onslaught, for it was dominated by a Turkish fort big and stout enough to house most of the townspeople alongside a garrison of thousands. Even the Ikhwan cavalry could hardly have forced its walls against defenders sure of their cause. But these defenders were far from sure and as the Ikhwan closed in, Prince Ali and his army fled. Thousands of townsfolk fled with them and as the mingled rabble of soldiers and civilians thrust their way in panic down the road to Mecca, the Ikhwan camel men fell upon them with zealous glee. When the rout was ended, there were 300 dead, Taif was looted bare and most of its buildings had been burned to the ground.

Hafiz Wahba, an able former schoolteacher of Egyptian origin who was in the King's foreign service at the time, provides a fascinating insight into the mentality of the Utaibah Ikhwan: 'I was told by a friend that when the *Ikhwan* first entered Taif and Mecca they smashed all the mirrors they found in the houses, not from lust for destruction but simply because they had never seen such things before. Any visitor to Khurma will see the results of such behaviour – perhaps a fragment of mirror on a wall, somebody's share of the loot – or a window acting as a door because Bedouin do not see the point of windows – or half a door instead of a whole one. Or there may be a quarter or a third of a carpet on the floor, because one big one has been cut up into fair shares.'[4]

With his army destroyed and Mecca itself now open to the marauding Ikhwan, the last vestige of Hussein's power was gone. The opposition to him of the Indian Muslims provided the British with a plausible excuse to drop him. Within a few days he was compelled to carry out the threat he had often used against the British in the past – to abdicate in favour of Prince Ali. At Jeddah, where Ali had finally taken refuge, the old man's private yacht, *Al Rahmatin – The Two Mercies* – was prepared to receive him and what remained of his British gold. He drove down from Mecca for the last time in a small convoy of shining new motor cars that he had only recently imported into the country. He sailed away to Aqaba and seven years of weary exile on the island of Cyprus and in Amman, a man betrayed by himself as well as his friends, the forgotten king of a vanished kingdom, full of bitterness right until his death in 1931.

Ali's succession lasted just long enough to ensure that the rest of the Hejaz was handed over to Ibn Saud in more orderly fashion. Worried by the reports of the Taif massacre and their effect on the Wahhabi reputation beyond Arabia, Ibn Saud hastened to rein in the Utaibah Ikhwan again before they could get to Mecca and lay waste the Holy Places in their zeal. Their commanders were instructed on pain of death to prevent any further looting or murder. When the vanguard did arrive at Mecca a few days after Hussein's departure there was no looting outside the palaces of the royal family. The Ikhwan did, however, purify the place in accordance with Wahhabi tenets, with the destruction of the idolatrous graves of 'holy men' and the confiscation of all musical instruments and human portraits.

When Ibn Saud arrived in person fifteen days later he, too, came as a pilgrim and led his Ikhwan commanders through the ceremonies of pilgrimage. It was the first time he had ever set foot in Mecca and he was determined to prove to the Muslims of the world that he was a worthy keeper of its blessedness. From the Muslim world, anxious about desecration, he invited delegations to inspect the two Holy Cities.

His restraint and piety paid off. At first, to be sure, the inhabitants of Jeddah were unconvinced and prepared themselves fearfully against Wahhabist siege and slaughter. To their initial horror, Wahhabi shells fell not only upon Jeddah but even upon the Tomb of the Prophet in Medina. Ali's army, now consisting largely of destitute West African pilgrims, was in no mood to fight and the defence of Jeddah lay in the hands of a motley bunch of mercenaries. 'Some abandoned American army lorries which a Syrian had picked up for £300 apiece were fitted out with thin metal sheeting and sold to the Hejaz Government for £2,300 apiece,' Reader Bullard, His Majesty's Vice-Consul in Jeddah, reported. 'Only one of these "armoured" cars was ever used. It was driven out of the barbed-wire defences by a Russian refugee and shot through and through by the Wahhabis...[the driver] obtained £50 as compensation besides a post-dated cheque of doubtful value for another £50. He went to Egypt, gave a party for his Russian refugee friends, and returned to Jedda penniless.'[5]

As the months went by and Ibn Saud still held back the Ikhwan from the last bloody assaults their commanders urged upon him, the Hejazis took courage. Their plight under Ali, after all, was far from happy. Their towns were cut off, the *Hajj* which normally made them rich was of necessity suspended, and their food supplies were running very low. 'Starvation and disease,' Philby wrote, 'were left to do their worst during the summer in the ill-supplied, over-crowded town, where the poor were reduced to begging for brackish water and sifting the dung of horses for undigested grains of barley.'⁶ Life under the tyrant of Nejd could hardly be worse and might even be better.

By December 1925, a year after the gruesome events in Taif, the people of the Hejaz were ready to accept the inevitable. Ali was persuaded to follow his father into exile and Ibn Saud arrived in Jeddah to accept the city's surrender. On 8 January 1926, he called together in Mecca all the sheikhs, merchant princes and imams of the province and there in the Grand Mosque, after Friday prayers, he was solemnly proclaimed in their presence to be the new King of the Hejaz. It was just twenty-four years since he had raised the Saudi flag again over the Musmak in Riyadh and now two-thirds of all Arabia was in his hands.

*   *   *

The capture of the Hejaz was the greatest triumph of Ibn Saud's career. No other part of Arabia was richer or more populous and none was remotely of the same significance to the non-Arabian world. It made him a king in consequence as well as in style, and the year after his proclamation as King of the Hejaz he became King of Nejd as well.

The Hejaz transformed the whole nature of his realm, adding to the simplicity of desert and oasis the worldly complexities of ancient cities and thriving ports as well as the international horizons of a great centre of religious pilgrimage. Truly Jeddah was the *Bilad al Kanasil* – the City of the Consuls. Never again would simple Wahhabism and the traditional ways of the desert provide a sufficient guide to the Saudi revival. Henceforth, no matter how hard the Saudis might try to accommodate the demands of the external world within their exclusive system, they would discover that, little by little, their system was infiltrated, debilitated and often overcome by the world.

The Ikhwan, with their special passion for the simple answers offered by God and the Sword, sensed this truth at once. They were horrified to see how often the worldly Hejazi merchants put the demands of Mammon before the commands of God, neglecting their prayers for the sake of commerce and profiting from trade in blasphemous indulgences. They were even more shocked to see their own Imam, Ibn Saud, turning a statesman's blind eye upon some of these impious practices. No sooner had they entered Jeddah, for example, than the city's venal merchants were protesting that if the Wahhabist prohibition on smoking was enforced they would be compelled to sacrifice immense stocks of tobacco at penal cost (£100,000 in Philby's estimation) to the businesses and their families. Ibn Saud actually heard the complaint with sympathy and allowed them to expand their stocks. Worse still, eventually the merchants were to pay him a slice of their reward as tax. Thus did expediency creep into the innermost sanctum of the Wahhabi state.

Of course, Ibn Saud could not disregard the feelings of his Nejdi supporters. New Wahhabi imams were appointed at the mosques while the Society for the Encouragement of Good and the Prevention of Evil was reformed to enforce a stricter regime of prayer. Consisting largely of zealous, and often ignorant, Nejdis, some of them Ikhwan, the Society established itself by force in the quarters of the various towns and imposed summary whippings on those who failed to close their shops at prayer time or smoked in public. Yet the power of the Ikhwan was limited. Although they took Mecca, and their leaders Sultan bin Bijad and Khaled bin Luwai even treated briefly with the Jeddah *Kanasil,* the Ikhwan bands were specifically prohibited from entering the port. Ibn Saud's chief concern was to reassure the Hejazi notables that they would be consulted under the new regime. He even held out vague, and opportunistic, promises that the Hejaz might become a Trust of all Muslims. For the Ikhwan, his path was clear from the end of 1925 when Hafiz Wahba was made Governor of Mecca in place of Sultan bin Bijad and Khaled bin Luwai. The godless Hejazis were even to tender advice in the administration of the new domain.

Soon after the fall of Mecca at the close of 1924, Ibn Saud promised the creation of a Consultative Council *(Majlis al Shura)* 'of the *ulema,* the dignitaries and the merchants' which would be 'an intermediary between the people and me'.[7] The Council was duly promulgated in 1926 in a decree which also promised a constitution. Although the constitution has still, over fifty years later, to materialize, the Council did play some minor role in the coming decades, at least in the Hejaz; and both Council and constitution were to figure prominently in any talk of reform.

The Ikhwan were restive. At the *Hajj* of June 1926, they picked a quarrel with Egyptian pilgrims which precisely echoed the intolerant deeds of their Wahhabi forefathers in Mecca over a century before and led to an estrangement between Ibn Saud and Cairo that lasted for a full decade. The focus of the row was an ornate camel-borne litter known in Arabic as the *mahmal*, which was traditionally sent to Mecca from Cairo every year. It was used as ceremonial transport for the tapestry covering (the *kiswa*) that was normally made in Egypt for the *kaaba* every year, the cube-like construction in the centre of the Grand Mosque which contains the black stone and is the focal point of the *Hajj*. It had come to symbolize first Turkish dominion over the Hejaz and, more recently, the tenuous claim of Egypt's King Fuad to the now defunct title of Caliph, and there had been tension even in Sherifian times. With the *mahmal* and its camel came, by tradition also, a contingent of Egyptian troops and a lethargic but noisy brass band.

The sight and sound of this colourful caravan among the pilgrims of 1926 was too much for some of the Ikhwan among the throngs at prayer. Throwing themselves upon the Egyptian escort, they started a battle in which several men were killed, more were injured and vast numbers of other pilgrims were immensely scandalized. Only the personal intervention of Ibn Saud's son Feisal prevented worse from happening; but his father's subsequent diplomacy could not avoid a breach with Egypt that lasted until King Fuad died ten years later.

The Ikhwan were becoming a challenge and there were times in the next few years when Ibn Saud's responses seemed uncharacteristically feeble and unsure. To a large extent the Ikhwan were his own brethren and they had served him well. But with as many as one hundred of their settlements scattered about the country by 1926, able to put 50,000 to 60,000 armed men into the field, they had gained a formidable strength and momentum of their own, and he could neither ignore nor disband them without conflict.

As a stopgap measure he decided, in 1926, to remove them as far as possible from the sensitive precincts of the Hejaz, where their military force was no longer needed, and let them expand their zeal once more among their own desert people, while he stayed behind in Mecca to supervise the incorporation of the Hejaz into his new kingdom. The Hejazis were reassured again, the rest of Islam came to see that Wahhabism in the Holy Places was not so fearful after all, and Ibn Saud's rigorous suppression of the banditry and extortion that had often been practised against pilgrims in the past encouraged a new influx of the faithful, whose money brought a faint flush of health to the Saudi treasury for the first time in Ibn Saud's career.

But the Ikhwan's discontents increased. Back home in Nejd they chafed in unaccustomed inactivity and sought new outlets for their fierce righteousness. With the Hejaz denied them, however, these were no longer easy to find. Ibn Saud had forbidden them to raid among his own people any more and although most of the country's frontiers were still officially uncharted, Saudi territorial expansion had almost reached its limit. The only rival independent power left in the peninsula now was that of the Imam of Yemen; and although some of the Ikhwan might have relished a chance to deal with his heretical theocracy as they had already dealt with the neighbouring Idrisis, the Yemen's rugged mountains and the fighting reputation of its tribes made that a perilous undertaking that Ibn Saud would not contemplate. Elsewhere, Saudi power was totally circumscribed by the great crescent of petty rulers under British protection, from Aden to Kuwait, and by the new Hashemite kingdoms in Iraq and Transjordan which also enjoyed British support.

It was upon these last that the Ikhwan finally vented their frustration. Their motives were a characteristic mixture of the spiritual and the material. Their fanatical contempt for all things Hashemite was undiminished and they were equally resentful of British influence which had not only set up these infidels in their false kingdoms – as they saw them – but had also persistently imposed ungodly restraints and compromises upon their own leader, Ibn Saud, in the interests of Christian power. Just as important, though, was the fact that without a new excuse for *jihad* the whole Ikhwan movement was doomed. This was less for the sake of loot than because, like all evangelical movements, it could only breathe in an atmosphere of proselytizing. Across the new northern frontiers, drawn up by Cox and accepted by Ibn Saud, the Ikhwan movement scented exactly the opportunity it needed to maintain its momentum.

Frontiers, by definition, are confining things, delineating territory and restricting people. But the desert is not a place of confinement; like the sea, its life, of necessity, deals in movement, space and fluidity. In search of grazing, a single tribe may cover thousands of miles in a single year. It may share some water wells with other tribes from far away and demand hospitality in return when it moves into their home territory. Regular caravan routes, too, span mini-continents of semi-emptiness, their only fixed points being the scattered cities or oases touched along the way.

Cox had tried to make allowances for this by his creation of neutral zones, but the fundamental conflict remained. If sovereignty and responsibility were to be established on either side of the lines he drew, the old fluidity had to be restricted. It was difficult for bedouin raised in desert ways to grasp the necessity for such a vital change, and in the absence of any ready means of policing the new frontiers they simply ignored them, happily roaming and raiding as they had always done over their own and other people's territory. For the Ikhwan the change was even harder to accept. To them the frontier was much more than an irrelevance: it was a sinful attempt to frustrate their holy mission, as well as a potential obstacle between them and still more riches. To have accepted it at all, even on paper, was something for which Ibn Saud could not be forgiven. To destroy it was to earn yet another ticket to Paradise.

In the year or two immediately after the 1922 agreement at Uqair, therefore, when the desert defence forces in the new Hashemite kingdoms were still rudimentary, what was supposed to be a peaceful line of division was in fact an area of confused and widening conflict, in which ordinary bedouin and the so-called shepherd tribes of the marches were caught between the alien demands of modern governments on one side and the menace of the Ikhwan on the other. Some of the Ikhwan's raids were spectacularly bloody and dramatic. Scores of men, women and children were butchered and thousands of sheep and camels were seized and driven back into Saudi territory.

On one occasion the marauders swept 500 miles across the desert to within an hour's camel ride of Amman, the new Transjordan capital, and were only prevented from rampaging through the town itself when a British military lorry encountered their column by chance and was able to escape in time to call up a counter-attack by armoured cars and aircraft from the capital. Had their raids continued on that scale, Ibn Saud himself must have been drawn swiftly into conflict with the British, after all, and Cox's frontiers might well have been overrun almost as quickly as they were drawn. But Ibn Saud's preoccupation with the Hejaz provided a diversion for Ikhwan energies and it was not until 1927 that they were turned once more upon the north, in default of other outlets. By that time, however, the British had organized more effective defences and the Ikhwan discovered that the cost of prosecuting their *jihad* was higher than ever before.

The tribes of Transjordan and Iraq were prompted by the British to form new defensive groups. In Transjordan, Colonel Peake – known as Peake Pasha to his men – recruited and trained the first bedouin regular troops in the Transjordan Frontier Force which later formed the nucleus of Jordan's famous Arab Legion. In Iraq, Peake's eventual successor as commander of the Legion, a young intelligence officer called John Bagot Glubb, did the same among the tribes along the edge of the Euphrates valley. Once more the old poachers of the desert were turned into gamekeepers; but whereas the Ikhwan were still using the old poachers' weapons of camel, sword and rifle, their opponents were backed by the growing strength of brand new armaments – the motor car, the radio and the aeroplane.

These were primitive enough, by modern standards – long-nosed Rolls-Royce armoured cars built on civilian chassis, crackling crystal sets, ancient morse tappers, and draughty, open-cockpit biplanes of painful slowness and inadequate range. But they were more than a match for camel raiding parties and no ordinary caravan could get far across a frontier without the knowledge or approval of the men who ran them. With their aid the British and their bedouin allies could tame the desert as it had never been tamed before. Against them, ultimately, the Ikhwan stood no chance; yet their dying struggles were more ferocious than ever.

Three men stood out among the Ikhwan leaders. First in skill and seniority was Feisal al Duwish, one of Ibn Saud's best generals and leader of the Mutair tribe, who were famous across Arabia for their special herd of black war camels, known as the Shurf, that they drove before them riderless into battle. Beside him were Dhidan bin Hithlain, chief of the perennially rebellious Ajman tribe from the Hasa province – they who had so often deserted Ibn Saud in times of crisis – and Sultan bin Bijad bin Humaid, chief of the Utaibah. Whereas the first two were regarded as opportunists, Ibn Bijad, whose authority stretched across the central desert between Riyadh and the Hejaz, was a man of great religious force, bigoted perhaps to Western eyes, but genuine withal.

By the beginning of 1927 this zealous trio was in a fever of discontent at Ibn Saud's continual retreat from the true path of piety in the Hejaz and their own inaction. In January, Ibn Saud was obliged to convoke a conference of some 3,000 Ikhwan in which the *ulema* ruled on their grievances. These ranged from such Christian and Satanic innovations as the telegraph network Ibn Saud was installing for reasons of tribal security to Feisal's visit to godless England, the King's failure to coerce the Shia and to prevent the 'infidel' Muslim tribes of Iraq and Transjordan grazing their flocks on Saudi land.

While the *ulema* were, in some respects, favourable to the trio's views, they ruled that the crucial issue of *jihad* lay with the discretion of the Imam, Ibn Saud. The response of the Ikhwan zealots came in September. They seized on a British attempt to build a new desert police post, at a place called Busaiyah, eighty miles inside the Iraqi frontier, claiming that it was a base for an attack on Saudi territory. Ibn Saud protested to the British; but a band of Mutair attacked Busaiyah themselves, killed all but one of the workmen and began a series of murderous raids all along the Iraqi frontier. The British became more anxious than ever because the raids posed a severe threat to the plans of the Turkish Petroleum Company to build a pipeline and railway.

Ibn Saud seemed powerless to stop the Ikhwan, for many bedouin approved of their raids, hated the British and resented Ibn Saud's apparent complicity in the growing attacks on their old freedom. Many, too, could not see why, after years of preaching *jihad* himself in order to expand and fortify his kingdom, Ibn Saud should now find it politic to say that holy war was wrong. Certainly, Ibn Saud's attitude seemed highly equivocal to the British. Relations were cordial enough between the King and Britain for the Treaty of Jeddah to be negotiated by Clayton, the former head of the Arab Bureau, in May 1927. This recognized Ibn Saud's independence and new frontiers. But when Clayton returned in the summer of the following year to discuss the question of police posts in Iraq, Ibn Saud insisted they should be dismantled – possibly because he had promised that to the Ikhwan. Glubb, for one, felt that Ibn Saud genuinely feared rebellion if he opposed the raids while his pride and ambition approved them. It was just so, fifty years later, when Saudi leaders spoke of Palestinian attacks. But the Ikhwan soon threatened to split the new Saudi kingdom from top to bottom, thrusting Ibn Saud and the British into each other's arms and tarring their old leader with the brush of betrayal and apostasy.

For two years the country was racked with uncertainty and sporadic violence while the Ikhwan's commandos continued to ravage the northern frontiers. But gradually the new defence forces of the British began to gain the upper hand there, while at home the mood was shifting. At a Congress in Riyadh in November 1928, which over 10,000 bedouin attended, Ibn Saud threatened to abdicate. Amid cries of 'We will have no one else' the King secured a promise of support against the rebels.

When Ibn Bijad arrived on the Iraqi border in February 1929 with

3,000 men intent upon yet another raid, he heard convincing reports
that his line of march northwards was already covered by a powerful
force of British aircraft, armoured cars and troops. The Ikhwan had
learned enough by then to know that if the reports were true, their
raid was likely to be a costly failure. Ibn Bijad hesitated; then,
unwilling to sacrifice the chance of loot, he turned his force south
again to attack and plunder the camps of other Nejdi tribes settled in
Hasa. It was a fatal error. As long as they had directed their war
against the British and their allies, the Ikhwan's claim to be the
spearhead of true Saudi virtue was difficult to dispute. But when they
showed fear of the heathen firepower and fell upon their fellow
Wahhabis instead, their support began to crumble.

Moving to Qasim that month, Ibn Saud recruited a new Saudi army
of townsmen and loyal Utaibah and met the rebel Ikhwan at a spot
called Sibilla, near Zilfi in the north – not far from the field of Jarrab
where Captain Shakespear fell in the ill-fated battle with the Rashid
fifteen years before. Two of the principal Ikhwan leaders were in the
field, Feisal al Duwish having joined Ibn Bijad and moved in from
Artawiyah. Ibn Saud was visited by Feisal al Duwish and tried to
persuade him to disband his men and accept the King's authority.
Some say Feisal spent the whole night in the King's tent, was
convinced by his old chief's arguments and went back afterwards to
his own side promising to secure the agreement of his partners. Some
say that he was adamant to the last and was merely buying time to
mount a surprise attack on the King's army. Whatever the truth of
that, it was Ibn Saud who struck first. With his whole kingdom now
visibly at stake, he could hardly afford to take a chance on men who
had already proved so stubborn and so ruthless. His army had the
advantage of numbers and at first light on 30 March he launched his
men upon the Ikhwan lines in what proved to be the last major
traditional battle between bedouin forces in Arabia. His brothers
Abdullah and Muhammad and two of his eldest sons, Saud and Feisal,
were in command of the four columns. Once more the war banners
streamed out and the war cries pierced the cold silence of the desert
dawn as the horses and camels pounded down upon the enemy camp.
Once more the field was covered in shrieking confusion as the first
charge swiftly dissolved into a thousand hand-to-hand combats. But
the surprise was too much for the Ikhwan. Feisal al Duwish was
caught at breakfast and seriously wounded. Ibn Bijad beat a retreat. In
less than half an hour it was virtually all over.

Ibn Saud's better instincts almost betrayed him. Moved by the pleas of Feisal's veiled women, he allowed the Mutair leader to return to Artawiyah to die. But the tough old man slowly and secretly recovered, and the Mutair were still with him to a man. Meanwhile, the ever restive Ajman rose in rebellion when in May Ibn Hithlain, who had stayed out of the battle of Sibilla, was treacherously murdered by Abdullah bin Jiluwi's son Fahd. As Hasa sank into turmoil as far as the Kuwait border, the Utaibah – even though Ibn Bijad had been captured and was safely in jail – attacked Ibn Saud's rear.

Although the Ikhwan rebels fought with tenacity, they could not match Ibn Saud's new power. Bitterly regretting his gentleness, he strengthened the new army with imported motor cars and somehow sent them across to Hasa, broken springed and stripped of most of their seats but still too fast and strong for the Ikhwan cavalry.

Feisal al Duwish played his last card by trying to entice the Sheikh of Kuwait himself to his side by offering him the return of the territory he had lost to Ibn Saud through the warring of the Mutair and Sir Percy Cox's fiat at Uqair. A few years earlier the ploy might conceivably have worked. But by the end of 1929 Kuwait also had suffered much from Ikhwan raids and, besides, the British defence forces established in Transjordan and Iraq had thrown their protective mantle over the Sheikh's little territory as well. Even if he had wanted to co-operate with Feisal, the British were now in a position to stop him. So he turned down the old man's proposal, and the Ikhwan's fate was sealed. Harried repeatedly by Ibn Saud's forces, unable to find new allies, Feisal al Duwish was forced at last to lead his shattered columns into Kuwait, not for the sake of pillage or conversion but for sanctuary and survival. The British who had been his enemies for so long were waiting for him with their armoured cars and aeroplanes; but they knew that Feisal could fight no more, and instead of demolishing what was left of his forces they politely suggested that perhaps the time had come for him to surrender. He was a stubborn old man and would not give in easily, but in the end he recognized that there was no escape. On 10 January 1930 he rode from his camp in the car of the British Political Agent in Kuwait, Colonel H. R. P. Dickson, to the local headquarters of the Royal Air Force and there, with a mumbled word or two of Arabic and a momentary, respectful inclination of the head, he ceremoniously handed over his sword to Air Vice Marshal Sir Charles Burnett. Then, turning for the last time to Dickson before he was taken away into British custody, he said: 'I hand my ladies to your personal charge, O Dickson, and from my protective honour to your protective honour.'[8] It was the highest compliment he could bestow – an acknowledgement of the honour of an enemy whose overthrow he had sought so fiercely and so long in the name of God.

With the collapse of the Ikhwan rebellion, the most fanatical, reactionary and expansionary elements of Saudi power lost their driving force. In late February, under the watchful eye of the British, Ibn Saud held a relatively amicable meeting with the Iraqi King aboard the sloop *Lupin* in the middle of the Gulf. What was left of the Ikhwan armies was disbanded and many settlements, including Ibn Bijad's base at Ghatghat, were first dispersed and then destroyed. The memory of the Ikhwan was to lie dormant until a new generation of zealots recalled their exuberant violence in Mecca, fifty years on. As for their leaders, Feisal al Duwish was transferred by the British to Ibn Saud's mercy, on condition that his life was spared, and when the two warriors met again at the King's camp they kissed noses in bedouin fashion with tears streaming down their cheeks. The ladies were given into the King's care and Feisal died eighteen months later. Three years after that the remaining survivors, led by the unrepentant Ibn Bijad, tried to escape from prison during Ibn Saud's brief war with the Imam of Yemen. They were re-incarcerated for their pains in an old Turkish dungeon in the fort of Hofuf in Hasa, and were never seen or heard of again.

Nowadays, even the name of Feisal al Duwish seems to have disappeared from the Saudi scene. Men of the Mutair tribe are said to remember him as one of their great leaders, but so fierce was the challenge he represented to Saudi rule that it is still thought politic not to speak of him in public. Meanwhile, by a typical quirk of fate, his most poignant and substantial memory is preserved far off in the misty uplands of Scotland. The sword he handed over to Air Vice Marshal Burnett was passed on punctiliously to the limbo of 'higher quarters' as part of the spoils of war until, beribboned with the red tape of bureaucratic ignorance and indifference, it was eventually rescued again by its first recipient as a personal momento of a great occasion. From him it passed on by inheritance to his daughter, and now it hangs in the hall of her home, at the foot of the staircase in a small, stone-built villa in Aberdeenshire, Scotland. It seems an appropriate domesticated end for this last relic of a stern and wild old man; for his submission was the first sign that the domestication of Arabia had truly begun.

8 The Coming of Mammon

1931 – 4

I tell you, Philby, that if anyone were to offer me a million pounds now he would be welcome to all the concessions he wants in my country.[1]

IBN SAUD

THIRTY years on from the capture of Riyadh, on 18 September 1932, Abdul Aziz bin Saud, sometime prince and sultan, always sheikh and imam, and – since 1927 – twice a king, decided on yet another title. Abolishing the cumbersome dual monarchy he had established after the capture of Mecca as King of the Hejaz and King of Nejd and its Dependencies, he proclaimed the unity of both under his own name in the Kingdom of Saudi Arabia. Superficially it seemed an act of unique presumption, implicitly announcing the absolute title of a single family to rule over most of the Arabian peninsula. But in reality it reflected the state of affairs that actually existed, for most of Arabia was now, in fact, a Saudi family fiefdom. That it should take the family name was merely a symbolic way of expressing the truth.

Ibn Saud, at fifty-two, had reached the zenith of his career. With the defeat of the Ikhwan his power was unchallenged from coast to coast. Internally, there was scarcely a flicker of rebellion or disaffection left. Externally, he was recognized as Arabia's premier ruler not only by his old friends the British, who sent a full minister to Jeddah in 1930, but also by many other foreign powers whose consuls in Jeddah had reported with approval his measures to restrain the Ikhwan and improve the pilgrim traffic. One of them, by a curious historical irony, was the Soviet Consul, whose Government had been the very first – even before the British – to recognize Ibn Saud as the new King of the Hejaz-Nejd in 1927. No doubt Moscow was mindful of its responsibility for many millions of Muslims in central Asia – but its action provided a curious echo of that of the Tsarist Government in 1903, when it sent its Consul from Bushire to Kuwait to become the first European ever to meet Ibn Saud. If the Kremlin had any hope of securing advantage from its promptitude in Jeddah, however, it was as quickly disappointed there as its predecessor had been in Kuwait. The Soviet Union was interested in the commercial possibilities of the Hejaz. However, an attempt to offload cheap goods on the market at Jeddah elicited a cry of protest from the Hejazi merchants and, no doubt, the British. The Soviet Minister, a fretful Tartar Muslim called Karim Khakimov, was quite unable to prevent Ibn Saud imposing a total embargo against Soviet goods in the Hejaz in 1928.

Around the frontiers of Ibn Saud's new kingdom there was also, by and large, a gratifying air of stability and peace. Although the old Saudi quarrel with the Hashemites in Iraq and Transjordan lingered on, even after the death of old King Hussein of the Hejaz in 1931, the destruction of the Ikhwan had removed the chief remaining danger of provoking a clash on the Kingdom's northern borders; and in the south and east, the wide barrier of the Rub al Khali and British control of the sheikhdoms beyond it were, for the present, effective obstacles to any misunderstandings. Only the Imam Yahya in Yemen remained to dispute Ibn Saud's supremacy – and his claims were soon settled by another convincing display of Saudi force.

The two kingdoms had been in a state of suspended conflict ever since the Saudi occupation of Asir in 1920. In 1926, the Idrisi lowland territory had become a Saudi protectorate and was annexed in 1930. In the course of 1931 and 1932, Yemeni fears of encirclement and eventual reduction by Ibn Saud, encouraged by Idrisi exiles, led the Imam to reassert his authority among the tribes around the northern and eastern fringes of his mountain heartland, in the provinces of Najran, Asir and the coastal strip called the Tihama. Historically, the Imam's claims at least to Najran were reasonably good, for most of the tribes here had paid tribute to him or his predecessors for centuries. But a clash with Saudi expansionism had become inescapable. For the Imam, too, had a special Muslim faith to propagate, but it was one of a radically different kind from the Wahhabis'. To the Yemeni Muslims of the Zeidi sect, who had ruled with and through their Imam for many centuries, he possessed an aura not unlike that of a medieval Pope among the Christian princes of Europe. He was accepted not merely as yet another of the Prophet's multitudinous descendants, but also as God's representative on earth – a heresy of blackest hue to all Wahhabis, whose rejection of any personification of, or even mediation with, the Deity is absolute.

Although both rulers tried to avoid open war and talks were held in the Imam's mountain capital of Sana in 1933, this religious schism helped to keep the contest alive. In March 1934, Ibn Saud sent his forces under his son Feisal into action. Honed by repeated warfare, they proved far too strong for the Yemenis. Within three weeks Feisal captured the Imam's chief port at Hodeida, commanding the road to Sana, having outmanoeuvred the ill-equipped Yemeni opposition all the way down the Tihama coast. As British, French and Italian warships were hurried to Hodeida in Europe's anxiety at this new shift in the balance of power, Ibn Saud prudently ordered Feisal to proceed no further, and he did not blockade the highlands because, as he later told a British visitor, this would have cut off the supply of *qat*, the narcotic leaf to which Yemenis are so devoted, which would have emptied the town. But it was too much for Yahya. In a famous message to his rival monarch, the Imam cried, 'Enough! Enough! Enough!' Ibn Saud responded with statesmanlike celerity, for he called off the six-week war at once and invited Yahya to negotiate a new frontier between them.

Legend has it that Yahya then proved even shrewder than Ibn Saud, for he is said to have replied to the invitation with a counter request, couched in the highest terms of Arab flattery, for the Saudi King to take matters entirely into his own hands and decide the new frontier for himself. By tradition this placed upon Ibn Saud an absolute obligation to be generous. The result was another statesmanlike gesture enshrined in the Treaty of Taif of 13 May, which gave back to the Imam nearly half his lost territories, including the southern portion of the Tihama coastal strip, leaving only the uplands of Asir, Najran, and the Idrisis' former base of Jizan in Saudi hands. Yemeni pride suffered, however, and two years later, according to Muhammad Almana, one of the King's interpreters, an attempt was made on Ibn Saud's life by two Yemenis in the Grand Mosque at Mecca. They were forestalled by the speedy action of his eldest son Saud, who hurled himself in front of his father.

With the Treaty of Taif, the last major territorial additions were made to the Saudi patrimony and another portion of its once non-existent frontiers was marked out – by Philby the next year, in what was to be his longest journey. Except for the final touches to the borders with the British-protected sheikhdoms, the last of which was still more than forty years away, the Kingdom of Saudi Arabia was established on the map of the peninsula more or less as we know it today.

The young Emir of Nejd had won over no more than perhaps 50,000 people. He was now the undisputed ruler of 1–2 million. What had been an isolated and almost unknown desert territory now spanned sea coasts and cities, mountains and oases, as well as the all-important Holy Places of Islam, to which so many millions of people around the world habitually turned their eyes, their prayers and their feet.

Yet at heart the Saudi state remained little changed. Its organization was still entirely personal, a traditional desert principality writ large. Except for Ibn Saud's personal great *majlis,* or council, where all his subjects were welcome to seek an audience with him and all received presents and, sometimes, a homily, there were no central governing institutions to speak of. The civil service hardly existed outside the Hejaz, beyond a handful of the monarch's personal advisers and a few secretaries, paid in periodical gifts of money and clothing. Apart from fifty or sixty negro bodyguards, the military forces were only whatever groups of armed tribesmen the loyal sheikhs and princes could raise at the King's command – the army had been disbanded after Sibilla. The entire income of the state, exiguous as it was, was regarded as the King's personal property. Fundamentally, like the medieval kingdoms of Europe, the King was the state because no other focus of loyalty or tribute existed beyond his person – and the ever-present threat remained that if his person should be removed or his reputation stained, the state he had created would disappear as well.

Luckily, Ibn Saud's personality had expanded with his years and his domains. He was fifty-two when he declared himself King of Saudi Arabia, and although he was past his physical prime, with his wide chest beginning to slip towards a widening belt, a limp from a leg wound that became more pronounced as he grew older, and a partially blind eye, he was at the peak of his moral power. Wisely, he had given up going into battle himself – after the defeat of the Ikhwan at Sibilla he never again took the field in person. But his great height and strength still gave him a physically commanding presence upon which age and experience had bestowed a commensurate air of casual but profound authority that impressed everyone who met him. He was still accessible to all his subjects, whether in his mud palace with its

four square towers – one for each wife it was said – or in Jeddah, or
travelling the country with what was now a regular caravan of motor
cars (the camels were not finally retired until 1935), holding open
court at every stop in a marquee of white canvas, floored with several
dozen Persian rugs and surrounded by all the state paperwork in large
wooden chests.

Ibn Saud held the reins of absolute power. His word was as strong
as ever his sword had been. As he put it once himself to another of his
foreign visitors, his rule was a pragmatic mixture of democracy,
diplomacy and sheer command: 'We raise them not above us, nor do
we place ourselves above them. We give them what we can; we satisfy
them with an excuse when we cannot. And when they go beyond
their bounds we make them taste of the sweetness of our discipline.'[2]

What really bound Ibn Saud to his people, however, and enabled
him to maintain what amounted to one-man rule over a steadily
widening kingdom, was his zealous attention to the family connection. The traditional network of first cousins, related through both
blood and marriage, was invaluable in providing loyal men of his own
generation and experience to act when necessary as his lieutenants.
Two of the most notable among early provincial governors were the
half-brothers Abdullah bin Jiluwi and Abdul Aziz bin Musaid,
Governors respectively of Hasa and Hail, the two most sensitive of the
King's early conquests. Both were of the same generation as the King,
both had been prominent among the famous forty companions who
stormed the citadel of Riyadh, and both subsequently had proved their
loyalty in a score or more of battles under Ibn Saud's banner. In such
hands their King could leave his expanding provinces without fear;
indeed, their devotion to his cause has echoed down through the
decades to the present day. In Hasa the office of Governor is now a
Jiluwi monopoly – the present incumbent is a grandson of the
fearsome Abdullah – and when Abdul Aziz bin Musaid died in 1977,
at the reputed age of ninety-three, his funeral was attended by every
one of Ibn Saud's sons who was in Riyadh at the time, from Crown
Prince Fahd downwards (King Khaled himself would certainly have
attended, for his mother was the great man's daughter, but he was in
London for medical treatment).

Other blood relatives of the King became advisers at his court. His
brother Abdullah, although twenty years younger than Ibn Saud, was
especially well regarded for his quiet wisdom, and by the time he was
twenty-five had become the King's chief family adviser. His cousin

Saud Kebir, of the once notorious *Araif,* had become his brother-in-law through marriage to his favourite sister, Nura, and was also among his stoutest supporters. But it was Ibn Saud's own prodigious activities in the marriage bed that did more than anything to strengthen the family bonds that held his kingdom together. Women, and talk of women, were always a special pleasure to him, and he was proud of his ability to keep numbers of them happy. He once told an English visitor that he had only three real pleasures in life: prayer, women, and perfume, in that order; and those who came to know him in his more intimate moments vouched readily for his enthusiasm for all three.

Ibn Saud never exceeded his limit of four wives at any time, although in later years, like the Prophet Muhammad in his mature years, he did – in Philby's words – 'allow himself a certain latitude in such matters not strictly in accord with the rules'.[3] Thus, besides his four wives, each established in a separate house with her own retinue of slaves and servants, he also maintained a house of his own run by four favourite concubines who enjoyed a status 'indistinguishable from that of wedded wives'.[4] He usually had four favoured slave girls as well to complete his regular domestic team and, in common with most other sheikhs and princes, he was in the habit of accepting as his natural due the night-time favours of any young girl who was presented to him by his hosts when travelling away from home.

All this was accepted Arabian custom; and perhaps the only surprising thing about Ibn Saud's exploitation of it was that he did not form even more liaisons and produce more offspring than he did. But it must be remembered that for the first two decades of his manhood, when his potency was presumably at its greatest, modern medical aids were still almost unknown in Saudi Arabia and miscarriage and disease ended many young lives prematurely. When the international influenza epidemic of 1919 reached the peninsula it killed three of the king's known sons, including his firstborn, Turki, as well as a number of his daughters and his beloved wife, Jauharah, who 'governed all his senses', as he was later, weeping, to tell Hafiz Wahba.[5] It seems likely that other, endemic, diseases killed many more either before or soon after birth so that they went unrecorded in the family tree. Even so, Ibn Saud's surviving issue was far more numerous than that of his predecessors, partly, no doubt, because his conquests brought him greater opportunity as well as need for political marriages, and partly, perhaps, because by the 1920s both doctors and modern drugs did become available at court. The doctors came chiefly from Syria, and were trained mostly in Ottoman schools of medicine. Their knowledge quickly gained for them a special position in royal counsels which, in a modified way, has continued ever since. Rashad Pharaoun, the Syrian doctor who presided over a chest of medicines that became increasingly dominated by aphrodisiacs of Turkish and German manufacture, remains a valued adviser to this day.

To his growing brood Ibn Saud was a stern but at the same time indulgent father, as most Arab parents are. All his surviving sons now seem to remember him with genuine affection although, in the nature of things, most of them saw comparatively little of him. Those who grew to manhood before the Second World War had a completely traditional Arab education. They learned to read and write with instruction from the Koran, they mastered the arts of camel- and horse-riding, and discovered how to swing a sword, aim a rifle and hunt gazelle and bustard with the Saluki and the falcon. But few of them were permitted to do work of state.

By the time Ibn Saud conquered the Hejaz, he had sufficient confidence in his two eldest sons, Saud and Feisal, to appoint them to permanent roles as his effective understudies. Feisal was named Viceroy of the Hejaz itself – a post for which he seemed best qualified by his diplomacy and early experience of the world during his tour of Europe, for those were attributes no one else in the Saudi family enjoyed at that time and upon which the Hejaz, with its international horizons, was likely to make many demands. Saud, who had never been outside the Kingdom, was kept at home in Riyadh as Viceroy of Nejd. Although lacking both the intelligence and the subtlety of his father and younger brother, Saud was a tall young man of charm and humour, indolent but free with his favours in matters of both money and love. His personality suited the desert tribesmen and as Viceroy he soon established an easy popularity among them that helped to confirm him as his father's naturally ordained successor. In 1933, while in his early thirties, Saud was nominated as heir for the acceptance by the people. In what was an intense historical irony, it was Feisal who stood in for the absent Saud at the ceremony of allegiance in the Grand Mosque in Mecca.

Surrounded and supported by his ever-growing family in this way, Ibn Saud lived a patriarch's life that seemed at first sight to differ from that of his ancestors only in its scale. But innovations had been entering Ibn Saud's domains after the First World War, especially the motor car, the aeroplane and the wireless. The first automobiles, or 'trombils' as the Arabs called them, actually arrived in Riyadh as early as 1924 – two upright Model T Fords which had been landed on the Gulf coast and were literally dragged by camel through the worst of the desert between there and the capital. In due course they were added to.

At first the cars were used chiefly for royal entertainments. On the firm, level ground that stretched for many miles around Riyadh, they could outpace the fastest horse and they were soon in regular employment for hunting expeditions. At the time it seemed good clean desert fun, for the game was plentiful, the cars were few, and the animals still had a sporting chance. Within a few years their bold new pastime became the means of massacre, so that now, when the young bloods of Arabia go 'hunting' in their General Motors trucks and Range Rovers, with modern repeating rifles and sometimes sub-machine guns by their sides, there are no more oryx and only a handful of ibex to give chase to.

As more cars arrived (and more drivers were imported to drive them), Ibn Saud was able to copy the British in Iraq and Transjordan and build the primitive motorized force that helped to crush the last of the Ikhwan. By that time, he was fully awakened to the possibilities of all the new heathen instruments of war and peace and was bent upon introducing them to his kingdom as fast as possible. He had to fight the conservative theologians every step of the way, as much as he had to fight the Ikhwan on the battlefield. The learned men greeted every novelty with outright hostility, even sending emissaries to the new wireless station at unexpected times to see if they could trap Satan on one of his visits there.[6] But their pious reservations were fated to be demolished by the temptations of convenience. It was difficult to maintain that the wireless and the telephone were Devil's magic when they were employed in such holy work as directing the forces of Ibn Saud and Wahhabism against the heretics of Yemen; even more so when Ibn Saud proved to the assembled *ulema* of Mecca that they could be used to transmit the words of the Koran through apparently empty air. Truly, that was a work of God – and as these curious inventions proved to be not without material convenience to the holy men, as well as to other people, they were soon accepted. Nevertheless, no non-Muslim worship was permitted on the Kingdom's soil and Western visitors came to Riyadh only at the King's behest, and in Arab dress.

By the early 1930s, a rudimentary wireless network, installed by Marconi through Philby's offices, covered all the major cities and most of the main frontier posts. The royal court had a substantial fleet of cars at its disposal and many merchants in the Hejaz had become such ardent buyers of cars and buses for the pilgrim trade that their cutthroat competition swiftly led to bankruptcies. A number of private electrical power plants had also been set up, to supply the royal palaces and some of the wealthier merchants. 'It is nothing,' the King would tell the marvelling bedouin, 'just a machine and some pieces of wire.'[7] Ibn Saud had even got the British to send him some aeroplanes. The aircraft arrived in September 1930 – four de Havilland biplanes, with British pilots and maintenance crews, which landed on a makeshift strip outside Jeddah to be greeted by the King in his marquee and an immediate clamour from all his young sons, cousins and nephews to be taken up for joy rides.

But these acquisitions, too, had a serious purpose, for both in civil and military use they helped to bind still tighter the disparate provinces of the Kingdom. Five years later, when the war in Yemen was over and Mussolini's Italian empire was just beginning its conquering thrust into Abyssinia, on the other side of the Red Sea, the first Saudi pilots were sent to Italy for training as the nucleus of the Saudi air force – to the intense anxiety of the British.

Yet such innovations cost money and were complex to administer. The appointment of his two eldest sons as Viceroys had been a tacit acknowledgement of the increasing complexity of government. True, he gave them as little independence as he could, but their titles implied a new breadth to the country that was difficult to reconcile with Ibn Saud's purely personal rule. Most of the impetus for this change came, inevitably, from the Hejaz, whose conquest not merely greatly enlarged the Kingdom but fundamentally modified its nature. Ibn Saud realized what the Ikhwan failed to see, that it was impossible to treat the Hejaz – like Hasa or Hail – simply as yet another extension of the Wahhabi state. Even if it had been possible to impose strict Wahhabism on the natives of the province, it certainly could not be done with the thousands of foreign pilgrims who came there every year. Nor could the foreign consuls in Jeddah who watched over the interests of their citizens involved in the pilgrim traffic – many of them from Christian imperial powers like Britain, France, the Soviet Union, and the Netherlands – be expected to accept the sort of restrictions commonly imposed on non-Muslims in Riyadh. A new public tolerance was not merely desirable, therefore, but essential if the Hejaz was not to prove more of an albatross than an ornament in the Saudi Kingdom.

Moreover, the requirements of the pilgrims, the consuls and the merchants who lived off them imposed fresh administrative demands on the Saudi court. Health services had to be maintained and extended if epidemics and diseases were not to become an annual accompaniment of the *Hajj* and a lasting discredit to Ibn Saud's kingdom. The fate of the French vessel *Asia,* which caught fire in Jeddah harbour in 1929 with 1,800 North African pilgrims aboard, was a terrible warning. Complex civil and commercial law also had to be administered. The Dutch bank, moneychangers and customs offices had to be supervised, foreign manners and languages had to be mastered or accommodated. The simple ways of Nejd were hopelessly inadequate.

To some extent this could be dealt with by simply maintaining the existing Ottoman structure, as taken over from the ill-fated King Hussein. But even that cost money and required supervision; and although the *Hajj* was self-financing to the extent that it generated trade for the state to tax, Ibn Saud's administration had neither the men nor the experience to handle the sums – and the problems – involved.

Little by little, therefore, after 1925 the skeleton of an administration was developed that went beyond the royal family and Ibn Saud's person. The Foreign and Domestic Courts in their little offices in the Riyadh palace, and a separate Foreign Ministry, employed both Arab and foreign business and professional men with skills to which few Saudis could aspire. As was only to be expected, the administration's growth was a profoundly haphazard affair. At one extreme it drew in old-established Hejazi merchant houses like the well known Alireza family (sometimes written as Ali Ridha or Aliridha) who became, almost from the moment of Ibn Saud's impoverished arrival in Jeddah, advisers and lenders to the court, thereby laying the foundations of the family's vast fortune. At the other extreme it included St John Philby, who had resigned from British Government service in 1925 and had become an even greater nuisance to his old employers than before. Although he never attained any official position at court, he was valued by Ibn Saud for his foreign contacts, worldly knowledge and forthright manner. He became one of the King's chief confidants and guides amid the mysteries of modern diplomacy and commerce. He was even, for a brief period in 1927, in charge of the royal stable of some 250 cars until there was an insurrection among the chauffeurs and mechanics.

Alongside these was a motley assortment of non-Saudi Arabs who came from half a dozen countries in search of power and profit as the King's new executive assistants. Some of them enjoyed only a brief reign, but several dug themselves in so well that a quarter of a century later, even after Ibn Saud's death, they were still among the royal family's most influential advisers. They included men like Fuad Hamza, a Druze sheikh from Palestine who secured almost instant approval as the King's chief executive in foreign affairs, under Prince Feisal; Hafiz Wahba, the Egyptian schoolteacher who became Ibn Saud's first Director of Education but was soon appointed instead as his Ambassador in London where he remained for so long that he eventually was honoured as the doyen of the diplomatic corps at the Court of St James; and – best known in the end – Yusuf Yassin, a young Syrian who came as a teacher to the King's family and quickly rose to be political secretary and, especially in later years, a notorious thorn in the flesh of all Westerners in the Kingdom.

Like everyone else around the King, these new recruits remained utterly dependent on his favour and his power; yet because they dealt with affairs in which the King himself was a novice, their opportunities for discreet deception were immense. It would be unfair to say that they were excessively corrupt. The three just named above were generally regarded as honest men and they all did the Saudi state some service. But it was in the nature of the state itself that it should be run as the King's personal fiefdom and, therefore, that any persons sufficiently trusted by the King should build their own, lesser, fiefdoms in his shadow. Conflict of interest between them and the state was not only unrecognized but, by definition, could not exist. If the King was the state, then so were they because they held their positions solely at his consent. Any favours they enacted or exacted were simply an extension of his own. So, as the King's responsibilities widened beyond his actual capacity to handle them, the myriad details of daily administration became a rich source of pickings for these new administrators.

For none were they richer than for the King's personal treasurer, Abdullah Suleiman.[8] He was an exception among the new royal entourage – a Nejdi born and bred from an old-established merchant family, he had spent a year or two in Bombay learning the secrets of trade and book-keeping before he became attached to Ibn Saud's household. There he proved intelligent and resourceful, especially in the fine and very necessary art of making a little go a long way. In charge of finance from 1932 onwards, he was the chief steward for the rest of Ibn Saud's life and did more than anyone except the King himself to maintain and expand it as a giant family business. Described by Philby as 'a frail little man of "uncertain" age but with something of the inspiration of the prophets in his soul',[9] he was, in fact, the only man apart from the King – ever excepting Ibn Jiluwi in Hasa – with any real power. Outside the King's own presence, where he kept a humble demeanour, he was treated almost like a king himself.

From his earliest days as ruler, Ibn Saud had always been poor; even when he took over the Hejaz and its generally lucrative pilgrim traffic and customs revenue, the proceeds quickly proved inadequate. As early as 1929, various advisers intimated to the British in Jeddah that a renewed subsidy might be welcome – in return for Ibn Saud curbing the tribes or refusing to have anything further to do with the cheap Soviet goods and petrol being dumped in Arabia.

That year, the great crash of the industrial economies produced a slump in the prices of agricultural goods in the main pilgrim countries of India and the Dutch Indies. Few could afford the £50 or so a head for the *Hajj*. From 132,000 souls in 1927, the *Hajj* dwindled to 38,500 in 1931, the revenues of the state were cut by one-third while a failure of pilgrim confidence brought the riyal to catastrophic depths.

Philby estimated that the Government was in debt some £300,000 – to the wireless and telegraph companies, the Government of India and such bankers as Abdullah Qusaibi. Officials were bringing in drafts against customs receipts and gold was being exported at an alarming rate. Particularly galling for Philby was Abdullah Suleiman's swoop on the petrol stocks and general goods of his company, Sharqieh. Eventually, at the end of 1931, Suleiman persuaded the King to impose a moratorium on further debts and a regime of principal and interest payments of 5 per cent per year.

The next summer, a mission headed by Feisal set off for Europe in the hope of finding means to alleviate the court's distress. In London, Fuad Hamza held somewhat unsatisfactory talks with City interests and the Bank of England, while Feisal took part in what were now familiar rounds of state. In Moscow, however, the Soviets proved more accommodating. Arriving at the western station on 29 May, to a reception and a band playing 'the Internazionale and various Arabian airs', whatever they were, Feisal met the Soviet leaders and was fêted by local commissars from Leningrad to the Caucasus. On leaving for home after touring the Baku oilfields, Feisal cabled from his steamer on the Caspian that he 'had been specially impressed by the oil derricks and the techniques of getting oil, a product so beneficial to humanity'.

The British soon learned what Feisal and Molotov had talked about. Not only was the Soviet Union willing to pass over a debt of some £30,000 for stocks of petrol and kerosene that Abdullah Suleiman had commandeered for the court, Fuad told the British Minister, Sir Andrew Ryan, on 13 June, but it would also make an immediate loan of £1 million. In return, Ibn Saud must lift the trade embargo and sign a commercial treaty and a treaty of friendship. Fuad explained that, of course, the King was opposed to the treaty of friendship but desperately needed money and if Britain did not provide £100,000 at once, he would be forced to accept the Soviet Union's offer. Sir Andrew was aware that Fuad was setting Britain off against the Soviet Union and had, no doubt, heard from the garrulous Philby the remark made by the King early in 1931: 'I tell you, Philby, that if anyone were to offer me a million pounds now he would be welcome to all the concessions he wants in my country.'[10] But Sir Andrew was worried enough to cable London that he was 'slightly perturbed'. 'The Soviet Minister appears to have adopted the Arab headdress permanently,' he wrote somewhat absurdly.

As it turned out, the loan and treaty were never signed. Ibn Saud, it appears at this stage, cared too much for British sensibilities, and Abdullah Suleiman, in Sir Andrew's view, detested Feisal and Fuad. Agreement was reached on lifting the embargo but the Soviet mission got no further and, in May 1938, at the height of the Stalinist purges, Karim Khakimov disappeared from Jeddah, never to be heard of again.

The financial crisis did not abate. It was not that Ibn Saud was personally greedy. Extravagance was part of his lifestyle. Every day there were hundreds, sometimes thousands, at his palace to be fed. As the Dutchman, D. van der Meulen, one of the most perceptive and least partisan observers of the court, wrote: 'He preferred the bedouin way of living...[But] wanted money to distribute among the poor of the towns, villages and tribes he passed on his way. He wanted to give royal presents to his visitors and to the men who served him...[And] Abdullah as-Sulaiman had his own ideas about state finances.'[11]

Abdullah Suleiman made some effort to curb spending. He once had to flee an enraged Saud, to whom he had closed the Treasury, and take refuge on a vessel in Jeddah harbour until the Crown Prince's wrath had abated. But he could not curb the King's extravagance or bring order to the chaos of his finances. As long as there was any money in the till the King would spend it in the conviction that discontent and dishonour would follow if he did not. Abdullah Suleiman was therefore obliged to hand over whatever the King called for with no record kept either of its receipt or its disbursement. The collection of taxes was farmed out to powerful individuals at a price, on condition only that they handed over a due share to the King's private purse. Corruption followed as night the day and a habit of mind was developed at the centre of Saudi affairs that accepted these vices as the natural accompaniment of power.

From time to time there were signs that Ibn Saud himself was worried by this cancer in his state, but they were too fleeting and too lacking in any real grasp of what was wrong to offer much hope of reform. He was, after all, only human; and for all the grandeur of his achievements he remained as the years advanced a true child of his time and place. Born in the desert to desert ways, he could not comprehend the world into which he had fought his way so boldly; and like so many desert conquerors before him, he nibbled its tempting fruit without quite knowing the poison that lay within. Yet there lay in store an irony greater than anyone could have guessed, for as the 1930s advanced and poverty and extravagance together wrought mischief with his Kingdom, a new source of money was about to be discovered that would, in time, wreak far more.

9 The Land of Promise

1932 – 8

Although the British originally had the concession and did some drilling, they abandoned it as unproductive. Then the Americans came along.[1]

PRINCESS ALICE, COUNTESS OF ATHLONE

ON the upstairs floor of Simpson's restaurant in the Strand in London, seated discreetly at a corner table, two men faced each other across the plates of pink roast beef and watery vegetables for which that establishment has acquired its worldwide fame, and settled, without knowing it, the fate of Arabia. The older and taller of the two, who was host, spoke with an American accent and had the smooth and confident bearing of a practised executive or official. He was Francis B. Loomis, a former career officer in the United States Government service who had once been Under-Secretary of State for Theodore Roosevelt and was now a consultant on foreign affairs for the Standard Oil Company of California.

The second man, short, square and full of restless energy, his grizzled beard and weathered tan so out of keeping with his grey City suit, was none other than Harry St John Bridger Philby, just back home in London after the completion of his grandest Arabian exploit – the exploration of the Rub al Khali.

On the face of it, the two seemed unlikely to have much in common; but Philby had accepted the American's invitation to lunch that May of 1932 on the personal recommendation of the U.S. Consul General in London and on the entirely natural assumption that Loomis, like thousands of other people at that time, wanted to hear about his travels. It was only after he had talked for some time, with his usual enthusiasm, of his discoveries in the Rub al Khali that he learned there was another purpose to their meeting. Did he, asked Loomis, quietly changing the subject, think there was any chance of anyone getting an oil concession in the territory of Ibn Saud and, if so, would he be willing to co-operate with Loomis in obtaining it for Standard Oil? Philby was equal to this abrupt change of tack. Unknown to Loomis he had been thinking about such a possibility for some time. His commercial agency, Sharqieh, was in desperate straits because of the King's failure to pay. He replied with a qualified yes on both counts; and so, by the time the port had succeeded the cheese, one of the most important of all American commercial enterprises overseas was launched.

The roots of this momentous meeting lay in those complex negotiations of the 1920s to decide which international oil companies should have access to the new Middle Eastern concession areas. In 1923, at the time Major Frank Holmes won his inconclusive concession from Ibn Saud, they were still unresolved. But five years of haggling later, they had resulted in a series of agreements between certain British, Dutch, French, and American oil companies to share out the territory of the old Turkish Empire between them and to transform the pre-war Turkish Petroleum Company into the Iraq Petroleum Company (I.P.C.), which it formally became in 1929, for the exploitation of specific areas in Iraq. As part of this general rearrangement of the Middle Eastern oil scene, two American companies – Exxon (then Standard Oil of New Jersey) and Mobil (then Standard Oil Company of New York – Socony) – secured between them a quarter share of the new I.P.C. alongside Anglo-Persian (which became Anglo-Iranian in 1935 and British Petroleum in 1954), Royal Dutch Shell and the new Compagnie Française des Petroles, thus becoming the first U.S. participants in what had been hitherto a European preserve. But the cost to them was acceptance of a cartel arrangement with their partners and some other American oil companies which greatly restricted their freedom of manoeuvre. Called the Red Line Agreement, this inaccurately but conveniently included virtually the whole of the Arabian peninsula within the former territory of the Turkish Empire and said that within that region – all appropriately outlined on the map in red – each signatory would have a veto over the actions of the others. Thus it was impossible for any one company to obtain or exploit a new concession without first offering a share in it to the others, and if one decided not to participate all the others were compelled to reject it too.

In the context of the late 1920s, it was not perhaps so restrictive a scheme as it now sounds. Massive new oilfields had been opened up in the Americas in the ten years after the First World War and when the crash of 1929 halted half the industries of the Western world – and, incidentally, bankrupted Ibn Saud – it was clear that the companies were facing a hopelessly glutted market as well as declining sources of capital with which to finance new exploration. There were, however, some companies which had not profited so much from earlier discoveries and which were still anxious to increase their sources of supply if they could, especially at a time when competition was low. One of them was Standard Oil of California, commonly known as Socal; and as it was not a party to the Red Line Agreement it was able to act in the Middle East where other major companies had tied themselves hand and foot.

For more than a year before Loomis's lunch with Philby, Socal had cherished a belief that there was a sporting chance of discovering oil in Hasa. There was, of course, no direct geological evidence to support them. The survey that Major Frank Holmes had commissioned there, under the concession he secured from Ibn Saud in 1923, had proved disappointing, and he had been unable to re-sell the concession to anyone else. Beset by lack of money and poor results, and with the rebellious Ikhwan making foreign prospecting anywhere on Saudi soil a hazardous task, Major Holmes had been forced to abandon the concession, leaving two years' rent unpaid, to compound Ibn Saud's financial difficulties. Shortly after the Holmes survey, Red Sea Petroleum Company, a subsidiary of Shell, gained rights to the Farasan Islands, off Asir, from Hasan Idrisi. Although they actually did some drilling in 1926–7 the prospect was poor and the rights lapsed when the Idrisi domains were annexed by Ibn Saud. The Dutch had also been asked to carry out a survey but finally refused when no money to cover costs was forthcoming.

Holmes had, however, been more fortunate on the island of Bahrain, where he had obtained a similar concession. After Anglo-Persian had turned down the chance of buying that from him he contrived to get Socal to take up the option, thus stepping outside the Red Line cartel whose members were bound by Anglo-Persian's refusal. Socal's preliminary surveys soon belied Anglo-Persian's gloomy forecasts and before long their engineers on the island were suggesting that there might be oil not only under Bahrain itself but also, and probably in larger quantities, under Hasa. If Jebel Dukhan (meaning 'Hill of Smoke') in Bahrain was a natural oil dome, so was that of Jebel Dhahran in Hasa and, if they knew anything at all about the oil business, its exploration was surely worth a try.

The difficulty was to get there. Holmes was their man and he was full of promises about introducing them to his old friend Ibn Saud. But, alas, his credit was exhausted there after his failure to follow through on the original concession and he could accomplish nothing. When Socal's very first test well on Bahrain produced a heavy flow of oil on 1 June 1932, confirming all their hopes, some more urgent action seemed required to extend their activities across the water before other companies found a way to forestall them.

But there was no government channel available through which to contact the King, for although the United States had recognized Ibn Saud's rule in the Hejaz, it had no traditional interest in the *Hajj* there and had not yet appointed a representative to join the other Western consuls and legations in Jeddah. Some other, private, channel would have to be found – and Philby provided the obvious answer.

When Philby resigned from British Government service in 1925, he determined to seek his fame and fortune in Arabia alongside the man who had become his hero, Abdul Aziz bin Saud. Already known and liked by the King, he soon secured some royal favours and was able to maintain himself by starting a commercial agency, which he called Sharqieh, in Jeddah. As a non-Muslim, however, he was handicapped in his new career. The local merchants were jealous of his special relationship with the King. Yet he could not visit the King in Mecca, nor travel anywhere beyond the immediate surroundings of Jeddah without the King's permission – and as the King was in the thick of his battle with the Ikhwan, whose detestation of foreigners was boundless, permission was rarely and grudgingly given. So his business and his ambitious plans for exploration both languished. Not until 1930 did his fortunes begin to turn. Then, with the Ikhwan defeated and only a small rebellion in the north-east to trouble him, Ibn Saud was able to embark on the modernization of his country and he turned to his old friend Philby for help. The favourable first result was a long-sought contract for Philby's company to supply a set of Marconi radio transmitters for the Government's use. He was not paid for years for that. The second result was the King's permission to embrace Islam.

Most of his friends in London and elsewhere were aghast at his apostasy. Yet for Philby the matter was one of simple common sense. He had no conventional belief in God so he had no sense of betraying any faith. On the other hand, he needed the freedom and the identification with Arabia that acceptance as a Muslim would confer. Above all, he wanted to belong. The wisest comment was a somewhat testy one attributed to Sir Andrew Ryan, the first British Minister in Jeddah: 'Mecca and Islam will give Philby the background which he has needed so badly ever since he quarrelled with the Government.'[2]

Philby was not disappointed. Welcomed promptly by Ibn Saud to the inner circles of his court, he became one of the King's closest confidants and, hearing his repeated lamentations about his lack of money – which Philby, of course, shared – was able to offer his services in seeking out foreign businessmen who might be interested in exploiting mineral concessions and water resources in the Kingdom. According to his own account, he told the King of just the man – a certain Charles R. Crane, an American philanthropist whose father had made a fortune out of sanitary fittings and who had once unsuccessfully advised President Woodrow Wilson on the intricacies of the post-war settlement in Palestine, as a leader of the so-called King-Crane Commission. Crane had since provided from his own pocket the money for windmills and modern road and water works in Yemen. Other accounts have it that Crane, whose hobby was to meet outstanding men, had been attempting to visit the King since 1926, and had even set out the next year only to be shot up by Ikhwan.[3] Whatever actually occurred, in early 1931 Crane was invited to meet the King in Jeddah where he readily agreed to send his chief engineer from Yemen, a tough Vermont Yankee called Karl Twitchell, to cast his practised eye over the Kingdom and see what exploitable minerals and water it might contain.

Twitchell's first reports were discouraging. The Hejaz offered little, he thought, in either water or minerals except for some ancient gold workings that might be revived with profit.[4] Hasa, when he got there, seemed to him – as to the men of Socal across the water in Bahrain – to be worth a closer examination as a possible source of oil. His advice to the King, however, was to wait upon Socal's results in Bahrain. if they proved promising enough, other companies would probably be willing to put money into Hasa. Reluctantly, and because he had already been disappointed in Holmes, the King accepted his advice. But on Abdullah Suleiman's advice, Twitchell was asked to go back to the United States to raise money for his gold mine proposals. According to the interpreter Muhammad Almana, Twitchell's report was also passed on to the British Government in London, to whom Ibn Saud felt he owed a moral (and, indeed, an actual) debt. London was uninterested in the oil prospect.

Meanwhile, Philby had returned to his own real ambition. He sought the King's permission to attempt the crossing of the Rub al Khali – a feat hitherto regarded as unthinkable if not actually impossible. At first the King was reluctant; there was still tribal raiding on the distant fringes of the desert and he did not want to risk untoward incidents. Then, in 1931, another explorer, Bertram Thomas, a British official in the service of the Sultan of Muscat, made a sudden dash by camel from the south Arabian coast clean across the eastern section of the Rub al Khali to the British-protected sheikhdoms of the Trucial Coast. To Philby this was a desperate disappointment. To Ibn Saud it seemed more like a deliberate insult for he believed that Thomas, representing a rival ruler, must have passed without permission through territory he claimed as his own. All the more reason, then, to send Philby off to show what could be done in the same trackless land under the aegis of Ibn Saud. So Philby went at last in January 1932, driving himself, his men and his camels to utter exhaustion for the next three months.

He emerged from this bleakest of journeys a hero in his own right, winning the gratitude of the King and the applause of the world. When he arrived in London that summer, he found himself the lion of society, the press and the lecture room. To Francis B. Loomis and the directors of Socal, he seemed heaven sent.

The lunch at Simpson's that resulted from all this disparate activity, however, did not bear fruit at once. Having heard from Philby that the King would certainly welcome an approach, provided that the terms were good enough, and that Philby himself might act for Socal as long as it did not interfere with his other plans, Loomis returned to the United States for consultation, while Philby set off in the opposite direction. Taking his wife, Dora – who had accepted his conversion to Islam and the King's present to him of a slave girl with private distaste but public composure – he enjoyed a grand and leisurely car tour through most of the capitals of Europe and the Middle East on his way back to Jeddah. By the time he got there in December Socal had acquired the services of Twitchell, whose geological observations confirmed their own and who could also act as guide and mentor in Arabia, if necessary. But Philby was the man with Ibn Saud's ear, and his desk was piled with cables enquiring urgently of his whereabouts. For in his absence Socal had decided to go ahead with an offer to explore Hasa and they were anxious to have Philby make the first formal approach to the King. Philby, meanwhile, advertised himself as a free agent in a letter to Anglo-Persian.

With the last embers of the Ikhwan revolt still burning in the north and bankruptcy staring Ibn Saud in the face, what was wanted by the King and Abdullah Suleiman was cash – and as fast as possible. They were encouraged by Socal's interest and delighted by the news of the strike in Bahrain which, as Twitchell had indicated, must improve the prospect of finding oil on Saudi territory. They also knew that Persia and Iraq were beginning to squeeze better terms out of the oil companies as new discoveries were reported and exports rose. So they pitched their terms high, asking through Philby for £5,000 a year in rental plus an immediate loan of another £100,000, all to be paid in gold. Socal was unimpressed. In view of their meagre information about His Majesty's terrain, they replied, the figures seemed quite burdensome. But as negotiation was obviously possible, they decided to send out their own man, a forty-year-old lawyer, whose round baby face looked fifteen years younger, by the name of Lloyd N. Hamilton. He arrived in Jeddah from Port Sudan aboard the little Egyptian steamer *Talodi* on 15 February 1933, accompanied by his wife Airy, and Karl and Nona Twitchell.

Jeddah had the ambience of a place that had remained unchanged for centuries and in 1933 it certainly looked exactly as Lawrence had first seen it twenty years before:

> It was like a dead city, so clean underfoot, and so quiet. Its winding, even streets were floored with damp sand solidified by time and as silent to the tread as any carpet. The lattices and wall-returns deadened all reverberation of voice. There were no carts, nor any streets wide enough for carts, no shod animals, no bustle anywhere. Everything was hushed, strained, even furtive. The doors of houses shut softly as we passed. There were no loud dogs, no crying children: indeed, except in the bazaar, still half asleep, there were few wayfarers of any kind; and the rare people we did meet, all thin, and as it were wasted by disease, with scarred, hairless faces and screwed-up eyes, slipped past us quickly and cautiously, not looking at us...The atmosphere was oppressive, deadly. There seemed no life in it. It was not burning hot, but held a moisture and sense of great age and exhaustion such as seemed to belong to no other place...a feeling of long use, of the exhalation of many people, of continued bath-heat

and sweat. One would say that for years Jidda had not been swept through by a firm breeze...[5]

Twitchell, of course, was already familiar with this silent, steaming place, but apart from him and Charles Crane scarcely any Americans had ever set foot there. Britain, and to a lesser extent France, were still the dominant foreign powers throughout the Middle East; and, in spite of a few hard-won breaches in the old order, like the entry of American companies to the I.P.C., firm 'no entry' signs still confronted American businesses which sought to expand their interests in territories where British political control was paramount. Even Socal had been compelled, by British regulations, to set up a company registered in Canada before it could begin operations in Bahrain.

The slow ways of the Saudi court meant long days of waiting, punctuated by long hours of polite conversation through interpreters (including Philby) over Arab coffee. As the weeks went by, the temperature and the humidity rose and tempers and suspicions rose with them. The ubiquitous Major Holmes put in an appearance with mysterious hints that he, too, wished to bid again for the concession he had once abandoned, but he soon disappeared when he was reminded of the rent – at £2,000 a year – he owed the King. More menacingly, the I.P.C. decided to take an interest, spurred on by its chief British shareholder, Anglo-Persian. At first they asked Philby to represent them in negotiations, but when he answered, rather belatedly, that Socal had already bespoken his services, they sent one of their own men, Stephen Longrigg, to bid on their behalf.

As the long talks and longer silences dragged on, it was clear to Philby that the King would prefer, if he could, to let the British company have the concession, although – in true Arab style – the Americans were always given the impression that Ibn Saud would naturally prefer them to have it. But the King's long reliance on the British was drawing to a close, in spite of himself, for reasons that now seem to epitomize much of Britain's subsequent decline in the world and America's irresistible rise. Britain's treaties with the King just after Shakespear's death in 1915, and in May 1927 to take account of the Hejaz conquest, recognized Ibn Saud was no petty potentate like the Gulf Emirs and made no mention of oil. As for Abdullah Suleiman, he was solely interested in the down payment, for he knew how much it was needed; and the British were not ready to put up anything like the money Abdullah Suleiman wanted.[6] They rejected the idea of a preliminary loan altogether and would only offer to pay their annual rental in Indian rupees instead of gold. But behind this apparently penny-pinching attitude were other, more serious, errors of judgement, reflecting an elderly complacency in British enterprise. For one thing, I.P.C., with the backing of its major British shareholders, was confident that it had enough oil in Iraq to keep its markets happy for as far ahead as it could see. For another – and in spite of the Bahrain discovery – it was still dubious about the possibility of finding any oil in Hasa. Indeed, one Shell director of the day was so convinced that a Saudi concession was a wild goose chase that he offered to drink all the oil that was ever found in the country. Their chief concern was to keep out their American rivals.

The Kingdom's need for cash was now so urgent that Ibn Saud could not afford to turn down any advantage, and after a month or two it was clear that the real negotiation was not about whether I.P.C. or Socal would win the concession, but how long Ibn Saud and his advisers could hold out in trying to squeeze the maximum out of Socal alone. It was not until May 1933 that the agreement was signed at last and, considering the comparative weakness of the Saudi position, with I.P.C. out of the running and their own treasury empty, it represented a considerable triumph for the tenacity of Abdullah Suleiman's bargaining – and, perhaps, for Philby's translations and his shrewd appraisal of just how far each side might compromise.

The terms were for an immediate loan of £30,000 in gold, with another £20,000 in eighteen months, plus a first annual rental of another £5,000 in gold and subsequent rentals of the same amount to be paid in agreed foreign currencies. There were some few further delays in getting hold of the gold. The depression had forced the United States to impose an embargo on gold exports that April. An application to export the sum of $170,327.50 in gold from the United States was refused by a young Under-Secretary of the Treasury named Dean Acheson and 35,000 English sovereigns had to be bought in London instead and shipped out to Jeddah on a P & O steamer. On 25 August 1933, just over a year after Philby's lunch with Loomis, and more than six months since Hamilton had arrived in Jeddah, they were carried into the Netherlands Trading Society, which housed the only bank in Jeddah, and counted out upon the table in the manager's office, one by one, all 35,000 of them, by the somewhat out of practice hands of Abdullah Suleiman.

What Socal had got for its money was title to explore a vast area only vaguely delineated but around four times the size of Britain, most of it barely known to anyone but its few scattered inhabitants and all of it devoid of even the most rudimentary aids to modern penetration or settlement. A secret annexe spoke of preferential rights to both Nejd and the Kuwait Neutral Zone in return for another £100,000 loan should oil be found in commercial quantities. What the California-Arabian Standard Oil Company (Casoc), the company Socal created in November 1933, did with its entitlement is now well known. Within weeks the first exploration parties crossed from Bahrain to set up their Saudi headquarters in the fly-blown little port of Jubail, while 1,000 miles away in Jeddah the administrators moved into Philby's beautiful and ramshackle house to maintain liaison with the Saudi Government. Within months, the first permanent camp was established in Hasa and all the paraphernalia of the oil industry was beginning to pile up on the Saudi shore. There were pipes and cranes, girders and drums, wrenches and dies, cars and lorries. There were also improvised shower baths and water closets, screen doors to keep the mosquitos out and electric fans to circulate the still, steamy air. There were medicines, too, and Flit guns for the flies, and electric lights to dispel the darkness, not to mention an aeroplane for survey work. By and by a whole new civilization had been created where, but a few months previously, there had been nothing but a few mud huts, the camels and the sea for as long as men had been there.

But the reward for which the Americans looked did not come easily. As the months passed into years, the money flowed out in ever-increasing quantities, the temporary headquarters were replaced by permanent camps with prefabricated, air-conditioned bungalows, and Jubail's treacherous creek was supplanted by a brand new pier at an insignificant bay called Al Khobar – but all this effort brought no return. The low hill of Jebel Dhahran, just inland from Al Khobar, looked as promising on the spot as it had done from across the water in Bahrain, but it did not yield the hoped-for oil. Christened the Dammam Dome, the structure was mapped and measured from every angle, sampled by the geologists and assessed by the engineers until it seemed that oil must be conjured out, whether it was there or not. But when drilling began in 1935 it brought little but disappointment. Here and there traces of oil or gas were found to raise hopes for a day or two. But the success of Bahrain eluded the drilling teams.

Dammam No. 1 was abandoned and Dammam No. 2 was spudded in. That was dry, too, and the teams moved to Dammam No. 3. Dry again – and so on, to Nos 4, 5 and 6. It was the end of 1936 when Dammam No. 7 was started and by now the cost was beginning to alarm the home office of Socal in San Francisco.

In that year, Socal, Casoc's parent company, invited Texaco (then called the Texas Company) to take a half share in the concession. Apart from any financial considerations, Socal needed additional distribution outlets for the prospective production. Texaco had developed these, but required access to new sources of oil. It was a marriage of convenience. At the same time, they formed a joint marketing company called Caltex, covering an area from Egypt to Hawaii, to sell any future output. It was not until 1944 that Caltex was renamed the Arabian American Oil Company (Aramco).

Dammam No. 7 proved a troublesome well from the start, with blockages when the sides caved in, delays in the supply of new drill pipes and the usual assortment of oilfield accidents. After a year it was down to 4,500 feet and had found not a sniff of oil. It was beginning to look as though the director of Shell had been right and that I.P.C. had done well to save their money by not pursuing the Hasa concession.

But on 4 March 1938, San Francisco received word from the drilling team that completely changed its outlook. Dammam No. 7 blew in with a steady flow of 1,585 barrels of oil a day. It was not a true gusher, such as had been seen in Iraq. But three days later, tests recorded a flow of 3,690 barrels a day. With nowhere to store the oil – although they did not know it, the oil was mixed with highly toxic hydrogen sulphide gas – the crew flowed it back into Dammam No. 1.

It might seem curious now, but the flow of oil was overshadowed for the Americans, at least, by a visit from minor English royalty. 'Alice was tremendous,' an Aramco account was later to say. 'She was a greater success than Dammam No. 7.'[7]

Alice was none other than Princess Alice, Countess of Athlone, grand-daughter of Queen Victoria, who in the spring of 1938 became the first member of her family to visit Ibn Saud's kingdom and the first European lady ever to cross Arabia from shore to shore.

This tale of imperial twilight begins, appropriately enough, at Ascot where Crown Prince Saud, on a visit to England in 1936, was being entertained to luncheon. 'Uncle George' was not well and

Queen Mary did not like the idea of speaking to the prince through an interpreter; so it was the sprightly Alice, then in her early fifties, who prattled to Hafiz Wahba and was duly invited to pay a visit. The Foreign Office, no doubt, welcomed this opportunity to show the flag as the storm clouds began to gather.

The royal party, including the Earl of Athlone and Lord Frederick Cambridge, arrived at Jeddah on 25 February 1938, and was entertained with style and charm by Feisal and the King. 'Arab dinners are not very bright as a rule,' commented the British Minister, Sir Reader Bullard, perceptively, 'but they are short. The Amir Faisal's dinner was exceptional in being long and gay.' At a tea party the next day, Ibn Saud himself was in excellent form and attempted to eat a piece of iced cake with a fork. 'He dropped his fork and he just sat patiently while the servants cleared away first the fork and then the victorious cake.'[8]

There were outings, too, to the oasis of Wadi Fatima where Philby 'with very pink feet in sandals' aired his topographical knowledge, and a picnic 'with fishing' where Abdullah Suleiman let everybody down by failing to arrive with the catch.[9] The King teased all his men, but especially Philby, without mercy.

Princess Alice's account of the journey across the peninsula, written for her grandchildren, reflects a similar innocent delight. Much space is given to the difficulty of, in her own words, 'powdering one's nose'[10] in the desert camps, particularly since she had chosen to wear voluminous and clumsy black robes. Forty years later, she was to advise 'Lillibet' to wear the same dress on her visit to Riyadh. Prince Saud had kindly installed two European-style bathrooms but otherwise Princess Alice had little good to say of his taste; his private apartments were 'garishly painted and full of clocks, glasses, horrid English wardrobes, lodging-house dressing-tables and hard divans or chairs right around the walls'.[11] But Saud's sister gave her and her maid more suitable clothing and she visited the influential Nura, now about sixty years old but still handsome and a power behind the throne.

There was bustard-shooting on the way to Hofuf and a silent dinner there with Saud bin Jiluwi, who had succeeded his father as Governor of Hasa in 1935. Princess Alice termed him 'very stodgy'. At the oil camp, 'Everyone was frightfully American, nice and hospitable'.[12]

They were also efficient. Just over a year later, in May 1939, a harbour and tankage had been built at the roadstead of Ras Tanura – which Twitchell had recommended as a port in his survey – and Ibn Saud was being entertained on the deck of the tanker *D. G. Scofield* by Caltex dignitaries. Then they went ashore, presents were handed out to the King and Abdullah Suleiman, and Ibn Saud reached out his bony hand and turned the valve through which his wealth was now to flow.

10 America Takes Over

1939 – 45

I hereby find that the defense of Saudi Arabia is vital to the defense of the United States.[1] – FRANKLIN D. ROOSEVELT

A MILE to the north of the mud walls of Riyadh, separated from the town by the undulating sand and gravel of the desert, Ibn Saud and several of the senior members of his family began to build in the 1930s the last of the royal Saudi palaces in the old Nejdi style. Until then the King had continued to rule from his smaller palace within the walls close to the old Musmak fortress-prison, but as his household and his responsibilities mounted, with ever more women and children to look after and ever more tribesmen from his new territories to honour, appease, cajole and feed, the sheer pressure of numbers forced him to look for new quarters. Luckily the money from the new American oil venture permitted him, for the first time, to plan something on a grander scale.

The Murabba (Square) Palace, with its square, fortified towers that rose at intervals from the outer wall, was by Nejdi standards an imposing structure. It rose sixty or seventy feet above the desert, with vast slab sides of red mud discreetly decorated with bands of geometrical impressions and a royal crest of crossed swords and palm trees over the main door. Inside, there was a public coffee hall fully twenty feet high, with three rows of plaster columns to hold up the roof, a big open hearth in one corner where the blackened copper pots were kept hot on the glowing charcoal, and bright cushions, rugs and elbow rests around the walls where guests waited for the call to the King's *majlis*.

Beyond lay a maze of corridors, stairs and courtyards, harem quarters, guest palaces, kitchens and offices which housed what seemed like the population of a small town. Massive negro slave guards lounged at the entrance to the *majlis* and the harem stairs. The divines sat cross-legged in the corridors, crooning to themselves the verses of the Koran. Groups of men in their desert robes and sandals huddled in conversation in every spare corner, picking toenails, or lay stretched in sleep in the cool shade of a courtyard. Royal children often ran shouting from room to room, and regularly from the *majlis* itself, where the King held his daily audience, the cry for coffee would come echoing through the halls, shouted from guard to guard like a password until the coffee-maker himself appeared, clashing a fistful of little handle-less cups in one hand to signal his arrival and flourishing his brass pot in the other.

It was here that the first American Government mission to Saudi Arabia was entertained by Ibn Saud one day in the spring of 1942. Only one of them had ever been in the Kingdom before, and he was by that time an old friend of Ibn Saud whom the United States Government had wisely appointed to lead the mission. He was Karl Twitchell, the geologist, gold-miner and oil company contact man. With him were the United States Minister to Egypt, Alexander Kirk, who was commissioned to open the first-ever diplomatic relations between Washington and Riyadh, and a small group of agricultural experts from America whom Twitchell had recruited with the State Department's aid and at the King's request to launch a scientific survey of the water resources of Nejd.

They were astonished and delighted by the scene that Twitchell introduced them to. In the *majlis*, besides the King himself – still an impressive giant of a man although he had now passed sixty and had begun, out of elderly vanity, to use black dye on his greying beard – were Ibn Saud's eldest son and successor, Crown Prince Saud, his brother Feisal, still Viceroy of the Hejaz, and the Finance Minister, Abdullah Suleiman.

Awkwardly, the Americans squatted in the unfamiliar desert robes the King had given them and listened to Twitchell and the interpreters exchanging elaborate Arabic greetings with the King. Eventually, they filed into another long room nearby where a feast had been prepared. Eight huge copper platters were piled with pyramids of rice. On top of each pyramid lay a whole roast sheep, and around them all the rest of the menu was spread with splendid disregard for either health or order – flat bread, roast chickens, boiled eggs, blancmange, tinned fruit, yellow custard, pink and orange soft drinks. It was a royal if bilious initiation into the timeless customs of Arabia; and if their visit had ended there, the Americans might well have believed that Ibn Saud's kingdom had somehow retained the secret of permanent insulation from the world around it. But that evening Crown Prince Saud showed them the other, changing side of the Saudi coin. After a Western dinner in his own palace, nearby, the lights were turned off in his salon, a folding silver screen was unrolled and, with a steady background hum from the new generators that supplied electric power to the palace, the Crown Prince personally demonstrated the very latest royal enthusiasm – a moving picture show. The Americans could make out through the scratches dancing on the screen a months-old British newsreel and the figures of President Franklin D. Roosevelt and Prime Minister Winston Churchill signing the Atlantic Charter on board the U.S.S. *Augusta* in August 1941. Their historic agreement included a promise that the war would produce no territorial changes 'that do not correspond to the fully expressed wishes of the people concerned'. But if, on the bosom of the Atlantic, all was sweetness and light between two great, equal allies, in Arabia their struggle had begun.

*   *   *

Ostensibly the American mission went to Riyadh to put new heart into the King. As international shipping losses mounted with the submarine war in the Atlantic and all available production of steel, lorries, pipes and drills was diverted to the urgent needs of battle, the Americans at Dhahran found themselves at the end of a steadily shrinking supply line. Soon they were forced to abandon most of their work and until 1944 they could do little more than operate at the merest trickle of 30,000 barrels a day, the equivalent of four minutes' production today.

This was a cruel disappointment to Ibn Saud who, with his usual thriftlessness, had fallen at once into the habit of spending his modest new oil income. To make matters worse, the upheavals of world war, the requisitioning of ships and the threat from the Italians across the Red Sea in Eritrea and Ethiopia, had closed down the annual pilgrim trade so that the Kingdom's other great source of money had dried up almost completely.

There were many in Ibn Saud's entourage who looked hopefully at first for a quick German victory in the war to remove the British from the Middle East. Notable among these were the growing number of advisers of Syrian or Palestinian origin who were deeply fearful that Jewish immigration into British mandated Palestine must eventually lead to the eviction of their fellow Arabs and the creation of a Jewish State. Ibn Saud himself had been content to take deliveries of Italian arms in 1938 and open relations with Hitler's Germany the next year when the flamboyant and likeable Dr Fritz Grobba, the German Minister in Baghdad, paid a visit to Jeddah. But to do the King justice, he never let the arguments of his markedly pro-Axis advisers deflect him from the judgement he had always held that Britain was the power best fitted to help him; and even in the darkest moments of British retreat, he forced his courtiers to listen to the radio news bulletins from London and to applaud whenever they gave news of a British victory. When St John Philby lectured him, with characteristic contrariness, on the inevitable German triumph – pointing to the rate at which Britain's merchant fleet was being destroyed as evidence that his countrymen could not possibly hold out for more than a few months – the King confided to other friends that Philby must be mad. In 1941 he even gave his tacit approval to Philby's temporary arrest in India by the British authorities when the Englishman was on his way from Saudi Arabia to the United States, where he proposed to convey the same bleak message to the American public through an extensive lecture tour.

Yet although Ibn Saud remained steadfast in his perception of the value of Britain's friendship, earning Winston Churchill's praise as a good friend in time of need, he was treading a tightrope between the expectations of his own people and the constraints imposed on him by Saudi Arabia's renewed isolation. In a single disastrous month, May of 1941, the Arab world watched shattered British forces evacuate Greece to the Axis, Field Marshal Rommel press relentlessly across the Western Desert towards Egypt, and a pro-Axis revolt menace British control in Iraq. Although the Golden Square revolt, so called because the four leaders were said to have been bought by Dr Grobba's gold, failed to open Iraq to Germany, the position of the Allies in the Middle East looked ever more perilous. Ibn Saud could count on advances from the oil company, food from Britain and the Empire and, from 1940, a British subsidy that quickly mounted to over £1 million a year. But these did not still the dissident voices hinting that he should squeeze more profit out of Britain's discomfiture, or else abandon the old alliance altogether.

When British diplomats had attempted, in the dark days of 1941, to interest the United States Government in taking on some of their financial burden in Saudi Arabia, Washington had refused. But by the time Twitchell, Kirk and the agricultural experts arrived in Riyadh in 1942, America had entered the war and oil supplies, especially the 'great prize' of Saudi reserves, to serve the Far East theatre and post-war needs assumed great importance. Furthermore, Washington was no longer content to leave Britain a free hand. In Iran, where Britain and the Soviet Union had collaborated to oust the allegedly pro-German Reza Shah in 1941, America was playing an increasingly active role. Its distrust of Britain was only marginally exceeded by its fear of the Soviet Union.

The Riyadh mission was to be the forerunner of a quick and effective campaign, abetted by the intrigues of the Saudis and Caltex, to replace the British as the King's chief protectors. The experts came officially at Ibn Saud's invitation, but the inspiration of the visit was attributed to Sheikh Abdullah Suleiman and the oil company. Caltex was professing deep anxiety about the amount of money it was advancing to keep the King afloat. This was to rise to $10 million by the next year, and Caltex dared not refuse further payments without ensuring that an alternative source of income was available to Ibn Saud. If the British were left to carry the burden alone, Caltex officials urged, there was a risk that they would ask for oil concessions in return and so recoup the ground they had lost in 1933 when I.P.C. failed to match the American offer. I.P.C. had already claimed the Hejaz, and the British might even persuade the King to throw Caltex out of Hasa. The company turned, therefore, to the U.S. Government, urging it to step in to protect America's national interest.

Officials of the State and Interior Departments were already alarmed by reports from oil company sources at home that America's domestic petroleum reserves were being run down too rapidly and might be exhausted within fifteen years. They could not risk having the richest potential overseas fields poached from them by the British. On the other hand, as the war consumed all Britain's capital stocks and foreign exchange, the British Government was even more urgently aware that its future solvency might well depend on how much oil it could keep in its own hands around the Persian Gulf, and since American companies had moved into Iraq, Bahrain, Kuwait, and Saudi Arabia, Britain's share of those reserves had been declining rapidly while the U.S. companies' share was to rise to 42 per cent by 1944. To help arrest that process, the British were certainly prepared to prop up Ibn Saud a little longer, however tiresome his spendthrift habits might be, in the hope of gaining post-war favours in his Kingdom. At the very least, Caltex had to be prevented from flooding the market for oil from sources under British control.

Thus, behind the screen of wartime friendship and co-operation on the battlefield, mutual suspicions mounted swiftly between the two allies in Arabia. While the British looked askance at America's agricultural mission and the subsequent dispatch of an engineering team which began to drill new water holes all around Riyadh, the oil company fanned U.S. suspicions with accounts of the arrival of a British anti-locust group in which, according to Caltex, petroleum geologists were said to be employed. The oil company also professed alarm over plans to open a British bank in Jeddah – only the second proper bank there – saying that Britain was forcing Saudi Arabia into the sterling area and would undermine Caltex's activities as an American firm.

By the beginning of 1943, these oil company whispers had reached the highest level. The chairman of Texaco and the president of Socal personally urged Harold Ickes, Secretary of the Interior and Petroleum Administrator for War, to frustrate the British and protect American interests by securing official lend-lease aid for Ibn-Saud. Ickes passed on their recommendations, with approval, to the State Department and the White House, and on 18 February 1943 President Roosevelt issued an Executive Order declaring the Kingdom vital to the defence of the United States. Accordingly, lend-lease funds would be made available forthwith to the Government of Saudi Arabia.[2] The great American takeover had begun.

The British were intensely suspicious and resentful as they saw the new relationship being built before their eyes in ways directly parallel to their own earlier attempts to bind Ibn Saud to their side. The Administration in Washington quickly made its supremacy plain. A quarter of a century earlier London had been the first and most natural port of call for the young Prince Feisal when he made his first trip abroad on behalf of his father. Now it was to Washington that he went, to meet the President. He arrived there in September 1943, accompanied by his younger brother Khaled (now King Khaled), and was greeted on his first evening by a White House dinner for forty people, including the Vice-President. There was no trudging about in the rain on this occasion, looking for a place to stay. The two princes and their party were put up with full state honours at Blair House, the official guest house, just across the road from the President himself, and when they left Washington to see something of the country they were given a private railway car to take them to Los Angeles and San Francisco. Hurriedly, the British invited them to stop in London on their way home, but it was somehow characteristic of Britain's incapacity to match America any longer that when the princes arrived with a magnificent, jewel-encrusted sword as a present for Winston Churchill from their father, the Prime Minister was not there to receive them. They were welcomed by Mrs Churchill instead – an agreeable enough gesture in British eyes, no doubt, but scarcely calculated to impress two young princes of the desert who had been raised in a tradition of absolute male supremacy.

In other fields, the contest was even more unequal. While the British were attempting to curb Ibn Saud's budget for 1944, the Americans weighed in with their subsidy. Although this was carefully set at a figure just equal to that of Britain, American arms and other supplies to the King soon exceeded anything the British could spare. Washington's plan to imitate Churchill and Anglo-Persian by buying into Caltex foundered because of opposition from Socal and Texaco, but the U.S. Government encouraged the oil companies to press ahead with preparations for post-war development of their Saudi resources on a scale that the British could not equal. There were to be new wells drilled, new areas explored, a new refinery constructed at Ras Tanura and a strategic new pipeline laid from Dhahran across Arabia and Syria to the Mediterranean coast. By 1944, when the German armies had been thrust out of the Middle East theatre and the tide of war had turned decisively in favour of the Allies elsewhere, the American Government felt confident enough to give Caltex permission to start work on these grand new projects and to release the materials. Money, supplies and Italian prisoners-of-war began to arrive for the half-abandoned Saudi camps. America's wealth and strength were visibly sealing the contract that was to make Saudi Arabia an American economic protectorate for as far ahead as anyone could see.

British officials protested vainly. In February 1944, Churchill was worried enough to cable President Roosevelt that Whitehall feared 'that the United States has a desire to deprive us of our oil assets in the Middle East on which among other things, the whole supply of our Navy depends'. Roosevelt replied promptly that he was equally anxious about reports 'that the British wish to horn in on Saudi Arabian oil reserves'.³ In Jeddah itself, the first permanent U.S. envoy, James Moose, fought a running battle with his British counterpart. Moose was succeeded late in 1944 by Colonel William A. Eddy, the son of a missionary and inured to Arab ways from his childhood in Sidon. His campaign against the British Minister in Jeddah, Stanley Jordan, was eventually to lead to the latter's removal.

Ibn Saud and the 'Syrians' were adept in exploiting these differences for their own benefit, whether over who was to repair the Riyadh road or who was to provide a financial adviser. One result of the King's manoeuvres, having pointed out to the British that the Americans were prepared to train his army, was the establishment in 1947 of a British camp outside Taif, Ibn Saud's summer capital, where a few British officers began the laborious business of making a modern force out of a few hundred undisciplined bedouin recruits.

But Britain was rowing against the tide of her own exhaustion and America's expanding power. Although Jordan had written in 1944 that Britain still enjoyed a 'preferential position' and the U.S. had 'little experience in Mohammedan countries',⁴ the arrival of such committed and, incidentally, anti-imperialist American Arabists as Eddy was to prove him wrong. As late as 1945, Ibn Saud was still to reject U.S. demands for a military airfield at Dhahran – primarily to assist in bringing back troops – because of British misgivings. But as the German armies began to retreat on all fronts throughout 1943 and 1944, the British were particularly exposed in the Middle East to all the resentments inspired by their past imperial record.

Ever since the First World War, the British had exercised either direct rule or some form of tutelage all the way from Aden through the Gulf to Iraq, Jordan, Palestine, and Egypt, and in most of these territories unhappy Arab memories of the betrayals involved in the Hussein-McMahon correspondence, the Sykes-Picot Agreement and the Balfour Declaration had never been suppressed. Indeed, the Germans had tried hard and with some success to exploit them, fomenting the Golden Square uprising in Iraq in 1941 and financing many anti-British factions in Egypt, Palestine and Syria.⁵ Now, with Britain plainly weakened by the war and Allied promises of self-determination to uphold them – underwritten by the terms of the Atlantic Charter – there was an undertow of Arab determination to wrest from the wreckage of the Second World War that genuine independence of which they believed they had been cheated by the deceptions of the First World War.

The British Government was aware of the dangers of this feeling, especially to its Middle Eastern oil interests, and sought to counter it by co-operation rather than by repression. Thus, in 1944, even while it was struggling to retain its foothold in Saudi Arabia against American pressure, it became the discreet midwife of a new League of Arab states in which, for the first time, individual Arab countries could come together in a search for political and economic unity. The advantage of such a League to Britain, it was thought, was that it would reduce inter-Arab rivalries – such as those of the Hashemites and the Saudis – and thereby make Anglo-Arab negotiation and co-operation easier. At the same time, Britain's association with any movement towards Arab unity might help to erase Arab recollections of imperialism.

Superficially, at least, it was a plausible notion. The difficulty was that such a current could not, by definition, be directed by a non-Arab power; yet neither could any one Arab government take the lead without courting the envy and hostility of all the others. It was, therefore, with some satisfaction and quiet hope that the British – after offering several public hints of their support – saw Ibn Saud and King Farouk of Egypt issue a joint appeal to their fellow Arab rulers to discuss the creation of a League of independent states in which joint Arab policies could be agreed upon. Even in this rudimentary and in many ways bizarre form, with an elderly and puritanical desert chieftain harnessed to a young and notoriously sybaritic urban monarch, it was enough to persuade the other Arab governments to take notice. Their representatives duly met in Alexandria in October 1944 and by March of the following year they were able to announce the birth of the Arab League.

British hopes that the new League might be employed to their advantage, however, were dashed at once. For the truth was that nothing held these disparate and quarrelsome states together save two things: the desire to end British rule and influence among them and to intervene in a growing conflict between Arab and Jew in Palestine. As these two goals were inextricably linked, through Britain's direct responsibility for the Palestine struggle, the League immediately became little more than an instrument for co-ordinating anti-British actions.

It is unlikely that Ibn Saud actually intended that to happen. As a devout Muslim, dedicated to the idea of Islamic unity, he was interested in creating some genuine Arab co-operation. Yet there had always been a deep ambivalence about the Anglo-Saudi relationship. From the days before the First World War when Whitehall steadfastly preferred to uphold the Turkish Empire rather than court Ibn Saud, through the long Anglo-Hashemite alliance and Britain's support for the minor sheikhdoms of the Gulf against Saudi pressure, there had never been a time when London and Riyadh had been wholly compatible. Now, as the contradictions and deceptions of past British policies in the Middle East ate ever more deeply into the crumbling framework of Anglo-Arab relations in general, the Anglo-Saudi differences were inevitably deepened too.

The emotional focus of these differences was Palestine where, in the closing months of the Second World War, a flock of thirty-year-old British chickens was coming home to roost. For many years Ibn Saud's interest in the affairs of Palestine had been marginal, for he had been far too concerned with establishing his own kingdom throughout the 1920s to have much energy or interest left to spare for the complaints of the Arab leaders under the British mandate there. It was not until the 1930s that he began to concern himself, like other Arab leaders, with the growing violence provoked by the continuing arrival of Jewish immigrants in a predominantly Arab land. In 1936 he attempted to mediate, along with the Kings of Iraq and Transjordan, to prevent further violence. By then, educated Palestinians and Syrians had come to weigh heavily in his growing entourage, including his chief political adviser, Yusuf Yassin, who hailed from Latakia on the Syrian coast, Rushdi Mulhas, his Palestinian political secretary who was also deeply opposed to British rule in his homeland, and Jamal Husseini, nephew of Hajj Amin Husseini who, as the Grand Mufti (chief Muslim leader) of Palestine, led the Arab resistance there to both Britain and the Zionists until he was obliged to flee to Berlin. Their strong personal interest in the developing Palestinian tragedy certainly helped to arouse Ibn Saud's sympathies. Yet in the end he needed none of their prompting to form his judgement. Islam was enough.

Although by Wahhabite standards Ibn Saud was now only moderately zealous in his devotions, he would have been false to the deepest roots of his nature if he had not deplored the systematic Judaeization of what Arabs universally regarded as part of their homeland. In the long debates on Palestine that absorbed his closest relatives and advisers at his *majlis*, he returned repeatedly to the Koran's guidance on the matter. The Jews, he would remind his audience with all the authority of a lifetime's study of the Holy Book, had always been the enemy of the Prophet and their history was full of rebellion against Allah. Did not the Koran say the Jews would listen to any lie (Sura V 42), and did they not utter blasphemy against the Almighty (Sura V 62)?[6] If they behaved like guests in Palestine they would, of course, deserve to be treated like guests; but if they drove away the Arabs from the land of their fathers and dared to claim it as the land of their fathers instead, they would surely stand accursed.

As Jewish immigration continued to mount, for all that, his opposition to any political concessions by the Arabs only grew more adamant. When a British Government commission of inquiry recommended partition of Palestine between Arabs and Jews in 1937, Ibn Saud joined all the other Arab leaders of the day in condemning the proposal out of hand, on the grounds – which seemed to all Arabs eminently reasonable – that as the Jews had no right to be in Palestine anyway, other than the title bestowed on them unilaterally by Britain, any partition must amount to Arab surrender. When the British convened a round table conference in London two years later, at which Arab and Jewish leaders were to study new proposals for negotiation, Ibn Saud sent Feisal with firm instructions not to give an inch; and when the prince duly joined his Arab colleagues in refusing to sit at the same table with the Jewish delegates, forcing the British to go fruitlessly to and fro between the delegations in a fashion now familiar to many later mediators, he signalled his wholehearted approval.

After the start of the Second World War there was a time when both Roosevelt and Churchill half believed that Ibn Saud might hold the key to an ultimate accommodation in Palestine, but events demonstrated that this was wishful thinking, induced by a characteristically eccentric initiative by St John Philby. Before his spell of preventive detention in 1941 – the British authorities were soon convinced he was harmless – Philby had busied himself with a scheme for Saudi-Zionist co-operation that combined cynicism and romanticism in roughly equal parts. Although he was himself a stout defender of Palestinian Arab rights, he had reached the conclusion – sensible enough in itself – that the Arabs could gain nothing by trying to put the clock back. He therefore proposed that Ibn Saud should be persuaded to take the lead in accepting a Jewish state in Palestine with a view to forging a brand new, pan-Semite alliance. But how to nudge his hero in the right direction? Well, Ibn Saud was short of money and the Jews were short of land, so, said

Philby, a swap should be possible. If the Zionists would give Ibn Saud £20 million, he would then be able to induce the Arabs to leave Palestine to the Jews and provide for their resettlement elsewhere, using his prestige and the acumen of the Jews to create a new, united Arab state under Saudi leadership.

If such a proposal had ever had the smallest chance of acceptance by the Saudi leader it would, no doubt, have been cheap at the price; and Philby peddled it earnestly to Dr Chaim Weizmann, the Zionist leader, who since the First World War had believed that the Arabs would accept a Jewish state in return for their independence. Philby took it to Churchill, who recommended it to Roosevelt, until it seemed to Philby with his usual cocksureness that it could not possibly fail. Yet fail it did, as anyone but Philby would have known it would, when the King, one day in late 1940, having heard Philby out with commendable patience, simply refused to discuss the matter further. The true effect of Philby's bizarre intervention was revealed three years later when, under pressure from Dr Weizmann, Roosevelt sent Eddy's cousin, an American Arabist called Colonel Harold Hoskins, to Riyadh to ask whether Ibn Saud would agree to meet Weizmann for talks about Palestine. The King, reported poor Hoskins, had exploded in dreadful wrath, declaring that he hated Weizmann personally because the Zionist leader had insulted him by offering a bribe of £20 million if he would accept Arab settlers from Palestine. Philby, said the King, had been the intermediary for this scandalous proposition and he had even intimated that the bribe would be guaranteed by President Roosevelt. As Hoskins reported later to the President: 'His Majesty said he had been so incensed at the offer and equally at the inclusion of the President in such a shameful manner that he had never mentioned it again.'[7]

So much for Philby and Zionist hopes of Ibn Saud. Yet with the war nearly over and the full extent of the Jewish holocaust in Nazi-occupied Europe becoming steadily more apparent, the conflict implicit in the original Balfour Declaration between the concept of a Jewish homeland and the rights of the existing Arab inhabitants of Palestine was approaching a climax. Jewish and American pressure on Britain as the administering power to admit the refugee survivors of the holocaust to Palestine was countered by a flat Arab refusal to give any more ground. Jewish terrorism, partly suspended during the war as most Jews fought with the Allies against Hitler, was reborn in Palestine as the immediate German threat to the Middle East receded, and Arab counter-terrorism revived along with it.

Caught in the middle of this deepening conflict, Britain's credit was sinking steadily throughout the Arab world, and at Ibn Saud's court the anti-Zionist passions of the King and his advisers powerfully reinforced more practical arguments for shedding the old British connection in favour of a new American umbrella.

There were signs of occasional regret in the King over this turn of events when he would say – with a touch of old man's sentimentality, perhaps – that the British had been his friends for so long that he did not want to turn his back on them now. But stronger than any sentimentality was that necessary and familiar bedouin eye to the main chance which dictated a prompt accommodation to the new facts of life; and by the time President Franklin D. Roosevelt decided that he ought to meet the Saudi ruler in person, in February 1945, on his way back to Washington from discussions at Yalta with his allies, Joseph Stalin and Winston Churchill, there was no longer any doubt in which direction those facts were pointing.

Roosevelt did not trouble to tell Churchill of his plans until the day before they left Yalta and, in the name of wartime secrecy, any hint of the arrangements for the meeting was scrupulously kept from the ears of the British Minister in Jeddah. Churchill was furious at seeing his old friend and rival so casually steal a march on him, especially with a ruler over whom they had already exchanged sharp words, and he promptly sent the Foreign Office whirling into action to make sure the Saudi leader set aside some time for him as well. Roosevelt, however, had gained the advantage with his prior invitation and it fell to the U.S. Navy, represented by the destroyer U.S.S. *Murphy* under the command of Captain B.A. Smith, to carry Ibn Saud away from his own Kingdom for the first time in fifteen years, to meet a foreign ruler. The last occasion had been the meeting with King Feisal of Iraq on board the *Lupin*. This time he was to sail under the American flag to meet President Roosevelt on board the cruiser U.S.S. *Quincy* in the Great Bitter Lake of the Suez Canal, midway between the Mediterranean and the Red Sea.

Getting him there with the necessary secrecy proved as difficult for the Americans as the protocol of the earlier visit had been for the British. Apart from the active and well trusted Colonel Eddy, only the King and a handful of his closest advisers were supposed to know anything about the trip, and the King had been advised not to travel with more than a maximum of ten servants and companions. The

Americans had not yet really learned about the customs of a desert ruler. When the day came to board the *Murphy* in Jeddah harbour, the King arrived with a party of forty-eight people, preceded by several dhows laden with vegetables, rice and 100 live sheep with which, he insisted, he would feed the entire ship's crew as well as his own retinue. In vain did Captain Smith and Colonel Eddy explain that the ship could not accommodate so many people and that its lockers were already stored with more than enough food for their brief journey. Lack of accommodation, said the King, was of no importance; he would of course sleep in his usual tent and his men were accustomed to bunking down wherever they could. As for food, no good Muslim could eat meat which had been killed more than twenty-four hours previously, so the sheep would have to come too.

Reluctantly the Captain agreed to the first part of this royal command and the King's tent was erected on the *Murphy*'s foredeck, with his simple, wooden throne in the middle and the rugs spread about for his daily *majlis*. The rest of the party found improvised sleeping quarters in the gun turrets, the wheelhouse and the gangways down below. The sheep, however, provided a slightly greater problem. Imagining his lovely ship ankle-deep in droppings and running with the blood of daily slaughter, the Captain at first was adamant in refusing to accept a single one. But the demands of statesmanship at length overcame his seaman's protestations. A compromise was negotiated by Colonel Eddy and seven sheep were tethered on the destroyer's fantail, with bundles of greenstuff to munch for the trip and a makeshift hearth on which to cook them as well.

Jeddah was left in a state of consternation at the party's departure, throbbing with bazaar rumours that the Americans had kidnapped the King; but by the time they arrived at their rendezvous with the President, three days later, the King's party and the ship's crew had come to terms with each other. Most of Ibn Saud's companions had managed to sneak into the crew's quarters to see the latest Hollywood movie starring Lucille Ball, a spectacle to delight deprived eyes, and most of the crew had tried their hand at eating the hot roast lamb and rice at a royal feast. When they parted, the King gave every man a present of $40 apiece, with gold daggers and swords for senior officers, and the Captain gave the King a pair of binoculars and a couple of submachine guns.

On board the *Quincy* this note of bizarre cordiality was maintained. Roosevelt was then within two months of his death and plainly a sick and exhausted man, but his famous charm was still in working order and he employed it for all he was worth on the King. Although he was a chain-smoker he forbore to light a cigarette in the King's presence out of deference to his religious feelings, and hid in the ship's elevator for a quick puff whenever he felt he could no longer go without.[8] When the King pointed out that he was now crippled with arthritis and could only walk with a cane, and showed a keen interest in the President's wheelchair, Roosevelt immediately made him a gift of the spare chair that travelled with him. The King was just as determined to be the perfect guest, and when he discovered that the exigencies of naval schedules and wartime convoy systems would prevent the President from enjoying an Arab dinner with him (a dispensation for which, we may suspect, Roosevelt was privately not unthankful), he insisted on having coffee brewed and served on the *Quincy* by his men as the minimum permitted gesture of Arab hospitality.

Even the issue of Palestine could not divide them, although for a time it came perilously close to doing so. The President was naturally concerned about the fate of the Jews in Europe and in spite of Ibn Saud's earlier outburst to Hoskins on the matter he still had hopes of persuading the King to use his influence with the Arabs in favour of more Jewish immigration to the Holy Land. With all his eloquence he described Jewish sufferings under the Nazis and invited the King's advice. Colonel Eddy, who was interpreting between the two men, described Ibn Saud's reply as prompt and laconic: 'Give the Jews and their descendants the choicest lands and homes of the Germans who had oppressed them.' When the President said the Jews would prefer to go to Palestine, Ibn Saud replied with unassailable logic, 'Amends should be made by the criminal, not by the innocent bystander.'[9] It was, he observed, the Arab custom to distribute the survivors and victims of war among the victorious tribes in proportion to their capacity to support this extra burden. In the Allied camp, he noted, there were now fifty countries. Compared to most of them, Palestine was small and poor, yet it had already been assigned more than its quota of European refugees; why should it now be asked to take still more?

Roosevelt was impressed in spite of himself, remarking afterwards that he had learned more about Palestine in five minutes with Ibn Saud than in all the arguments and memoranda he had ever had from his staff.[10] With a politician's instinct for the soothing promise, he gave the King two personal undertakings which were soon to become notorious for their apparent betrayal. He, President Roosevelt, would never do anything which might prove hostile to the Arabs, and the U.S. Government would make no change in its basic policy in Palestine without full and prior consultation with both Jews and Arabs. On that the two men parted, satisfied that they not only liked but understood each other. That there was, in fact, a fundamental misunderstanding still between them, with Roosevelt speaking only for himself while the King supposed that the President was committing the honour of the United States Government, was only to be revealed after the President was dead.

By contrast, when Ibn Saud met Churchill three days later, relations between them were no more than mutually respectful and on occasions were downright strained. Like Feisal's visit to London after his American tour, the meeting had been arranged in haste chiefly so that Britain might continue the pretence of rivalling America in Saudi eyes. But the reality, as both men knew, was that the pretence was now too threadbare to deceive anyone. Ibn Saud had even felt obliged to ask Roosevelt if there was any American objection to his accepting Churchill's invitation – and the broad magnanimity with which the President gave his kind permission must have told the King that the President no longer feared any real competition there. Nor did he have to. The two men met in the Grand Hotel du Lac, a gloomy roadhouse on the shores of a reed-fringed stretch of water in the Fayoyum oasis, fifty miles south of Cairo. It was a place more accustomed to gin-drinking, duck-shooting parties of British officers on leave and weekend assignations for illicit couples from the haut monde of Cairo than to the affairs of high statesmanship, and Churchill responded to its louche ambience more naturally than the puritanical Ibn Saud.

Characteristically, the Prime Minister declined to follow Roosevelt's tactful self-denial in the presence of the King. Instead he announced boldly that 'if it was the religion of His Majesty to deprive himself of smoking and alcohol I must point out that my rule of life prescribed as an absolutely sacred rite smoking cigars and also the drinking of alcohol before, after, and if need be during all meals and in the intervals between them'. The King, according to Churchill, 'graciously accepted the position',[11] but a frisson of disapproval must certainly have crossed his mind. In any case, the two found little to say to each other, and when they parted the King seemed dissatisfied with his reception. It appears also that his temper was not improved by

Rashad Pharaoun's careless loss of the great medicine chest. Although the King went home on a British cruiser – allegedly intended to impress him by contrast with the smaller destroyer placed at his disposal by the Americans for the journey north – he reported to Colonel Eddy afterwards that he had not enjoyed himself. 'The food was tasteless; there were no demonstrations of armament; no tent was pitched on the deck; the crew did not fraternize with the Arabs; and altogether he preferred the smaller but more friendly U.S. destroyer.'[12] Colonel Eddy was clearly enjoying himself.

Some months later there was an apt postscript to all this when the parting presents profferred to the King by the two Allied leaders at last arrived in Saudi Arabia. As was customary with truly distinguished visitors, the King had given to each of them a collection of jewels, gold daggers, swords and harem clothes worth many thousands of dollars. Churchill in return had been able to produce only a box of moderately expensive perfume, but on the spur of the moment, as if embarrassed by the inadequacy of his present, he told the King he would send him 'the best motor car in the world'. He meant, of course, a Rolls-Royce, but in 1945 those glamorous cars had been out of production for nearly six years and the British Ministry of Supply had to make a lengthy search before a pre-war model was discovered in mint condition, languishing in a dealer's garage. It was fitted with a special throne and did not arrive in Jeddah until 1946. Then, driven carefully across the desert to Riyadh by two members of the British Legation staff and one of Ibn Saud's former drivers, it was left in the courtyard of the Murabba Palace for a final brush and polish before the formal presentation to the King the next day. But before Cyril Ousman, the odd-job engineer who had become British Vice-Consul, could hand over the vehicle, the King came upon it in the courtyard and took an instant dislike to it. His one remaining eye observed with baleful contempt that, being British, the steering wheel was on the right-hand side. As he always liked to sit in the front of his cars, especially when hunting, this meant that he would have to sit on the driver's left. But the Arab place of honour was invariably on the right. Thunder-browed at such British carelessness, he told his younger brother Abdullah that he could take the car away. The men from the Legation formally presented the car the next morning, but the Rolls-Royce soon vanished from the palace. For the next thirty years it was lost in Abdullah's garage, gathering dust and rust, until it was rescued after Abdullah's death in 1977 as a potential exhibit for an Abdul Aziz al Saud museum.

No such indignity awaited President Roosevelt's present. Although he was dead before it arrived in Riyadh, it was obvious at once that it was a hit – a twin-engined Douglas DC-3 airliner with an American crew on free loan for a year. When the ill-fated Rolls arrived, the DC-3 had already been flying Ibn Saud and his family around the country for several months and the American Legation in Jeddah was happily negotiating a long-term contract for Trans World Airlines to take over when the crew's free year was up and organize the first-ever scheduled Saudi civil airline. Thirty years later, when Churchill's Rolls was only a museum piece, T.W.A. and Saudi Arabian Airlines would still be in business together running some of the busiest routes in the Middle East. The era of American hegemony had begun.

11 Mammon Triumphant

1946 – 53

And therefore, at the Kynges court, my brother,
Ech man for hymself, ther is noon oother.
CHAUCER, The Knight's Tale

IF the First World War was the time when the modern Middle East was born, the Second World War was the period in which it came of age. By the time Germany and Japan had surrendered, the French empire in the Middle East had been destroyed and British power was broken. Nationalism had become an irresistible challenge in many Arab countries.

A whole set of new threats to traditional imperial power emerged from the Mediterranean to India. In Greece, a communist-inspired civil war was in progress that promised – with the tacit complicity of Moscow – to establish Russia's first outpost of direct influence on the Mediterranean shore. In India, for so long the key to British interest in Arabia and the Gulf, nationalism was on the brink of victory over the Raj. In Turkey the Soviet Union was also demanding concessions, and in Iran it launched a new version of the old Tsarist drive southwards by establishing a puppet Soviet in the northern province of Azerbaijan.

Devastated and impoverished by the war, the British were now beyond resistance. Within eighteen months of the war's end their retreat from empire began to resemble a rout. In a single week of February 1947, London announced that it would like to withdraw from the Palestine mandate, informed Washington that it could no longer afford to maintain its support of the Government in Athens against the communist guerrillas, and declared that it would complete an unconditional withdrawal from India by the end of June – thus abandoning in four months what had been so jealously guarded for two centuries.

Into much of the vacuum the Americans moved with power and decisiveness. The Soviet troops were pushed out of Iran by the threat of force in 1946, and in March 1947 President Truman announced his famous 'doctrine' establishing that the defence of Greece and Turkey was vital to the interests of the free world.

In Saudi Arabia, which had declared war on Germany just in time to be eligible for admission to the new United Nations Organization, the United States' replacement of British power was all but complete. Only one issue threatened to break the new bonds that had been forged by men like Colonel Eddy and the officials of Aramco, as Caltex was now known. This was Palestine. Even before the war had ended, President Roosevelt's rash promises to Ibn Saud were cast away by his death and by the different attitudes of his successor, Harry S. Truman. At the Potsdam Conference, where the Allied leaders met in the summer of 1945 to apportion the fruits of victory in Europe, Truman was already pressing Winston Churchill to authorize the immediate immigration of more Jewish refugees from Hitler's pogroms in Europe. When Churchill was defeated in a general election before he could reply, Truman took up the same cause even more strongly with the new Labour Government under Clement Attlee, urging the admission of 100,000 Jews to Palestine forthwith. Attlee's reply expressed the British Government's deep reluctance to be harried into any further quarrel with the Arabs, particularly those leaders like Ibn Saud whose control of oil output could spell life or death for bleak, post-war Britain.

The U.S. State Department argued, with all the conviction its growing corps of Arabists could muster, that open support of Zionism was bound to damage American interests, while the Defense Department, supported by the oil companies, pointed with alarm to the continuing depletion of America's domestic oil reserves and the corresponding need to secure good relations with the Arabs for the future. But in Congress and the White House those arguments carried little conviction. In November 1945, after listening to a State Department recital of the damage to American prestige in the Middle East already caused by his pronouncements, President Truman dismissed his officials with the now famous words: 'I'm sorry, gentlemen, but I have to answer to hundreds of thousands who are anxious for the success of Zionism; I do not have hundreds of thousands of Arabs among my constituents.'[1]

For all the pretensions to co-operation voiced in the Arab League, the Arab states were united only in the abstract.—in rejecting the idea of more Jewish immigration and the corollary of a Jewish state. In practical terms they were at sixes and sevens. Ibn Saud had once argued that a continued British mandate was preferable to a triumph for the Jews or an expanded state for the Hashemite Abdullah in Transjordan. But other members of the Arab League, in fighting to throw off the rule of Britain everywhere, forgot that they needed British support against the Jews; and in their violent denunciations of American pressure for immigration, they actually stirred genuine emotions about Jewish suffering and helped ensure that President Truman could not retreat, even if he had wanted to, from his personal commitment to the Jews.

The Arabs failed totally to marshal the one effective power that nature had suddenly thrust into their own hands – the power that the world would later come to know and fear as the 'oil weapon'. But at this time they still lacked the expertise, both political and technical, with which to organize boycotts, shutdowns or embargoes, and the Arab states chiefly concerned were also precisely those in Arabia and the Gulf where, with the exception of Iraq, modern administration was weakest and the demand for money most acute. Of nowhere was this more true than the Kingdom of Ibn Saud.

For the United States, Ibn Saud's attitude was as critical as it had earlier been for the British. Aramco and the State Department continued to assure him and his advisers that they would do all they could to limit American support for Zionism. But President Truman was not to be deflected, and when Feisal went to Washington in 1946 to try to secure from him an affirmation of Roosevelt's undertakings he got nowhere. Yet, even as late as the autumn of 1947, when Britain finally dropped the whole Palestine crisis in the lap of the United Nations, formally announcing that it was washing its hands of the mandate, Feisal in New York at the General Assembly seemed to hope that President Truman would ultimately feel bound by his predecessor's promise to do nothing without Arab consultation. It was wishful thinking. When the Assembly met to debate Palestine's future, the American vote was cast decisively in favour of the kind of partition plan that the Arabs had repeatedly rejected, to set up both a Jewish state on the better part of the soil of Palestine and an Arab state on the remainder. According to a sympathetic biographer, it was both a stunning political blow and a personal affront to the prince, who had actually assured his fellow Arab delegates beforehand – on the strength of his conversations with American officials – that they need have no fear of such an outcome.[2] Moreover, he had to endure the stinging insults of Zionist supporters in New York. Feisal told the U.S. Minister in Jeddah, J. Rives-Childs, that he would have broken relations with Washington if he had been in a position to take such a decision.[3]

The State Department would, no doubt, have welcomed some signs of effective Saudi opposition to support its arguments. Aramco came to its aid with a declaration that it would have to stop work on the great Trans-Arabian Pipeline (Tapline) stretching from the Gulf to Sidon on the Mediterranean, causing the Secretary of Defense, James Forrestal, to declare in shocked tones that unless the U.S. had access to Middle East oil, 'American motorcar companies would have to design a four-cylinder motorcar sometime within the next five years'.[4] The Arab League, too, reached a tentative agreement that it should deny pipeline rights to American oil companies until Washington's policy was changed, and the Prime Minister of Syria went so far as to visit Ibn Saud in Riyadh to try to persuade him to take punitive action against Aramco.

If the King had been a younger man or if Palestine had been an issue that personally affected his subjects, like the loss of a traditional grazing ground, he might perhaps have acted more vigorously. But though he continued to berate the Jews in his *majlis* and would listen to broadcasts from Jerusalem with tears pouring from his eyes, he left it to the Arab League in Cairo and the abysmal incompetence of Arab states more closely involved. Thus, when the first full-scale Arab-Israeli war began, after the British had completed their Palestine scuttle and the United States had led the world in recognizing the independent state of Israel, the two most significant reactions in Saudi Arabia were a decision by Aramco to resume work on the Tapline project and the Saudi Government's settlement with the company of a lengthy dispute over payment of oil royalties. Ibn Saud did, however, dispatch a brigade of his new army to the front. By the time the war ended in the armistice of 1949, with Israel established in frontiers that were to last until 1967, it was clear that the Arab economic power that both the British Government and the U.S. State Department professed to fear was a fiction. The emptiness of these Arab threats was to set a damaging precedent.

Towards the end of 1949 Saudi Arabia reopened a territorial dispute that had lain dormant for a dozen years. Secretly and conceptually, the Kingdom had never renounced its aspiration to assert its hegemony over the greater part of the Arabian peninsula, excluding only Yemen, the coastal plains of the Sultanate of Oman, and the Hashemite kingdom of Jordan up to the northern point of the Wadi Sirhan, about fifty miles from Amman. Saudi Arabia's land borders with only Iraq, Kuwait and Qatar had been precisely delineated and were internationally recognized. It was a situation that the King was happy to live with. He would have liked to regard the whole of the southern littoral of the Gulf as part of his ancestral domain. More specifically, there was the Buraimi oasis. Situated in the undemarcated area where the historic claims of the Kingdom, the Sultanate of Oman and the Emirate of Abu Dhabi converged, its inhabitants had paid tribute to the Saudi emirs from 1853 and again from 1869 when it was under full Wahhabite occupation. On this and other related issues there was a tangle of dubious precedent justifying Saudi assertions. The Kingdom's claim to Buraimi would probably have remained quiescent if it had not been for the intrusion of the oil companies into the desert regions and the waning of British power.

Early in 1949 Aramco exploration parties began to probe into the coastal regions east of Qatar across the Riyadh Line and into what the British Government regarded as the rightful territory of the Emirate of Abu Dhabi, its protectorate. Ibn Saud had never accepted the demarcation, an amendment of one formulated in 1935 and proposed by the Foreign Office in 1937. But in October 1949 the Saudi Government in the name of the King formally put forward a claim to what was generally accepted as four-fifths of Abu Dhabi's land – given what appeared to be Ibn Saud's apparent acquiescence to the Riyadh Line. Included was a broad corridor to the Gulf adjacent to Qatar, Buraimi, and also the Liwa oasis, together with its surrounding settlements. Liwa was the ancestral home of the Al Bu Falah, to which the ruling family of Abu Dhabi belonged.

Whatever Ibn Saud's belief in Saudi rights, it is very unlikely that he, with his health and concentration failing, would have taken the initiative in asserting the claim. Just who supplied the inspiration remains unclear. Probably it was Yusuf Yassin, perhaps in collaboration with Fuad Hamza and Khaled Ghargani, a Libyan, who were now the King's closest advisers. A forthright Arab nationalist in his younger days and something of an Anglophobe, Yusuf Yassin would have relished the chance of discomfiting or humiliating imperialist Britain – quite apart from the opportunity it gave him to ingratiate himself further with the King. Certainly, he became the chief advocate of the scheme. But the suspicion that Aramco might have been responsible for planting the idea in Yusuf Yassin's brain cannot be ruled out. Aramco adopted the Saudi cause with alacrity and enthusiasm. George Rentz, the distinguished Arabist who headed the company's oriental department, welcomed the opportunity to accumulate an archive of historical material and to win favour with the Saudi authorities for the future. He and his colleagues lost little time in delving into arcane research for material to be used as backing for the claim. They combed the desert for co-operative tribesmen. They compiled exhaustive tax lists and family records to support the Saudi case. 'Aramco at this time was more Saudi than the Saudis,' one contemporary observer commented.

Once committed, Ibn Saud could not retreat. Not only ancestral rights but also his honour, self-esteem and position as a great Arab monarch were at stake. The evidence is that he became increasingly preoccupied with and confused by the dispute, and more intransigent. At the same time he would make protestations of his great friendship for Britain which, he would say, was as precious to him as his family's. Mournfully the King protested that his devotion to Britain was no longer reciprocated. In his decrepitude Ibn Saud simply could not understand why his old friends could not see the justice of his claim.

From the start Feisal was a vigorous champion of the cause. Both he and Ibn Saud would insist in diplomatic exchanges that the issue was not one of territorial expansion or material gain but tribal loyalties. The claim was 200 years old and the people in question owed loyalty to the Saudi Emir even when he was exiled in Kuwait. Feisal went so far as to argue that the Sunni tribesmen of Oman, chafing under the rule of the 'heretic' Sultan Said bin Taimur Al Bu Said, who belonged to the unorthodox Ibadi sect, wanted to pay fealty to Ibn Saud.

The King was pleased that the Duke of Edinburgh, serving as a Lieutenant on board H.M.S. *Chequers*, accompanied by Admiral Sir John Power, Commander-in-Chief of the Royal Navy's Mediterranean Fleet, on board his flagship H.M.S. *Surprise*, paid a courtesy visit to Jeddah on 17 January 1950. But he saw as a threat the announcement that a 'Persian Gulf Frontier Force' was to be formed. Two years later the Trucial Oman Levies, as the force was named, was created. One reason for its formation was the threat of Saudi aggrandizement. Another was to check more effectively the trade in slaves destined for the Kingdom and the amorous proclivities of Ibn Saud's sons, not the least his designated successor. The return to power of a Conservative Government and Winston Churchill as Prime Minister in the British general election of 1951 brought hope of an amicable compromise. Eventually a conference was held in Dammam in February 1952, but it collapsed and was adjourned *sine die*.

By then the dispute focused heavily on the nine villages of the Buraimi oasis where, the British said, the Sultan of Oman and Sheikh Shakhbut bin Sultan al Nihayan of Abu Dhabi enjoyed sovereign rights over three and six respectively. On 20 September 1952 a small Saudi force of forty men led by Turki bin Utaishan entered Hamasah, one of the villages claimed by Oman. The operation was only made possible by the logistical support provided by Aramco. The Saudis were blockaded by the British and deadlock ensued. Ibn Saud said that he would rather see Turki bin Utaishan dead than withdraw. Towards the end of the year, Raymond Hare, the U.S. Ambassador, proposed – on instructions from the State Department – that the conflicting parties should agree on a standstill so that there would be no escalation of the dispute. Ibn Saud was in favour of American mediation, an indication of the extent to which Saudi Arabia was falling under the influence of the superpower, but Washington was unwilling to contemplate a role as intermediary.

In April 1953 Ibn Saud, reluctantly and ambivalently, agreed to the principle of arbitration in the course of an audience with G. C. Pelham, the British Ambassador, but only on the basis of 'historical rights'. The diplomatic stalemate would continue for more than a year.

*   *   *

The House of Saud's feud with the Hashemites continued unabated in the early 1950s. Suspicion and animosity were directed chiefly at Iraq, the more powerful of the two Hashemite monarchies. Relations were hardly improved when Ibn Saud, who had given refuge to Iraq's former Prime Minister Rashid Ali Gailani, exalted him in the summer of 1951 to an official adviser. Rashid Ali had been the leader of the Golden Square revolt in 1941 and had forced the Prince Regent Abdul Illah and his pro-Western right-hand man Nuri Said to flee abroad before the monarchy was fully restored by British military intervention. Rashid Ali's presence at the Saudi royal court could be seen as an implied threat or a defensive measure; the Iraqi authorities for their part appeared to do nothing to stop tribal raids across the border into Saudi Arabia in the autumn of 1951. Abdul Illah made an overture early in 1953, but it proved wholly counter-productive. Ibn Saud was enraged by his proposal that they should hold talks on his yacht in the Gulf. The King regarded the invitation from a dynastic foe, who was not even his peer, to meet in an Iraqi vessel and a neutral venue as a flagrant insult.

In Jordan, as Transjordan had been known since 1947, the formidable King Abdullah was murdered in July 1951 and was succeeded by his son Talal. Towards the end of 1951 Ibn Saud invited him to Saudi Arabia in a diplomatic gesture of reconciliation. The young Hashemite monarch was taken not only to Riyadh but also to Jeddah and Mecca in the Hejaz, his ancestral domain. Ibn Saud showered him with gifts, but refused a request for a loan. Talal was suffering from a growing mental illness and his reign was short. In August 1952 he was deposed in favour of his eldest son Hussein, aged sixteen.

Active Saudi diplomacy of a less conventional kind concentrated on Syria. Ibn Saud wanted its friendship, or at least good will, as a counter to the threat of Iraqi encroachment on his realm. At the same time he was concerned about the expansionist dream harboured in Damascus of a 'Great Syria' including not only Lebanon but also Jordan, part of whose territory – up to the northern end of the Wadi Sirhan – he regarded as rightfully belonging to his dynasty. Conversely he feared a Jordanian attack on Syria. An additional, urgent reason for courting Syria was the construction of Tapline. Transit dues had to be negotiated and a general agreement on the passage of oil to Sidon on Lebanon's Mediterranean coast had to be ratified. The early flood of oil revenue made available plenty of money for Yusuf Yassin to interfere in the political manoeuvrings of his native country, which was to gain a unique reputation for instability and the frequency of its coups d'état. It was an operation in which Yassin revelled.

Less secretively, in its first public exercise in dollar diplomacy, Saudi Arabia decided in the summer of 1949 to accommodate the Syrian Government of Colonel Husni Zaim by raising a $6 million loan. Zaim was assassinated in August by Colonel Sami Hinnawi, but Saudi Arabia was anxious to oblige his successor and pressed on with its plans. Yet no sooner had the sum required been raised in the U.S. than $4.5 million of it was used by the Saudi Government to settle accounts with merchants to whom it owed money. The pledge was renewed when Hashim Atasi, Prime Minister of Colonel Hinnawi's new Syrian regime, visited Riyadh in November. Syria, it was agreed, would repay the loan in commodities, particularly cereals of which she had a surplus at this time. There were other reciprocal conditions. Abdullah Suleiman then approached Aramco for an advance to finance the credit. Concerned to ward off mounting pressure for bigger revenues and also to settle the question of Tapline, Aramco complied. A first tranche of $2 million was paid by the Saudi Exchequer and the Tapline agreement signed in January 1950. The second instalment was not paid. The most likely reason was the Saudi Exchequer's shortage of cash and growing indebtedness. The balance of the Aramco advance, it seems, was spent to defray the royal court's prodigal spending. But Ibn Saud could not ignore his commitment. After the visit of the Syrian Chief of Staff to Riyadh in the summer of 1950, the balance of $4 million was handed over. This time the money was lent by the Netherlands Trading Society (now the Algemene Bank Nederland N.V.) against future receipts for oil.

Saudi Arabia also sought to maintain the best possible relations with Egypt, always a cardinal policy aim for the Kingdom. Despite his differences with Britain over Buraimi, Ibn Saud had advised King Farouk to reach a settlement of the dispute over U.K. occupation of the Suez Canal Zone. But when in October 1951 Nahhas Pasha, the Egyptian Prime Minister, abrogated the Anglo-Egyptian Treaty, his action was applauded in Saudi Arabia. It seemed almost like a good seal of approval when Egyptian newspapers and magazines, which were full of anti-Western and more specifically anti-British propaganda, were allowed to circulate in the Kingdom. But by then the ageing Ibn Saud would have been too comatose to interest himself in such peripheral matters. His Anglophobe advisers were probably responsible.

With the decline of Britain and defeated, discredited France, the U.S. had asserted itself as the predominant influence in the Middle East. But as a critic and castigator of 'imperialism', a guise in which the world's leading power could indulge for some years after the Second World War, the U.S. was correct and cautious in its dealings with the Kingdom, a strange political entity quite remote from American ideals of liberalism and democracy. The approach was one of an enlightened benefactor promoting American material interests as well as more idealistic, though confused, concepts.

In practice, both Aramco and the State Department appreciated the fact that Saudi Arabia did not suffer from the feelings of inferiority, cultural confusion and love-hate so common amongst those peoples who had been colonized or dominated by the European powers. The more perceptive amongst them were aware of the converse: Saudi contempt for the Unbeliever or those who were not privileged by their possession of and proximity to the holiest shrines of Islam, an arrogance more evident in the conquering Wahhabis of Nejd than the cosmopolitan Hejazis. But Saudi xenophobia was only partly due to an innate sense of spiritual superiority; more defensively, it arose also from an embarrassed bewilderment at being confronted by a technologically more advanced civilization, many of whose ways contradicted – though not in terms of mercenary gain – those traditional ones which sustained the Kingdom. Saudis appreciated the outward manifestations of Western advancement of which the U.S. was now the pioneer. In the Kingdom of the late 1940s it seemed epitomized, at two extremes of spending power, by Cadillacs and Coca-Cola.

St John Philby about this time was heard to comment of the Americans that they overcame Saudi xenophobia by acting as if it did not exist. Relations were considerably eased by their buoyant self-confidence, unaffected egalitarianism, and tolerant acceptance of local 'foibles' that Europeans of a more colonial frame of mind would regard as perverse bloody-mindedness. American benevolence was typified in these days by the array of services provided for the local community from the earliest days by senior Aramco employees based in Dammam. Indeed, the instrument of American foreign policy in Saudi Arabia tended more and more to be Aramco rather than normal Government channels. This was not appreciated by the Aramco accountants in New York.

This American influence did not result in any dominance of a quasi-colonial nature or even an easy relationship. Rather, it was a tangled and often strained inter-dependence. Despite Ibn Saud's and Feisal's genuine preoccupation with the question of Palestine and the American-Zionist connection, in the early days these issues were relatively minor. Immediately, the friction was over money and profit, and the conditions governing access to and disposal of the precious oil. As Aramco built up its production after the Second World War, Ibn

Saud, who in 1931 had been unable to pay what it owed the Soviet Union for imported oil, experienced an explosion of wealth of which he could never have dreamed. Revenue paid by Aramco to the Government increased from about $10 million in 1946 to over $50 million in 1950. But it was not enough to satisfy the aspirations of an embryonic government or – more relevantly – the tastes of a greedy, extravagant royal family which proliferated by the year.

Ibn Saud, the royal family and their advisers, as well as the merchant community catering for the extravagance, seemed to assume that the exponential increase in income enjoyed in the first three years would continue indefinitely. Production had leapt from 164,000 barrels a day in 1946 to 476,000 barrels a day in 1949. In that year the market slackened and the rate of increase slowed, but the Saudi hunger for money did not.

The first taste of oil wealth merely whetted the royal family's appetite and greed, which fed upon itself. No proper accounting system existed nor would one have been comprehensible or even acceptable to Ibn Saud who, in his dotage, did not appreciate and was not told of the seriousness of the problem. Accustomed to the traditional, hand-to-mouth means of financing the needs of the Kingdom, Abdullah Suleiman had neither the will nor the ability to assert any control. Any attempt to do so would have been vigorously resisted by the princes, who looked upon God, in the form of Aramco, to provide. Frequently drunk, the Finance Minister paid only the most perfunctory attention to his responsibilities. By the summer the pay of members of the Army and Police Force were no less than four months in arrears. Units stationed at the military headquarters were demoralized and restive. The currency was on the verge of insolvency.

In this situation Ibn Saud's persistent demands for more revenue took on a greater, more acerbic urgency in keeping with the mounting irritability of his last years. Indirectly they were intensified – and their satisfaction assisted – by a change in the ownership of Aramco. With the encouragement and approval of the United States, which wished for a wider ownership in Aramco, Exxon (then named Standard Oil of New Jersey) and Mobil (still called Socony at this time) started negotiations in May 1946 with Socal and Texaco, who were acting as one company – Caltex – in the Middle East. They were amenable to an approach by their two bigger rivals mainly because of the opportunity the merger could present of opening up new markets for Caltex and expediting the development of its concession. Negotiations were concluded in March 1947 with the result that Exxon obtained a 30 per cent share and Mobil 10 per cent of the operation, leaving the balance shared equally between the two Caltex partners.

The arrangement was an apparent breach of the Red Line Agreement in which the shareholders of the Iraq Petroleum Company had bound themselves not to acquire concessions independently within the boundaries of the old Turkish Empire without the consent of the other signatories to the Agreement. It was immediately challenged by the other partners of the Iraqi group. The objections of British Petroleum (then still the Anglo-Iranian Oil Company) and Shell were bought off by Exxon and Mobil's commitments to purchase from them considerable volumes of crude under long-term contract. Having sought British legal counsel, the Americans argued that Compagnie Française des Petroles (C.F.P.) and Calouste Gulbenkian had no right of objection because their shares in I.P.C. had been declared forfeit during the war. They had been pronounced 'enemies' because they had operated in France under Nazi occupation. C.F.P. and Gulbenkian embarked on a costly legal suit that was called off at the eleventh hour in November 1948 after Exxon had undertaken to finance a large expansion of the Iraq Petroleum Company's facilities and, in addition, Gulbenkian received a generous grant of oil. The merger was accepted by all and the Red Line Agreement became a dead letter.

A row erupted at the first board meeting of the reconstituted Aramco on 27 January 1947. The new entrants Exxon and Mobil lost no time in insisting that Aramco should sell its oil at the general world rate of nearly $1.50 a barrel rather than at a price of just over $1 as had been the practice of the Caltex partners. Their main concern had been increasing sales and their market share. Acting as one, the two companies argued that selling Aramco oil at the prices charged by others would only restrict Saudi output, slow the development of its resources and damage relations with Ibn Saud. The prolonged battle was won by Exxon, strengthening its leading market position in the process, when in June 1948 Caltex agreed to raise the price of Saudi oil to $1.43. The profits made by Aramco and dividends paid to the four shareholders went up accordingly, but no extra benefit accrued to the King's Exchequer and the realm. Ibn Saud made known his displeasure. His growing irritation was not soothed by the conclusion in 1948 of a supplementary accord altering the terms of the original concession. Under it, the company raised its royalty payments by another 5 cents per barrel. In addition it relinquished its on-shore part of the concession in the Neutral Zone shared with Kuwait, agreeing also to pay $2 million in annual rental for the off-shore area until oil was discovered, and thereafter $2 million in annual royalties (Safaniyah, the world's largest off-shore oilfield, was discovered by Aramco in 1951).

Aramco had further cause for alarm at the turn of the year. Aminoil, the group of independent U.S. oil companies whose biggest shareholder was Phillips Petroleum, reached agreement with Kuwait on exploiting a concession covering its half of the Neutral Zone. Included in the terms were a down payment of £7.5 million and better royalties than Saudi Arabia received from Aramco. Ibn Saud was in favour of the same company taking his share of the territory, making it a joint concession. But the American tycoon J. P. Getty trumped Aminoil with a down payment of £9.6 million and royalties of just over 50 cents per barrel. And so the Saudi part of the Neutral Zone went to his company, Pacific Western, in February 1949. With the consent of the two host Governments, Getty's company and Aminoil agreed on a joint exploration effort. By then Ibn Saud knew the details of Venezuela's agreement with its concessionaires, concluded in November 1948, which gave it a 50 per cent share of the profits.

Reacting with alarm, Aramco raised its royalty payments from 20 cents a barrel to 30 cents in an attempt to ease the pressure. But Ibn Saud was not satisfied. Aramco's relations with the King and his advisers became fraught, as one incident early in the summer of 1950 showed. Aramco had been seeking repayment of the $6 million advance to Saudi Arabia to finance the loan originally promised to Syria. Far from trying to mollify the oil company, an Egyptian lawyer acting for the Ministry of Finance took the offensive, telling an Aramco representative that the Saudi Government, as a sovereign government, was not bound by the terms of the concession agreement. This was not the point at issue but the truculence of the adviser chilled the spines of Aramco's chiefs. They remained outwardly unflinching but were deeply apprehensive. 'Incidents of harassment and sabotage against Aramco brought to the forefront the question which was always in the background: how much could the companies stall in meeting the Government's revenue demands without losing the concession?'[5]

Unknown to the royal court in Riyadh, an atmosphere of crisis was building up in Aramco's headquarters in New York and the State Department in Washington which had become alarmed at the threat to their strategic interests. A meeting of company executives and Administration officials was held on 11 September 1950 under the chairmanship of George McGhee, Assistant Secretary of State for Near East Affairs. One course of action considered to placate the King was for Aramco to further relinquish areas of its concession, allowing the Saudi Government to auction them off to other companies, of which there were many desperately anxious for a foothold in the Kingdom. An argument against this possible solution was the implications that it would have for the Aramco companies elsewhere in the region. A bigger constraint was the U.S. Administration's concern that oil prices should be held down to assist the recovery of the war-devastated economies of West Europe and Japan. The conference then examined the more drastic option of conceding a 50 per cent tax on profits. Inevitably, the companies would try to maintain their margins, which would be pared by half, with the result that consumers would have to pay a higher price for Saudi oil.

Another meeting presided over by McGhee took place on 2 November 1950. By then the U.S. Government and Aramco were even more fearful about the future of the invaluable concession. The consensus was that a 50:50 arrangement on the Venezuelan model might satisfy the Saudis. Clearly, Ibn Saud was now aiming for nothing less, and on 4 November 1950 he issued a decree imposing a 20 per cent tax. That in itself was technically a unilateral breach of the concession agreement, but Aramco chose to disregard it as they desperately sought a final solution proposed by themselves. It was decided to offer a 50:50 profit split, though not before the Inland Revenue Service had been persuaded to relieve them of double taxation by granting credits against American obligation in respect of payments made to the Kingdom (as it granted to American companies operating in Venezuela).

The agreement between Aramco and Saudi Arabia was finally signed in Riyadh on 30 December. Ibn Saud had won a notable victory – perhaps his greatest off the battlefield. Within a year the practice whereby the host took 50 per cent of the profits had become the common pattern in the Gulf. Royalties were set at 12 per cent but credited against tax. The system of tax credits was granted to other American companies operating concessions abroad throughout the world.

The immediate benefit felt by Ibn Saud was that his revenues from Aramco rose from $56.7 million in 1950 to $110 million in 1951. This development in no way ensured smooth relations. Within months the Ministry of Finance was wrangling over financial responsibility for the protection of Tapline and the company's contractual obligations in this respect. At the end of December, Abdullah Suleiman, accusing Aramco of a variety of malpractices, launched into a bitter dispute with the consortium over the means of calculating the sterling portion of Aramco's payments, which in 1950 had amounted to 25 per cent of the total but had gone up to 37 per cent in the first ten months of 1951. The 1950 agreement included a formula whereby sterling was valued against gold and the dollar at the International Monetary Fund's rates. Saudi ire was caused by the fact that this method seemed to put a considerable, unjustified premium on sterling as compared with transactions on the local free market. In the face of Abdullah Suleiman's wrath, which seems to have been partly feigned and no doubt was fired by alcohol, Aramco gave way and agreed to amend the formula in favour of the Government. Bickering continued on a number of other points as Suleiman and his colleagues sought to ensure that its Hashemite rival was not receiving better terms from the Iraq Petroleum Company.

A serious cause for complaint was the practice of setting 'posted prices' – the official rates asked from third-party customers, as opposed to those at which producing ventures sold to their owners. Iraq's 50 per cent income tax had always been based on this reference. When Aramco set a posted price of $1.75 a barrel compared with the $1.43 charged to the shareholders, Ibn Saud naturally wanted a share of the difference. His determination on this score hardened when Aramco increased its posted price in 1953 to $1.93. It increased its 'off-take' price at which the four shareholders bought from Aramco to give the Saudi Government some compensation, but denied it a half share of the difference between production costs and the posted price. Such was Ibn Saud's frustration that, despite the mutual antipathy between the two royal regimes, an Iraqi oil delegation was welcomed to Riyadh in June 1953. But not until 1955 did Aramco concede the improved terms sought by Saudi Arabia. Meanwhile, though the implications were not fully appreciated at the time by the ailing King, Getty and Aminoil had discovered the commercially viable oilfield at Wafrah in the Neutral Zone in 1953. Production began the following year. More important, Aramco found in 1953 the Ghawar deposits, which proved the biggest field ever discovered and twenty-seven years later still contained more reserves than all those of the United States.

Whilst constantly pressing Aramco for more money, Ibn Saud and the royal court turned to it for a multitude of services. For its part the company, ever anxious to improve the climate of relations with them, was eager to gratify their wishes. Perhaps the most extravagant of them, a whim of the King's that cost $50 million, was a railway. During a visit to Egypt in 1946 to cement his new understanding with King Farouk Ibn Saud had taken a journey by rail – the first he had ever made. Aramco was ordered to produce something similar from Dammam to Riyadh. The oil company hired the American company Bechtel to build it and work was completed in 1950 with a golden spike driven in before the King in his wheelchair at the Riyadh railway terminal. The cost, as usual was offset against oil revenues. The deductions to defray the debt caused friction until a system was agreed early in 1952. Aramco itself was able to make some use of this futile venture for freight, though the Railways Authority tried to impose an extra high tariff on the company.

Bechtel, which had laid Tapline for Aramco and built the Ras Tanura refinery, seemed as well entrenched and as permanent a fixture as the oil company in the late 1940s. It was the principal public works contractor for the Government and was involved in building the Jeddah and Dammam harbours, the airports there and at Riyadh, the electrification schemes for the two cities, the resurfacing of the Mecca-Medina road, and other building programmes. Bechtel had negotiated what looked to be a very cosy arrangement whereby it was paid its costs plus a guaranteed profit. It was much resented and was referred to as *maktal*, 'the place of killing'. Towards the end of 1951, however, the company decided that its involvement was more trouble than it was worth. It was owed $1.8 million by the Saudi Exchequer when the arrangement was terminated. The U.S. State Department was said to have regretted Bechtel's decision. The company continued, however, to carry out work on behalf of Aramco. It was replaced by another U.S. concern, Michael Baker Inc., but its relationship with the Saudi Government was only short lived.

As early as 1952 an extension of the railway to Jeddah was considered. But far more important for connecting the country's far-flung population centres and binding the unwieldy Kingdom together was the build-up of Saudi Arabian Airlines, which had come into being in 1945 with a fleet of three DC-3s, including the one given by Roosevelt to Ibn Saud. From the start T.W.A. provided management and technical services. In 1952, the airline bought six secondhand

Convair Skymasters as well as some Bristol airfreighters from Britain. Coca-Cola, an important symbol of the American way of life which was not then blacklisted by the Arabs, had a well established bottling plant (Pepsi-Cola's entry into the market was prevented in 1951 by the shortage of water in Jeddah).

Commercially, however, the Kingdom was not an American preserve by any means and within the ruling hierarchy there were evidently some influential voices anxious that it should not become one. Even though Ibn Saud had a hearty contempt for the French they were chosen to set up in 1951 two munitions factories at Al Kharj, while in 1950 Philby had been successful in winning for two British companies, Braithwaite and Thomas Ward, the contract for the Jeddah-Medina road. There was a pro-German lobby which was believed to include Abdullah Suleiman, but it had to contend with official anger over reparations being paid to Israel. Nevertheless, Siemens in 1952 was awarded the first big telecommunications contract. Then a German combine called Govenco replaced Michael Baker Inc. in 1954.

The U.S. State Department, with the great strategic asset in the forefront of its mind, wanted to increase American engagement, both to establish a binding relationship and also to modernize the archaic Kingdom. An Export-Import Bank (Eximbank) loan of $15 million had been extended as early as 1946 and a further $10 million in 1950. In the latter year the Federal Reserve Bank of New York agreed to help with the stabilization of the currency by making available dollars placed by Saudi Arabia with it as collateral for lend-lease silver previously made available to the Kingdom. The U.S. Administration enthusiastically complied with a Saudi request for advice on the revision of the Kingdom's tariff system by sending customs experts and accountants, responsible directly to the King and royal court, to assist with the audit of the realm's finances. At the United States' initiative rather than Saudi Arabia's, an agreement on technical co-operation under a Point Four aid programme was signed. Under it an American mission was quickly dispatched in 1951. One of them, a pilot, was shot at while making an aerial survey over Riyadh and subsequently jailed. Undeterred, Washington sent a mission to study ways of instituting a more rational monetary system.

Towards the end of 1950 the U.S. restarted negotiations on the use of Dhahran as a military base and, in 1951, gained a five-year lease in spite of Saudi apprehension about foreign interference and intrusion.

The Kingdom obtained in return an American commitment to supply weapons and not only to train the Saudi Army but also to build up a small Air Force and Navy. The British military mission that had come in 1947 was asked to leave, though the Vampire aircraft ordered on its recommendation were delivered in 1952. Saudi Arabia's relationship with the U.S. was more tangled than ever. Its special nature was given singular recognition when Secretary of State John Foster Dulles visited Riyadh in 1953.

12 The End of an Era

1949 – 53

You may my glories and my state depose,
But not my griefs; still am I King of those.
　　　SHAKESPEARE, *Richard II*

AS early as 1946, during his visit to King Farouk of Egypt, Ibn Saud had shown distinct signs of fatigue and ageing. From 1947 onwards he presented a sad spectacle of decline. The crippling pain in his knees became so acute that he put aside the rubber cane with which he had walked with increasing difficulty for several years. Instead he took to a wheelchair – a larger replica of the one given him by President Roosevelt aboard the U.S.S. *Quincy*. His eyesight grew worse and by 1949 even wearing his thick-lensed glasses he was half blind. Tiring easily, he would spend progressively more of the day dozing and less attending to the affairs of state in his declining years. Most distressing of all to Ibn Saud was the loss of his potency. The aphrodisiacs supplied through the good offices of the U.S. Embassy were not sufficient to restore his sexual powers. Increasingly cantankerous and unapproachable, he began to withdraw into a cocoon of solitude. Despite the advancing age of his older sons, he was as imperious as ever, not allowing them to sit down in his presence unless he had bidden them to do so. For company he favoured his sixth-born son Muhammad, who would regale him with stories of his hunting, a pastime that Ibn Saud could no longer enjoy.

In those last years of his reign, as oil revenues flowed in and almost as quickly evaporated, an atmosphere of disintegration and uncertainty prevailed. The whole edifice of government, such as it was, centred upon Ibn Saud. His rule was personal and control over his realm lay

very much in the dominance of his presence. The system of government
had evolved in an *ad hoc* manner. It was a matter of people and their
relationships to him. The 'Royal Cabinet' was an informal body which
included Abdullah Suleiman, Saud bin Jiluwi, and Ibn Saud's advisers,
of whom Yusuf Yassin and Khaled Ghargani were the most important
in his later days. The number of Ministers could be counted on the
fingers of one hand. Ibn Saud had appointed Feisal as Foreign Minister
in 1930 and Abdullah Suleiman Finance Minister in 1932. Subsequently
he had established a portfolio for Defence in 1944 and the Interior in
1951. Mansour, the King's twelfth-born son and one of his favourites,
was the first Minister of Defence and Aviation. He died in Paris of
kidney trouble in 1951 and was replaced by his full brother Mishaal, Ibn
Saud's eighteenth-born son. Abdullah al Feisal, eldest son of the
Foreign Minister and the first second-generation prince to hold office,
became Minister of the Interior, with responsibility for Health,
amongst other things. But there was still no apparatus for government,
let alone for the disbursement of petroleum revenue for the benefit of
the King's subjects.

In the spring of 1953, three more departments of state were
established and placed under the responsibility of three of the old King's
more promising sons. Fahd was appointed to take over the newly
created Ministry of Education. Born in 1921, he was Ibn Saud's
eleventh son. Sultan, Fahd's eldest full brother, born in 1924, at this
time was given charge of the Ministry of Agriculture. He proved to be
relatively hard-working and pugnacious, but lacked Fahd's subtlety.
Talal, twenty-third born son, who was then only in his early twenties,
was entrusted with the Ministry of Communications. He had vied with
Mansour, until the latter's death, as Ibn Saud's favourite by virtue of
the great beauty, grace and intelligence of his mother Munaiyer, who
was reputed to be a concubine of Armenian origin. Talal was born in
1931 and named after another son by the same wife who had died earlier
that year. A spoilt child, Talal had even at that point been exposed to
modern, external influences. A Ministry of National Economy was also
set up but made part of Abdullah Suleiman's Ministry of Finance.
There was still no formal framework for the Ministers to meet
collectively, however, and the broadening of the bureaucratic base did
not immediately result in greater ministerial harmony. No sooner had
Talal taken office than he was arguing with Mishaal as to who was
entitled to run Saudi Arabian Airlines and, in consequence, have the
power to commandeer aircraft for royal or personal convenience.

Nowhere could the inadequacy of Ibn Saud's rule be more damaging than in finance. For him there would have been no contradiction between trusting Abdullah Suleiman's loyalty while knowing how much he had profited from his long tenure of office. Nor was there any trace of conscience on the part of the Finance Minister, who was heard openly to boast of being the richest man in the Kingdom. His unique position owed much to the skill and resource with which he had been able to raise, husband and juggle funds as early as the conquest of the Hejaz. In an article deploring the general decay and decadence in Saudi Arabia Philby could describe the keeper of Ibn Saud's Exchequer as 'the ablest administrative brain in the country'.[1] But that organ was not the instrument it once had been, not the least because of the influence and ravages of alcohol. In 1949, when the Kingdom found itself almost insolvent, Abdullah Suleiman was in full decline, a 'worn out old toper', according to one contemporary British diplomat. Yet his ability to assert any control over the purse strings would have been severely limited by the King's failing grasp of the finer points of administration and an implicit collective agreement that he should not be told how serious the financial position was. Moreover, the country's revenue, of which four-fifths came from oil royalties, was looked upon as the royal family's. Despite his high standing, there would have been little that Abdullah Suleiman could do to curb princely demands upon the Exchequer. The same diplomat quoted above recalled how on one occasion when Muhammad ran out of money he sent a servant to fetch more from the Finance Ministry at gun point. Nor did Suleiman's chief assistant, Muhammad Sorour Sabhan, a man of slave origins, have the disposition or character to help him assert a measure of fiscal discipline. In this chaotic situation Hendrick Entrop, the manager of the Netherlands Trading Society, was the other most important person in Government circles – an unpaid and unofficial adviser. His organization was the Kingdom's main banker, often a substantial creditor, and handled oil royalties and the mint, which would arrive in sacks, to be counted and then stored under the stairs in the Dutch concern's premises.

Control over spending was improved by Najib Salha, a Lebanese, who for a while effectively displaced Muhammad Sorour Sabhan as Abdullah Suleiman's right-hand man. At the turn of the year the arrears of pay for the Army and Police had been cut from four months to two months. In the spring of 1951 Najib Salha drew up some proposals for budgetary reform. The King showed little interest or enthusiasm, rather the contrary, and referred the matter to Saud, who rejected most of them. The fact was that Salha's modest ideas about imposing any kind of a budgetary system did not endear him either to senior princes, including the heir presumptive, or the merchant community of the Hejaz who had profited hugely from the initial spending spree. Both categories invested most of their gains abroad in Egypt or Lebanon.

The money owed by the Exchequer to businessmen servicing royal extravagance began to mount in 1950. Already the practice had started whereby debts were sold off at discount, normally to those who exercised more influence with Abdullah Suleiman. It was the smaller merchants who theoretically lost out, but they generally covered themselves by setting a price for their services that took account of this circuitous method of obtaining payment in reasonable time.

Notwithstanding higher oil receipts and monthly instalments resulting from the agreement with Aramco at the end of 1950, the Saudi Treasury was desperately short of cash. In March 1951, requiring $6 million to pay for the French munitions factories, Abdullah Suleiman approached the Banque de l'Indochine, which had lately opened in Jeddah. It refused to advance the sum required. Enraged, Suleiman spread the rumour that the French concern was going to have its permission to operate withdrawn, causing – as he intended – a run on the bank. He was baled out by Aramco, which advanced him $2.5 million against future taxes, and a $3 million loan from the Netherlands Trading Society, in addition to $6 million or so of revolving credit which the Exchequer had with the Dutch banking house.

In the early spring oil production slumped from the peak of 713,000 barrels a day in February to only 400,000 barrels a day because of marketing difficulties. Nevertheless, Ibn Saud agreed that the money set aside for the princes' individual royal stipends should be raised, though regularized under a new system of fixed personal budgets. About this time Abdullah Suleiman reverted to his custom of the penurious early 1930s of personally authorizing all the Government's disbursements. It was a practice guaranteed to delay payment, especially since his working day could be curtailed to as little as an hour and a half because of the temptations of the bottle. The merchant community was in a state of agitation and worry. One trader claimed to be owed the equivalent of £7 million. But because he was illiterate and kept all his accounts in his head he could not press his claim strongly and was totally at the mercy of the Finance Ministry.

While Muhammad Sorour Sabhan and Najib Salha engaged in internecine war, Abdullah Suleiman became so desperate that he cabled Dr Hjalmar Schacht, who had been Governor of the Reichsbank under Hitler and was now an independent consultant,[2] and Dr Zaki Abdul Motaal of the National Bank of Egypt, for assistance. Schacht was engaged in Indonesia and declined to offer his services. Motaal responded positively, but having had a look at the fiscal mess returned to Cairo after a few days.

Although most of Najib Salha's budgetary proposals were rejected, he did win approval for an Income Tax Law that was decreed by Ibn Saud early in 1950. It laid down reasonable rates for salary earners, self-employed businessmen and companies. From the start, members of the royal family and the *ulema* were exempted. Other Saudis immediately objected, pleading on religious grounds that it was incumbent upon them as Muslims only to pay *zakat* (alms tax) to charity. The wrangle went on for the best part of the year before the Ministry of Finance agreed that Saudi merchants should be completely exempt from the new income tax provisions. In addition to their corporate liability, foreign companies were made to meet their Saudi employees' tax obligations other than *zakat*.

The law contributed nothing to the budget for the fiscal year running from June 1950 to May 1951, which set spending at 490 million riyals (the equivalent of $125 million). The outline published was a meaningless gesture towards modernity that gave no real indication of priorities or likely disbursements. It was also largely notional, but revealed the unabashed intention of devoting no less than 313 million riyals out of the total 490 million riyals, or nearly two-thirds, to 'State Palaces, Princes and Royal Establishments'. The main headings were Defence with 89 million riyals (including 48 million riyals for the National Guard and tribal subsidies), General Debt (the servicing of it) at 70 million riyals, Development and Constructional Schemes at 91 million riyals, and Miscellaneous at 26 million riyals. The Government's accumulated borrowing continued to grow.

The budget for 1952–3 gave a far more detailed breakdown.'Riyadh Affairs' – meaning Royal Palaces, Princes and Royal Establishments and this time also tribal subsidies – was allocated 152 million riyals of the total expenditure envisaged, or just under a quarter. Defence was to receive almost as much. Nearly one-sixth, 100 million riyals, was set aside for debt repayment. One positive aspect of the budget was the 33 million riyals allocated to the Health Department, for which Abdullah al Feisal was responsible.

Later that summer, Ibn Saud felt confident enough about the Kingdom's financial security to issue a decree to abolish the pilgrims' dues. In making the announcement it was revealed that about $55 per head had been levied on the faithful making the *Hajj* – a large amount for all but the affluent. The loss to the King's Exchequer would be in the region of $5 – $6 million over a full year. Pilgrims still had to pay nearly $20 charged for the upkeep of the Holy Places and also the quarantine tax of almost $15.

In the course of 1953 the Kingdom's financial situation worsened again in the absence of the restraining influence of Najib Salha. He had departed in December 1952, disappointed at not having been appointed Abdullah Suleiman's deputy. He returned to Saudi Arabia in July 1953 and was made Assistant Deputy Minister of Finance, but left two weeks later because he had not been given effective control of the economy. At that point he estimated the Kingdom's debts to be something between $150 million and $200 million.

Saudi Arabia's monetary system may have been archaic, but at least the country had the hardest currency in the world, in contrast to the chaos and disorder of its finances. From 1935, when the first Saudi riyals, whose value was based on their metallic content, were minted, it had been on the silver standard. The only fiduciary element in the indigenous currency had been the base metal coins called quirshes, twenty-two of them to the silver riyal, denominated in ones, halves and quarters. Supplementing them were British gold sovereigns, circulating predominantly in the Hejaz and used mainly for the bigger transactions. Indian rupees and, to a lesser extent, Egyptian pounds were also a means of payment, particularly in the coastal areas and around the time of the *Hajj*. The Kingdom was on a strange bimetallic standard. Arabians, not the least Ibn Saud who opposed the idea of paper currency, had a strong preference for gold and silver coins with an intrinsic value. So much did they respect the concept, that 'King's Heads' (George V) carried a premium over 'Queen's Heads' (Victoria) on the correct assumption that the former were less eroded by the passage of time.

Up until 1950 about 150 million silver Saudi riyals had been minted and were in circulation. In the following three years another 127 million had to be minted to cater for the rapid growth in economic activity. Of one order for 25 million made at the end of 1950, no less than 7 million, it was noted, went straight to the royal family in Riyadh. The versatile and amenable Netherlands Trading Society acted as the Government's fiscal agent up until 1952, while the foreign exchange dealers Al Kaki and Salem bin Mahfooz Company helped with local payments and the transfer of funds within the country. (In 1952, the firm founded a fully-fledged bank that was to become the National Commercial Bank, the biggest in the Kingdom, thirty years later.)

In 1950 the King approved a proposal to purchase Saudi golden coins. The contract was given to the Banque de l'Indochine. One million were minted in France and arrived in the Kingdom early in 1952. An order for another one million was placed in the summer of that year. The system worked because the state's oil receipts accounted for a large and ever-growing proportion of total economic activity. Yet, despite the soundness of the currency, it was a confused and haphazard one. Exchange rates fluctuated wildly. From 1945 to 1953 the rate for the Saudi riyal varied from 89 to 40.75 against the British sovereign, and against the dollar from about 3.2 to 5. At the centre of these big swings was the relationship between the British sovereign and the dollar where the range over the eight years was in the region of $10 to $20, and the value of silver on the world market. As its price rose in 1949–50, anything from 2 to 5 million riyals were reckoned, conservatively perhaps, by Arthur N. Young to have been smuggled out of the country monthly.[3] A great deal more were probably hoarded.

The banks in Jeddah and Dammam were concerned almost wholly with trade financing and the attendant, potentially very profitable, problems of foreign exchange. Apart from the Netherlands Trading Society, the Banque de l'Indochine, and from 1952 Al Kaki and Salem bin Mahfooz, the British trading company, Gellatly Hankey, had also engaged in banking activities until these were taken over by the British Bank of the Middle East. Commerce with the outside world was immensely complicated. Because of the exchange rate fluctuations, Abdullah Suleiman could not have budgeted with any accuracy at this time even if he had been capable of it and royal whims had made it possible.

Arthur N. Young, the expert provided by the United States under the Point Four Agreement, was a man of supreme tact, patience and persuasiveness. He arrived in the Kingdom in 1951. The brief of his mission included fiscal administration. Whilst his efforts in this sphere may have been reflected in the presentation of the budget of 1952–3, he achieved far more dramatic results on the monetary front. Ibn Saud approved the establishment of the Saudi Arabian Monetary Agency (S.A.M.A.) in May 1952 in accordance with Young's recommendations. Soon afterwards another American, George Blowers, formerly of the State Bank of Ethiopia, was appointed as its first Governor and proved to be as politically adept as his fellow American. After Young's departure early in the year on completion of his mission, exchange rates began to fluctuate rampantly again. One of Blowers' first actions on his arrival was to order another 20 million riyals to meet the requirements of the Government, pilgrims and speculators. S.A.M.A. opened its doors early in October with powers to handle what was termed the Government's 'Royal Account'. This gave it initial control of most of the country's dollar and part of its sterling earnings, in addition to the issue of the metallic currency. S.A.M.A.'s vault was about 70 foot by 70 foot square and 8 feet high, with capacity to store 30 million silver riyals and several million sovereigns. Blowers recognized that the people's deeply rooted adherence to money with intrinsic value must delay the introduction of paper currency, which was, anyway, undesirable from the point of view of monetary stability. Thus, on the basis of the sovereigns ordered from Paris, which were put into circulation in tranches of 200,000 a month, a rate of 40 riyals to a sovereign was set and skilfully maintained by S.A.M.A.

Policy aimed at depreciating silver against the dollar and sterling worked. Then in the summer of the following year Saudi Arabia's first paper scrip was subtly introduced. S.A.M.A. announced the issue of Pilgrims' Receipts that could be exchanged for Saudi silver riyals or foreign currency. Their immediate purpose was to relieve, literally, the pilgrims of the burden of carrying quantities of heavy coins. It was evidently meant to be an educative process. There was no certainty that these quasi-travellers' cheques would continue in circulation – but they did, serving for the next eight years as the Kingdom's first bank notes. Ibn Saud was probably too old and tired to be aware of the experiment, or – if he was – to care.

Despite the blinkered profligacy of the royal family, financial mismanagement, and the administrative obstacles placed in the way of foreign contractors – not the least payment delays – there was haphazard progress towards building what the country needed. Palaces were the most conspicuous form of expenditure, draining the resources of what in most places elsewhere would be regarded as the public treasury but in Saudi Arabia remained very much a privy purse. Nevertheless work, for the most part financed by the Eximbank loans, proceeded on ports, roads, water supplies, electrification, drainage, and telecommunications. Education remained completely neglected and left in the hands of men of religion – enlightened by the Koran but obscurantist in many other respects. But the need to supply health services was recognized, which owed much to the efforts of Abdullah al Feisal.

Inevitably, the concentration of the unco-ordinated attempts at development were the main urban centres. For the bedouin, who probably constituted more than one-third of a population hardly in excess of three million souls, life was little changed from what it had been a century before – unless they happened to roam in the vicinity of the oil fields and find employment. In brash contrast to the fumbling and bemused efforts of a traditional Arabian regime and, for the most part, an isolated society, was the rapid expansion of the oil industry. This dynamic force, galvanized by the American work ethic, the profit motive and the wonders of modern technology, gave big opportunities for established merchants to enrich themselves and others entering business for the first time to make fortunes. In the late 1940s, Suleiman Oleyan, who thirty years later was one of Saudi Arabia's billionaires, was earning 300 riyals a month driving a truck for Aramco. The oil industry also spawned a proletariat exposed to disruptive and radical influences in the midst of a theocratic and medieval environment.

Aramco corporately and its American employees individually had continued to exude anti-colonialist sentiment and enlightened benevolence. Torn between the financial interests of the shareholders and the exacting demands of Ibn Saud, the management sought to ease the friction by satisfying royal requirements and whims. Saudi Arabian Government Services, an Aramco off-shoot, bought food for the palace and even provided chefs. They purchased and looked after vehicles on behalf of the royal family. Aramco administered the Al Kharj experimental farms from which most of the food produced went to the royal palaces. Left to itself the company might have included these expenses as part of overall production costs, but in deference to the board of directors in New York they were deducted from royalties.

The rapid build-up of operations was such as to strain the management of any company. As far as human welfare was concerned, the immediate priority was to ensure that expatriate American personnel were provided with home comforts comparable with those enjoyed back home in Texas or California. Somehow the company neglected to pay equal attention to or observe the mood of their growing labour force.

Aramco had entered an under-developed and thinly populated backwater. From 1946, when the minimum wage was three riyals a day, it had started to provide medical services and housing. The assumption seemed to be that it was enough to provide remuneration, employment and living conditions for locally recruited labour or immigrant Arabs, including Palestinian refugees, superior to anything enjoyed by others in the area. Thus, it came as a shock when a group of Saudi and other employees presented a petition in May 1953 demanding higher wages, better housing, and more privileges.

The contrast between the affluence of the green township on the Jebel Dhahran above and the dormitory accommodation for the workers below was a stark one and a source of resentment. So, too, was the time spent in transportation to ever more remote sites. Leaders of the protest formed a committee. Aramco was not allowed to recognize or deal with labour unions, but held informal consultations. The Saudi Government stepped into the vacuum by setting up a Commission of Inquiry. After talks with the Aramco management it recommended improved conditions. Members of the strike committee, however, refused to answer its questions and were thrown into Dhahran's insalubrious prison early in October.

The eruption of protest the following day appeared to be spontaneous. Demonstrators attacked a police station. A U.S. Air Force bus and company vehicles were stoned. It soon became clear that the anger was directed less against the American presence in Hasa than the regime. Five thousand troops were drafted into the area, primarily to guard the installations. But both the soldiers and the police wanted to avoid a confrontation. By mid-October, 17,000 out of the 19,000 work force were on strike. However, all but a fraction of the strikers returned to the company's offices in the third week of the strike to collect their pay, and this signalled a general return to work. Twelve leaders of the protest were exiled to their native towns and villages. Aramco was thankful that Saud bin Jiluwi, the autocratic and brutal Governor of Hasa, was out of the country. Repression on his orders could have led to grave, long-term repercussions.

There was no evidence of external direction, organized cells or ideological motivation. But Aramco itself felt duly chastened. It was a frightening experience so soon after the explosion of popular and nationalist sentiment in neighbouring Iran had shown how vulnerable an oil concession could be.

After the disturbing eruption Aramco undertook a study of the working conditions, living standards and general aspirations of their Saudi employees, who then numbered over 10,000. It was found that their average salary in 1952 was 414 riyals a month, then the equivalent of $110. Nearly 90 per cent of them owned their own homes and about 600 possessed a motorcar. Though over half of them were under twenty years old, no less than two-thirds of them were married and had two children. But 45 per cent of those with families lived in the company dormitories and had to travel back to their villages at the weekend. One important factor behind the strike may have been the fact that Aramco was laying off workers. With most of the basic work on infrastructure and plant completed, Aramco was running down its labour force from a maximum of 24,000 reached in 1952 to nearly half that level nine years later. In the list of preoccupations concerning the company's indigenous work force, the survey identified education and training, job security and opportunity for advancement well before salary levels, which were numbered seventh in priority.

Change in the face of Western influences and technological innovation was not limited to Hasa, or relatively cosmopolitan Jeddah which, chafing under Wahhabite domination, had long been exposed to the outside world. In the capital, Riyadh, wealth was undermining those fundamentalist values and stern observances that had inspired the formation of the Kingdom and continued to be the main rationale for it. Increasingly, the religious façade of the Kingdom seemed a sham maintained by the dynasty to excuse itself from attempting even modest reforms and to preserve for itself exclusive privileges. There was evidence of decadence and double standards everywhere. Avarice and graft appeared to be the order of the day. It seemed to some cynics that if the hands of all thieves, defined in the broader sense, were to be cut off there would be very few Saudi hands left in the Kingdom.

In 1951 Abdullah al Feisal was largely responsible for the legalizing of football. It had been banned previously, not only because the *ulema* disapproved of Saudi youths besporting themselves with their thighs naked, but because soccer clubs were suspected of being a cover for subversive political activity. Film projectors proliferated in palaces and houses. Talal had an open-air cinema in the grounds of his opulent new home which was very audible to passers-by. Even more seriously, alcohol was eroding traditional values. For the Wahhabite more than any other sect of Islam, the consumption of alcohol is an abomination and a manifestation of 'Satan's handiwork'. Ibn Saud had been magnanimously liberal by his own strict standards in giving foreigners a special dispensation to import liquor in the 1930s. Yet the forbidden juices that became available tempted many Saudis other than the ageing Abdullah Suleiman. Apart from professional bootleggers, there were more than enough foreigners, like one notorious engineer employed by the U.S. Embassy in Jeddah, willing to cater profitably for their tastes.

Of those last years of Ibn Saud's life bitter St John Philby wrote: 'It is difficult to assess how far the old King was aware of the deviations of certain members of his family. Some cases were too notorious to escape his notice and severe displeasure; and one son sentenced to life imprisonment for an unsavoury murder obtained his release only at the amnesty celebrating his father's death. Others boasted of their debauches with wine and strange women, and seemed to think that their father was unaware of their doings.'4

The son in question was Mishairi, the twenty-fourth born of Ibn Saud's sons. In that era of confused but more relaxed transition, Saudis, including princes, mixed more freely with foreigners than they did three decades later. Mishairi was a personal friend of Cyril Ousman and his wife Dorothy. A former N.C.O. in the Royal Army Service Corps, Cyril had settled in Saudi Arabia in 1929 as a freelance engineer and supervised the operation of a small steam desalination plant – known as the Condenser, or *Kindusah* in Arabic – on the Jeddah foreshore. He was engaged as the British Vice-Consul and had a flat in the old British Residence, a two-storey, rectangular building constructed round a courtyard.

The Ousmans had known Mishairi, who was born in about 1932, since he was a child, as they had other sons of Ibn Saud. One afternoon in November 1951 an incident occurred when he visited their flat. Cyril refused Mishairi a drink saying that he had had enough already. The Vice-Consul also told the prince, whose business had gone awry, that he could not possibly facilitate a favour requested of him, whereupon Mishairi drew a revolver. Dorothy pulled the gun from him. That night he left peaceably enough, but four days later returned, telling Dorothy that he had come to kill her husband. She told him not to be so 'silly' and hustled him away, slamming the door. Having locked it, she roused her husband. They barricaded the door with a wardrobe. Mishairi returned with a shotgun and blasted several shots through the door. Cyril, meanwhile, decided that he should escape to seek help and prepared to jump from a window. He was confronted by Mishairi who had anticipated his move and shot him in the stomach. Cyril fell back through the window, killed instantaneously.

Ibn Saud knew and liked Cyril. On hearing of the murder, he was quoted as saying, 'Where is Mishairi's head?' The young prince hid for two weeks until the King's anger abated. It was widely believed that he took refuge in Feisal's house. When he reappeared he was arrested. Dorothy was asked to choose, as she was entitled under *sharia* law, between demanding the death penalty or blood money. She wanted to opt for capital punishment but was dissuaded – probably by her former colleagues – and settled for a gift of £20,000 instead. When informed that this would be treated by the British Inland Revenue as one year's income and taxed, she left for South Africa.

It is not clear how significant this affair was in Ibn Saud's decision to decree total prohibition of alcohol. The fact is, however, that not until September 1952 did he impose a ban on transactions and the consumption of liquor by non-Muslim foreigners. Big employers, particularly Aramco, were apprehensive that the measure would lead to resignations and difficulties with recruitment. Any hopes of the regulation being relaxed or loosely applied were soon dashed. Cargoes of alcoholic drinks en route at the time of the ban were transhipped or returned to the consignor. As existing stocks dwindled, approaches for leniency were made to the Saudi authorities, but to no avail. The effect of the ban was to increase the price of illicitly imported alcohol and restrict its consumption to the more affluent. After some unfortunate experiments in distilling, and in recognition of human frailty, Aramco drew up and circulated amongst its non-Muslim employees a guide on how safely to manufacture intoxicating drinks at home.

From 1950 to 1951 Ibn Saud's grasp over the affairs of his realm weakened rapidly and he delegated all but supreme authority to Crown Prince Saud who, in anticipation of his father's death and the possibility of some dispute over the succession, applied himself with enthusiasm and vigour to his responsibilities. For his part, the ailing King seemed anxious to build up Saud's image and confidence. In 1950 he sent the Crown Prince to represent him at the *Hajj*. He did the same again in 1952. On the latter occasion Saud stayed on to re-organize the administration of the Hejaz, the fiefdom of Feisal who was its Viceroy, placing his own men in positions of importance. An overhaul of the system inherited from the Ottoman Empire may have been overdue. Restructuring it along the lines of Nejd, where Egyptians, Syrians and Lebanese had all made their influence felt, was not necessarily the ideal solution.

With zeal Saud sought also to assert the decree on prohibition and to invigorate the Society for the Encouragement of Good and the Prevention of Evil. He was frequently heard to complain about the decline in morality, even though his own life style was becoming noticeably more sybaritic. His efforts on behalf of Wahhabite orthodoxy scarcely endeared him to many citizens of the Hejaz, where he spent four months of the later summer and early autumn of 1952, but the province was made to feel the stamp of his authority.

Saud stayed only a few weeks in Riyadh before proceeding to Hasa, accompanied by Abdullah Suleiman. Eagerly the merchants of the region flocked to greet them. Freely accessible to the people, Saud distributed largesse and talked of new projects. He managed to convey the impression of a leader with a traditional, common touch and modern, reforming intentions. Then, in the spring of 1953, the Crown Prince cut a dash on a tour of Arab countries. In Syria alone he was said to have given away cash and gifts worth 3 million riyals. In August Ibn Saud transferred to him the title of Commander-in-Chief of the Army.

A final consolidation of Saud's position came with the King's decree of 10 October 1953 establishing a *Majlis al Wuzara* – a Council of Ministers. The Crown Prince should draft the functions, responsibilities and organization of each ministry, the decree laid down. It did not specify whether the King or Crown Prince should be vested with the role of Prime Minister but said that there should be a deputy to act in the absence of the premier. Furthermore, the Council of Ministers was to meet once a month and decisions were to be taken by a simple majority. It did not take practical shape until the spring of 1954.

Ibn Saud's decree was a very belated attempt to bequeath the juridical basis for some kind of institutional structure to replace his personal rule, but it was a skeletal and unproven one. Another and perhaps equally serious failing on the King's part was that he had not delegated more authority to his elder sons. They had not been allowed to achieve any real stature under his dominating presence. That was particularly true of Saud. He had developed a regal manner that also conveyed benevolence and solicitude, but there was very much more vanity and arrogance than real self-confidence in his posture.

In the summer of 1953 the Crown Prince's position seemed unchallenged. Clearly it was the will of Ibn Saud that he should be the next King. Faisal, the next in line, was generally acknowledged to be more able, however. His long experience as the Kingdom's envoy to the world at large had given him a knowledge and interest in international affairs that was the frequent subject of comment by foreign ministers and envoys who met him. Foreign policy was considered to be his only sphere of influence. Feisal had stood by without apparent protest as Saud reorganized the administration of the Hejaz. In the last years of his father's reign he seemed both apathetic and unambitious – but also enigmatic. The fact that he seemed reluctant to apply himself to hard work in those years was generally attributed to ill health.

The main speculation centred on Muhammad, Ibn Saud's favoured companion, who spent much of his time in the desert with the tribes. There was a dark suspicion that he might raise them in revolt in a bid to seize power.

There were doubts whether any successor could hold together the far-flung realm that was the creation of Ibn Saud's valour and personality as well as good fortune. It was feared that the Hejaz might secede, and the Kingdom's main military strength was shifted from Nejd to the Hejaz in 1952–3. Such was the general uncertainty in 1953 that even the possibility of Saud bin Jiluwi carving out an independent state in Hasa was not ruled out by anxious foreign diplomats and oil men.

Ibn Saud's issue of the decree setting the legal framework for establishing a Council of Ministers was his last act as King. Soon afterwards he suddenly sickened and was taken to Taif in the hope that the cool, clear air would revive him. Sensing his death was near, the royal family converged on his summer palace on the great escarpment overlooking the coastal plain and Jeddah. They did so partly out of respect but also in anticipation of a dispute over the succession. Saud had been declared heir apparent as long ago as 1933, but in the recent years of the King's decline and dotage no conclave of senior members of the family had confirmed the choice, nor apparently had it received any subsequent blessing from the *ulema*. Nothing could be certain, however. As it was, Ibn Saud seemed to recover about the turn of the month.

Still seeking assiduously to project himself as a noble successor, Saud felt confident enough about the King's health to travel on 8 November to Jeddah where he was received with full military honours, including a brass band. But on the following day the King suddenly suffered a relapse. The news was relayed early that morning to Saud, who hastened back to Taif to be by the bedside of his father. The Crown Prince arrived too late to see him expire. The great Ibn Saud died in the arms of Feisal.

It was a tense moment as the Crown Prince arrived at the gates of the palace. It was defused as Feisal rose to greet him and acclaim him as King. Other members of the royal family hailed him as such, although there was certainly one strong silent voice. Pushing through the throng, Feisal then took a ring from one of the dead King's fingers and presented it to Saud in a gesture of loyalty that seemed to allay all doubts about the succession. Reciprocating, the new King passed it back to Feisal, declaring that he was now Crown Prince and heir apparent. Mecca Radio broadcast the news of the King's death at noon. A perceptible sense of grief was widespread throughout the Kingdom, though it was not openly expressed. Feisal flew back to Riyadh with the body, which was loaded onto a truck and buried in an unmarked grave outside the city with the minimal ceremonies practised by the Wahhabite sect.

A decade later it would seem appropriate that Feisal should have both held the King in his arms as he died and also laid him to rest. But for the moment the new King and Crown Prince showed an equal concern about expressing the solidarity between them. As representatives of foreign states paid their condolences to Saud a few days later in Jeddah, he made a point of bringing Feisal forward to shake their hands. As they drove to prayers the following Friday, Feisal sat in the front seat of the wide limousine. Also thrust into prominence at this ceremony was Saud bin Jiluwi, observers noticed. Amongst the foreign dignitaries present was Colonel Gamal Abdul Nasser, strongman of the Egyptian regime that had overthrown King Farouk in July 1952. It was a smooth succession. Only Muhammad, one of the most powerful princes and the third eldest of Ibn Saud's surviving sons, refused to swear the oath of allegiance.

The new King was bloated by his majesty and determined to display it to the world. When a photographer from *Life* magazine arrived at

Jeddah airport without a visa, he was nevertheless ushered within two hours into the royal presence. Saud was anxious that his ascendancy should be portrayed to a readership of many millions throughout the world.

13 The Years of Ozymandias

1954 – 8

My name is Ozymandias, king of kings:
Look on my works, ye Mighty, and despair!

                                SHELLEY

THERE is a tradition among the people of Arabia that to destroy a dead man's house deliberately is to dishonour his memory. If his sons do not live in it after him, it is better to leave the house empty for time and the elements to do their work. Like so many traditions in Arabia now, this one is under constant attack as rising land values and the demand for modern amenities place disused old property at a premium; and in the old core of Riyadh, Jeddah or Hail, the wholesale destruction of traditional housing has become commonplace in recent years. But there has always been a great reluctance to knock down more princely homes. In the centre of Riyadh, empty palaces dominate some of the choicest land, losing a corner here to a new road, another there to someone's new villa.

There were, in all, ten monarchical palaces built or at least started in the 1950s – in the capital, in Jeddah, Mecca, and Medina, in the pleasant juniper-clad hills above Khamis Mushayt in Asir, and on the Red Sea coast. The most remarkable example can be seen in the centre of Riyadh's wealthy suburb of Nasariya. A pink-washed palace big enough for a sun king, built only twenty-five years ago yet already all but a vacant ruin. Persian carpets by the hundred covered its acres of marble floor. Crystal chandeliers by the score lit its giant halls and stairways. In its gardens, vast fountains played by night and day, imported cage birds trilled among orange trees, and jasmine and thousands of rose bushes made the desert bloom. Turkish tiling and

French brocade hangings, giant Spanish lanterns and Chinese vases as big as a man, Rosenthal dinner services for five hundred or a thousand people and gold and silver plate by the ton. Nearly £10 million went into its construction and sumptuous furnishing in the mid-1950s. More precious oil revenue was spent on a surrounding complex of satellite palaces, harem quarters, a mosque for the royal family, a school for its progeny, a hospital for its household, and a barracks for the bodyguard employed to protect it. Around all this was built a pink-washed wall, fifteen feet high and seven miles long, with a triumphal archway for an entrance.

The arch is now the centre of a busy traffic junction. All but a mile or two of the pink wall has disappeared. But the central palace remains, a symbol of extravagance of epic proportions. The Khozam Palace in Jeddah still stands, though its massive rooms are carpeted with dust and shattered crystal. So, too, does the beach palace on the Red Sea; the poor of Jeddah resort there on Fridays to picnic.

The Ozymandias who built these forlorn monuments was Ibn Saud's designated successor and eldest surviving son, Saud bin Abdul Aziz bin Abdul Rahman bin Feisal al Saud, who was proclaimed King of Saudi Arabia on his father's death in November 1953. Their neglected fate precisely echoes the destiny of their creator. His reign lasted eleven years, but his existence was expunged from official recognition. Saud exceeded even Ibn Saud in the number of his progeny. He fathered at least fifty-two sons and fifty-five daughters, apart from those borne by concubines out of wedlock. None held any office of state after his reign ended. Even in private conversation Saudis are reluctant to talk about him. But Saud entered his inheritance with every appearance of advantage. His father had built a country that was united and at peace in a way that Arabia had not known since the time of the Prophet. He claimed its throne at his father's express wish and with his father's immense prestige behind him after several decades of preparation for the job. He was born on the very day that Ibn Saud took Riyadh in 1902. From the time of the great Hejaz victory in 1926 he had been his father's official deputy, first as Viceroy of Nejd. He was declared Crown Prince in 1933, the year after Ibn Saud proclaimed himself King of Saudi Arabia. Many doubted the wisdom of Saud's designation as the King's successor. For Ibn Saud to choose his successor at all was a grave departure from desert custom which required that true leaders be chosen by consensus, based on the quality of their performance. Yet Ibn Saud singled out his eldest surviving son no less than twenty years before he died.

Saud lacked his father's energy and authority. He was popular with most of the bedouin and he had an aptitude for tribal politics, though this tended to be expressed in the marriage bed. But he wanted too much to be loved to be any sort of disciplinarian. As a commander he was no more than indifferent, never showing either the dash or the determination that distinguished the young Feisal or, for that matter, Muhammad, conqueror of Medina. As an administrator Saud rarely, if ever, did more than assent to whatever seemed the line of least resistance. He was plagued, too, by physical deficiencies for most of his life. Thick lenses shielded eyes that were weak from birth. Because of fallen arches he had a slow, slightly clumsy gait. In time he suffered stomach ailments – the cause of which only became apparent later. Like his father, he became obsessed with failing sexual potency which led him into the hands of quack doctors – Swiss, German, Austrian, and even a Turk who claimed to be the Waltz King of Europe. They prescribed diets, pills and injections which drained him of most of his remaining strength.

Primogeniture was not a principle observed by the House of Saud. But in the 1930s Ibn Saud was in search of international recognition. Especially, he wanted to impress foreign governments and their consuls in the Hejaz with the stability he had imposed upon Arabia. To be able to name a Crown Prince, in accordance with alien customs rather than those of his own family, must have seemed to him the sort of gesture they would appreciate, symbolizing the existence of an orderly succession that would ensure the survival of the Kingdom even if he were to die abruptly himself. That would have been one reason for the nomination of Saud. Another, perhaps more important, was because he was the least contentious choice. Ibn Saud certainly had no illusions about his capabilities or his life style, but in the forefront of the old King's mind must have been awareness of the categorical need to avoid any divisions within the family over the succession that could well have tempted Ibn Jiluwi to secede.

Saud at first seemed to be a promising successor who, if he could never have unified the country and created the Kingdom, at least seemed capable of preserving the family patrimony. Even before his succession he had created an image of himself as a well-intentioned man, anxious to bring about both reform and modernization. Having inherited and probably been partly responsible for the creation of the statutory framework for a system of government, he gave substance to it a few months after his accession to the throne. In the spring of 1954 he established the first Council of Ministers provided for by his father's decree in the previous October. Those holding the seven portfolios already in existence retained their positions, including Abdullah Suleiman, despite the new King's dislike of the old retainer and Feisal's contempt for him. Ibn Saud had commanded that the Finance Minister should be kept on after his death. In addition, a Ministry of Health was set up under Rashad Pharaoun, the physician of Syrian origin who had long served Ibn Saud, and a Ministry of Commerce under Muhammad Abdullah Alireza, a leading member of the Jeddah trading company Haji Abdullah Reza and President of the Union of Chambers of Commerce. But in the summer of 1954 he resigned after a blazing row with Feisal and was replaced by Muhammad Sorour Sabhan. In addition, Yusuf Yassin and Khaled Ghargani were appointed Ministers of State in a twelve-man council under the premiership of the King. Outwardly, at least, the Kingdom had a governmental structure.

At the same time, Saud decreed that Riyadh should become a modern capital, with the evident intent that it should capture some of the power and prestige that had slipped away from Nejd in the previous years to the advantage of Jeddah and Dammam. By 1956 an ambitious building programme was under way to provide accommodation in Riyadh for the new ministries in buildings that were for Saudi Arabia at that time grandiose in concept along the new tree-lined boulevard leading from the old city to the airport. Several new state secondary schools were opened. In 1957 Riyadh University was inaugurated and the first favoured young Saudis were awarded scholarships under royal patronage to pursue their studies abroad.

But Saud was, in fact, hopelessly ill prepared for the responsibilities he had to face as King. He became almost at once a prisoner of people more clever and cunning than himself. Nearly all his experience was of the traditional desert life among the tribes – a life distorted, however, by the rapid increase in wealth of the previous few years. Of the world which made that wealth possible Saud knew very little more than his father. His only foreign tongue was a few words of English, and although he had been abroad a few times he had nothing like the overseas experience of Feisal. He had never been to the United Nations or held diplomatic conversations with important world statesmen. He knew few foreign businessmen, and the oil industry was a closed book to him. Being a king meant enjoying his rights of marriage and concubinage, accepting the plaudits and petitions of the bedouin tribesmen as often as possible and spending the oil revenues with as much ostentation as could be. But of modern administration, trained civil servants, ministerial structures, and constitutional responsibilities he was ignorant and his kingdom still entirely innocent. The Council of Ministers amounted to little more than a façade concealing Saud's preference for private advice and a cabalistic system of intrigue. The entire financial recipts of the Government, the mounting proportion of it from petroleum revenue that had risen from $56.7 million in 1950 to $234.8 million in 1954, was still regarded as the royal family's and, more specifically, the King's income. Despite improvements in the drawing up of the Finance Ministry's budget, at court there was virtually no control over the money allocated to the privy purse.

The modern world seemed to be rushing in apace. But so, alas, was the modern world's greed. The royal court under Saud became a forcing house of waste, decadence, corruption and intrigue as a growing throng of princes and self-serving advisers plunged their hands into the bulging coffers. Towering apartment blocks in Cairo and vast villas in Lebanon soon provided the visible proof of their material success and lack of confidence in the regime. Secret Swiss bank accounts disguised evidence of the rest. The Council of Ministers was a positive innovation but did not prevent the contracts for development projects being inflated by commissions.

All the way down the list of royal Saudi officials, from the chamberlain's office to the junior cooks, bribery and graft took its daily toll. Wages and bills went unpaid for months while those who should have paid them pocketed the money.[1] Expensive palace fittings were 'lost' and sold back again to the royal commissariat from shops in Riyadh and Jeddah owned by the King's 'advisers'. Food disappeared from the palace kitchens in the same way, and as the annual kitchen bill for Saud's ten palaces was $5 million there were rich pickings to be made from such simple robbery.

Contractual and financial irregularity were endemic, the rule rather than the exception. It was most dramatically highlighted by the agreement reached in 1954 between the Government and Aristotle Onassis, the Greek shipping magnate. The original deal was signed in January but amended three months later in April. Its extraordinary provisions could only benefit Onassis and the Saudi interests who facilitated it. Onassis's main concern was guaranteed employment of ten oil tankers that he was to have chartered on long-term contract to

H. St John B. Philby in 1946: Jack
to the English, Abdullah to the
Saudis

ritain's last fling: a British military mission in Jeddah in the early years of the war. Major-
eneral Sir E.L. Spears with Ibn Saud and Yusuf Yassin

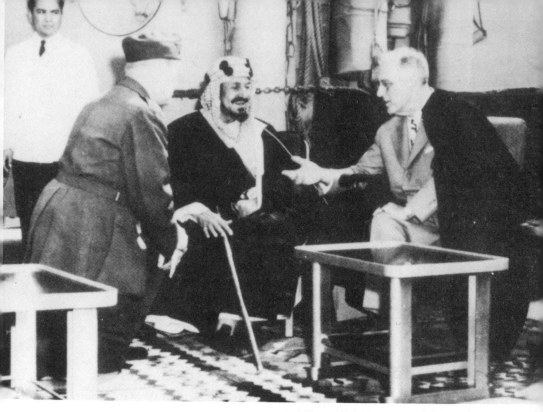

America in command: President Roosevelt and Ibn Saud aboard the U.S.S. *Quincy* in the Great Bitter Lake, 12 February 1945

Feisal at the United Nations General Assembly in December 1947 at which the U.S. shocked the Arab delegates by voting for the partition of Palestine. The man with the cigar is Jamal Husseini, Ibn Saud's former adviser, and Ali Alireza, later Saudi Ambassador in Washington, is second from right. Far right is Camille Chamoun, the Lebanese Christian leader

Feisal and Khaled at the helm of a U.S. Coastguard vessel in San Franscico Bay, September 1943

Old enemies: the first meeting, on 29 June 1948, of Ibn Saud and King Abdullah of Transjordan, who was ousted from the Hejaz by Saudi troops in 1924-5. Only the issue of Palestine could have brought Abdullah to Riyadh

Mishaal, eighteenth-born son
of Ibn Saud, in Washington in
November 1951 making his
first trip abroad in his mid-
twenties as Minister of Defence

Saud and the 'bulging-eyed
tyrant', Imam Ahmed of
Yemen (seated third from left),
listen to a speech of welcome
at the Egyptian Embassy in
Sana, August 1954

Anglo-Saxon Marine, Shell's transport subsidiary. The plan fell through because an executive in the oil company was found to have an interest in the broking firm setting up the arrangement. Thus, at a time when the tanker market was slack, it was with relief that Onassis, who on one of his trips to Jeddah by yacht was accompanied by Maria Callas, concluded an agreement with Saud and Abdullah Suleiman that allowed him to establish a limited private company, registered in the Kingdom, called the Saudi Arabian Tanker Company (Satco). The crucial aspect of it came in the amendment – with which Aramco immediately took issue when the terms of the deal were revealed. It contained terms giving Satco preference in shipping crude oil and refined products from both Saudi terminals and Sidon at the end of Tapline. Onassis undertook to put a minimum of 500,000 dead weight tonnage under the flag of the Kingdom – but would also give privileged treatment to any additional capacity flying its colours. The only restriction on this almost exclusive right was that vessels owned by Aramco prior to the end of 1953 should have first option on carrying Saudi oil. It was laid down that the rates should not be less than the 'internationally recognized' ones (those announced by the London Tanker Brokers' Panel), but these were defined as those prevailing between March 1952 and March 1954. Because of the global tanker over-capacity and even more depressed market conditions in prospect it was, indeed, a generous concession. Onassis could not lose.

Once details of the deal, a quasi-monopoly contrary to all the ideals of free trade, became known there was a furore throughout the shipping world. Other oil companies unaffected by the agreement implemented an undeclared boycott against Onassis. He was sued for $14.2 million by Spyridou Catapodis, a Greek living in the south of France who, amongst other dubious activities, pimped for prominent Saudis. Onassis, it emerged, had guaranteed him a substantial sum in return for an introduction to Muhammad Abdullah Alireza if the deal materialized through the latter's good offices. In the event, Alireza was instrumental in negotiating the agreement with Abdullah Suleiman. Spyridou Catapodis was paid nothing and at one point Onassis tried to dispense with his services by bringing in Hjalmar Schacht as intermediary. The court proceedings also revealed that Alireza stood to gain a down payment of $1 million, and seven cents for each ton shipped, with a guaranteed minimum of not less than $168,000 a year. Onassis, it was established, signed the contract with an ink fabricated to evaporate from the paper – which it did.

Eventually, the Saudi Government agreed to independent arbitration which took place from 1956 to 1958 in Paris and Geneva. For Aramco the immediate point at issue was whether they could be compelled to use tankers not freely chosen by them. More fundamentally, the sanctity of the 1933 concession agreement was at stake. Many of the learned affidavits emphasized just how detrimental to the Saudi national interests the Satco agreement was. For instance, Edmond Gérard d'Estaing, President of the French Committee of the International Chamber of Commerce, commented: 'If implemented the 1954 agreement will lead to an immediate and progressive reduction in sales of petroleum by Aramco and accordingly in Treasury receipts of the Government of Saudi Arabia in accordance with the convention of 1933.'[2] The opinion of Professor J. R. M. van den Brink, former Dutch Minister of Economic Affairs, was: 'It needs only a little imagination to predict that the Onassis agreement, especially in the long run, will cause great advantages to Mr Onassis and great disadvantages to the Saudi Arab Government.'[3] In August 1958, the arbitration panel found in favour of Aramco. By then many governmental changes had taken place. But even though the proceedings had clearly shown that the Satco agreement was not in the interests of the Kingdom as a whole, there was common resolve thereafter never to accept any contractual clause providing for independent, international arbitration.

Personal jealousies and factional strife thrived in the sickly atmosphere of the royal court as individual and family groups schemed to win the King's favour. With truth at a premium and dishonesty in the chair, those with the glibbest tongues climbed fastest up the ladder of success. None climbed faster than a certain Id bin Salem, a native Nejdi of no previous royal connection, who started life as a mechanic in the palace garage. He became the royal pimp and supplier of alcohol, and in the last few years of King Saud's reign the chief focus of power behind the throne. He provided for the King's whims and tastes. To foreign diplomats, Aramco men and, not the least, senior princes like Feisal and Muhammad, he seemed to be virtually the unofficial premier, supervising Saud's appointments, monitoring his conversations, authorizing his signature.

Imprisoned in this atmosphere of the seraglio, Saud himself grew progressively more irresolute. A quarter of the royal income was spent importing military equipment, much of which the unschooled Saudi soldiers did not know how to use and some of which was never unpacked from its crates. A royal yacht was purchased for $2 million and sold again for a quarter of that price because nobody could be found to sail or maintain it.⁴ Most of its equipment had been pilfered by the King's own men. The turnover in royal gifts consumed more millions when the palace concluded an agreement with a Swiss firm to produce a special gold watch with the King's head portrayed on the face that became the standard reward of every visitor. Within a year or two they were such a common sight throughout the Middle East that they had become known, derisively, as 'Mickey Sauds'.

The King's living arrangements acquired a gargantuan vulgarity. Apart from his extravaganza at Nasariya he maintained nine other palaces around the country and moved frequently from one to the other with a household of several hundred people in tow. On desert safari among his bedouin tribal friends he never moved with less than fifty vehicles and he held court in a special travelling tent palace big enough for a circus ring. His personal desert sleeping trailer, run up for him at a factory in Tulsa, Oklahoma, was reputed to have cost nearly $400,000 and contained a pale green drawing room with gilt armchairs and sofa, a pale blue bathroom with gold fittings, and a royal blue bedroom with a king size velvet bed placed beneath wall to wall ceiling mirrors. At every stop he was honoured with fresh young 'brides' from sheikhly families wishing to demonstrate their loyalty. Often he obliged by taking them to his velvet bed beneath the mirrors, but he did not always consummate the marriages. Such was the impetuous Muhammad's impatience with the extravagance of Saud and his closest kin that as early as 1955 he put forward a plan for Saud to be deposed and succeeded by Feisal. It was a premature move, but discontent was rife in the Kingdom. In the previous year there had been an isolated mutiny in one army unit and an officer was executed. Subversive, communist-inspired pamphlets were found circulating in Hasa in 1955. Anti-monarchical slogans were even found painted on palace walls in Riyadh. This simmering unrest was undoubtedly stimulated by the influx of Egyptian, Syrian and Palestinian professionals, technicians and workers. In the course of the year known or suspected troublemakers were deported by the score, but in the long term Saud only seemed to be compounding the problem posed by foreign and Arab manpower by his short-lived decree forbidding young Saudis to receive secondary education abroad.

One, at least, of the old favourites decided quite early that he could stand it no longer. Even before the death of Ibn Saud, St John Philby had been growing restive for some years over the obvious corruption at court, and in two of his recent books he had been outspoken about the way money was destroying all the old Arabian virtues. As long as Ibn Saud lived, however, his position was protected by the King's personal respect for him. Once the old man died his back was open to the knives of intrigue. Characteristically, the changed circumstances only made him more critical. In a series of lectures to Aramco staff in Dhahran, delivered in February 1955, he spoke with fierce candour about the decay that he saw around him. It was the excuse his many enemies at court were looking for and a few weeks later they waited upon him in Riyadh with King Saud's personal order that he must leave the country. Before the end of April he was gone, driven across the Jordan frontier under military escort. 'I think,' he wrote wryly to a friend at the time, 'I must be quite the only person who has ever left the country with no share of the spoil.'5

In foreign as well as domestic affairs, the same sad course was followed from early promise to eventual near disaster. King Saud began well, with a popular visit to Egypt in the spring of 1954 to restore the tentative alliance his father had made with King Farouk. The partnership between Saud and Nasser, who now led the revolutionary corps of officers, headed by Colonel Muhammad Neguib, who had overthrown Farouk less than two years earlier, was even more bizarre than the first. Although Farouk was as dissolute as Ibn Saud was puritanical, they had at least the common interest of monarchy to unite them. Now it was the Saudi partner who was dissolute and the Egyptian who was the puritan. Saud was seeking to protect his inherited privileges. Nasser represented the hot breath of revolution. But they had a common tactical interest. Both wanted to capture the rising Arab mood of anti-Western militancy that had been released in the wake of Israel's triumph in establishing a Jewish state in Palestine. For Nasser it was a means to his first aim of getting the British out of Egypt. For Saud it offered a way of increasing the pressure on the British over Buraimi. For both it was also a potential lever against their common rivals – the British-sponsored Hashemite regimes in Jordan and Iraq.

With such enthusiasm did Saud adopt Nasser's policy of 'positive neutrality', which rejected all alliances with non-Arab powers, that in 1954 he expelled the U.S. Point Four mission that had been welcomed three years earlier. The King also became paymaster for the Egyptian President's propaganda. At first he seemed to have backed a winner, for there was a genuine populist Arab emotion that Nasser soon learned to exploit. With strikes in Aden, riots in Jordan and demonstrations in Iraq – financed by Saudi cash and inflamed by Nasser's words – the British and their protégés were soon under mounting pressure; and Saudi policy in the Buraimi dispute grew steadily bolder.

Finally, in July 1954, through U.N. auspices, Saudi Arabia and Britain agreed upon the terms of arbitration of the dispute. At the end of the year the composition of the tribunal was announced. The chairman was Charles de Visscher, a Belgian. He was assisted by a Pakistani and a Cuban. Saudi Arabia's nominee was Yusuf Yassin and Britain's Sir Reader Bullard, who were to act as independent judges. An interlude of nearly nine months elapsed before it actually convened, a period that gave Yusuf Yassin ample time to purchase the loyalty of the tribes in areas around Buraimi and Aramco to prepare the massive Saudi memorandum. The British written submission was much shorter and more selective – complacently omitting, according to one Arabist employee of the Iraq Petroleum Company (Abu Dhabi's on-shore concessionaire), some of the more convincing material. Both sides presented a bewildering mass of esoteric scholarship centring basically on the traditional allegiance of the tribes and the recipients of their tribute, on which the outcome of the case would be decided. There was no certainty that the arbitration would go Britain's way. On the contrary, the Saudis, thanks to Aramco's scholars, seemed as if they had a good chance of establishing their claim to sovereignty over what had been something of a no-man's-land.

Eventually the tribunal met on 11 September 1955 in Geneva. It was short-lived. The British became convinced that Yusuf Yassin was trying to suborn the Pakistani member of the tribunal. There is no doubt that he was. Moreover, Charles de Visscher caught him instructing those presenting the Saudi case and Bullard resigned, charging that Yassin was 'in effective control of the conduct of the proceedings on behalf of the Saudi Government and is representing the Government on this tribunal rather than acting as an impartial arbitrator'.[6] The arbitration was never resumed. In October the British Government published evidence claiming to prove the extent of bribery in the area of the oasis, including one payment of £30 million. Later that month Sheikh Turki bin Utaishan was evicted by the Trucial Oman Levies (without the help of Aramco, he and his men were incapable of making their way home and had to be repatriated via

Bahrain, courtesy of the R.A.F.). The eviction was condemned by Saudi Arabia as an act of aggression, and John Foster Dulles spoke reprovingly of the shock to what he called Saudi 'public opinion'. Under pressure from the U.S., Britain had been previously restrained from evicting the Saudis. 'Hands off Buraimi' became a new theme in Arab nationalist rhetoric.

For Saud the whole affair was a humiliation, and for Yassin, who had assured him that Britain was a spent force, it was a serious miscalculation that cost him much credit at the royal court. The King was confused and wanted to retaliate. Abdul Rahman Azzam, his Egyptian adviser, was in favour of an appeal to the U.N. Security Council rather than any hostile action. Saud decided to do what he could to discomfit the British. About the only option open to him was to finance and intensify support for the revolt raised by the Imam Ghalib and his brother Talib against the Sultan of Oman. The campaign was orchestrated by Saud bin Jiluwi, the tyrannical Governor of Hasa. Little expense was spared in providing arms and training. Yet by early 1958, one year before the rebellion was finally crushed, Saud was responding positively to the efforts of Sidney Cotton, a multi-millionaire Australian entrepreneur who was interested in an oil concession, to mediate between Britain and Saudi Arabia. Saud complained to Cotton that he had inherited and not created the Buraimi dispute. He said that he would not mind losing the case before an international tribunal. Clearly he wanted a settlement even if Saudi Arabia had to renounce its claim. But Selwyn Lloyd, the British Foreign Secretary, insisted that there could be no discussion of the dispute before diplomatic relations were resumed.

Beguiled by his sycophantic courtiers and advisers, Saud was happy to cultivate his image as an honest Arab leader menaced by British imperialism, and yet more money flowed out of his purse to eager political supplicants from Arab nations who were ready to sing the right song over Buraimi on his behalf. But Nasser was doing even better. Precociously adept at political improvisation and diplomatic manoeuvre, and with the assurance of Saudi financial support, he moved with growing confidence from one triumph to another. With discreet American support, he forced the British to evacuate their last military bases in Egypt, ending over seventy years of virtual British occupation. Then, in 1955, in defiance of both Britain and America, he rejected Western plans for a Middle Eastern defence agreement, commonly called the Baghdad Pact, and denounced the Hashemite regimes of Iraq and Jordan as Arab traitors when the first agreed to join the Pact and the second showed signs of following in its wake. He demanded freedom for Algeria from the French and sheltered Algerian nationalist exiles in Cairo. He challenged the new Israeli state by sending Palestinian *fedayeen* across the borders from Gaza to raid Israeli farmlands; and when Washington tried to bring him to heel by refusing his request for arms supplies, he signed a secret deal with Czechoslovakia to equip the Egyptian forces with Soviet weapons. At the Bandung Conference in April 1955, the first major conference of developing countries, he was fêted as one of the leaders of the new 'Third World', ranking equally with India's Jawaharlal Nehru, Indonesia's Sukarno, and China's Chou En-lai.

The Arab rulers were divided by Nasser's success. King Hussein, who had succeeded to the Jordanian throne in 1952 when his mentally deranged father, Talal, was deposed, Crown Prince Abdul Illah and King Feisal, grandson of the first King Feisal of Iraq, looked more nervously than ever towards each other and the British for protection. In Syria, on the ,other hand, the republican regime leaped quickly aboard the Nasser bandwagon. Led by President Shukri Kuwatli, an elderly, silver-haired, smooth-tongued son of a family that had previously been Ibn Saud's agents in Damascus and originally recommended Yassin as an adviser, it turned the Saudi-Egyptian axis into a tripartite alliance. There was little doubt, however, where most ordinary Arab sympathies lay. In Palestinian refugee camps, in growing city slums, in liberation movements, universities and new Arab middle-class homes from Rabat to the Gulf, Nasser was suddenly the man to follow. Enterprising merchants sold his picture in the suks and his clamorous denunciations of 'imperialism' became the staple fare of political discourse in the coffee houses of a dozen Arab capitals as Cairo Radio's 'Voice of the Arabs' – *Sawt al Aarab* – broadcast them to an eager and admiring audience. Even in Saudi Arabia, Nasser's words and deeds made their mark. Although their echoes were still muted there by the strength of tribal tradition and the depths of desert indifference to city-bred politics, the new wage labourers of the oilfields and the growing class of Arab immigrants who had been attracted by the prospect of easy money – many of them Egyptians or Palestinians themselves – followed his career with pride and hope, made only the greater by its apparent contrast with the corruption they saw around them.

Warning signals began to appear here and there that the encouragement of Nasser might be dangerous for Saudi Arabia. More and more he talked of revolution as the Arab necessity, and spoke of Arab oil as the sinew of Arab strength. A three-day strike by Aramco workers in the spring of 1956 showed where that sort of thing might lead. Ostensibly it was no more than a protest against some changes in the company's transport system, but when the strikers began to widen their demands for better treatment all round the Government decided to take no chances. The strike was suppressed by force, 200 men were arrested and three were publicly beaten to death on the orders of Saud bin Jiluwi.[7] The King subsequently issued a decree making strikes a punishable offence. Aramco and other contemporary observers had no doubt that the troubles were essentially a political protest against the regime. There was no evidence of external direction or organized cells. But the trouble was fomented and led by one Nasser Said who subsequently fled the country and formed an opposition group abroad called the Arabian Peninsula People's Union.

Later that year the true measure and menace of Nasser's magnetism was demonstrated to Saud and the royal court – and by then it was impossible to offer any public challenge to it. The occasion was the Suez crisis when Nasser nationalized the Anglo-French company that still owned the Suez Canal and defied the world to stop him.

Ordinary Arabs everywhere were enraptured by his boldness and even hostile Arab governments like those in Jordan and Iraq were forced to rally behind his banner as he rejected Anglo-French threats and American pressure to rescind or modify his decision. When he arrived in Dammam that September to persuade Saud that the time was near when Aramco oil supplies might have to be used as a weapon against the West and more Saudi cash might be needed to underwrite the defence of Egypt, tens of thousands of Arabs turned out to greet him in the biggest demonstration the Kingdom had ever seen. Saud, at his side, was utterly overshadowed. For that day, at least, Nasser was King. There must have been, at that moment, acute concern in the ruling hierarchy as the extent of the Saudi dilemma was made apparent.

But so close did Saud and Nasser appear to be at this point that Howard Page, a senior Exxon executive, convinced Eisenhower that the King held the key to solving the Suez crisis. Accordingly, a top-level mission headed by Robert Anderson, Secretary of the Treasury, was dispatched to Riyadh for talks with Saud and his entourage. Their task was nothing less than to get Saud to accept the Suez Canal Users

Association for operating the waterway. The mission was coolly received. Feisal was not impressed by Anderson's somewhat convoluted talk about the great advances being made by the U.S. in the field of nuclear energy and his assertion that it could satisfy West Europe's energy requirements. The reply sent back to Eisenhower was apparently non-committal. Punishing Nasser would be of no avail, it said. If the U.S. wanted peace in the region it should use its leverage on Israel to repatriate or compensate Palestinian refugees. Feisal took aside one member of the mission, William Crane Echard, and insisted that the message be conveyed personally because the Saudis feared that Britain, France and Israel were planning to attack Egypt.[8]

Until Nasser's visit to the Kingdom the Saudis had assumed that money would buy him much as it bought tribal loyalty. But his reception made it obvious that he had established an independent source of power through his own actions, and Saudi money was only helping to make a Frankenstein's monster that might turn upon those who created it. Yet in the full flush of the Suez crisis, when the whole Arab world was compelled to take Nasser's side, there could be no question of Saudi help being withdrawn. On the contrary, as the crisis marched on to its violent denouement, with Israel's invasion of Sinai, with Anglo-French aircraft bombing the outskirts of Cairo and Anglo-French forces seizing half the Canal, Saud was compelled to go still further along Nasser's path, if only to protect himself. By the first week in November, Egyptian naval teams had blocked the Suez Canal with sunken ships and Syrian army engineers had blown up the pumping stations on the I.P.C. pipeline from Iraq to the Mediterranean, to demonstrate to Britain and France that Arab oil could be denied to them. Within the Kingdom itself, saboteurs set off explosions in the French-run munitions factory. The situation left the King with only one course: he broke diplomatic relations with Britain and France and announced that all Saudi oil shipments to the Anglo-French aggressors would be banned forthwith. By the end of the month there was petrol rationing in both Britain and France and the world was faced with its first Arab oil embargo.

The Saudi action was far from a decisive gesture. Oil was still being shipped to other countries and it was an easy matter then for the oil companies to divert tankers on the high seas to British and French ports so that not much was lost either to the Saudi Treasury or to British and French consumers. The closure of the Canal and the damage to the pipelines was of more immediate consequence, since they meant that oil for any European destination now had to make the long haul round the Cape; and in the age before the supertanker that created immediate shortages that were reflected in rationing and higher prices.

The cost of being in the vanguard of Arab liberation was mounting. Expectations aroused by the earlier distribution of Arab money were higher than ever in the aftermath of Suez. But the advantage now lay almost entirely with Nasser. Riots and demonstrations in his name became regular happenings from Lebanon to the Gulf.

Hashemite Jordan bent with the prevailing wind. Under radical pressure the young King Hussein in March 1956 had dismissed Glubb Pasha, the respected British commander of the Arab Legion and bane of the Ikhwan. In a risky exercise in constitutional democracy in October he allowed a radical government led by Suleiman Nabulsi, a nationalist devoted to Nasser, to emerge and establish itself following what was considered a fair general election. Negotiations for the termination of the Anglo-Jordanian treaty were soon under way and the summer of 1957 set as the date for complete military withdrawal. Even Iraq's staunchly pro-Western regime felt forced to withdraw base facilities for the R.A.F. In January 1957 Saud joined with Egypt and Syria in an 'Arab solidarity agreement' whereby three countries undertook to compensate Jordan for the loss of its British subsidy. Thus, Saud found himself financing his family's old and bitter dynastic rival. He also had obligations to Nasser, who had appealed for cheap oil the previous year. The Kingdom had agreed to accept payment in Egyptian pounds for the required supplies, which gave per barrel revenue in real terms only one-third of what other customers paid.

But now a new factor entered the turbulent Arab scene – a growing fear of the new Soviet threat. Since Nasser's first Eastern arms agreement the Soviet Union had moved swiftly to reinforce its new foothold in the Arab world. The Baghdad Pact had been virtually demolished by the British action at Suez. Soviet arms and experts were flowing into Cairo and into Damascus as well, and the Soviet Union had agreed to take over the construction and financing of the new Aswan Dam. The United States reacted in alarm to this evidence that the most populous and powerful of all Arab states had now fallen under Soviet influence and it determined to cut Nasser down to size. The means it chose was a campaign to strengthen the conservative regimes of the Arab world through what was soon christened the Eisenhower Doctrine. Promulgated in a message to Congress from

President Eisenhower on 5 January 1957, the new policy rested on a formal American promise to provide military and financial aid to any Middle Eastern countries 'requesting such aid against overt aggression from any nation controlled by international Communism'.[9] It was, in short, the language of the Cold War brought into the heartland of Arab politics and, as such, it was in direct – and intentional – conflict with Nasser's own post-Bandung concept of 'positive neutralism'. Inevitably, therefore, it was read by Nasser as a deliberate provocation and he launched an immediate onslaught on the Eisenhower Doctrine and on any Arab leaders who voiced sympathy with its purpose.

From the beginning of 1957 inter-Arab discord became more intense as Nasser's propaganda offensive swept across the Middle East; and with the Hashemites of Baghdad on one side of the conflict and Nasser in Cairo on the other, it was inevitable that the focus of the struggle was soon identified as Damascus. If Baghdad could sway the regime in Syria to its side – or, more realistically, engineer its replacement by another more sympathetic to the pro-Western, conservative outlook of the Hashemites – King Hussein's Jordan and President Camille Chamoun's Lebanon could be linked in an anti-Nasser bloc stretching throughout the Fertile Crescent from the Mediterranean to the Gulf. If Cairo secured Syrian support, on the other hand, Chamoun and Hussein would be isolated and Baghdad would be reduced to relative impotence.

As this grand conflict took shape in 1957 King Saud's former position as Nasser's ally became untenable. As an absolute monarch of the most traditional kind he could not ignore the threat posed by Nasser's call to revolution. As the ruler of a state supposedly dedicated to God's service he had to resist the prospect of an Arab, Islamic alliance with the 'atheistic communism' of the Soviet Union. As the recipient of American aid and money, both through Aramco and the United States Government, he could not support Nasser's onslaught on Washington's policies. And purely as a practitioner of *realpolitik* he was compelled to resist the implicit Nasserist threat of a unified Arab command that would extend from Cairo to Baghdad right along his own northern frontiers.

Belatedly – and not without some heavy American per-suasion – Saud recognized the long-term threat to his regime from Nasser. Suddenly he veered from the path of 'positive neutrality', turning precipitately in the process towards the United States. He could not have conformed more obligingly with the strategy evolved by Eisenhower and Dulles to detach the Kingdom from its loose alliance with Egypt and Syria formed in the summer of 1956.

The U.S. Administration was concerned about Nasser, with the Kremlin's backing, becoming 'head of an enormous Moslem confederation'. Eisenhower wrote in his memoirs: 'To check any movement in this direction we wanted to explore the possibilities of building up King Saud as a counterweight to Nasser. The King was a logical choice in this regard; he at least professed anti-Communism, and he enjoyed, on religious grounds, a high standing among all Arab nations.'[10]

The dangers of such a strategem were not fully apparent at the time to either the U.S. President or the Saudi King, though they should have been. American backing at a time of radical ferment within the Kingdom and throughout the Arab world might only place the Saudi regime in greater jeopardy than it already was. Conversely, it was absurdly unrealistic of the U.S. Government to believe that such an inept ruler as Saud could be a bulwark against communism.

Immediately, however, these aspects of what seemed like a volte-face were not as obvious as they would become in retrospect. Because of the tough line being taken by Eisenhower to bring about Israeli withdrawal from the Gaza Strip and Sharm el Sheikh in the winter of 1956–7, the standing of the U.S. amongst Arabs was as high as it would ever be over the next decade and a half. The turning point came with Saud's state visit to Washington at the end of January 1957. The event started inauspiciously. As he sailed into New York harbour with his enormous retinue, he was greeted by a 24-gun salute from a U.S. Navy vessel – and a resounding snub from Robert F. Wagner, the Mayor.

The White House, the State Department and the U.S. Navy had informed Wagner that extraordinary courtesies were to be extended to the King. The Mayor, a Catholic himself but very sensitive to the feelings of his large Jewish constituency, refused to roll out the red carpet or meet the King because of anti-Semitic remarks alleged to have been made by him. As American Naval aircraft flew out to greet the Arab King, who by then had trans-shipped from the liner *Constitution* to the frigate U.S.S. *William A. Lee*, the Mayor called in reporters to tell them that Saud was 'not the kind of person we want to be recognized in New York City'. Paying due regard also to his voters of Irish ancestry, Wagner added that Saud was 'a fellow that says slavery is legal, that in the Air Force you can't have any Jewish boys, and that a Catholic priest can't say Mass'.[11] Wagner was correct. In Saudi Arabia, slavery was as much a part of life as it had been in the deep south of the U.S. before the Civil War. Wahhabite sensitivity insisted that no Jews could serve with the U.S. Training Mission, and any Christian ceremony was banned in the Kingdom. Wagner was duly applauded by the *New York Post* for a 'heartening display of self-restraint'. The newspaper commented: 'What is not debatable is the grotesqueness of staging lavish testimonials to a ruler who practises the most primitive forms of oppression and intolerance.'[12] The only official welcome received by the Guardian of the Holy Places was provided by some fifty unsympathetic policemen, mainly of Celtic origin, and devotees of an absolutist Pope. Their main concern was the King's swarthy bodyguard wielding daggers and submachine guns. Saud and his entourage ensconced themselves expansively in the Waldorf-Astoria, occupying seven floors of the hostelry.

Eisenhower, however, was so keen to win over Saud that he had already planned to depart from normal custom by going to the airport to meet his guest. There he hailed him as a great Arab leader. By coincidence, on the very same day the House of Representatives approved by a large majority the draft declaration of the Eisenhower Doctrine authorizing him to intervene by force if necessary against any overt act of aggression in the Middle East region at the request of the country attacked. Appreciating that U.S. and Western strategic interest in the Kingdom far transcended Wagner's petty political preoccupations, Eisenhower's Administration arranged for the route from the airport to Blair House to be lined with military guards of honour and bands. No less than 20,000 people, including many of the Saudi citizenry in the U.S., as well as idle and curious American onlookers, were reported to have witnessed his arrival. *The Times* correspondent observed that they 'took a ready liking to the King's young son sitting gravely on his father's knee'.[13]

The infant's presence there was not an accident. The twenty-fourth born of the lusty King's recorded legitimate progeny was crippled with poliomyelitis and needed treatment. Impossible though it was for the King to identify and remember the names of all his progeny, he had a particular affection for Mashur. He had been booked into a Jewish-endowed hospital in New York, a plan which could have had embarrassing repercussions had it become public. And so, on Aramco's recommendation, a reservation was made with the Walter Reed Hospital in Washington. Encouraged by his eager advisers, Saud decided to capitalize on his son. His chef José Arnold wrote: 'The pictures of Prince Mashur, from his first appearance on his father's lap during the ride into Washington, caught the imagination of the American public. Even the most cynical newspaperman was unable to discourage the response of the American people to the paralysed little boy. The hospital party for him was an international social event for children. His big round innocent eyes and his mischievous smile, featured in every major newspaper in the United States, inspired more favourable publicity for Saudi Arabia than any planned propaganda campaign could have possibly achieved, and completely offset Mayor Wagner's attempt to discredit the visit of King Saud.' As for the old-fashioned rustic charm of Blair House, one of the King's coffee boys commented that the U.S. President must 'be a poor man because he had such an old place for his guests to stay in'.[14]

Yusuf Yassin and Muhammad Sorour Sabhan conducted the detailed negotiations with Dulles. In line with the Point Four aid agreement concluded in June 1951 the U.S. undertook to strengthen Saudi Arabia's Armed Forces, supply modern weapons and give economic aid with prime consideration being given to the expansion of Dammam port. Saud and his delegation requested arms deliveries worth $300 million, including F-86 Sabre jets. This figure was regarded by U.S. Administration officials as excessively high. The Saudi request for the military equipment to be paid for by grant aid was turned down, but the U.S. agreed to make the provision of training programmes free, at an estimated cost of $50 million over five years. Saud then publicly endorsed the Eisenhower Doctrine, describing it as 'a good one which is entitled to consideration and appreciation by Arab governments'.[15] On his return the King conferred in Cairo with Nasser and Kuwatli in an attempt to assure them that he was doing his best for the Arab cause. Back in Riyadh he entertained the Shah of Iran. But in April Saudi Arabia duly agreed to a five-year extension of the rent-free lease for use of Dhahran base. The understanding that no Jews should be involved in the training programmes was believed to have been reconfirmed privately.

When in Washington Saud was careful when confronted by television cameras to pay lip-service to Nasser. Of greater significance, perhaps, was his meeting with Prince Abdul Illah of Iraq. Together with Charles Malik, the Lebanese Foreign Minister, they held private consultations on the Middle East. Thus, the ground was laid for an entente with the Hashemite regime of Iraq. Saud paid a state visit to

Baghdad in early spring 1957. Then in April King Hussein of Jordan, with American and Saudi backing, overthrew the radical parliamentary government of Suleiman Nabulsi elected the previous year. Saud's flirtation with the 'progressive' Arab forces was over. By the summer of 1957 Saudi Arabia was effectively in the conservative camp with Iraq and Jordan, as well as Lebanon.

In April 1957 a number of subversive Palestinians and also, it was reported, Saudis, were rounded up in Dhahran and Riyadh. They admitted to being under the direction of Colonel Ali Kashabah, the Egyptian military attaché in Jeddah. The malcontents were said to be involved in a plot to assassinate Saud. Wholesale deportations of Palestinians and Egyptians followed. The affair was reported in the Beirut press and subsequently confirmed by Western diplomats in Jeddah. Saud was heard to complain of the rank ingratitude that he received in view of his £40 million paid out in subventions to Nasser. A few days after the round-up, President Shukri Kuwatli of Syria travelled to Riyadh accompanied by Anwar Sadat, Nasser's Islamic specialist and one of his leading anti-Western propagandists, and Abdul Latif Baqouri, Minister of Labour. Their purpose, it was believed, was to give Nasser's personal reassurance that he had no knowledge of the plot.

That spring the Saudi Government abruptly stopped the arrangement with Egypt whereby it had agreed to supply the country with oil payable for in soft Egyptian pounds rather than dollars to ease Egypt's foreign exchange difficulties. This decision may have been made for financial rather than political reasons – the Kingdom also announced its intention to withdraw its financial support of Jordan – but once again Saudi Arabia came under heavy verbal fire from Cairo.

Relations with Syria, meanwhile, had deteriorated rapidly from mid-1956. Towards the end of June the Saudi Ambassador to Syria was withdrawn after the publication of remarks attributed to Khaled Azem, the Syrian Minister of Defence, who attacked Saud and Hussein as men 'drowned in a sea of contradictions' and 'tools of America'.[16] The Times correspondent in Beirut acknowledged that Azem was the doyen of the Syrian Army but with a blind naivety about the nature of the King: 'Probably only a representative of the younger generations, emancipated from religious orthodoxy, would thus dare to attack King Saud, whose authority as keeper of Islam's Holy Places should not be underestimated.'[17] In August Azem signed an economic and technical accord with the Soviet Union. No sooner was the ink dry on the agreement than the Syrian Government expelled three American diplomats alleging that they were plotting to overthrow the regime. The charge may well have been justified.[18]

Amidst well-founded fears about drastic U.S. action to prevent the establishment of a 'Soviet satellite', Saud and his self-serving advisers sought to distance themselves from Washington, advising Jordan and Lebanon to do the same. In his anxiety to bridge the rift at the heart of the Arab world Saud visited Damascus in September where he publicly condemned aggressive moves against Syria, uttered fulsome expressions of brotherhood and gave assurance of Saudi support for the country's independence. In the U.N. General Assembly, Ahmed Shukairi, the Kingdom's delegate, said: 'Who is in power and who is not in power in Syria is the concern of Syria alone.'[19]

Saud was bidding to rally the Arab states behind him and return to the path of 'positive neutrality' by using his Washington connection to defuse an intensifying crisis. But in October, Nasser, having reached a secret agreement with Damascus the previous month, regained the initiative by dispatching troops to Latakia to show Egypt's solidarity with Syria against the perceived threat from Turkey and Iraq. Syria made it clear that it was no longer interested in Saud's mediation. On 1 February 1958 the merger of Egypt and Syria was announced subject to a plebiscite that predictably returned a 99.0 per cent vote of approval in both countries.

The formation of the United Arab Republic was a development that caused grave qualms for the Saudi regime which traditionally sought the best possible relations with Egypt while exercising a measure of influence over Syria. A fortnight later it had reason to tremble. At a press conference in Damascus on 5 March, Lieutenant-Colonel Abdul Hamid Serraj, the Syrian Army's Chief of Intelligence, claimed that he had been offered and had accepted a bribe of £1.9 million to arrange the assassination of Nasser by blowing up his aircraft. The sum was made up in three cheques worth £1 million, £700,000 and £200,000. They were cashed at the Damascus branch of the Arab Bank and the money was, it was said, set aside for development projects. For good measure Serraj passed around photostats of the cheques.

The go-between was said to be Assas Ibrahim, one of Saud's many fathers-in-law. He was quoted as saying that the U.S. was aware of the plot. The State Department had no comment to make on the allegation. From Riyadh there was only a stunned silence. That in itself seemed an implicit admission of guilt. A conciliatory statement or a denial might have helped defuse the tension. Indeed, in anticipation of one Radio Cairo called off its propaganda barrage of charges and insults for two days. The only response was a news item on Mecca Radio saying that Saud had ordered a strong guard to be placed around the Egyptian Embassy in Jeddah to protect it from the anger of the people. Saud was now exposed to the most violent vituperation and vilification. According to a later account from Cairo quoting 'responsible informants', which was reported in the *New York Times,* Saud had sent President Nasser a letter conceding that he had indeed spent large sums of money in Syria over the past few months. The purpose was to counter the growth of communist influence, the letter was said to have declared.[20] The Saudi Government announced the formation of a committee to look into the affair but nothing more was heard about it. The evidence was just too damning.

Colonel Serraj's allegations were well founded and Nasser's indignation righteously based in fact. Saud, now enfeebled by ill health, probably would not have had the daring or guile to initiate the plot. The finger of suspicion inevitably points to Yusuf Yassin who still rankled over his family's sequestered property in Syria. As at the Buraimi arbitration tribunal in 1955, Yusuf Yassin's tactics proved to be totally counter-productive, ripping to shreds the veil of civility covering the deteriorating relations between the two countries. Later it was claimed that the C.I.A. was well aware of the scheme and tried to warn the King of the 'trap' into which he was falling. 'The man who we had hoped might eventually rival Nasser as an Arab leader was in grave trouble,' wrote Eisenhower.[21] His blunder might well have proved fatal for the dynasty.

14 A House Divided
1958 – 62

Musing upon the king my brother's wreck
And on the king my father's death before him.

T. S. ELIOT, *The Waste Land*

SOME kind of polarization of the conservative and radical regimes of the Arab world was the inevitable result of the chain of events that brought about the merger between Egypt and Syria in February 1958. The outwardly cordial relationship which the Kingdom had been able to enjoy with President Nasser of Egypt up until 1956 was bound to be strained by the new conjunction created by the formation of the Baghdad Pact, the Suez War, the proclamation of the Eisenhower Doctrine, the political turmoil in Damascus during 1957 and finally the creation of the United Arab Republic. Even a leader as feeble-witted as Saud had been forced to recognize that the principles for which Nasser stood were totally at variance with those of his dynasty.

The radical forces at work in the Arab world were sufficiently and alarmingly present in the Kingdom for the C.I.A. to conclude that the possibilities of a coup there were very real. Saud had only himself to blame for the measure of discontent within his realm. However, its existence made the maintenance of the best possible relations with Nasser a cardinal aim of foreign policy. Despite his visit to Washington early in 1957, it would still have been possible to make a more plausible pretence at pursuing the path of 'positive neutralism'. As it was, his wobbling effort to tread the middle road did not in any way convince Nasser. It was exposed totally by the ham-handed attempt to influence events through bribery.

Nasser must have grimly observed that Saud could afford to lash out £1.9 million in bribes but, pleading financial difficulties, had stopped the cheap oil committed to Egypt in the wake of the Suez War – for a lesser saving – and had also informed Jordan of the Kingdom's inability to pay the £5 million subvention committed for 1958. By March of that year Saudi Arabia was, indeed, on the verge of bankruptcy. The financial crisis dated back to 1956 when oil revenues levelled off at about $340 million, but Saud and his Government – if it could be characterized as such – had gone on spending as if the bounty had continued to grow as it had previously. The Saudi Arabian Monetary Agency (S.A.M.A.) had been given powers to audit revenue and expenditure at its foundation in 1952, but these had been successfully eroded when the Ministry of National Economy was set up in 1953 and made an adjunct of the Ministry of Finance. Since the first budget had been published for 1947 – 8 they had appeared only intermittently and bore little relation to reality. By 1956, Abdullah Suleiman could no longer control the situation and resigned after a blazing row with Feisal. He was replaced by Muhammad Sorour Sabhan, then regarded as one of the Crown Prince's men. He proved equally incapable of enforcing any discipline over royal extravagance.

By the beginning of 1958 the Government was in debt to the tune of 1,800 million Saudi riyals, the equivalent of $480 million at the official rate of exchange, of which 700 million riyals was owed to S.A.M.A. International banks refused credit so that Aramco was asked to underwrite further loans against future revenue early in 1958. With the rise in the world price, most of the silver riyals had left the country and gone out of circulation in the middle of the decade. Gold and foreign exchange reserves had fallen to a point where they covered only 14 per cent of the Pilgrims' Receipts issue which was depreciating fast. On the free market the rate fell from the official one of 3.75 per U.S. dollar to 6.25. The entire money supply as it stood at the end of 1955 was exceeded by the amount of credit advanced by the commercial banks in 1956 and 1957. Inflation was roaring. Capital poured out of the country putting the balance of payments even more out of equilibrium.

There were doubts about the King's physical and moral fitness to rule even amongst those as degenerate as himself. As early as 1955 foreign observers whose business it was to know such things were aware of his growing addiction to alcohol and partiality to Cointreau. It was a well kept secret – even from the eyes and ears of Nasser's intelligence service – until 1962. By 1958 what were politely referred to in court circles as the 'irritating liquids' were compounding a multiplicity of ailments, not the least his stomach troubles, his high blood pressure and his sagging legs which, because of poor blood circulation, could hardly carry the weight of his body. At a time when monogamy was becoming fashionable outside the royal family the size of Saud's stable of concubines and the number of his progeny, particularly those of dusky hue, had become a cause for comment and contempt at home as well as abroad.

Urgent action was needed to preserve the patrimony of Ibn Saud. Amongst the elders of the family and the senior princes a consensus finally coalesced that the King must either step down in favour of Crown Prince Feisal or hand over authority to him. But a full nineteen days elapsed between news of the alleged £1.9 million bribe paid to Colonel Sarraj as an inducement to assassinate President Nasser and the final confrontation, indicating both disagreement and agonizing over what should be done. Resolution of the problem was not easy. It necessarily involved consultation with the *ulema*. Islam respects authority deriving from legitimacy, which Saud unquestionably had. The country was, as St John Philby had written, 'nothing if not royalist to the core'. Saud's half-brothers had sworn a vow of allegiance to him. But according to the *sharia*, he who exercises authority does so only with the blessing of his subjects. In the opinion of the majority of the sons of Ibn Saud he might have appeared to have forfeited that authority. But courteous, genial and generous as ever – to those who could penetrate the wall of predatory advisers surrounding him – Saud was still much loved within the royal family. Moreover, full regard had to be paid to his following amongst the tribes. As open handed with money and far reaching in his conjugal affiliations as Feisal was prudent and by that time monogamous, the King was far more popular with the majority of the inhabitants who mattered. In the context of traditional Arabia, Saud was a 'politician' of flair compared with Feisal. The Crown Prince's inhibitions were another weighty factor. He took his oath of loyalty particularly seriously and was anxious that the King should not be forced to concede authority. It would create a bad precedent and risk splitting the House of Saud.

Feisal had brooded darkly over Saud's conduct of affairs at home and abroad in the course of 1957, during which he is believed to have become totally disillusioned about the successor to Ibn Saud's throne. Although he was First Deputy Premier and Foreign Minister, the Crown Prince had been on the periphery of events or out of action for more than a year because of ill health. For six months he was in the

U.S. where he had two operations. On his journey home he stopped in Cairo and spent all January there still convalescing. Alarmed by Saudi Arabia's drift from the path of 'positive neutrality', Feisal had four meetings with Nasser in which he attempted to ease the growing strain in relations. At this point he would have been well aware that pressure might be placed upon him to choose between continuing to play whatever role the King assigned to him or a more important one chosen by the family.

Muhammad and Fahd, third and seventh of Abdul Aziz's surviving sons, took the initiative in pressing for Saud's abdication in favour of Feisal. Behind them was the weight of Khaled and Abdullah – both of them beloved of the bedouin – as well as Fahd's six full brothers. Feisal would go no further than agreeing to the transfer of executive authority to himself. He calculated that the King would have no choice but to accept an ultimatum to this effect. It was duly delivered on 24 March 1958 as Saud sat between Feisal and Abdullah bin Abdul Rahman, the brother of Ibn Saud, who had played a crucial role in bringing about the compromise. The King's own poor health was believed to be one reason for his immediate, unquestioning submission and his failure to anticipate the move to divest him of executive power. He had made no attempt to rally the tribes to which he had distributed so much of the country's precious oil revenue. Saud had no one to turn to and no option. He agreed to hand over all actual powers to Feisal and, after some haggling, to the wording of a decree saying as much.

The Crown Prince had the credentials to deal with the crisis. In terms of knowledge of regional and global affairs he was a man of the world amongst adolescents. He had represented Ibn Saud in 1939 at the futile Arab-Jewish conference in London called by Britain to work out a solution to the problem of Palestine. In the same capacity he had attended the 1946 – 8 sessions of the U.N. that grappled with the fate of the mandated territory. Feisal was not only conversant with Arab nationalism but also understood and identified himself – to the extent that supreme loyalty to his dynasty allowed – with the expression which it had taken with Nasser's nationalization of the Suez Canal. Unlike the dithering Saud, he had been prompt to applaud Nasser's action in nationalizing it. Earlier he had been quick to declare his opposition to any Arab state joining the Baghdad Pact. At this point Feisal was on good terms with Nasser and appreciated the need to come to terms with his charismatic personality.

In retrospect it is strange to recall that the Crown Prince was then

referred to in the columns of Western newspapers as an Arab
nationalist and a Nasserite. Within a couple of years, when the
constitutionalist movement was flourishing and espoused by several
leading princes, the epithets conservative or reactionary would be
conferred upon him. No more than any of his kith and kin did he want
to change the nature of the Saudi system or weaken the family's
domination of it. But, quite apart from the pressing need to save the
economy from ruin, Feisal was aware of the fact that concessions to
progress would be required if the antiquated structure was not to be
endangered by the increasingly strident call to revolution coming from
Cairo.

Austerity is the quality most frequently associated with Feisal. In
reality his vulpine face betrayed in part a record of past dissipation.
Once in the 1930s Feisal and Khaled had enjoyed themselves in the
pursuit of un-Wahhabite pleasure so blatantly in neighbouring
Kuwait that the Emir had had to ask them to go home. And Feisal
was by no means adverse to the temptations that Western cities could
afford. In 1945 he was the accredited delegate of his country at an
inaugural session of the U.N. held at Central Hall, Westminster,
London, but when the historic hour came Feisal was conspicuous by
his absence, to the dismay of his hosts. The police traced him to a
house of ill repute in Bayswater. It was only after the removal of a
non-malignant tumour from his stomach in 1957 that he renounced
alcohol forever, thereafter adhering to an unappetizing diet of boiled
food and – his sole addiction – an endless succession of cups of tea
which were enough to strain the bladders of some visitors detained by
him for long conversations. In private he was once a chain-smoker
who could show signs of irritation at being deprived of a cigarette
during a public audience. This may have accounted for his
disconcerting habit of scratching his crotch in public. Towards the end
of his life, however, he cut down to five cigarettes a day before
renouncing them completely.

Feisal never kept concubines and had only three wives throughout
his life. The first was Sultana bint Ahmed al Sudairi, daughter of one
of Ibn Saud's right-hand men, and the sister of Hassa, the most
famous of the first King's wives, who had borne him no less than
seven sons – the eldest of them Fahd. Sultana had given birth to one
son, Abdullah, in 1921. Having assisted his father in carrying out his
duties as Viceroy of the Hejaz, Abdullah al Feisal had been appointed
Minister of the Interior in 1951. His second wife, Haya bint

Muhammad bin Abdul Rahman, mothered Khaled and Saad. The marriage was dissolved in 1940. Thereafter Feisal's affections centred on his third wife, Iffat bint Ahmed al Thunayan, his favourite, whom he had married in 1932 or 1933 and who eventually was to become known, after the Crown Prince's accession to the throne, as the Queen. The union was the result of romance rather than arrangement. She gave birth to Feisal's other five sons – Muhammad (born in 1937), Saud, Abdul Rahman, Bandar, and Turki. In addition there were nine daughters in all by the three wives. Feisal sent his sons abroad for secondary and higher education, to Princeton's private and very expensive Hun School and to Princeton University, and even gave his daughters some schooling, unlike Saud, only one of whose progeny reached the secondary level. Not the least of Iffat's contributions to her husband's life came in the person of her younger brother Kamal Adham, who became his closest confidant and adviser.

For reasons of policy Feisal would have eschewed large palaces of the sort that Saud squandered the country's oil revenues on. In reality, however, his own chosen lifestyle was conspicuous for its lack of ostentation and luxury. And his dedication to work was such that, combined with his reluctance to delegate, it would in later years be almost counter-productive. From about 1958 he developed an almost unvarying routine, to the extent that those around him said that one could tell the time from it. Regardless of the sunrise, the traditional indicator that competed with G.M.T. for observance, he would arrive at his office at 9.40 a.m., take tea at 11 a.m., and at 12.15 go through the ritual of noon prayers. He then presided over lunch, which was open to the family, and slept briefly before praying again and going back to his office. Fifteen minutes before sunset Feisal would be driven to the desert to say more prayers. After dinner he visited either of his two elder sisters in their homes before returning again to work until 10.30 p.m. From 11 p.m. he would sit with his daughters, sons and closest half-brothers. He listened to the news and retired to bed at 1.20 a.m.

Feisal's physiognomy, which was progressively worn by arduous work and heavy responsibility, reflected an inner puritanism and a stern integrity. One of the most striking features of his gnarled face – the mean-looking downward twist of his lips – seemed symptomatic of his deep cynicism, till he smiled, when his face radiated kindness and charm. Feisal believed in no one and nothing – except God. As time went by, the gates of Paradise diverted his attention more and more.

Announcement of the decree by Saud making over full powers to Feisal was enough to defuse the tension with Egypt. Nasser welcomed the transfer of authority – though not explicitly – through the Cairo media and regarded it as a triumph for himself. Within a month, on 28 April, the Saudi Arabian Government had issued a statement on foreign policy reasserting the country's 'positive neutrality' from which Saud had deviated so grievously through his meddling in Damascus. With oblique reference to the blunder, it deplored 'the storm raging over the Arab states' as contrary to the dignity of the Arabs and of benefit only to 'the enemy who lurks among them'. In an unmistakable gesture to Nasser, Saudi Arabia promised to return goodwill to Arab brothers 'whenever it receives a response to such good intentions'. With reference to the agreement with the U.S. reached early in 1957, the statement defensively asserted that there was 'no American base in Dhahran and Dhahran is...not a depot for [American] military weapons'.[1] Looking further afield, Feisal held out hopes of a resumption of diplomatic relations with Britain provided that the situation arising from the British 'aggression' in Buraimi and other disputed areas could be solved by arbitration, and with France when it ceded Algerian independence. The only real diversion from the line of positive neutrality was the implicit exclusion of communist countries from the Kingdom's orbit.

Egypt responded to Feisal's conciliatory words, and relations improved through the summer. The measure of détente even survived the grave shock to the House of Saud from the coup in Baghdad on 14 July 1958 when the young King Feisal, his uncle the Crown Prince Abdul Illah and most of the Iraqi royal family were slaughtered in cold blood, together with Nuri Said, the Premier.

Saud was badly shaken. Feisal, however, remained cool and aloof. When the Saudi delegation at the U.N. asked him for instructions he told them to say nothing. Amidst the furore, he believed, rightly, that the safest policy for the Kingdom was one of non-alignment. As U.S. Marines landed on the beaches of Lebanon and the British sent aid to the surviving Hashemite monarch in Jordan, the Crown Prince stayed silent and neutral. In August he travelled to Cairo to see Nasser and after nine hours of talks felt able to say: 'Thank God our relations are so good now that if there was a cloud in the sky it has passed by.'[2] The horizon was to remain relatively clear for the next two years.

Dealing with the country's economic problems was as big a priority as placating Nasser. Feisal was ready with a prescription. The previous autumn, when he was in New York, Feisal had met Muhammad Sorour Sabhan who told him all was well with the state's finances and the cover for the currency was still 100 per cent. Despite his illness and his prolonged absence from the country, the Crown Prince had too much sense to believe him. Though effectively out of office in 1957, Feisal set in motion a mission from the International Monetary Fund. Its members were Anwar Ali, a Pakistani, and Zaki Saad, an Egyptian. The latter was a friend and compatriot of Abdul Rahman Azzam, an exile from Nasser's revolution in Egypt, who had been Saudi representative in the Geneva arbitration over the Buraimi dispute and was then serving the Kingdom as a political counsellor in New York and assisting its delegation at the United Nations. His daughter Muna subsequently married Feisal's second son Muhammad. It did not need any great degree of expertise to see that the Government's gross over-spending and indebtedness was at the root of the troubles. Zaki Saad was particularly forthright. The King once asked plaintively how he could be expected to reduce his spending when he had so many palaces to maintain. The Egyptian economist retorted curtly: 'Blow them up, Your Majesty.'

On the very day of the decree that placed the Government in the hands of the Crown Prince, Zaki Saad had arrived in Riyadh to be his financial adviser. One of Feisal's first actions was to inform Muhammad Sorour Sabhan, who was at Geneva at the time, that his services were no longer required. The holder of the King's purse strings had been far too judicious about the preservation of his post and the influence deriving from it to attempt any exercise at financial restraint. When Feisal visited the Treasury he found only 347 riyals there. The currency coverage was down to 12 per cent, mostly in gold and silver that had been pledged against borrowing from abroad. Zaki Saad remained for a while and even persuaded Saud against taking an expensive foreign trip, but Feisal asked Anwar Ali, a more tactful man, to stay on and run S.A.M.A. The partnership lasted until 1974 when the Pakistani died.

Feisal was quick to obtain some legal safeguards to limit the scope for Saud, his sons and advisers to make mischief and to prevent unwarranted hands from dipping into the state's coffers. The still shaken King assented to a new charter for the Council of Ministers to replace the one dating from 1953 which had, in effect, made it an adjunct of the royal *diwan*, or court. The 1958 decree laid down that all Government revenues must pass through the Ministry of Finance and National Economy. Henceforth funds were to be paid out only after authorization by the Council of Ministers. It in turn would be answerable to the Prime Minister who was given sole responsibility for asking the King to dismiss any one of them. Under the statute, the Premier's resignation meant that all other members of the Council of Ministers should climb down with him and a new one be established. The decree made it a criminal offence for any member to sell and let property or buy and hire it from the state – a fair indication of the kind of malpractices which had been taking place. The King was left with the power to sign decrees, however, and therefore by implication a veto.

In practice, the balance of power was an uneasy one. Feisal had reliable supporters in Fahd, who was Minister of Education, Sultan, who had replaced Talal as Minister of Communications, and his own son Abdullah al Feisal, who was Minister of the Interior. But Fahd bin Saud, a son of the King, who had replaced Mishaal as Minister of Defence, continued to hold the post although he knew little of military affairs and the Army was said to be less than happy with him. Another of his scions, Saad bin Saud, who was barely twenty years old, was Commander of the tribal levies known as the National Guard, increasingly referred to by foreigners as the 'White Army'. It had been built up as a counterweight to the regular Armed Forces and its main task was to protect the ruling family from dissidents at home.

It seemed doubtful that Saud would be prepared to act merely as a constitutional monarch. The prospects for such compliance did not look good. Only a few days after ceding power, Saud let it be known to a number of people, including President Camille Chamoun of Lebanon, that he was still in charge and that the Crown Prince was his proxy. Relieved of more onerous responsibility Saud was soon to embark on a campaign of spending the considerable resources at his disposal in a bid to win the hearts and minds of his subjects.

Feisal, meanwhile, set about restoring the Kingdom's finances. His first act was to slash existing appropriations. When some ministers demurred, the Premier denied all the departments any money at all for a few weeks. Ruthlessly, he suspended work on projects and payments, including those owed to the Government's creditors. Then in June the Crown Prince began to administer the first dose of medicine prescribed by the International Monetary Fund. An emergency exchange rate was applied to replace the import controls imposed in 1957 that had failed to stem the drain on the country's reserves. Foreign currency was made available at the official rate of 3.75 riyals to the U.S. dollar for essential consumer goods and capital equipment. The free market provided for other imports and its ability to cater for demand was governed by the volume of funds made available by Anwar Ali at S.A.M.A. At the same time, Feisal tightened restrictions on imports to the extent that only the humblest of motor cars could be purchased from abroad. It was a severe measure for a country where American limousines were already a status symbol. Despite the fiscal austerity, he introduced in August a Pensions Law, at a fairly heavy immediate cost to the Exchequer, designed to rid the administration of retainers over the age of sixty and thus alleviate the Government payroll in the longer term.

The iron hand on the till was first seen to good effect in the budget for the Hijri year 1378 – 9 which lasted from 11 January to 30 December 1959. It balanced revenue and expenditure at 1,355 million riyals, the equivalent of $375 million at the official rate of exchange, of which 12 per cent was set aside for debt retirement. Payments to the King – under the heading of 'Private Treasury' – were amalgamated. Including the sum of 190 million riyals set aside to sustain the tribes and bind them with affection – but to be spent according to the King's own discretion – the total was 236 million riyals compared with 292 million riyals in the previous year.

There were some signs of economic recovery. As early as the summer of 1959 Feisal was able to ease the curbs on the import of motor vehicles, though not sufficiently to meet demand. Despite continued stringency, the Crown Prince, no doubt aware that indebted officers might be dangerously discontented ones, improved the salary scales of the Armed Services. After an increase in oil production of nearly 10 per cent in 1959 and the prospect of further expansion in 1960 it was possible to set the budget for the latter year at 1,597 million riyals, up 17 per cent on the previous one. The share allocated to debt servicing rose to nearly 15 per cent. The funds set aside for new projects, including the extension of the Grand Mosque at Mecca, was doubled, but to only a very modest 110 million riyals, accounting for just 7 per cent of the total. The allocation for the privy purse, the Private Treasury, was raised by 12 million riyals to 248 million riyals. The greater amount was probably explained by a rationalization of accounting procedures and the inclusion of items that were not in the consolidate fund for 1959.

The second phase of the stabilization programme came into effect early in 1960 when the riyal was devalued at 4.50 to the dollar. Import restrictions were abolished, but for social reasons it was decided to continue subsidies on imported food grain, livestock, drugs and medicine, vegetable oil, and powdered milk. S.A.M.A. was authorized to issue the Kingdom's first proper paper money, as long as it was backed 100 per cent by gold, in place of the discredited Pilgrims' Receipts. As confidence recovered, the Agency began to exchange gold sovereigns for Saudi riyals. The first World Bank mission to visit the Kingdom in 1961 commented that the execution of the programme had been fast and effective. One of its own recommendations was quickly followed up when a Supreme Planning Board was established in 1961. By the end of 1962, when the Kingdom faced its next big external challenge in the form of the Yemen civil war, the Government had cleared its debts and maintained full cover of the currency issued.

In the five years from 1957 to 1962 oil production surged ahead by over 50 per cent from a rate of 1.02 million barrels a day (about 50 metric tonnes) to over 1.64 million. The increase more than compensated for the loss resulting from the oil companies' cuts in posted prices which shocked Saudi Arabia, Venezuela, Iran, Iraq and Kuwait into forming the Organization of Petroleum Exporting Countries (O.P.E.C.) in 1960. The Government's oil revenues rose from $296.3 million in 1957 to $409.7 million in 1962. Thus a base for steady growth was laid and it would not be undermined in the future.

Many tribal chiefs and their followers did not welcome Feisal's austerity programme, nor did the new class of merchants who had been enriched by the big contracts made possible by the King's extravagance. In addition, there was evidence of a growing political consciousness amongst Army officers, inspired by Nasser, many having received training abroad. They had little reason to respect Fahd bin Saud, the Minister of Defence, and resented the greater devotion given by the Kingdom to the National Guard. Moreover, the Crown Prince's cuts in defence expenditure were not appreciated.

Within the bosom of the royal family itself, there was a group of young princes, half-brothers of Saud and Feisal, who had developed liberal views of their own. They began to press for a reform of the traditional system and faster social change, not the least to advance themselves more rapidly in the ranks of the family. Also, there was the whispered agitation of the King's advisers, most notably the rat-like Id bin Salem and Saud's unprepossessing sons, who together had most to recover and gain from the return of the King to full power. In the last analysis they probably did most to persuade the King that he could and should win back his prerogatives. But it was paradoxically the liberal princes who provided the catalyst.

For mainly opportunistic reasons Talal responded most readily to what seemed the irresistible forces set in motion by Nasser. Quite apart from his income from the privy purse, Talal was rich in his own right because of an inheritance from his mother. Almost certainly the Levantine in his upbringing accounted for the fact that he was the most sophisticated, by the standards of the outside world, and showed an entrepreneurial streak earlier than any others. His drift away from tradition was probably encouraged by his first marriage to a daughter of Riad Solh, the Lebanese politician and perennial Premier. Talal was also one of the brightest of Ibn Saud's sons. At the age of nineteen he had been appointed Comptroller of the Royal Household and three years later Saud made him the Kingdom's first Minister of Communications. In that capacity he had quarrelled violently with Mishaal, five years his senior, who was at the time Minister of Defence and Aviation. The dispute might have had something to do with Talal's connection with certain dissident elements in the Army who were rounded up in April 1955. At any rate, his connection with them cost him the Ministry of Communications and he was dispatched to Europe where he served concurrently as Ambassador to France and Spain.

Talal returned in 1957 with what were for a traditional regime dangerously radical ideas about constitutionalism and representation in Government. To varying degrees at the turn of the decade they were shared or toyed with by his own full brother Nawwaf, Badr and Fawwaz who were the offspring of Abdul Aziz's fifth wife Bazza, Abdul Mohsen and Majid. It is difficult to say to what extent they saw reform as necessary for the dynasty's survival or as a means of leap-frogging up the ranks where respect for seniority by birth was rooted. However, they found a ready following amongst the foreign-educated intelligentsia, professionals, administrators and businessmen, especially in the Hejaz which still rankled under Nejdi predominance. No doubt the constitutionalists also evoked some secret sympathy in the officer corps.

Understandably, the categories that constituted an embryonic middle class resented a scheme of things that gave them little opportunity to reach positions of power. The most prominent of such commoners was Abdullah Tariki, Director-General of the Department of Petroleum and Mineral Resources who had enjoyed higher education in the U.S. He was the would-be scourge of Aramco, second only to Perez Alphonso of Venezuela in the formation of O.P.E.C. and an international figure because of his pronouncements about the rights of the oil producing states. Shortly after the transfer of executive power to Feisal in April 1958 Tariki claimed exultantly: 'We in Saudi Arabia have just taken a step forward to a constitution. Eventually this country will become a constitutional monarchy.'³ But the Crown Prince had no intention of moving in such a direction. He did not even exalt Tariki, to whom he had given support in his vain battles with Aramco, to ministerial rank despite the importance of the oil sector. The Department of Petroleum and Mineral Resources remained under the Ministry of Finance over which Feisal had taken a firm grip. However, although the hopes of the constitutionalist movement were not fulfilled, it gained ground, and towards the end of 1959 began to assert itself. By then the reformers were coming to the conclusion that the restoration of a weak Saud and the ousting of Feisal might help them to realize their objectives. For their part, the myopic monarch and his sons calculated that the liberals might be used to gain lost prerogatives. It was General Said Abdullah Kurdi, Saud's Chief of Intelligence, rather than the insidious Id bin Salem, who is believed to have convinced him of the advantages of such an alliance.

The ground had been well prepared. Saud was not idle for long after his humiliation in March 1958. Relieved of the burdens of office he embarked upon a series of peregrinations through the country displaying majesty, munificence and magnanimity. As Feisal tightened the purse strings, he disbursed bounty on a scale never known before and the Saudi press dutifully recorded his acts of generosity. He had a style that caught the popular imagination in the rural areas.

By the summer of 1959 Saud was exhausted. His blood pressure was so high that it amazed foreign physicians, who advised him not to ascend above a level of 1,000 metres. Yet the King felt reinvigorated by his stay in his new Al Hariyah Palace at Taif in the mountains during the high summer, though he was beginning to suffer from severe stomach trouble brought about by the 'irritating liquids'. Thus, he embarked upon a Mediterranean cruise in a chartered Greek vessel with a party of 300 having earlier steamed with some trepidation through the Suez Canal. At Ismailia the King was honoured by only a cool, perfunctory greeting from a minor Egyptian

Government official. Saud proceeded from Venice for medical treatment at Baden-Baden and Bad Nauheim where a restaurant so delighted him and his entourage of 300 that he decided to have a replica built in the harem patio in the Nasariya. On his return journey he was entertained by Nasser. The Egyptian leader evidently calculated that some show of forgiveness might help to boost the King to the detriment of Feisal.

The following year Saud was once again roaming the country with his lavish motorized desert caravan undertaking good works of the kind that should have been undertaken by the Government. Again the Saudi press and Radio Mecca hailed the nobility of the 'Guardian of Religion and Prime Defender of Lofty Precepts'. He participated with the leading pilgrims at the ceremonial washing of the *kaaba* during the *Hajj*. It was he, not Feisal, who took the publicity for entertaining Muhammad V of Morocco and the Emperor of Ethiopia on state visits earlier in the year. When Saud visited Medina in November 1960 'the enthusiasm of the people knew no bounds', reported *Al Bilad* on the 11th. On the 12th of that month the newspapers sang his praises with poems.

Saud was behaving in the grand patriarchal manner of Arabia – in his interest rather than the country's. Spurred on by his advisers and sons, he was mounting a bid to regain power in the only manner that he knew. Dipping deeply into his privy purse, the genial, flabby King did little to better the Kingdom by his impulsive hand-outs for minor projects, but the sum total of his benevolence was to create a feeling of optimism and even euphoria that was not in any way justified by economic realities. From the near bankruptcy of 1957–8 to the hopeful beginning of 1960, a policy of austerity and retrenchment was the only course open to a responsible Prime Minister. The effect of the King's benevolent jamboree was to make Feisal look mean and put him on the defensive. It was a measure of his exasperation with Saud's showmanship that he felt constrained to get published early in 1959 the essence of a report by Zaki Saad in which he attributed the economic crisis and the ensuing austerity to the extravagance of the royal family.

Towards the end of 1959, it appears, Saud began to flirt seriously with the idea of an alliance with the reform-minded princes. They, too, had made an impact. Nawwaf had given away part of his income and some properties. The press applauded him for his good works which, one newspaper said, must surely win him a place in

Paradise. Talal donated a hospital and a mobile clinic to Riyadh. He was cutting a dash in public, gaining an image for progressive modernity and increasing in popularity amongst the educated commoners in the main towns. After Feisal abolished censorship early in 1960 the press aired the idea of constitutional reform. The movement gained momentum, as did Saud's cause. During a visit to Cairo in May, Nawwaf spoke of 'a tendency to establish, for the first time in Saudi Arabia, a constitutional assembly to prepare the state's first constitution'.[4] It was clear that Saud was rallying 'liberal opinion' to his cause. By the summer he was actively thinking in terms of the creation of provincial assemblies half appointed and half elected, representatives of which would form a central parliament to advise the Council of Ministers and to approve the budget.

Finally Saud felt strong enough to force the issue. When on 19 December Feisal presented the budget for the new financial year, the King refused to sign it on the pretext that the details were not itemized. In a note protesting that the conflict was one of principle and not a personal quarrel, Feisal declared: 'As I am unable to continue I shall cease to use the powers vested in me from tonight and I wish you every success.'[5] He was careful not to hand in his resignation. Nevertheless, Saud regarded his compliance as such. Fahd refused to admit its validity. The King still felt the need of Feisal's support. But his offer to reconfirm him as Crown Prince and give him the title Deputy Prime Minister was studiously ignored. Feisal withdrew from the political scene and was not to speak to the King again for another seven months.

Saud formed a new Council of Ministers. As Minister of Defence he appointed Muhammad bin Saud, regarded as one of the least incompetent of his sons, whom – it was widely suspected – the King wanted to nominate as Feisal's successor. Talal, Abdul Mohsen, and Badr, the twenty-second-born son of Ibn Saud, were given, respectively, the portfolios for Finance and National Economy, the Interior, and Communications. Abdullah Tariki was promoted to head the new Ministry of Petroleum and Mineral Resources. Ibrahim Suwayl, a career diplomat, who had been serving as Saudi Ambassador in Baghdad, was made Foreign Minister. On the very night that the formation of the new Government was announced, Talal told the Beirut newspaper *Al Hayat* of his hopes that 'the future will prove that the new Government will serve the Kingdom in all its efforts and will implement the necessary reforms'.

Confusion and – for the reformists – disappointment followed. Talal had his constitution ready. It provided for a National Assembly of not less than ninety members and no more than 120, of which two-thirds would be elected and the others nominated by a joint committee of princes and ministers. The parliament would have the right to propose bills, including ones dealing with finance. Under it the King was given power to veto legislation and also to dissolve the Assembly with the proviso that new elections would be called within three months. On 24 December Radio Mecca announced that the draft had been submitted by the King to the Council of Ministers. Three days later, the Director of Broadcasting and Publications denied that any such statement had been made – though it had been monitored abroad. The liberal princes knew where they stood when the King's address to the new Council of Ministers omitted any mention of constitutional reform. Talal felt betrayed but did not give up.

Restored to executive power, the King embarked on a flurry of activity in other respects. 'I am determined to work with all the strength I have to ensure comfort, dignity and prestige for the people,'[6] he declared. A budget that was basically Feisal's was balanced at 1,720 million riyals and quickly decreed. With a healthy increase of revenue anticipated, it was possible to triple the allocation for new projects to nearly 291 million riyals, not including the electrification scheme for Riyadh which was a special priority and entered separately. The most significant feature of this part of the budget was the 20 million riyals set aside for 'the enlargement of holy mosques' and 'Mecca projects'. It is not known whether these were unchanged from Feisal's draft, but contemporary observers suspected that Saud had made a shift in the allocations to consolidate the support of religious leaders. Appropriations for defence and the National Guard were slightly lower at 241 million riyals, as was the one for the privy purse. At 248 million riyals, however, it still accounted for 14 per cent of the budget and more than education, health, labour and social affairs combined.

The King promised strict fiscal discipline. At the same time he eliminated dues on foodstuffs, building materials and industrial tools. On 7 January 1962 he presided over the first meeting of the Supreme Planning Board whose chief was Talal. The King then abolished the real estate taxes levied by the municipalities. Radio Mecca reported that the towns and villages of the Hejaz were praising the King's munificence. It commented on his interest in building roads, schools and hospitals. But although more resources were available, progress was in reality slower than it might have been because of Talal's belief in planning concepts. The World Bank, together with other U.N. experts, were in the process of drawing up proposals.

Soon the alliance between the King and Talal was under strain. In February the young prince raised the question of the constitution. Totally confident that they would disapprove of any such notion, Saud referred him to the *ulema*. They told Talal bluntly that the constitution of the country was the *sharia* and there could be no other. For someone who had committed himself to the reform of the system and spoken so much about it, the rebuff was a hard one. His failure to make any progress with constitutionalism did not enhance his prestige with those who saw themselves as progressively minded. Distrust and disagreement were such that the new regime was crumbling after little more than two months. Early in March the King issued a decree prescribing prison sentences for ministers – his chosen servants – found guilty of 'deliberate fabrications in attempts to alter the royal order'.[7] It was assumed to be aimed mainly at Talal, as well as Abdul Mohsen and Badr to a lesser extent. One incident at the turn of 1961–2 showed how rapidly authority in the Kingdom was disintegrating. Yusuf Yassin approached Aramco with a proposal that it should advance a loan of $14 million to the Syrian Government. His evident purpose, again, was the recovery of his family's sequestered property. Aramco declined, saying that such financial assistance was the responsibility of the Saudi Arabian Government.

Saudi Arabia had entered into a new era of seedy court intrigue. On this front Talal gained a few victories. He resented the King's reliance on two commoners, officially appointed as royal advisers at the end of the previous year. They were Feisal Hegelan, a diplomat who had served in Washington, and Abdul Aziz Muammer, a former Director of Labour with strong left-wing views, who had been imprisoned for a while in 1955 accused of having connections with subversive elements and whose presence was equally resented by Feisal and his supporters. Under pressure the King dispatched them abroad as Ambassadors, Abdul Aziz Muammer to Switzerland and Feisal Hegelan to Spain.

He scored again when the former Finance Minister Muhammad Sorour Sabhan returned from exile in Cairo soon after Saud's resumption of executive power and offered his services as Chief of the Royal Court, a post that happened to be vacant. This was opposed by Talal, as well as by Abdullah bin Abdul Rahman and Khaled, who both privately abused Saud in the coarsest terms known to the Arabic language. The King rejected Muhammad Sorour Sabhan's proposition probably because of his demand that Id bin Salem, the King's procurer of women and drink, should be banished from court.

Another victory for Talal was the appointment to Chief of the Royal Court of the neutral Nawwaf. He was appointed against the wishes of Saud's sons – the 'Little Kings' as the Egyptian newspaper *Al Haqaiq* derisively called them at the time. However, little more than two months later Nawwaf resigned over a minor decree, believed to have been instigated by Saud's sons without his knowledge, which he claimed infringed the authority of the executive. His departure left Id bin Salem to reign supreme at the *diwan*.

Talal's connection with the wider Arab world made him a force to be reckoned with and one that the King was reluctant to break with, particularly as he feared Talal had discovered records in the Ministry of Finance and National Economy indicating his complicity in fiscal irregularities and might expose them. Articles in *Al Anwar*, the Beirut daily newspaper that consistently took a pro-Egyptian line, alleging that Feisal would be the recipient of part of the commission involved in the concession agreement with Japanese interests for oil exploration in the off-shore waters of the Kingdom's share of the Neutral Zone, were believed to be the work of Talal, who was a friend of the paper's editor, Said Freyha. And at the end of March Talal had seen Nasser in Cairo, intending, evidently, to give the impression that he was the real power in the land, and, as a progressive, had unique access to the Egyptian leader.

Breaking point came, however, when Talal gave a press conference in Beirut on 14 August and made statements that deeply angered the King. He said that a 'national council' was being considered by the Council of Ministers. This was untrue unless he was referring only to Abdul Mohsen, Badr and himself. Talal acknowledged his quarrel with Feisal but dismissed it as only a difference of opinion over the system of government. Such public airing of dissension in the royal family could not be tolerated. Talal also mentioned the approaching expiry of the U.S. lease on Dhahran air base. That was considered an unwarranted intrusion into foreign policy. Worst of all was an aside about the King. 'That man,' he said, has 'so far been behaving himself.'[8] Saud was furious. He sent a cable reprimanding Talal and calling him home. The prince flew to Jeddah and lingered there for a few days before Abdul Mohsen and Badr persuaded him to see Saud.

Talal refused to apologize and threatened to take no part in the Council of Ministers unless the King redeemed his 'promises' about the constitution and representative government. Enraged, Saud promptly dismissed him. For good measure he fired Abdul Mohsen and Badr as well.

The King and the 'Free Princes' had used each other for their own ends. The alliance and its rupture, however, damaged beyond repair the chances of their ambitions being fulfilled. Talal erred grievously in thinking that he could manipulate Saud. Even more fatally for his career in public life, the prince completely alienated Feisal who, he should have realized, was the man of the future.

Saud seemed almost as isolated as he had been in the spring of 1958 with no one of weight within the royal family to turn to. Ambitious as ever, Nawwaf, who had now veered to the middle of the road in Saudi politics, accepted Saud's offer of the Ministry of Finance. Feisal bin Turki, the posthumous only son of Ibn Saud's first male child who died in the influenza epidemic in 1919 and was a full brother of Saud, was appointed Minister of the Interior. Feisal bin Turki was distinguished by the fact that he was the tallest member of the family. He was to hold office for only seven months and later drank himself to death. The Ministry of Communications was entrusted to a commoner, Atallah Saad. The Council of Ministers could not have been thinner. The King was well aware how weak his position was. Earlier in the summer when they were both staying in Taif, Saud had made an approach to Feisal. In a letter he had plaintively asked why he would not co-operate. The Crown Prince replied that he had no intention of doing so 'as long as you treat the Kingdom as if it were your own estate'. Under pressure from some senior princes, including the redoubtable Muhammad bin Abdul Aziz, Saud called on Feisal in Taif on 8 August 1961 in search of a reconciliation. It was their first meeting since Saud's refusal to sign the budget on 19 December of the previous year. The Crown Prince said bluntly that he would only co-operate if the whole of the Council of Ministers resigned and his government was restored. Saud refused to capitulate. It was a subject of some comment that Feisal did not pay the King the courtesy of seeing him off at the airport on the following day. However, despite his distaste for Saud which almost amounted to revulsion by 1961, the Crown Prince was not able to maintain his splendid isolation much longer. The Government was disintegrating and events abroad again threatened to bring the fabric of the House of Saud crashing down.

Earlier in the year the King had showed political sense in informing the U.S. that Saudi Arabia did not intend to renew the American lease of the Dhahran facilities, for the arrangement exposed the Kingdom to radical propaganda. In July he had reacted with uncharacteristic decisiveness in dispatching a brigade of troops to Kuwait when its treaty relationship with Britain ended and Iraq threatened to gobble up the state. As it happened, Baghdad's claim implicitly embraced the greater part of Hasa, including its oilfields. But in September another event took place that Saud, showing renewed signs of exhaustion and physical decline, could not cope with. In the early hours of the 18th of that month, tanks rolled again in the streets of Damascus, senior Egyptian officers and officials were bundled out of bed to be dispatched home still wearing their pyjamas, and Syria seceded from the United Arab Republic.

Saud dithered nervously. On 10 October Feisal flew to Riyadh for consultations with him. Saudi recognition of the new Syrian regime was announced on the same day. By the end of the month the verbal truce with Egypt came to an abrupt end as Nasser, deeply wounded by the failure of his great experiment in Arab unity, unleashed his propaganda machine against 'the palaces of reaction in Riyadh and Amman'. Muhammad Heikal, the editor of *Al Ahram*, alleged that Saud's liberal disbursements had undermined the union. Later he was to write how Saud had admitted to Nasser in 1967 that he had spent no less than £12 million to break up the union.[9]

For the second time Saud was forced to yield executive power to Feisal – though only with great reluctance and after dogged resistance. On 15 November there was a dramatic session of the Council of Ministers at which Abdullah Tariki alleged that one half of Kamal Adham's commission from the Japanese oil concession was to be given to Feisal and one quarter to his wife Iffat. Kamal Adham, Feisal's brother-in-law, was agent for what became known as the Arabian Oil Company and had legitimately received a handsome lump sum from the Japanese and a contractual promise of 2 per cent of its net profits. Before shuffling out of the chamber Saud is said to have terminated the proceedings by asking the Petroleum Minister to find out if the percentage could be transferred to the Treasury.

Later that night Saud collapsed with acute stomach pains. Before departing for treatment at Aramco's hospital in Dhahran he made an extraordinary attempt to reassure his subjects of his physical fitness to govern. He insisted on his favourite wife, the fat Umm Mansour, being brought to him. With the physical support of four slave girls he successfully had intercourse with her. The glad tidings about his virility were spread about the capital by Id bin Salem who, by Saudi standards, maintained an extraordinarily close relationship with Umm Mansour. In hospital, where he was examined by a number of specialists, the King was found to be suffering from severe internal bleeding caused by alcohol. The staff found it impossible to prevent his kinsmen bringing in more of the 'irritating liquids', bottles of which were found underneath his bed. It was agreed that he should go to the Peter Bent Brigham Hospital in Boston for surgical treatment. Before leaving he sought, with the help of Id bin Salem, to retain some kind of control through a regency council of five princes of his own choice. Yet the weight in favour of Feisal acting as his deputy was such that the enfeebled King had to acquiesce. But the Crown Prince undertook to make no changes in the Government during Saud's absence. The King left for the U.S. with a large entourage. In Boston he was operated on for cataracts in both eyes and then underwent an operation on his stomach. The party ran up bills of $3.5 million. Aramco paid on the understanding that the company should be reimbursed one year later.

In Saud's absence Feisal presented a new budget projecting expenditure at 2,085 million riyals, including 400 million riyals for development. The defence appropriation rose from 243 million riyals to 322 million riyals. There was also a significant increase in the allocation for education, up from 149 million riyals to 171 million riyals. Yet the administration was hamstrung. As he convalesced in New York and then Florida, Saud tried haphazardly to direct affairs of state from afar and sent messages to the ministers whom he had appointed. There was bickering in the Council of Ministers as Nawwaf accused Abdullah Tariki of spreading rumours that he had been criticizing Feisal behind his back. Feisal heard of the Minister of Petroleum's allegations about the commission paid by the Japanese and – with what appears to have been righteous indignation – called him to account. Then, on his own initiative, Tariki announced that an investigation had been authorized and instructed the Arabian Oil Company to cancel the commission even though he had been advised that it was legal.

Saud arrived home on 7 March 1962 intent on appointing a new Council of Ministers under himself. The senior princes, who had not been tainted by close association with the King over the previous fourteen months, told him in no uncertain terms that Feisal should form a government of his own. Excluded from it were Nawwaf, Suwayl, and Tariki. Feisal took over as Foreign Minister, as well as Premier. Musaid bin Abdul Rahman was appointed Finance Minister. The tenth son of Abdul Rahman al Saud, he was the uncle of the King and Crown Prince, though only 39–40 years old at the time. A man of undoubted honesty and purity in his personal life, he would render great service for many years to come in controlling the Kingdom's finances. Ahmed Zaki Yamani became the new Minister of Petroleum and Mineral Resources. The new council was still something of a compromise. Muhammad bin Saud stayed as Minister of Defence and Feisal bin Turki as Minister of the Interior.

Tariki's fall was an event of some significance. He was the first of the Kingdom's native born sons to become a 'technocrat' of international renown, or – as far as the oil companies were concerned – notoriety. He was exceptional for his lucidity, honesty and idealism as well as his courage. During the 1950s he had been quick to espouse the revolutionary idea put forward by Frank Hendryx, an American oil adviser, that a government might be released from or able to override contractual obligations as a result of 'changed circumstances'. Tariki's role in the formation of O.P.E.C. in 1960 was without doubt his outstanding achievement. He was also the moving spirit behind the first Arab Petroleum Congress in 1959 and a star performer at subsequent sessions.

Tariki negotiated a deal with Japanese interests that gave Saudi Arabia a 56 per cent share in the profits from the concession awarded in Saudi Arabia's half of the Neutral Zone shared with Kuwait. Justifiably, he was vociferous in reminding Aramco, Saudi Arabia and the world of the better terms obtained by Venezuela from the companies operating in its territory. Yet he did not improve on the Kingdom's. Most vitally he failed to settle the pricing of the crude pumped to the coast of Lebanon through Tapline. On that issue the Saudi Government had an excellent case. It received the same revenue for oil shipped from Ras Tanura in the Gulf while the company sold the same commodity at Sidon on the Mediterranean seaboard at a higher price and made more profit.

Within sensitive Saudi political circles Tariki was to prove his own worst enemy and his own undoing. Perhaps this slightly built, prematurely greying demagogue was before his time. In vital respects he was, certainly, naive and tactless. Worse, he knew much less about the oil industry than he and others thought. With no justification he harped upon how Saudi Arabia had been denied its just deserts by the 'profits' which, he alleged, Aramco made on freight. He also emphasized the importance of realized rather than posted prices – until the former sagged as supply swamped demand and the companies unilaterally cut per barrel revenues in August 1960. In earlier days Feisal had backed him in his militancy. However, the fact that Tariki had been appointed by Saud to the Council of Ministers in December 1960 as the country's first Minister of Petroleum and Mineral Resources and his association with the 'Free Princes' were probably unforgiveable to the Crown Prince. As it was, Tariki's fate was no doubt sealed by his public complaints about the Arabian Oil Company's deal with Kamal Adham.

Feisal could be as vindictive as he could be forgiving. Tariki first sought exile in Cairo where he resisted attempts by the Egyptian authorities to make him serve their propaganda machine. Subsequently, later in the summer of 1962, he proceeded to Beirut where he established himself as an oil consultant, as ready as ever to expound in public on petroleum policy. He published an article saying that the Prophet had been the first socialist. This was the occasion for Feisal putting pressure on President Chamoun of Lebanon to force him out of the country. Tariki then took refuge in Kuwait.

Tariki's dismissal brought to power a successor who was to become far more illustrious. Ahmed Zaki Yamani had been legal adviser to the Council of Ministers during Feisal's spell in power from 1958 to 1960. The Meccan-born jurist was to prove as astute, charming and calculating as Tariki was gauche, ill-humoured and disingenuous. From the start Yamani was, in the words of an O.P.E.C. colleague, 'a man of strategy and realism'. When he took office Yamani was by no means unknown to Aramco. In 1961 the company had been faced by a crisis seemingly trivial but of serious import. A not very significant businessman of Hasa had for some time been demanding that Aramco should sell more of its oil to him. As was its right, the company refused. It became alarmed, though, when the merchant took the matter to the *sharia* court, a step that put in question the company's concession agreement and might have invalidated it entirely. Aramco called in the honey-tongued lawyer who smoothly sorted out the problem. As a former Secretary-General of O.P.E.C. put it, 'He got more out of Aramco because he got on with Aramco.' Within one year Yamani had successfully negotiated a retroactive settlement of the dispute over the Sidon price differential, and took credit for it. Behind the scenes, however, the U.S. Government, in its concern for the stability of the Kingdom after the outbreak of the Yemen civil war, used its influence to persuade the Aramco partners to concede more. It brought a down payment of $160 million and better terms for the future.

From the beginning of 1962 relations between Saudi Arabia and Egypt deteriorated steadily and became very tense in the spring just before the time of the *Hajj*. Cairo alleged that a Saudi vessel had fired upon Egyptian fishing boats in the Red Sea. Riyadh ordered the return home of Egyptian workers, which had started in the previous year, to be expedited. Egypt complained about harassment of its nationals making the *Hajj*, who were told that their Egyptian pounds were unacceptable and could not be changed for riyals. One shipload did not disembark at all after the Saudi authorities had delivered a sharp insult. It was the old problem of the *kiswa*, the woven covering for the *kaaba*, which had been the centre of the riot in 1926. The Saudis now refused to accept it on the grounds that the fabric was of too poor quality to cover the cube around which the pilgrims make a seven-fold circumambulation. By mid-summer verbal warfare had reached a strident pitch.

Apprehension and uncertainty at the royal court was compounded by the activities of Talal. The prince incurred the King's displeasure when he cabled Nasser on 23 July congratulating him on the tenth anniversary of the Egyptian revolution. A few weeks later he announced in Beirut that he and his half-brothers Abdul Mohsen and Nawwaf had decided to free their concubines and slaves. Talal revealed that he had thirty-two of the former and fifty of the latter. The event was enough to precipitate a meeting of the Council of Ministers in Taif. After it, Saud was infuriated but impotent. Talal's passport was withdrawn on 16 August while he was in Beirut. His half-brothers, Fawwaz and Badr, together with their cousin Saud bin Fahd, handed in theirs in sympathy. On the same day Talal, dressed in a cream-coloured suit, gave a press conference. He explained that there was nothing personal about his conflict with Saud and Feisal. It was simply a question of democratic government. In some form, he believed, it was feasible for the Kingdom, despite the high level of illiteracy existing there. Talal envisaged a system suited to the country's needs, and social justice not of left-wing form but 'in a moderate form'. Nawwaf was quick to make his peace with the King and Crown

Prince. But three days later, Talal, Fawwaz, Badr, and Saud bin Fahd were provided with passports and left the country. They settled in Cairo, after Feisal had again put the squeeze on Lebanon. Later they were joined there by Abdul Mohsen.

The errant members of the royal family were from the start exploited by the Egyptian propaganda machine. Talal claimed that the Egyptian leader made the formation of a league of 'Free Princes' a condition for letting the five stay in Cairo and that they were discredited by being forced to go there. Certainly, Feisal wanted no opposition in the Kingdom. But the open breach in the ranks of the royal family could not have come at a more embarrassing time.

15 The Fox and the Lion

1962 – 7

Arabia should be at leisure to fight out its own fatal complex destiny.
T. E. LAWRENCE

THE death of the Imam Ahmed of Yemen, who had succeeded his father Yahya in 1948, was announced on 20 September 1962. Cairo Radio triumphantly declared that Saud and King Hussein of Jordan were just as dead but not yet buried. The event was enough in itself to cause qualms in Saudi Arabia where the bulging-eyed old tyrant's successor, the Crown Prince Muhammad Badr, was looked upon with grave misgivings because of his association with President Nasser of Egypt and dangerously liberal ideas. But the House of Saud was more than ever concerned with the principle of legitimacy and the King was quick to send warm greetings to the new Imam on his succession.

In 1956, when the Kingdom had been treading the path of positive neutrality with most conviction, Saud had joined with Nasser and the Imam in a mutual defence agreement under which they undertook to come to each other's aid in the event of an aggression against any one of them. The three states had in common, at least, hostility to the British colonial presence in South Arabia and the Gulf. But perhaps the most important result of the pact was the opportunity afforded Nasser to charm, influence and deceive the Yemeni Crown Prince. It was at Nasser's suggestion that he made the journey in 1958 which resulted in trade pacts with the Soviet Union and China as well as the purchase of Soviet arms paid for by a $3 million gift from Saud. More ominously, the way was opened for the dispatch of an Egyptian military mission and the training of young Yemeni officers in Egypt.

In 1958 Nasser issued an open invitation to other Arab states to join a wider Federation of Arab States based on the Egyptian-Syrian merger. He was embarrassed that the first and only response came from the despotic old Imam, a move for which the Crown Prince was largely responsible. But it did enable Nasser to further revolutionary goals in the Yemen. Saud had been sufficiently concerned about the Imam's adherence to the Federation to dispatch an envoy to Sana in an attempt to change his mind. His representative was kept waiting nineteen days before receiving an audience and then was rebuffed.

With the secession of Syria from the United Arab Republic the Federation of Arab States evaporated, the Imam sent Egyptian teachers back home and then alienated Nasser by publishing a provocative piece of doggerel condemning Nasser's socialism as atheistical and asking, 'Why do you pollute the atmosphere with abuse?'[1] Relations with Saudi Arabia had improved accordingly. Yet the damage had been done in Yemen, which was ripe for revolution.

Within a week of the new Imam's accession, on the night of 26 September 1962, a group of young army officers trained the guns of the very tanks bought by Muhammad Badr six years before on to his palace and reduced it to rubble. For a fortnight the new Imam was believed to be both dead and buried beneath the fallen masonry. It was widely believed that the Egyptian Chargé d'Affaires in Yemen, who was well aware of the conspiracies brewing, gave the signal. Cairo had long been nurturing the overthrow of what seemed the most fragile regime in the Arab world. Almost confused at what they had done, the lieutenants who had taken the initiative turned to Brigadier Abdullah Sallal, the 'progressive' whom Muhammad Badr had naively appointed Chief of Staff, and asked for his co-operation. Sallal gave it on condition that he should be President and he was duly proclaimed as such on the day after the coup.

The Brigadier and his collaborators were uncertain about the implications of their success except that to maintain it they must turn to Egypt. Within four days Egyptian troops had arrived by air together with the leaders that Cairo had groomed to lead Yemen into a new era, including Dr Abdul Rahman Beidani, Muhammad Mahmoud Zubairi and Mohsen Aini. More disembarked at Hodeida on the Red Sea on 6 October. Whether or not Cairo knew in advance about the timing of the coup, it gave Nasser a chance to salvage pride and prestige lost as a result of the break-up of the United Arab Republic. Nor could such a tactician have ignored the possibilities of undermining the British presence in South Arabia through control of Yemen, and if one is to take Nasser's propaganda at its face value then he also saw the overthrow of the Hamiduddin dynasty as a means of bringing down the monarchical regime in Saudi Arabia.

The Kingdom was almost friendless and very feeble. In the Arab world it could only count on the lukewarm friendship of King Hussein of Jordan with whose dynasty the House of Saud had only recently become reconciled. Amongst the conservative Islamic states, Pakistan and, to a lesser extent, Iran, were sympathetic to its predicament. The U.S. was regarded as an ambiguous ally. In Feisal's view, American friendship was irritatingly – and inexplicably – compromised by the preoccupation of the 'new frontiersmen' with not being identified with anachronistic regimes and keeping on the best terms with Nasser's Egypt to prevent it from slipping into the communist orbit. With Algeria's attainment of independence, the Kingdom had in 1962 agreed to resume diplomatic relations with France. It had none with Britain. That country's Conservative Government was anxious about Egyptian expansionism and concerned to shield the South Arabian Federation, which it had formed out of its Aden Protectorates, from the forces of radicalism, but diplomatic contacts over the previous four years had always foundered upon Saudi insistence that renewal of these ties was dependent on a settlement of the Buraimi dispute. Saudi Arabia looked and felt very isolated.

The state had been strengthened by the financial retrenchment supervised by Feisal and its rising oil revenues. But it still seemed too absurdly archaic to withstand the new challenge presented by Nasser. At that time its military capability was almost non-existent and in so far as it existed was regarded as more a danger to the regime than an obstacle to any invader. The regular Armed Forces were deliberately prevented from becoming too strong. In 1962 the loyalty of the Army and Air Force was suspect. Only the paramilitary National Guard could be relied upon.

Feisal reacted to the news of the overthrow of the Imam of Yemen with typical caution and deliberation. He was in New York for the session of the U.N. General Assembly when news of the coup broke. He met there with Prince Hassan bin Yahya, who had been representing Yemen at the U.N. and had been proclaimed Imam when Muhammad Badr's death had been announced. Feisal told him that the Saudi Government was sympathetic to the Royalists but though it recognized the legitimacy of the Hamiduddin it could not extend aid. Like the rest of his clan, Feisal had no love for the dynasty despite their shared interest in preserving traditional regimes and common fear of Egyptian revolutionary ambitions.

From the start Feisal was anxious to avoid any action which might provoke Nasser unnecessarily. Despite the mounting propaganda aimed at bringing down the Saudi monarchy he was sticking as long as possible to his policy of accommodating the Egyptian leader. He had two great advantages over Nasser: he was unflappable by nature, and he also knew at first hand the Yemen, its tribes and its territory. Feisal was shrewd enough to realize, even at this early stage, that Egyptian interference could be self-defeating. Feisal took the line that the people of Yemen should be allowed to decide their own form of government and he adhered to it throughout the conflict. He was equally consistent in his belief that diplomatic avenues should be used exhaustively to bring an end to the conflict and bring about Egyptian withdrawal.

When Feisal set out for the U.S. in September (he arrived before the coup in Sana took place) his main objective had been to protest to Kennedy about his support for Nasser, whose regime was the biggest recipient of American aid in the Arab world, and convince him of the folly of the 'new frontiersmen' in trying to contain Arab radical power through friendship with Egypt. Feisal was well aware of the long-term importance of the Kingdom's oil to the U.S. and sought to persuade Kennedy that America's best interests lay in Saudi Arabia and not Egypt. To him the idealistic Administration's attempt to keep on good terms with both camps in a polarized Arab world was inexplicable. When Feisal had lunch with the American President on 6 October he urged him strongly not to recognize Sallal's regime. Receiving no assurance he went away perplexed and not a little angry.

In the meantime, Prince Hassan bin Yahya and the Royalist Foreign Minister Ahmed Shaami set off for Jeddah and met Saud on 1 October. There, the mood was one of apprehension. The majority of the Council of Ministers was in favour of coming to terms with Sallal and six out of the eight commoners in it submitted a memorandum recommending recognition of the Republican Government. Saud was convinced that the coup was the work of the Egyptian intelligence, regarded it as a mutiny against a legitimate ruler and, rightly, saw it as a threat to himself. He was sufficiently well informed to know that the Republican Government's control over Yemen was minimal and, therefore, the Royalist cause was not dead. With characteristic impetuosity the King offered help, although both he and Hassan agreed that there should be no question of direct Saudi intervention. Thus, in Feisal's absence he decided, with the support of the senior princes and his own advisers, secretly to give money and weapons. Officially the country's stance was one of non-involvement.

No sooner had the decision to aid the Royalists in Yemen been taken than three Saudi air crew on a mission to fly supplies destined for them to the border town of Najran went instead to Cairo where they were promptly given Egyptian uniforms and ranks. On the following day they were joined by two more pilots in a training aircraft. As a result, the Saudi Air Force was grounded and two dozen officers were dismissed in a subsequent purge. Quickly realizing the threat to Saudi Arabia implicit in the Yemeni coup and in accordance with the defence treaty concluded two months before, King Hussein dispatched a squadron of Hunters to Taif to bolster morale. 'Let us bomb Sana,' said Saud in one of his more exuberant moments. The Commander-in-Chief of the Jordanian Air Force and two pilots promptly defected to Egypt, forcing the Hashemite monarch to recall the rest of his aircraft.

On 23 October, Talal, who a week before had renounced his title, announced plans 'to establish a national democratic government and to leave the people free to choose the kind of government they prefer'. The regime, he proclaimed, was 'steeped in backwardness, under-development, reactionary individuals and tyranny'. At the beginning of the month Talal had predicted the disintegration of the Saudi Kingdom into its four main regions – the Hejaz, Nejd, Hasa and Asir. He was confident, too, that his supporters would rise up in revolt. Other more impartial observers also contemplated the possibility of the state being dismembered.

When on 17 October 1962 Feisal was asked by Saud to form a new government he acceded to the request on condition that he should be given carte blanche to rule and also power to deal with the question of the princes' stipends. On 31 October he appointed a new Council of Ministers, sacking the commoners who had been in favour of appeasing Nasser and also the sons of Saud. He brought in some of the strong and resolute members of the family. Whilst retaining the post of Minister of Foreign Affairs, Feisal appointed Khaled as Deputy Prime Minister, Fahd as Minister of the Interior, and Sultan as Minister of Defence. Musaid bin Abdul Rahman kept the Finance and National Economy portfolio.

Egyptian air raids directed at the border towns of Najran and Jizan convinced Feisal of the need to aid the Royalists as well as to break off diplomatic relations with Egypt on 6 November. Financial support began at the rate of 11 million riyals a month, apart from money disbursed directly to tribal chiefs whom the Saudis wished to influence. However, a large proportion was syphoned off into various Saudi and Yemeni pockets so that probably less than 50 per cent, whether in the form of gold or guns, was eventually used for the purpose originally intended. The lion's share, in a ratio of about two-thirds/one-third, went to the eastern front through Najran where the fat but tough, dare-devil Khaled Sudairi was Governor rather than through Jizan to the initially less important western front where the Imam Muhammad Badr was operating. The imbalance was caused largely by the fact that Khaled was more aggressive in his demands than Muhammad Sudairi, Governor of Jizan. Generally, however, Saudi policy was a haphazard one with aid being directed at individual Hamiduddin commanders. The method of distribution tended to exacerbate their differences which badly weakened the Royalist effort.

Such was Saudi Arabia's feeling of beleaguered isolation that Kamal Adham had for some time been making overtures on behalf of Feisal to Britain. Hafiz Wahba, who had been Saudi Ambassador in London, had also been doing what he could to bring about a resumption of diplomatic relations with Britain. The Kingdom had not yet established its own external security service. It needed British assistance in building up its own intelligence apparatus and covert assistance generally. Moreover, Britain was still ensconced in South Arabia to which the tribes of east Yemen enjoyed easy access through the state of Beihan. Despite the contacts in New York and U.N. efforts to mediate, the two countries were no nearer resolution of the Buraimi dispute, but the way towards diplomatic reconciliation was eased by the decision of the British Government not to recognize Sallal's regime.

Washington had put heavy pressure on London to approve the Nasserite manifestation in Sana. But Sir Alec Douglas-Home, the Foreign Secretary, had wilted in the face of strong opposition to recognition of Sallal's regime from Mr Duncan Sandys, the Colonial Secretary, Mr Julian Amery, the Minister of Air, and other right-wingers in the Conservative Government. Of crucial importance was the mission of Lieutenant-Colonel Billy McLean, the M.P. for Inverness, who as Sir Alec's personal emissary travelled through

Royalist-held territory and established emphatically that in terms of control of territory, the support of the people, and prospects of permanence, Sallal fell pathetically short of the normal criteria for recognition.

Agreement on the resumption of diplomatic relations between Britain and Saudi Arabia was not announced until the end of January 1963, but by then Kamal Adham had obtained the intelligence link-up which he preferred to the U.S. Central Intelligence Agency, and several million pounds' worth of gold and light weapons had been airlifted from an R.A.F. station in Wiltshire in an operation organized by M.I.6. They were then transported onwards to Beihan, whose Sherif was well rewarded for passing on the weapons and booty to the tribes in Yemen. Only a small percentage of the rifles, machine-guns and bazookas were probably ever pointed at Egyptian troops. Yet they were considered for morale a cheap investment (by 1964−5 an old surplus Lee Enfield costing £7 10s. in the U.K. could fetch £50−£60 in the Middle East). Feisal's trusted and shrewd adviser, who was now chief of the newly created Foreign Liaison Bureau and was also given general supervision over the General Intelligence Directorate, the internal security service, was nevertheless quick to develop a special relationship with the C.I.A. also, and with the French Service de Documentation Extérieure et de Contre-Espionage.

In the early part of 1963 the flow of British aid in kind was stopped, although subsequently some shipments were made without even the Joint Intelligence Committee knowing about them. Beihan remained open for Saudi-financed convoys and also Western mercenaries. The colonial authorities in Aden regarded them with a benignly blind eye. The Saudis kept a very watchful eye on British activities on the new republic's southern border − anxious to know all the time what they were doing. With or without the knowledge of their Saudi paymasters, M.I.6 were in 1965 to organize parachute drops of supplies to the Royalists in Yemen from Israel in one of the more bizarre and esoteric episodes in the history of the period.

Feisal knew that the danger to the future of the family oligarchy came as much from within as from without. His anxiety can be seen from the alacrity with which he announced on 6 November 1962 a ten-point reform programme. Only one point − the last one listed − had any immediate impact on the country itself or, for that matter, outside it. That concerned the abolition of slavery, which Saudi Arabia's representatives at the U.N. had repeatedly said no longer existed in the Kingdom. It was a fiction impossible to maintain after Talal gave so much publicity to the freeing of his own. It was an embarrassing anachronism and a sensitive, as well as an obvious, target for Cairo's propaganda salvos.

As far back as 1936 Ibn Saud had banned the import-export trade in human souls, but a steady flow had continued under the cover of the *Hajj*. Thousands of pilgrims, particularly from West Africa, did not return home. Some were sold into bondage by their parents to Arab 'missionaries' before they ever left Africa. Others sold themselves into slavery either because a Saudi master offered a more secure subsistence or because they were unable to make their way home. There had also been a steady natural increase in the slave population, which proliferated, notably, in Saud's teeming ménage. In 1962 the number in the Kingdom probably amounted to 30,000, or one in every hundred of the indigenous population. Feisal had owned slaves, but he never claimed to have bought any. Although his Government declined in 1956 to subscribe to the U.N. Supplementary Convention on Slavery, Feisal had manumitted his slaves. Later he complained that none of them wanted their freedom and insisted on staying in his household. The fact was that many slaves felt themselves relatively privileged citizens. The status could even offer the prospect of enrichment. Muhammad Sorour Sabhan had achieved positions of great importance, becoming Finance Minister under Saud, and had consequently amassed a reasonable fortune.

The abolition of slavery was resented by traditionalists, including learned men of religion. They could point to the text of the Koran to show that the Prophet had approved it. Princes of the House of Saud had invested heavily in human flesh and blood. Compensation was offered, and within a year the equivalent of over $10 million had been paid out and 10,000 slaves freed, according to Radio Mecca.[2] Frequently, however, former masters and slaves shared the money and the old domestic situation continued.

The other points in the reform programme covered legal, administrative and social reform. Feisal promised a fundamental law and an expansion of the old Consultative Council, the *Majlis al Shura* established in 1926 and made up of twenty-four *ulema*, dignitaries and merchants, to advise on the government of the Hejaz and to act as an intermediary with its people. By 1962 it had long been defunct. Feisal expressed the hope that enactment of the necessary legislation 'will not be delayed', but eighteen years later it still had not been promulgated.

Other pledges in the reform programme took a long time to
materialize. But the five years following its publication did, indeed,
witness improvements in the administration, a marked increase in the
proportion of funds allocated to public investment, the emergence of a
more rational approach to planning, steady economic growth and
better standards of living. With the inexorable rise in oil production
and revenues, it should hardly have been otherwise. Time, however,
would prove just how slowly the spirit of the programme as a whole
and its specific pledges were to be fulfilled.

A year later, in November 1963, new regulations were issued for
the formation of a provincial government, including the formation
of councils, but the question of their establishment was to remain
indefinitely no more than a worthy intention. From this point
onwards, however, Nejd became known officially as the Central
Province, Hasa as the Eastern Province, Hejaz as the Western
Province, Asir as the Southern Province, and the tracts of land
adjoining the borders of Jordan and Iraq as the Northern Frontier
Province.

The reform programme at least reflected the good intentions of
Feisal, who had been privately assured by President Kennedy that the
U.S. would help Saudi Arabia in as much as it helped itself. As the
National Security Council and the State Department pressed for
recognition of Sallal's regime, Kennedy sought to reassure Saud and
Feisal further and in a letter to the Crown Prince dated 25 October
1962 he pledged 'full U.S. support for the maintenance of Saudi
Arabian integrity'.[3] The Administration then tried unavailingly
through diplomatic channels to bring about Egyptian withdrawal
from Yemen on the one hand, and a cessation of aid from Saudi Arabia
and Jordan to the Royalists on the other. It was a formula – as simple
to propose as it was difficult to implement – that was to be pursued in
vain for nearly five years.

In December 1962 the U.S. Air Force, with facilities at Dhahran
restored to it (even though the 1957 agreement had not been
renewed), overflew Riyadh and Jeddah to bolster Saudi confidence.
Help was given with the erection of some anti-aircraft batteries on the
Yemen border. Then, on 19 December, the U.S. formally recognized
the Yemen Arab Republic, having obtained from Sallal an undertaking
to honour existing treaty obligations – an implicit reaffirmation of the
1934 Treaty of Taif – and another from Egypt expressing its intentions
to withdraw troops from the country. Egypt's proviso was that Saudi

Arabia and Jordan, which had no troops there, did the same. The Republic's credentials were accepted at the U.N., and Sallal had international blessing. Thus fortified, he was proclaiming within a week that he had rockets with the potential of reaching the 'palaces of Saudi Arabia'. The U.S. Air Force again overflew Jeddah and the American Navy sent a destroyer on a courtesy call. Feisal received permission to publish Kennedy's letter, but it was small compensation.

The Crown Prince was bitterly resentful at Washington's reluctance to unequivocally support him rather than Nasser. Moreover, Kennedy had made conditions for the recognition of the Yemen Republic without establishing any mechanism whereby they could be fulfilled. On 6 January 1963 the Saudi Government announced preparations which could only be regarded as symbolic defiance. It appointed a Supreme Defence Council to establish military training centres, started enlisting new recruits, and mobilized the irregulars who formed the bulk of the National Guard. Feisal then dismissed Saad bin Saud from the command of the National Guard, replacing him with the trusted Abdullah. A few days later Egyptian aircraft attacked Najran again.

Time was to prove that Nasser was as hopelessly ill advised about the chances of Sallal subduing Yemen as he was about the prospects of the coup in Sana leading to a revolution in Saudi Arabia. For this, much of the blame rested upon the shoulders of Anwar Sadat, the member of the United Arab Republic's Revolutionary Command Council to whom Nasser delegated Yemen policy. The impetuous Sadat embarked upon the campaign enthusiastically, unlike harder-headed colleagues such as Zakariah Mohieddin who later was to incur Nasser's displeasure because of his opposition to the costly adventure. Sadat leant heavily for advice on such Egyptian protégés as his brother-in-law Dr Abdul Rahman Beidani. Beidani understood little of the sectarian realities of Yemen. Nevertheless, the Egyptian expeditionary force got off to a good start as more troops poured in to double its size from 20,000 to 40,000 men. After a visit by Field Marshal Abdul Hakim Amer to Sana, the 'Ramadan offensive' was launched. An armoured force swept in a wide arc north-east of Sana through the Jauf region, which was secured, westwards into the trackless deserts beyond the highlands – and perhaps through Saudi territory – to capture Marib and Harib, the latter a town on the border of Beihan.

Even before the offensive Feisal had asked the U.S. to arrange some kind of mediation. In March it arrived in the form of Dr Ralph Bunche, wearing the mantle of the U.N. in which he had won plaudits for his role in helping to bring about the 1949 Arab-Israeli armistice agreement. Dispatched after consultations between the U.S. State Department and U Thant, the U.N. Secretary-General, he visited only Yemen and Egypt, avoiding Saudi Arabia lest he be embarrassed by meeting any Royalists there. The demonstrations of Republican enthusiasm that greeted him at Taiz, Sana and Marib were well organized. Dr Bunche was evidently impressed also by his talks with Sallal, Amer and Nasser. He was satisfied that the Republican Government was in control of the country. He seemed convinced by the sincerity of Nasser's willingness to pull out his military forces in exchange for a Saudi stand-off.

Dr Ellsworth Bunker, an old State Department troubleshooter, was then duly sent to Saudi Arabia to offer U.S. air cover in the form of eight fighter aircraft and a commitment by Washington to persuade Nasser to withdraw his troops in exchange for a Saudi agreement to stop giving the Royalists aid. 'With sustained patience and consummate skill, Ambassador Bunker worked with Cairo and Riadh until he was able to secure from the United Arab Republic and Saudi Arabia an agreement for mutual withdrawal,' wrote John S. Badeau,[4] who was U.S. Ambassador in Cairo at the time and a leading proponent of conciliating Nasser. In fact, Bunker enraged Feisal by offering American favours, including experts to advise on the construction of a television system for the Kingdom, in return for promises to reform and develop his country.[5] More patronizing and provocative still was the emissary's suggestion that a disengagement would give the Crown Prince an opportunity to deal with 'unrest and rebellion' in his own country. This was particularly galling as in February a cache of arms weighing several tons had been discovered in the hills north-east of Jeddah the morning after a night parachute drop by the Egyptian Air Force. But there was no sign of co-ordination with subversive Saudi elements – if any existed.

Nevertheless, Feisal accepted the stationing of the eight American fighter aircraft on Saudi soil under what was known as 'Operation Hard-Surface'. His understanding evidently was that they would protect his air space, although the instructions of the force were simply to defend themselves while on patrol. Only after receiving a telegram from Kennedy guaranteeing Egyptian good faith did Feisal enter into the Pax Americana sanctioned under a U.N. seal towards the end of April 1963. Included in it were provisions for a buffer zone twenty kilometres either side of the border and a U.N. observer mission. Major-General Carl von Horn, its Swedish commander, had grave doubts whether either Egypt or Saudi Arabia seriously intended to honour the terms of the disengagement agreement.[6] Only the day after it came into force Amer told von Horn that Egypt, initially at any rate, would not pull back from the buffer zone and would leave some security forces there indefinitely to ensure the survival of Sallal's regime. Omar Saqqaf, Under-Secretary at the Saudi Ministry of Foreign Affairs, left a much better impression. He gave an unqualified assurance, on Feisal's word of honour, that Saudi Arabia had ceased to send the Royalists arms, ammunition, money and men.[7]

The Royalists and their sympathizers were anyway probably well stocked with equipment for a guerrilla war. However, during this period supplies unauthorized by the Crown Prince did cross the border. Von Horn had no doubt that Sultan – 'a volatile and emotional young man' – and other princely figures were responsible for such breaches of the agreement.[8] The Egyptians, in the meantime, withdrew two weakened divisions and replaced them with full strength ones. Their Ilyushin-28 bombers continued their attacks on dissident tribes and then towards the end of May 1963 struck Jizan again, forcing most of the town's population of 40,000 to flee in terror. In the mountains of Yemen the fighting continued with no appreciable gains being made by either side.

In July the first report that the Egyptian Air Force was dropping gas bombs on Yemeni villages in Royalist-held areas was published.[9] For four years confused debate and even more indignant outrage centred on the alleged atrocities that seemed confirmed by witnesses of undoubted integrity such as Lieutenant-Colonel Billy McLean and Colonel David Smiley, who was a British mercenary advising the Imam's forces, as well as journalists and officials of the International Red Cross, including André Rochat, head of its Yemen organization. There was no doubt of the suffering inflicted by the raids on hapless villagers, many of whom died in agony. Fragments of canisters were sent back to Britain, but scientists at the Government scientific research centre at Porton Down in southern England were unable to establish that toxic substances had been used of the kind banned in warfare by the Geneva Convention. It was thought that death by burns and respiratory damage was probably caused by phosphorous or other incendiary materials.

Von Horn's force was desperately undermanned and ill equipped to survey 6,000 square miles of 'heat-crushed desert and mountain'. He resigned on 20 August 1963 to be replaced by his Italian deputy, Pier P. Spinelli. Finally, U Thant withdrew the military members of the mission and only a few dozen civilians were left.

After less than a year of fighting it was already clear that no simple military solution was possible in Yemen and the war was bleeding Egypt rather than Saudi Arabia dry. Feisal was thinking in terms of a compromise whereby the Royalists and Republicans would sort out their differences and Nasser was under pressure to honour the American-negotiated agreement, not the least because of warnings from the U.S. that they might cut off aid if he did not. But El Rais – 'the leader', as Nasser was known – could not contemplate the consequences of withdrawal – the collapse of his protégé Sallal and the loss of Egypt's foothold on the oil-rich Arabian peninsula.

Nevertheless, as 1963 drew to a close, his attention focused on an Israeli plan to divert the headwaters of the River Jordan and thereby deprive Arab farmers of their chief source of fresh water. This challenge to Arab nationalism threatened to provoke a flare-up between Syria and Israel. Unwilling to be drawn by his revolutionary rivals into a battle which he could not win, Nasser wanted to make the question of the Jordan River a collective responsibility. Thus, in a spirit of reconciliation, he issued an invitation to the heads of thirteen states to attend the first Arab summit in Cairo in January 1964. The meeting resulted in the formation of a meaningless joint military headquarters under Amer. More constructively, it signalled a new era of co-existence between Egypt and the conservative Arab governments, paving the way for a resumption of diplomatic relations with Saudi Arabia.

The summit also brought to a head the long power struggle between Feisal and Saud. The King represented his country at the conference, and was duly embraced by Nasser on his arrival at Cairo airport and treated with the respect befitting a monarch. His participation at the summit was his first public act of significance since he handed over the Government to Feisal in October 1962.

Under pressure from the rest of the family to absent himself from the Kingdom, Saud had spent much of the intervening period abroad. In March 1963 he flew to Lausanne for medical treatment. He then proceeded to Nice where his retinue of 160, including wives, concubines and children, occupied eighty suites at the Negresco Hotel at an estimated cost of £20,000 a week. In defiance of his doctors and the susceptibilities of his stomach, the King consumed half a dozen plates of crushed bananas and ice cream daily. Accustomed as they were to extravagance and eccentricity, the people of Nice observed with amazement Saud's exotic entourage. The royal aircraft, a de Havilland Comet IV purchased for £1.3 million in 1962, was sent back to Geneva to pick up 300 pieces of baggage and yet more members of the party. On its return it crashed into the Monte Matto, fifty miles north of Nice, killing the British air crew and all those on board, none of whom were members of Saud's own immediate family. Poor maintenance or pilot fatigue were the probable causes. The royal party overstayed its welcome at the Negresco and the management had to billet 130 American tourists in other hotels.

From Nice Saud went on to Paris for further medical observation at the Clinique Violet, staying in the clinic nearly one week. His departure to Riyadh after nearly two months in Paris was delayed by a last-minute purchase of four hundredweight of fruit, which had to be rushed from Les Halles to the airport, and the disappearance of a car carrying some of his womenfolk, which took a wrong turning. In less than three weeks the King was on his way back to Europe complaining of stomach pains. A few days after his arrival in Vienna on 12 May 1963 he went into hospital and spent some days under an oxygen tent. A duodenal ulcer was diagnosed. Saud then rented a villa at Hinter Brühl, near Vienna, and his harem took up residence at a nearby sanatorium. By mid-summer about 100 of his courtiers had arrived in Austria, and several of his progeny had purchased no less than four substantial properties. The world at large began to speculate that he intended to abdicate and that Austria was his chosen exile. However, questioned about Saud's plans, Abdul Moneim Aqil, the King's Chief of Royal Protocol, said that the King was in daily short-wave contact with the Government and ruling by radio. Was it true that he had relinquished effective power to his brother? 'Certainly not. He is King and everyone knows that,' answered Aqil. After so much embarrassing publicity the consensus of the Al Saud was that it would be better to have Saud back in the Kingdom than abroad. He returned home in mid-September determined once more to rule rather than reign.

Saud had already been angered by Feisal's decision in his absence to post infantry and tank units of the Royal Guard, the King's private army, on the Yemen border in the vicinity of Najran. Tension increased during the autumn when Feisal drew up a budget for the following fiscal year reducing the allocation for the royal family. Co-existence between King and Crown Prince became even more strained when the former, as part of a deliberate campaign, began to disburse money amongst such influential tribes as the Harb, Shammar, and Utaibah. Then the King and his sons tried to rally support to bring about Feisal's resignation and recovery of control of the Kingdom for themselves. Their efforts were an abysmal, miscalculated failure. All the princes of any importance sided with the Crown Prince. The hotter and stronger heads like Muhammad, Abdullah and Sultan now pressed for Saud's abdication in favour of Feisal, demanding that it should be brought about by force if necessary. By this time the Crown Prince's contempt and hatred for his elder half-brother knew almost no bounds. He would never mention the King by name, referring to him as 'that man' or 'him'. He was resigned to the fact that Saud would have to be formally stripped of all power, but insisted that the transition should be a peaceful one with the blessing of the religious leaders, and also appear to be the responsibility of others.

Ranks were closing against Saud within the royal family. The faction arrayed against him was strengthened late in the summer of 1963 when Talal made his peace, tacitly and indirectly, with Feisal through the mediation of his mother Munaiyer. The princely defectors had been encouraged by Nasser to establish an 'Arab National Liberation Front' that was allowed half an hour a day on Radio Cairo to beam programmes to Saudi Arabia. But they soon became disillusioned and embarrassed by the content of the broadcasts made in their names which were dictated by Egyptian propaganda chiefs. Privately assured that their assets would be unfrozen and that they would not be punished, the four brothers and their cousin returned home in the autumn.

Hostility towards the King was such that on the night of 15 December 1963 Saud deployed some 800 troops of his Royal Guard around his Nasariya Palace. The National Guard, now firmly under the command of Abdullah and loyal to the anti-Saud faction, made countervailing moves. Saud was put in his place by the delivery to him of a document bearing the names of the leading princes and religious leaders reaffirming the position as it had existed since October 1962, namely that his prerogatives did not extend beyond the right to sign decrees and to be consulted on important matters. He was also left in command of the Royal Guard.

Saud remained head of state and as such went to Cairo as the Kingdom's representative at the first Arab summit meeting. The Egyptian leader's flattery helped to restore his self-esteem and Saud returned to the Kingdom determined once more to be a constitutional monarch. His ego was soon deflated. Early in March 1964 he was excluded from the consultations with Amer and Sadat who came from Cairo for talks on a reconciliation between Egypt and Saudi Arabia. In fury Saud wrote to the Crown Prince on 13 March demanding that he be treated with the honour befitting his rank, that his prerogatives should be fully restored and two at least of his sons should be given positions in the Council of Ministers. He went as far as to threaten to turn the Royal Guard's artillery onto Feisal's villa. The letter was returned by Muhammad, who strode through the portals of the Nasariya Palace and threw it at the King's feet. It is also said by some who were close observers of the unfolding drama that the prince even threatened to run his sword through Saud's corpulent frame. Such action would have been entirely in keeping with Muhammad's notorious temper.

Crisis point had been reached. The National Guard was put on full alert. To preserve appearances of noble detachment Feisal went to Jeddah for a while. One night three of the 'Little Kings' sat on the hillside jeering at the encampment of the National Guard which had taken up positions on the Muttlah Ridge and cut the main route into Riyadh. In the morning three empty gin bottles were found where the princelings had sat. The King stayed cocooned in the Nasariya Palace with his advisers and household, hoping that the tribes would come to his assistance. According to one intelligence estimate, he had distributed £30–£40 million in bounty amongst them over the previous six months. But the tribes did not move. Nor could Saud look to the National Guard for salvation.

A well calculated policy of subsidies had secured the loyalty of the Kingdom's most effective fighting force to Abdullah who had taken command of it early in 1963 and was, in turn, totally loyal to Feisal. He had to be restrained by the Crown Prince who refused to permit the National Guard to overthrow Saud and continued to insist on a 'constitutional' solution. At this critical juncture his policy was one of masterly inactivity. Having returned from Jeddah, with dignified aplomb he drove daily on his way to the office past the Royal Guardsmen who were becoming increasingly nervous. They sheepishly saluted him. 'You will see,' he told one foreign confidant, 'the people will laugh at them.' On 26 March 1964 a deputation of religious leaders and tribal sheikhs visited Feisal and declared their moral support for him. Later that night the Crown Prince sent a message to the King informing him that henceforth the Royal Guard would be answerable to the Minister of Defence. Officers of the force accepted the dictat and swore allegiance to Feisal. Major-General Othman Humaid, its commander, was already firmly committed to him.

On 29 March a dozen of the *ulema* issued a *fatwah*, or judgement, declaring that the King was unfit to govern. It was an archaic 'constitutional' device discovered and utilized by Feisal, who was far less passive in the whole affair than he wished to appear and than legend has it. Yet undoubtedly the initiative behind the crucial move came from Muhammad, with support from Abdullah and Fahd and his full brothers. Their uncle Abdullah bin Abdul Rahman also played a crucial role as an intermediary with the theologians. Doubts were expressed as to whether the decision was valid because it had not been issued in the name of all the *ulema*. Effectively, however, a bloodless transfer of power to Feisal as Viceroy, if not Regent, had been sanctified. Decrees were issued on 30 March backed by a letter from sixty of the senior princes quoting the *fatwah* of the leading *ulema*. Saud was deprived of all executive, legal and administrative powers. His private allowance was slashed by half to 183 million riyals, or the equivalent of $40 million. The King appeared to have been finally reduced to a figurehead.

The situation, however, remained anomalous. Saud was still an embarrassing anachronism to Feisal and his supporters who were struggling against the forces of revolution at home and abroad. Moreover, the moribund body refused to lie down. Stimulated by his corrupt clique, Saud started once more to intrigue. He continued attempts to win over the tribes of Hasa. There were also suspicions at this time that he was making his own contacts with Egypt. It was rumoured that he was plotting the assassination of the Crown Prince.

After the events of March, Feisal remained rooted in Jeddah apart from two forays abroad to the Arab summit in Alexandria in September 1964 and the meeting of non-aligned states in Cairo in October. With the new fiscal year on the horizon he had to prepare a budget on his return. Again he proposed a cut in the allocation for the King. He even attempted to persuade him and his sons of the necessity for it. The measure was of more political than economic importance because already increased oil revenues were defying the Government's ability to spend and the country's capacity to absorb them. Feisal explained the position to the senior princes and let events take their course. Towards the end of October he set out from Jeddah on a meandering journey punctuated by visits to tribal sheikhs and a big gathering of them at Taif. By the time he arrived in Riyadh an overwhelming consensus of the royal family, the *ulema* in its totality and – last and least, as a cipher – the Council of Ministers, had decided that Saud must abdicate. On 2 November Feisal swore on the Koran to govern in accordance with the precepts of Islam and the traditions of the country. His inhibitions overcome and his respect for legitimacy satisfied, Feisal was at last King. He gave Id bin Salem forty-eight hours to leave the country. Abdul Aziz Muammer, the former Director of Labour whom Feisal was convinced was a communist, was recalled home from Switzerland and foolishly obeyed the order – he was flung into prison in Hofuf and not released until ten years later, a physical, almost blind wreck. As for the deposed King, his successor said: 'Saud is our brother and we shall do our best to ensure his comfort.'[10]

Early in 1965, accompanied by a huge retinue apart from his harem and teeming progeny, Saud departed for exile in Greece, grudgingly recognizing Feisal as King before he left.

Not until March 1965 was the name of the new Crown Prince announced. The nomination of Khaled surprised many observers at the time because his uterine brother Muhammad was the third surviving son of Abdul Aziz. Within the House of Saud, however, no firm principles relating to the succession had been established. Generally, the practice – whether the leadership passed to a brother or son (the latter having been the usual procedure) – had been to choose the eldest acceptable candidate. But the rule of primogeniture had never been adopted. In his younger days Muhammad had certainly been ambitious and in the years immediately prior to his father's death it had been widely suspected that he was planning a bid for the throne. He was not only one of the more competent and intelligent of the senior princes but had a forceful character. Yet a year at least before the deposition of Saud he had been forced, perhaps reluctantly, to renounce any firm expectation of becoming the next in line. He did so because the consensus in the ruling hierarchy was that he should be passed over. For this there were understood to be two related reasons. Though normally of a pleasant and humorous disposition, he was liable to break out into almost uncontrollable rages. Though no alcoholic, Muhammad, it was well known, indulged in heavy drinking bouts. The two weaknesses often coincided. There was also some apprehension that he might fall prey to self-serving advisers. 'We do not want to be ruled by women and drivers,' was one princely comment heard at the time – a reference to Saud and Id bin Salem.

An understanding that Khaled rather than Muhammad should succeed is believed to have been reached in the autumn of 1963 when the conviction that Saud should be deposed was hardening. In terms of ability and aptitude, Fahd, the seventh surviving son (with Nasir and Saad senior to him), might have appeared a more suitable choice. Khaled by contrast, with his lack of interest in affairs of state and fondness for hunting, had no ambition. Yet he was the best possible compromise as far as the all-important consensus was concerned – more acceptable to the grudging Muhammad, and to most of the ruling hierarchy preferable to Fahd who tended to keep his own counsel and was regarded as aloof.

*   *   *

Resolution of the power struggle within the Saudi royal family happily coincided with détente in the confrontation between radicals and conservatives in the Arab world. The Kingdom could breathe more easily in the era of reconciliation ushered in by the Cairo summit conference in January 1964.

Despite the setbacks suffered by the Yemeni Royalists in the early part of 1963, the Republicans had failed to gain anything like control of the whole country. The relatively sophisticated Egyptian military machine had been contained and Nasser's enthusiasm for the adventure somewhat drained, even though he could in no way renounce his commitment to the revolution. Kennedy had been assassinated – to the delight of the Royalists – and scepticism about aid to Nasser was growing in Washington. Early in 1964 Feisal could regard with equanimity the departure of the 'Operation Hard-Surface' squadron after a row with the U.S. State Department over the disengagement agreement.

In July 1964 diplomatic relations between Saudi Arabia and Egypt were formally resumed. Feisal chose as his Ambassador Muhammad Alireza, the prominent member of the Jeddah merchant family who had been involved in the Onassis affair. Abdullah Suleiman apart, he had been the first commoner to be appointed to the Council of Ministers, in which he served as Minister of Commerce in 1954. His appointment was unrelated to his role as go-between in fixing the Satco deal prior to taking his office. An able man, he rendered signal services to his country. Swept away by nationalist fervour, Alireza had departed in 1956 to fight the Anglo-French forces in the Suez Canal Zone, thereby endearing himself to Nasser. Subsequently, he had become identified in Feisal's eyes with the constitutionalists. They had fallen out, but Alireza had made his peace with the Crown Prince when he took control of the Government in the autumn of 1962. Now Feisal needed him for a most sensitive assignment. As he set out for Cairo the sole instructions given to the envoy were to improve relations with Nasser. It was in an atmosphere of intense suspicion that he took up his post.

In the Yemen, the Egyptians made a last desperate bid to gain a decisive upper hand and encircle the Imam before the Arab summit conference set for 5–11 September 1964 at Alexandria. The offensive failed miserably in its objective. Nevertheless, Feisal hesitated to attend the meeting because he objected to what he regarded as attacks upon him by Muhammad Heikal, the editor of *Al Ahram*. On his arrival Nasser greeted him with a cool handshake in contrast to the warm embrace extended to those like Sallal with good revolutionary credentials.

Discussions at the summit concerned mainly a joint programme of action to counter Israel's scheme to exploit the waters of the River Jordan and the establishment of a Palestine Army as part of the joint military command. The gathering was also notable in that the use of oil as leverage against Israel was seriously discussed at a high level for the first time. But Feisal was not enthusiastic. 'The consolidation of the Arab struggle in the Occupied South and Oman' was also on the agenda and British imperialism there ritualistically condemned. Yemen was discussed privately between Nasser and Feisal who stayed on after the formal pan-Arab gathering had ended.

Nasser could be very charming if he wanted to be. He could also make a distinction between affairs of state and personal relations of a kind that Feisal in his dour struggle to preserve the Al Saud estate did not appreciate. He was not amused when El Rais pointed at a sultry-looking Sadat and said in jest: 'This man is responsible. Take him to court.' Encouragingly, however, the remark indicated Nasser to be on the defensive. In vain the Egyptian leader mobilized President Aref of Iraq and President Ben Bella of Algeria to use their persuasions to get Feisal to recognize Sallal's regime in Yemen. Feisal let Nasser know that Saudi Arabia was not committed to the restoration of the Hamiduddin dynasty, but stuck firmly to the position that the people of Yemen should be left to settle their own destiny without the tutelage of Egypt and the distorting presence of its military forces. For his part Nasser conceded that the regime in Sana should be more broadly based and include representation of those tribal elements which had been so bitterly resisting the occupying army. They had reached an understanding rather than an agreement. The Lion and Fox of Araby embraced before the latter departed.

Royalists and Republicans met at Irkwit on the Sudan's Red Sea coast and agreed on a ceasefire to come into effect on 8 November 1964, and a conference of 168 Yemeni notables was planned to take place in the town of Haradh starting on 23 November. But on 10 November the Egyptian Air Force was in action again.

Perhaps there was provocation from Royalist tribes. Some prominent Saudis, too, had a vested interest in seeing the conflict continue. A more fundamental cause of the resumption of hostilities was Egyptian alarm at the implications of the Irkwit meeting. Moderate Republicans of the sort who might make a reconciliation work were not enthusiastic about attending the proposed peace conference and taking instructions from their Egyptian mentors.

Once again the Egyptians tried to capture or drive out the Imam. Their forces rose to a peak of 60,000 men. But their probes into the mountains were ineffectual. Meanwhile, even Sallal's sponsors began to appreciate that he had no future. The Republican Government was beginning to disintegrate. Nasser's politically ham-handed satrap had progressively alienated moderate nationalists opposed to the Imam. In March 1965 a number of such Republicans visited Jeddah for secret talks convened by Kamal Adham about the formation of a compromise regime. As they had done at Irkwit, they insisted on the exclusion of the Hamiduddin.

To the Saudis the situation was somewhat disturbing. The Egyptians might, in their frustration, invade the Kingdom and mount an offensive across its flat desert territory to encircle the Imam and completely strangle the supply routes to the Royalists. Troop concentrations suggested that that was the intention and it seemed the only way of dislodging the Imam and his followers from their mountain fastnesses. During the first half of 1965 apprehension about such a possibility grew. Saudi Arabia's military defences were lamentably inadequate. Feisal, however, kept his head. He complained to the Royalists that he was not getting value for money on the military front. They, in turn, grumbled that they were receiving insufficient aid. 'My own view was that the Saudis, for their own reasons, were giving the Royalists just enough help to prolong the war but not enough to win it outright,' observed Colonel Smiley.[11]

In the wake of the coup in Yemen the Saudi regime had been slow to do anything about improving its military capability. Despite continuing differences over Buraimi, diplomatic relations with the U.K. had been resumed in 1963, and this made it possible for Feisal to request the dispatch of a British military mission to assist with the training of the National Guard. But his concern was more with internal than external threats. Abdullah, Commander of the National Guard, enquired about a weapon suitable for destroying armoured cars – making it implicitly clear that the main preoccupation was defending the regime from the regular Saudi Army. A film presentation of the Vigilant wire-guided anti-tank missile was duly arranged and shown by a team from the British Aircraft Corporation. The reaction displayed a certain naivety, according to one witness. 'Very good,' commented a National Guard Colonel, 'but where will we find the men prepared to attach the wire?' Abdullah asked whether the missiles could be fired from camels. Only if the National Guard was prepared to expend a beast for each one fired, replied a B.A.C. man wryly. 'Of camels, we have plenty,' retorted Abdullah.

After the shock caused by the defection of the pilots in 1962 the Saudi Government proceeded with painful caution to implement plans for providing the Kingdom with a fig leaf of air cover. The obsolete Vampires acquired from Britain in the early 1950s and the F-86 Sabres provided by the U.S. after Saud's visit to Washington in 1957 were not operational. The 'Operation Hard-Surface' fighters of the U.S. Air Force had departed in January 1964.

The first modern aircraft to be ordered were five Hercules C-130s. They were badly needed to transport the troops to one of several potential points of crisis. The deal was assisted by Adnan Khashoggi, an obscure 26-year-old businessman who within a decade was to achieve international fame and notoriety. Having obtained a retainer from Lockheed – a modest $2,000 a month – he quickly succeeded in arranging the transaction for the C-130s and was rewarded with 2 per cent of the sales price plus $41,000 per aircraft. But they were incidental to the Kingdom's desperate need for an air defence system. Negotiations on the purchase of one began in 1965. Competition for the order for combat aircraft reached a critical point by the late summer of that year. The three main contenders were Lockheed's F-104 Starfighter, Northrop's F-5 Tigers, and the British Aircraft Corporation's Lightning F-53s. Ten years later the hearings in the U.S. Senate Foreign Relations Committee were to reveal how tough, oriental bargaining over price was counterbalanced by haggling over commission which, by contrast, inflated the cost to the Saudi Government. Lockheed and Khashoggi looked to have the decisive advantage not the least because the U.S. was responsible for nurturing the Royal Saudi Air Force under the 1952 agreement. The Pentagon recommended the F-104. Northrop, meanwhile, had Muhammad, the powerful elder of the royal family, in its pay.

Pushing the interests of B.A.C., without the support of the British Government, was Geoffrey Edwards who several years before had campaigned to sell Saudi Arabia a U.K. air defence system. The indefatigable Edwards claimed to have made no less than seventy-one journeys to the Kingdom and to have spent £84,000 of his own money,[12] a figure he later raised to £180,000,[13] including loans, in his endeavours. As it happened, Sultan wanted to buy American aircraft anyway. Moreover, the Lightning was an interceptor that would have to be adapted for the ground attack capability also wanted by the Royal Saudi Air Force. Even so, it would be an inferior one compared to the American rivals. In the event the choice was made not by Riyadh but by Washington and London.

As part of a deal to offset the cost to the U.K. of buying American F-111s, Robert McNamara, U.S. Secretary of Defence, agreed to put his weight behind a joint bid including Lightnings and the American Hawk missile made by Raytheon. Also involved in the British part of the package, worth £145 million, was radar, the sub-contract for which was awarded to Associated Electrical Industries and Air Work Services which took responsibility for maintenance, training and management. Sultan would have preferred a deal under which the U.K. Government took responsibility for overall supervision. The compromise resulting from political considerations of no direct concern to the Kingdom was a very unsatisfactory one for the buyer and would later strain Anglo-Saudi relations. Fortuitously, it earned Edwards over £7 million in commissions.

Apprehension about Saudi Arabia's vulnerability was such, however, that the Government decided upon a smaller, interim arrangement. Edwards was quick to respond to the challenge in organizing the emergency programme, code-named 'Operation Magic Carpet'. The £16 million contract involved the purchase of six Lightnings and six Hunters already in service with the R.A.F., together with eight launchers for thirty-seven Thunderbird air-to-ground missiles drawn from British Army stocks to be based at Khamis Mushayt, a town in the southern highlands of the Asir within easy range of Najran and Jizan and also in a commanding position facing the South Arabian border. Personnel provided by Air Work Services and the British Aircraft Corporation took charge of the servicing and maintenance of the aircraft. The firm also provided ex-R.A.F. pilots, who became mercenaries by the simple expedient of being seconded to Edwards as his employees. Their salaries were raised from a modest £4,000 a year to £10,000 when it was established that they would be expected to undertake combat duties. The contract was negotiated in one long session lasting almost twelve hours. When it was signed Sultan wept openly. Within a week he was complaining about certain of its terms. By the time the aircraft and missiles were fully operational towards the end of the summer of 1967 the danger to the Kingdom, which they were intended to counter, had passed.

Political and military impasse led to a meeting between Feisal and Nasser on 24 August 1965, when they signed the Jeddah Agreement. This was intended to be the framework for all future efforts to solve the intractable problem of Yemen. Both signatories had much to gain from the détente. Trembling at the prospect of a direct attack, Saudi Arabia gained time to improve its defences. Nasser, increasingly preoccupied with Israel, needed to extract himself from the problem of the Yemen war as much as he could and save his face in doing so. In exchange for Feisal's commitment to stop aid to the Hamiduddin, Nasser agreed to a deadline for the withdrawal of Egyptian forces which was set for 23 November 1966.

The Jeddah Agreement laid down an agenda and terms of reference for a conference aimed initially at the establishment of a transitional government 'representing all the national forces and people of authority in the Yemen'. The formula gave detailed substance to the understanding reached at the Alexandria summit conference in September 1964. Yet as a framework it was very much less than adequate in ignoring the future of the Hamiduddin – and, for that matter, Sallal, who in October was 'retired' for nine months to Cairo. The ceasefire was maintained but the Royalist and Republican delegations quibbled over the agenda and the terms of reference when they met at Haradh towards the end of November 1965. After only three plenary sessions the conference broke up on 24 December and never reconvened. As it expired, Nasser became angered at the improvement of Saudi Arabia's relations with Iran marked by Feisal's talks with the Shah earlier that month and their joint call for an Islamic summit conference.

Feisal had entered into the Jeddah Agreement in good faith, but also with justified scepticism. Any good intentions that Nasser might have had about withdrawing his troops disappeared with the announcement by the Labour Government in London in February 1966 that the U.K. would leave its Aden base by 1968. Addressing a rally at Suez on 23 March, Nasser proclaimed that the Egyptians would remain in Yemen indefinitely until the 'revolution' could defend itself. He warned again that Egypt would strike at 'the bases of aggression'. It was the start of what he described as his 'strategy of long breath'. Freed from colonial rule, the south of Yemen seemed a more promising base for expanding Egyptian influence.

For fifteen months from the date of the British announcement about the evacuation of Aden the struggle between Feisal and Nasser was the central feature of the Middle Eastern scene, overshadowing the looming Arab-Israeli conflict until a few weeks before it erupted. The U.S. felt it necessary to reaffirm Kennedy's commitment to safeguard the territorial integrity of Saudi Arabia.

In June 1966 Feisal paid a state visit to the U.S. Like Saud nine years before, he was snubbed by John Lindsay, the Mayor of New York, who cancelled a dinner for the King out of respect for his electors. In answer to a question about the Arab boycott, he had replied: 'Unfortunately, Jews support Israel and we consider those who provide assistance to our enemies as our own enemies.'[14] For good measure, the State Governor, Nelson Rockefeller, announced that he would not be making a courtesy call. Feisal, who had been spat upon and verbally abused by New Yorkers in 1946, knew what to expect. Secretly, he was not displeased by the incident which only seemed to prove the purity of his Arabism. It was at his insistence rather than President Johnson's that the joint communiqué omitted any further American reassurance about the maintenance of Saudi territorial integrity. Overall, his performance was a dignified one that could be seen as laying the basis for the Kingdom's future attempts to weaken the U.S.-Israeli axis.

Feisal despaired of breaking the Saudi-Egyptian impasse over

Yemen. For him, Nasser rather than Israel was the devil incarnate and the Egyptian leader's revolutionary creed as sinister a carrier of Marxist plague as Zionism, an extraordinary identification and an obsessive conviction that Feisal had first given public expression to in 1962, in response to Egypt's barrage of abuse. Cautiously and cunningly on the defensive, however, the King was careful to refrain from attacking Nasser personally and giving unnecessary provocation. In Yemen, at least, he could see that with minimal Saudi help to the Royalists and the gold-lusting, ungovernable tribes, the Egyptians would be unable to impose their own will and establish a stable and representative regime acceptable to themselves. But disillusioned with and irritated by the Hamiduddin, Feisal brought pressure to bear upon them to broaden the basis of their regime. More actively than ever he contemplated moderate alternatives, whether or not they included members of the family of the Imam, who had retired to Taif because of ill health – and also Saudi reluctance to let him return out of respect for Sallal's confinement in Cairo.

Anxious to prevent the flickering embers of war from flaring, the King resisted the plans of the Imam's cousin, Prince Muhammad bin Hussein, who had emerged as the Royalists' most dashing and successful commander and was responsible for operations in the eastern sector, to mount a big offensive. They quarrelled and Saudi payments were suspended in the autumn of 1966. By contrast and blatant contradiction to the Jeddah Agreement, Nasser unleashed the black-jowled Sallal who returned to Sana in August that year in a frenzied attempt to terrorize the much diminished territory under his control in an orgy of executions and repression. Intensified bombing by the Egyptian Air Force on defenceless villagers in areas beyond the reach of Sallal's hairy hand merely underlined the frustration of Nasser and his faithful servant.

As subversion and violence in South Arabia grew, Feisal looked with increasing gloom in that direction and with well-justified apprehension about the chances of the Federation of South Arabia, nurtured with such painstaking and futile care by the Colonial Office, surviving after British withdrawal. Feisal still sulked over the refusal of the U.K. in 1963 to allow leaders and members of the South Arabian League, the moderate group founded in the mid-fifties seeking independence for Aden and the South Arabian Federation, which was a particular protégé of Kamal Adham, to return home. They had sought refuge in the Kingdom when the movement was banned in 1960 by the British authorities. Permission for them to go back to Aden was finally given towards the end of 1966. By then the urban liberals, who had never had a power base, had been hopelessly overtaken by events. The fact was hardly appreciated by their Saudi paymasters in their preoccupation with countering the Egyptian-backed Front for the Liberation of the Occupied South Yemen.

In November 1966 Feisal received a group of Ministers of the Federation led by the Sherif of Beihan, who had helped the Royalist cause so much in the early days of the Yemen war, and promised substantial financial aid. Yet, with the negative passivity that was to characterize Saudi policy in the region for many years to come, he did little or nothing to help consolidate the position of the Sultans before their pathetic collapse in the summer of 1967.

Feisal became more querulous towards Britain whose Labour Government he tended to see as a stalking-horse for Marxism. He believed that his main Western allies, the U.S. and Britain, albeit well meaning, were misguided and blinded to the threat of communism. He came to see himself as a lonely bastion standing against a remorseless Marxist tide. And so the Guardian of the Holy Places set about mobilizing the forces of Islam. Following the call for an Islamic summit conference made jointly with the Shah at the end of 1965, Feisal visited Jordan, Sudan, Pakistan, Turkey, Morocco, Guinea, and Mali in the course of 1966. The heads of state of Kuwait, the Cameroons, Somalia, Niger, and Sudan were invited to the Kingdom.

At the end of 1966 the Constituent Assembly of the Muslim World League convened in Jeddah with Sheikh Muhammad Sorour Sabhan, Saud's former Finance Minister, recalled from the murky mists of the past and forgiven, as its Secretary-General. With reference to Yemen, it called on Muslims everywhere 'to denounce this murderous war in which a Muslim kills his Muslim brother' and deplored the persecution of the Muslim Brotherhood in Egypt. There was talk of nineteen countries, representing 450 million Muslims, attending a summit – which was eventually to be convened in 1969. Beneath a pious and fraternal guise, the strategy was a political one designed to counter Nasser's ambition to dominate the Arab world. As such, it was not a success, but nevertheless was sufficient to enrage further the Egyptian leader who denounced the call for an Islamic pact as an imperialist conspiracy and a plot to resurrect in another form the Baghdad Pact.

As Feisal rallied religion to his cause, Nasser and his allies summoned the forces of revolution to overthrow the House of Saud. In its assault on the 'agent King' and the iniquities of his regime, the Voice of the Arabs recounted the heroic resistance of the Arabian Peninsula People's Union, the opposition group formed by Nasser Said, the instigator of disturbances in the oilfields, after he had been forced to flee the country in 1956. Radio Damascus promoted the cause of the Saudi National Liberation Front, a creature of the Baathists in Syria but of no account. Confusingly, Radio Sana, the mouthpiece of Nasser's satrap in Yemen, claimed the existence of another group – the Popular Front for the Liberation of the Arabian Peninsula. Towards the end of 1966 there were reliable, verified accounts of bombs exploding: three went off in Riyadh, and another near Dammam was detonated outside the beach palace of Saud bin Jiluwi, the tyrannical, feudal lord of Hasa. (He was to die of natural causes in March 1967 to the relief of his Shia subjects, whom he called 'dogs', a very vicious form of abuse since in Islam the animal is considered *haram*, or unclean.)

There was no evidence that any of these organizations had an indigenous presence within the Kingdom. The trouble was caused by infiltrators from Yemen. Early in January 1967 a number were captured. They were charged with responsibility for planting bombs at the residence of Fahd, at the Nasariya Palace, at the Zahra Hotel, and under a bridge over which the King was scheduled to pass. Seventeen Yemenis who had confessed to carrying out or planning acts of sabotage were beheaded in public in Riyadh on 17 March. Contrary to custom the executions took place before the end of the *Hajj* as a warning to any other 'pilgrims' contemplating acts of subversion. Other Yemenis were summarily deported.

Ammunition for Nasser's propaganda broadsides came suddenly and ignominiously in the grotesque form of the genial Saud. Restored to a semblance of life by a long Mediterranean cruise, the tottering ex-King arrived in Cairo from Athens in December 1966 with an entourage of 130 and nearly eight tons of luggage to be welcomed by the man whose past abuse would have made a reconciliation impossible for anyone less shameless. Saud came just in time to save his property from the sequestration orders that led to the seizure of Saudi property in Egypt, including Feisal's assets which were assessed at nearly $10 million.

There seemed no depths of treason to which Saud would not go in his attempts to regain his irretrievably lost throne. After a meeting with Nasser, Amer and Sadat, he protested that he had not abdicated and that the *ulema* had acted illegally in procuring his deposition. A willing subject to the Voice of the Arabs, he proclaimed, 'I strongly denounced the presence of troops in our dear homeland, especially since I liberated the country from the presence of the American military base in Dhahran.'[15] The C.I.A., he alleged, had been instrumental in his deposition and the U.S. Air Force had prevented the tribes of Hasa from coming to his rescue in 1964. Saud donated money to the families of the executed Yemenis and lent Egypt $10 million from his own private fortune. Then on 23 April 1967 he flew to Sana to extend full recognition to Sallal who reciprocated by greeting him as the 'legal King of Saudi Arabia'.[16] Saud gave the Republican regime $1 million.

Nasser's aim was to create a climate of unrest within the Kingdom as well as to force Feisal to weaken his stance on Yemen. That, too, would have been the purpose behind the intensified bombardment on Jizan and Najran which climaxed in a last crescendo from 10 to 14 May. By the time Saudi Arabia had registered a complaint at the U.N., Egypt, Syria, Jordan, and Iraq had placed their forces on the alert against Israel.

Tension had been mounting in the Arab world since the punitive Israeli raid on the West Bank village of Samu in November 1966. Nasser had ordered the emergency U.N. force in Sinai to leave and closed the Gulf of Aqaba to Israeli shipping. Saudi Arabia could not remain aloof. Feisal was treaty-bound to Jordan and had, after the Samu attack, pledged military support to Hussein. On 25 May, the Kingdom placed its own forces on the alert.

Feisal knew the conflict might gravely destabilize the established order in his country. At the turn of the month there were several explosions directed at the American presence in the country, including one – with little more force than a firecracker – on the wall of the U.S. Embassy compound. They were impotent protests but nevertheless a disturbing reminder of the possible repercussions for the Kingdom of an Arab-Israeli war. But Feisal had no choice but to declare his support for Egypt and make what gestures of solidarity he could.

About six hours after Israel had launched its attack on 5 June the King told his people on the radio to prepare themselves for the 'decisive battle'. On 6 June, with the Egyptian Army in full retreat towards the Suez Canal and Jerusalem about to be taken, the King addressed a public rally at Riyadh racecourse, saying, 'We consider any state or country, supporting or aiding Zionist-Israeli aggression against the Arabs in any way as aggression against us.' Rousingly he cried: 'To *jihad*, citizens! To *jihad*, citizens! To *jihad*, nation of Muhammad and the Islamic peoples.'[17] But the battle cry of the Wahhabites had no substance nor, indeed, much conviction. A Saudi brigade moved ponderously towards Jordan. Some of the senior princes may have been relieved to see it out of the country. It arrived in Jordan after King Hussein's troops had been driven across the river to the East Bank and not long before Israel finally accepted a ceasefire, having succeeded in its objective of occupying the Golan Heights.

Despite Feisal's reservations about using oil as a weapon, Saudi Arabia joined the other Arab producers in suspending oil supplies to the U.S. and U.K. after Radio Cairo had alleged that American and British aircraft had participated in the surprise attack which left the greater part of the Egyptian Air Force destroyed on the ground. The charge was patently absurd. Nasser and King Hussein soon dropped it, but the widely held Arab belief that the U.S. and Britain had colluded with Israel in the launching of the offensive lingered on. Feisal never paid any lip service in public to this comforting myth and categorically rejected the 'big lie', but politically he had to join a collective Arab embargo against the U.S. and Britain.

As it was, the imposition of the embargo was not sufficient to allay radical feeling within the Kingdom. In Dhahran anti-U.S. riots erupted on 7 June, a day referred to by Aramco men as 'Rock Wednesday'. A mob made up mainly of students from the College of Petroleum and Minerals stormed the military part of the airfield and wrecked the American officers' club as well as living quarters nearby, which had been hastily evacuated at the approach of the rampagers. It then moved on to attack the American Consulate. Shutters were torn down and windows broken, but the building was not penetrated. The only casualty was an Algerian student who fell from the flagpole as he ripped down the Stars and Stripes, breaking his leg. The unruly mob then invaded the perimeter of the Aramco senior staff camp, wrecking a number of houses after politely asking any remaining occupants to get out of their homes. The mob tried and failed to storm the telephone exchange. By then the National Guard had intervened. The rabble, its anger satisfied, was dispersed without bloodshed. No American citizen was attacked or injured. On the same day in Riyadh several thousand people surged down University Street. Police brought up trucks behind the crowd which eventually was broken up as demonstrators at the rear were hauled into the vehicles. The experience was an unnerving one for the regime. Oil production stopped from 7 to 13 June as the bulk of the Saudi labour force absented itself from work.

The ceasefire accepted by all the combatant states on 10 June was rejected by Feisal. He had nothing to lose. Furthermore, the Guardian of the Holy Places was more pained than any other Arab leader except for the gallant little Hashemite about the Zionist occupation of East Jerusalem and the Al Aqsa Mosque, the third most holy of Islam's shrines. Deliberately and prudently, albeit not very heroically, the King of Saudi Arabia had avoided direct involvement in a war which could only have brought humiliation to his soldiery or – even worse – insurrection at home. Feisal had no sympathy for Nasser whom he blamed for the débâcle and the devastating blow to Arab pride. The Lion of Araby had got his just deserts as far as he was concerned. Feisal's attitude was well summed up in one terse comment he made to an Arab diplomat: 'If someone throws stones at a neighbour's windows he should not be surprised or complain if the owner comes out and beats him with a stick.' His lip curled as he delivered this withering judgement.

Saudi Arabia was anxious to resume oil supplies to the U.S. and Britain for political and financial reasons. Feisal stuck to his view that oil should not be used as a weapon. He argued that the Arabs should use it to build up their economic strength. 'We should not be misled by ideas that communist friendship has imbued in some people's minds,' commented Radio Jeddah on 6 July. The following day it reported that 'citizens in various parts of the country' had asked the Government to reconsider the ban on oil exports to Britain and the U.S. The Egyptian press retorted that the Kingdom would have a fiscal surplus of 40 per cent that year and, by implication, could easily afford to withhold supplies. Iraq pressed for a total ban on exports for three months and thereafter a selective boycott of the U.S., Britain, West Germany, and Holland.

The embittered relations between Nasser and Feisal remained the main obstacle to the convening of an Arab summit. However, the deadlock was eased at the Arab foreign ministers' conference in Khartoum at the beginning of August 1967 when Mahmoud Riad of Egypt proposed a formal end to the Yemen conflict. The King knew, anyway, that he had the whip hand over Egypt which, with its economy in ruins and its armed forces shattered, could not afford anything less than complete withdrawal.

Arab heads of state, with the exception of those of Algeria and Syria who refused to have any truck with the 'reactionaries', met in the Sudanese capital at the end of the month. They agreed that there should be no recognition of, nor reconciliation with, Israel, but the defeated confrontation states were permitted tacitly to seek, through intermediaries, an honourable settlement. To enable them to remain steadfast and not capitulate, money was needed. Thus, the deal was struck. Kuwait would grant £55 million, Saudi Arabia £50 million, and Libya £30 million each year in quarterly payments to provide Egypt with a subvention of £95 million, Jordan with £40 million, and Syria with £5 million. The cost to the three conservative oil producers was a heavy one with the subventions amounting to some 20 per cent of their annual revenue at the time. However, Nasser had been made beholden to them and thereby underwent something of an enforced conversion to moderation.

The critical meeting between Feisal and Nasser took place in the house of Muhammad Mahgoub, the Sudanese Prime Minister, on 31 August. Bargaining to the end, El Rais proposed that Egypt and Saudi Arabia should agree on a regime in Yemen mutually acceptable to both of them with the proviso that, to save his face, neither the Imam nor Prince Hassan bin Yahya should be a part of it. Knowing his strength, Feisal rejected out of hand any such imposed solution, suggesting instead that the parties to the conflict call a conference and settle the future political organization of Yemen themselves. Nasser conceded defeat and committed himself to withdrawing all Egyptian troops by the end of the year. Feisal pledged an end to Saudi aid to the Royalists. He could well afford to do so because the Egyptians were already pulling out fast. An enraged Sallal was not consulted and by the end of the year had been overthrown while he was absent on a visit to Moscow. And so Feisal emerged the victor from a struggle that had lasted nearly five years. In the end the decisive factors had been Nasser's folly and the armed strength of Israel. Yet those factors should not in any way be allowed to detract from Feisal's resolute and skilful handling of the sustained challenge to the Kingdom. He had gained the upper hand well before Nasser allowed himself to be drawn into conflict with Israel.

16 Uneasy Interlude
1967 – 70

Brothers, there is one small remark I would like to make. I keep hearing repeated the words 'Your Majesty' and 'seated on the throne' and such like. I beg you, brothers, to look upon me as both brother and servant.[1] – KING FEISAL

THE House of Saud emerged from the five-year ordeal of the Yemen war·much fortified. It would be idle to speculate as to what form of anarchy or authoritarianism, disintegration or dismemberment, reaction or revolution, might have filled the vacuum if the monarchy had failed to survive the revolutionary threat from Arab socialism and Marxism. Mercifully for the dynasty, as well as its subjects and the West, the recall of Feisal at the end of 1962 initiated a new era which laid a basis promising future stability.

Saudi Arabia, as a political entity, was in 1967 still virtually synonymous with the royal family. A society essentially tribal in nature, its sense of nationality was still very embryonic though, if anything, strengthened by Egypt's strenuous efforts to undermine the system of government. However, Saudi Arabia had begun to assume the shape of a modern state with which its inhabitants could identify. Mounting oil revenues, wise disbursement of them by Feisal, and a small measure of social reform introduced by him were the main factors in this process.

Oil production increased by 70 per cent from 1962 to 1967 when it reached 2.8 million barrels a day (including Saudi Arabia's share of output from the Neutral Zone) and generated revenue of $909.5 million compared with $409.7 million five years earlier. The Government was much improved under Feisal but still very inadequate to cope with the surge of funds at its disposal. There was also a desperate shortage of qualified labour to undertake the kind of tasks that the Arabian of bedouin origin disdained because they were manually demeaning. The problem was compounded by political factors. All but 900 of the 7,000

Egyptians working in the country in 1961 had departed by 1964. There was a further drain on trained manpower as a result of the eviction of Palestinians and Syrians suspected of having affiliations with left-wing movements. As for hewers of wood and drawers of water, the exodus of Yemeni workers by volition and compulsion after the coup in Sana had been such that work on some projects stopped for a full two years. The regime, meanwhile, had become very sensitive about the paucity of subjects governed by itself. A census undertaken by a British concern in 1962 – 3 was a somewhat tentative exercise given the large proportion of pastoral, nomadic folk. It estimated a population of only 3.3 million and was suppressed.

Development budgets bore little relation to actual expenditure which was known to be consistently below appropriations. (Not until 1976 did the Government publish actual figures retrospective to 1970.) Nevertheless, it is clear that the rate of public investment more or less doubled over the five-year period. The private sector's rate of investment had gone up even more sharply, reflecting the climate of confidence created by the new King's management of the state. Officially it was estimated that national income rose at an average annual rate of 8 per cent from 1962 to 1967. Overall the performance was a fulfilment of the pledge made by Feisal in his ten-point programme to the effect that economic development – together with the financial revival which was taken care of by rising petroleum production – would be the Government's 'prime concern'.[2]

Road construction forged ahead. Over twenty cities and towns within the Kingdom were serviced by Saudi Arabian Airlines, which had built up its operations to an efficient level with technical, administrative and managerial asssistance from Trans World Airlines. Progress with port construction had been less satisfactory as the surge of imports drawn in by the economic boom congested capacity at Jeddah and Dammam. With a healthy profit margin guaranteed by the Government, the private sector had responded to the escalating demand for electricity by doubling capacity. A country-wide survey of water resources covering all the Kingdom except for the Rub al Khali was under way. Work had started on Saudi Arabia's first modern desalination plant near Jeddah. Ever larger sums allocated to the agricultural sector were evidence of the Government's good intent, but its growth lagged disappointingly behind all others.

The General Petroleum and Minerals Organization, or Petromin as it became known, had been established in November 1962 as a state oil corporation and a vehicle for industrialization. In 1967 Petromin started operating a new refinery outside Jeddah built by the Japanese company Chiyoda and took over from Aramco the marketing of petroleum products within the Kingdom. In line with the ten-point programme, the Foreign Capital Investment Code was decreed early in 1964 and provided the basis for joint ventures in the future, although it produced no immediate bonanza of Western technology and equity participation. Feisal showed serious intent about a more coherent development of the Kingdom and the diversification of its economy when he established in 1965 the Central Planning Organization to supersede the moribund Supreme Planning Board. In 1967 oil still accounted for 60 per cent of the country's Gross National Product.

The majority of people lived at no more than subsistence level, however. Nutritional standards were low even by the criteria of the developing world. Disease was widespread, not the least because of the lack of proper sanitation, supplies of pure water, and knowledge of the simplest rules of personal hygiene. Programmes carried out by the World Health Organization and Aramco had begun to eradicate malaria. A large-scale programme launched in 1957 had begun to reduce the incidence of cholera, a hazard made more dangerous by the annual *Hajj*. But at this time it was believed that 80 per cent or more of the population suffered from trachoma, the cause of King Saud's poor eyesight. Syphilis, contracted or congenital, was widespread. Many citizens were debilitated by dysentery, enteritis, and other intestinal problems which accounted for half the high number of deaths in infancy. In agricultural areas bilharzia was a serious problem and throughout the Kingdom tuberculosis was rife.

Justifiably, King Abdul Aziz, when he established the Department of Health in 1951, had decreed that it should be second in importance only to the Ministry of Finance. But not until the next decade was more meaningful progress made with free public medical service initiated by Feisal. In terms of construction the achievement was impressive. By 1968 there were eighty hospitals in existence with 5,952 beds, compared with twenty-nine hospitals with 2,617 beds in 1958. Some 200 dispensaries for out-patient treatment had been established and 300 health centres in villages, staffed by male nurses. The expansion exceeded the human resource required for it. In 1967 there were only 663 doctors, 1,153 male nurses, and 637 female nurses and midwives serving these facilities. Pakistanis, Egyptians, Syrians, Lebanese, Ethiopians, and Greeks made up the greater part of the staff. The Government had opened three schools for the training of Saudi nurses, but it had not yet felt able to establish a medical school. Because of the Koranic strictures against cutting open dead bodies, aspiring doctors were sent abroad to study – as if the contravention of Holy Writ was less serious abroad than at home.

Up until 1962 social security had been left to the tribe or extended family, religious charity, and royal benevolence. Feisal's reform programme recognized implicitly that such traditional forms of support were not sufficient for an oil-rich state. Within weeks of its issue, the Kingdom's first Social Security Law had been introduced, giving a minimum annual dole of 360 riyals a year to those over the age of sixty, the incapacitated, orphans, and women without means of support, with extra benefits for any dependent children. Homes for the old and disabled, orphanages, reformatories, and schools for the blind, deaf and the dumb were established.

Labour legislation 'protecting the workers' class against unemployment' was foreshadowed in Feisal's ten-point programme. Catering for the able-bodied did not prove easy. By 1962 the set minimum wage was still only 5 riyals a day ($1.11 at the then prevailing exchange rate) for those working in establishments of ten employees or more, of which there were relatively few anyway. Workers had never been permitted to form trade unions or to enter into collective bargaining. New legislation eventually decreed in 1969 established a system of arbitration committees as well as revising and improving working regulations. Only about 20 per cent of the economically active received salaries or wages. The rest were self-employed as nomads, farmers, shop-keepers, craftsmen, or individuals rendering sundry services. Many of them received less than 5 riyals a day, quite apart from casual labourers on the bread-line. Development inevitably led to a drift of people from rural areas to urban centres. Theoretically, the combination of economic boom and labour shortages should have ensured well-paid jobs for all. Too frequently, however, Saudis did not have the rudimentary qualifications, aptitude or inclination required. As the Kingdom became more and more dependent on expatriate skills, unemployment of nationals became a problem. Not until 1962 was the Kingdom's first vocational training school opened in Riyadh. By 1964 others were functioning in Mecca, Jeddah, Dammam, Buraidah, Hasa, Qasim, and Medina. But the promised legislation aimed at dealing with unemployment did not materialize for seven years. It proved impossible, meanwhile, for most concerns to fulfil the requirement that 75 per cent of their manpower should be Saudis.

Complaints by Saudis about lack of employment opportunities and growing reliance on imported manpower highlighted how economic growth had out-paced social change, particularly the educational attainments of the Kingdom. One of the most formidable constraints on the development of its vast and sparsely populated territory was an abysmally low rate of literacy, which was no more than 5 to 10 per cent at the beginning of the 1960s. Feisal was most acutely conscious of this shortcoming. Significantly, in drawing up his first budget for 1959 he had, despite the austerity programme, increased the allocation for education by 24 per cent as he reduced appropriations for the privy purse and defence by one-fifth and one-third respectively. Thereafter, ever larger sums were devoted to the nourishment of the brains of the Kingdom's children, rising from 114 million riyals in 1959 to 674 million riyals in 1968. The statistical evidence suggests that this vital sector grew at a faster rate than national income.

Quantitatively, the performance was impressive. From 1958 to 1968 the number of primary schools in the state system nearly tripled to 1,544. The number of pupils attending them rose from 91,787 to 329,127. The child-teacher ratio fell from 28:1 to 23:1. The curriculum concentrated heavily on the learning of classical Arabic, religion and Islamic history, but was an improvement on the traditional *kuttabs* where boys, from the age of six, had the Koran drummed into their ears by the local Imam until they knew it by heart and learnt nothing else. In 1958 there had been just one secondary school, though classes had been added to a number of intermediate schools to cater for those wanting to study beyond the age of fifteen. Ten years later more than 6,000 were receiving education at this level. The number was very low but represented progress.

It proved easier to build new schools and colleges than to ensure the quality of teaching and learning in them. Standards were not high. With more lucrative options open to them elsewhere, Saudis were not enthusiastic about taking up careers in education. They soon became heavily out-numbered by expatriate Arab teachers, except in primary schools. And even at that level they were in a minority by 1967. Despite the financial inducements, there was a limit to the number of competent teachers from abroad who were prepared to come to the Kingdom. Most of the Egyptians had left in the early 1960s. The most easily available alternatives were Palestinians, whom the authorities regarded with suspicion.

Higher institutes of learning had more places than there were Saudi applicants and were taking in other Arab students. Riyadh University had expanded considerably and had an enrolment of more than 4,000 students in thirteen faculties by 1967. The munificent former King had also founded the Islamic University of Medina as an international seminary, modelled on Cairo's Al Azhar, to propagate the Prophet's message throughout the world. More meaningfully for the country's physical development, the College of Petroleum and Minerals had been established at Dhahran in 1963, later to be upgraded to full university status. The King Abdul Aziz University in Jeddah came into being in 1967. By then there were more than 1,000 Saudis studying in the U.S. and half as many again in West Germany, Lebanon, Britain, Egypt, and elsewhere. The preference for foreign institutions so worried the Government that a few years later it ceased giving scholarships for first degrees abroad.

Saudis familiar with the outside world tended increasingly to choose their brides from more cultured Arab countries. That fact alone seemed to indicate the need for girls' education. The demand for it was such that several dozen private establishments had sprung up before 1960. The Government's decision in that year to start schools for girls was, nevertheless, a revolutionary and even daring one in a society where many saw such an innovation as anathema or a threat to the virtue of their daughters. To allay the conservatives' fears about the morality of the new schools, Feisal placed them under the supervision of the Grand Mufti rather than the Ministry of Education. Extra care was taken to ensure that true belief would be inculcated into innocent minds. Chaperones were provided to protect the girls physically.

Feisal was aware that even the most modest reform could provoke a backlash from arch-conservatives and bigots. Point six of his programme made elliptical reference to the problems faced. After fulsome reference to the need to maintain Islamic values, he promised 'to reform the Committees for Public Morality in accordance with the Shari'a and Islam's lofty goals, for which they were originally created, and in such a way as to extirpate to the greatest extent evil motives from the hearts of people'.[3] In practice, this piece of double-talk meant that he wanted to curb the excesses of the *mutawain*, or religious police, under the direction of those committees dedicated to full observance of the Prophet's dictates as the Wahhabites understood them. But Feisal could not say as much. Any Saudi wanting to see the Kingdom come to terms with the modern world could only be embarrassed by or resent the activities of these baton-wielding zealots who felt so violently constrained to force shops to close at prayer time, break dolls in the *souks* (as representatives of the human form), knock cigarettes from people's mouths in the streets, beat the exposed calves of European women, and even break into homes to smash record-players or to search for illicit liquor.

For many young Saudis who had tasted the freedom and delights of other civilizations, their own country was beginning to seem like a temple of boredom and backwardness. The programme obliquely acknowledged this fact with its promise to provide 'innocent means of recreation for all citizens'.[4] Despite the distaste with which the religious elders and the devout viewed even masculine torsos and thighs, it proved fairly simple to encourage sport. But the opening of public cinemas was out of the question. When in the summer of 1963 some enterprising individuals set up clubs to show films, the irrepressible *mutawain* soon got wind of the abominable initiative and put an end to it.

It had been something of a breakthrough when in 1961 or so male singers had first been heard on indigenous air waves at a time when the inflammatory salvos of Ahmed Said on the Voice of the Arabs was interspersed with the seductive tones of the great Umm Kalthoum. Then, to the horror of the bigots, a woman's voice was heard in 1963 for the first time on Radio Mecca. A deputation of religious leaders and lawyers protested to Feisal. The King coolly retorted that the Prophet himself had been enchanted by the voice of the poetess Al Khamsa. He added that they might soon be seeing women on television. In the face of bitter resistance, T.V. was introduced in 1965. Every effort was made to avoid offending susceptibilities. There was a long series of 'tests' showing nothing of substance, let alone the human form, before any proper programmes were beamed. Anything remotely resembling a love scene was cut from foreign feature films. Even Mickey Mouse was not allowed to be seen giving his spouse an affectionate kiss. Nevertheless, Khaled bin Musaid, a grandson of Ibn Saud, led a group of religious fanatics in an attack on the television station in Riyadh. Khaled was mentally unbalanced and had mystical delusions.[5] He had received treatment at a clinic in Vienna in 1961 and 1963. The group's attempt to storm the T.V. station and destroy the transmitter was foiled by the police. But the young prince and his companions refused to disperse even when an emissary from the King came with an invitation to come and discuss their grievances with him. At this point Khaled drew a revolver and was promptly shot dead by the senior policeman present. Feisal ruled that there was no question of his half-brother Musaid being entitled to blood money. The officer had only been doing his duty. It was an incident that would be recalled ten years later in far more dramatic circumstances.

The next year, 1966, the problems of edging Saudi Arabia into a new era were illustrated, embarrassingly, to the King by Sheikh Abdul Aziz bin Baz, the blind (from the age of sixteen) and old Vice-President of the Islamic University of Medina. Shocked by heresy taught at Riyadh University, the learned man felt that it was his duty to take issue with the theory of the solar system first propagated by Copernicus over three centuries earlier and bitterly contested by Christian theologians at the time. Baz wrote an essay that was published by two newspapers. The sage felt obliged to refute the theory that the earth rotated round a fixed sun. 'Hence I say the Holy Koran, the Prophet's teaching, the majority of Islamic scientists and the actual fact all prove that the sun is running in its orbit, as Almighty God ordained, and that the earth is fixed and stable, spread out by God for his mankind and made a bed and a cradle for them, fixed down firmly by mountains lest it shake.' Anyone who believed otherwise would be guilty of 'falsehood toward God, the Koran and the Prophet',[6] he warned. On hearing of the essay Feisal was so angered that he ordered the destruction of all undistributed copies. It was too late to stop the Egyptian press from picking up the text and quoting it with triumphant relish. Controversy raged in Saudi Arabia. Four months later Baz denied having said that the earth was flat but reasserted that it was static and the sun revolved round it.

In general, and in contrast to the hopes raised by Feisal's accession to power in 1962, the atmosphere was as socially repressive as ever. In the autumn of 1967, for instance, he felt forced to issue a royal proclamation warning the youth of Saudi Arabia against laxity in dress and behaviour. Under pressure from the *ulema*, the sale of Christmas trees was banned. Government officials were instructed to engage in group prayer at the appointed times of day and threatened with penalties if they failed to do so. The age of segregation in schools was standardized and set at nine. All girls older than that were forced to wear the veil. One minuscule advance was the publication in 1968 of a weekly supplement for women by one of the Riyadh newspapers. The cover of the first edition carried a picture of Queen Elizabeth II. It would have been impossible to have featured one of Queen Iffat, or any other Saudi woman for that matter.

It was increasingly open to question whether the King would have resisted the Grand Mufti Sheikh Muhammad bin Ibrahim al Sheikh and the *ulema* more vigorously even if he had felt able to. The five years of struggle and the heavy burden of an office on which all authority centred had taken not only a physical toll. The arteries of his mind were hardening. Since the reform programme had been pronounced in 1962, there had been no mention of the 'fundamental law for the country' or the expansion of the Consultative Council that had been pledged. In 1963, regulations about provincial government had been decreed, whose aim appeared to be to unify the country more effectively rather than to provide for local autonomy. Subsequently, there was some official talk about a more popular form of participation in provincial government. In the event no new institutions were established and the whole idea was quietly discarded.

The programme had laid great emphasis on the creation of a Ministry of Justice to organize the administration of the courts and a Supreme Judicial Council to reconcile modern legislative requirements, which the Prophet could not have foreseen, with the *sharia* law. The Ministry of Justice eventually came into being in 1970, but the Supreme Judicial Council did not emerge until the summer of 1975 – after Feisal's death. The Civil Service Law promised in 1962 had been ready for at least three years before Feisal signed it in 1971. The movement towards reform was effectively dead.

Feisal could not totally cure extravagance by the princes or exploitation of business opportunities provided by their privileged position. Not the least of his achievements, however, was to stop ostentatious expenditure and behaviour of a kind that could only create envy and resentment. He insisted on living in a modest grey villa on the Medina Road. On the outskirts of Jeddah the Al Hamra Palace, one befitting a monarch of his stature, was being constructed for him. But when it was completed he refused to occupy it, insisting that it should be used as a state guest house. His own austere example and the awe in which he was held were enough to inhibit other princes. In his reign, the building of anything that might be identified as a 'palace' came to a halt. He forbade the import of Cadillacs because of their association with the self-indulgent luxury of Saud's reign and the publicity given to it in the press abroad. The ban was to account later for the unexplained delay in the conclusion of an order for armoured cars being negotiated by Abdullah for the National Guard. The objection, it emerged, related to the name of the U.S. manufacturer – the Cadillac Gage Company of

Warren, Michigan. The deal went through when it was explained that the company had nothing to do with General Motors' limousines.

As Feisal became more autocratic, he showed a growing tendency to defer decisions and procrastinate. The sheer weight of work grinding him down may have been partly responsible. However, his immobility on the political and social front was more directly related to his apprehension about subversion. Nasser's humiliating setback in June 1967 had afforded some relief. Implied, if not agreed explicitly, in the deal struck at the Khartoum summit conference in August 1967 was an undertaking on the part of the Egyptian leader to cease trying to undermine the House of Saud. Even so, the King, with justification, remained suspicious of Nasser. Moreover, radical forces were on the ascendant in the Arab world. Following British withdrawal, a Marxist regime had established itself in South Yemen. In Iraq, militant Baathists seized power in July 1968. In 1969 there came the shock of revolutions within a few months of each other in Libya and Sudan. The civil war spluttered on in North Yemen, where an uncertain situation would not be resolved to the moderate satisfaction of the King until 1970.

Feisal and the senior princes felt beleaguered. The King became more and more fearful about the threat to his own regime. As the Soviet grip on Egypt tightened, he complained that U.S. support for Israel could only force Arab states into the arms of the Soviet Union and spread the virus of revolution to countries under conservative systems of government. This sea of troubles around Feisal made him less inclined than ever to contemplate any political evolution of the kind that Western diplomats felt was necessary if an upheaval in the Kingdom was to be averted. He may have been right. At this stage any meaningful movement towards popular participation in government might have merely furthered the expression of discontent that, undoubtedly, existed in worryingly large proportions.

Far from making any gesture to the forces of progress, Feisal felt increasingly disposed to take refuge behind the barricade of tradition and to concentrate the decision-making process more firmly than ever in the hands of the Al Saud and, more particularly, his own. The composition of the Council of Ministers formed by him in 1962 had expressed in institutionalized form the royal family's strong grip on power. Including the King, five of its thirteen members were from the House of Saud and held the key appointments. Three others, including the Grand Mufti, were from the Al al Sheikh, the regime's Wahhabite conscience, to whom the royal family was closely related by ties of marriage. Apart from the collateral branches of the latter, the Al Thunayan, Al Jiluwi, and Al Sudairi, it was by 1970 the only family into which the Saudi princes and princesses were allowed to marry.

In the eight critical years from 1962 to 1970 only two changes of portfolio had taken place and these both involved commoners. In 1967, Yusuf Ibrahim Hajiri, the Minister of Health, was sacked because of a cholera epidemic that might have been more efficiently handled. For three years Sheikh Hassan bin Abdullah al Sheikh, Minister of Education, filled the gap. Under pressure from the zealots who held him responsible for the offence to Allah given by the nascent, emasculated television service, Jamil Ibrahim Hegelan was removed from the Ministry of Information, but his transfer to the Ministry of Health in the summer of 1970 was the King's oblique riposte to the bigots. Yet such was the King's mood that it came as something of a surprise when another commoner, Ibrahim Anqari, was appointed to head the Ministry of Information in that year. Several princes were known to be eyeing the post with some interest. Against expectations, Muhammad bin Ali al Harkan, also a commoner, was appointed in 1970 to be first Minister of Justice, when many had expected the new position to be placed in more privileged hands.

Administrative improvements had been brought about under Feisal, but direction of the state remained a very personal affair centred upon his person. Important matters of government, especially those concerned with foreign policy, were made in princely conclave where no minutes were kept. Commoners like Ahmed Zaki Yamani were there to be consulted for their expertise when it was required. In seeking and maintaining a consensus of the powerful princes, the King did not dispense with advisers. Rashad Pharaoun, the blue-eyed Syrian doctor who had kept his hands very clean under Saud, was the last survivor of Abdul Aziz's northern Arab recruits. Abdul Rahman Azzam was retained to help with the outstanding problem of Buraimi and the border dispute with Abu Dhabi. The King's right-hand man was Kamal Adham, his brother-in-law. A man of great intelligence and infinite cunning, he did not arouse the hostility of the princes, as Saud's coterie had, and did not excite resentment of the business opportunities that his position opened up. Feisal also had faith in Omar Saqqaf who was promoted to the position of Minister of State for Foreign Affairs in 1967, though he was little more than a golden messenger. With no more than an oral briefing from the King he would be dispatched on diplomatic missions and could, in turn, be a useful conduit to the throne for diplomats, whose whisky he much appreciated. One recalls how in the mid-1960s he opened all the Foreign Ministry mail personally.

Despite Khaled's nomination as Crown Prince, he had developed no interest in participating in the government of the country. His preference for falconry, desert life and tribal confabulations was as strong as ever. According to well-informed diplomats serving in Jeddah at the time, there was a crisis within the royal family towards the end of 1969 when he indicated that he would like to be relieved of the responsibility of succeeding to Feisal. In doing so, he raised again the question of Muhammad's position. The eldest surviving son of Ibn Saud had accepted the consensus that he should be passed over, but a condition for his acquiescence and that of the ambitious Abdullah had been that Fahd did not move into second place in the line of succession. The King was not alone in being alarmed at the prospect of a serious split between the traditionalist senior princes and the more modern-minded camp centred around the sons of Hassa bint Sudairi. Feisal, Abdullah, and Abdullah bin Abdul Rahman, in particular, leant heavily upon Khaled to do his duty. He was forced to agree not to renounce his position as heir apparent for the sake of preserving the House of Saud. Muhammad remained a formidable and feared force in the royal family. His seniority by age continued to be respected and Khaled would defer to him in private and cede precedence at table.

Abdullah had maintained and intensified the loyalty of the National Guard to the royal family. His achievement in doing so was all the more remarkable on account of one particular handicap. He was a rare phenomenon amongst Arabs because of his stammer, normally a serious shortcoming given the importance attached by Arabs to fluency of speech. His love of falconry and equestrianism was well known. At home, where in private he chain-smoked, his favourite pastime was playing pool. As Commander of the tribal levies, whose primary function was to protect the regime, he had established himself as a key figure. His mother hailed from the great Shammar tribal confederation and he had a natural rapport with the bedouin.

Abdullah bin Abdul Rahman ('Uncle Abby') was still a very influential voice in the family, essential in the consensus-reaching process. His brother Musaid bin Abdul Rahman had moved even closer to Feisal's ear. A man of blinkered religious conviction, which was not belied by his private conduct, he performed steadily and honestly as Minister of Finance from 1962 to 1975, though he never produced any figures relating to actual expenditure.

Much more promising for the future of the House of Saud had been the rise to prominence of Fahd and Sultan who had shown themselves loyal and committed supporters of Feisal in his struggle against Saud. In the formation of the Council of Ministers in 1962, Fahd had been made Minister of the Interior at the age of forty-one, and Sultan appointed Minister of Defence and Aviation at the age of thirty-eight. The evidence was that the King looked to them for advice as much as he did to the arch-conservatives. They could be said to represent the 'liberal' face of a King whose inflexibility was becoming more pronounced by the year. With a high quotient of intelligence, the two full brothers were blessed as well with an innate worldly wisdom. They were more aware than most of their kith that money alone could not buy security, defence and development.

Fahd, second only to Feisal, stood out by reason of his intelligence and flexibility which seemed to make him a champion of moderation in the House of Saud. He had acquired a grasp of foreign affairs and modern realities generally. His taste for the pleasures that the West could offer and the relative ease with which he adapted to its ways have helped what must have been a process of self-education for one who had received only traditional court schooling. His pronounced pro-American leanings were apparent. But Fahd was also prone to indolence. He was very much less than assiduous in his attendance of the time-consuming consultations within the royal family, apparently preferring to choose his own company. His detachment was a cause of distrust amongst the traditionalists. Ibn Saud's eleventh-born son showed none of the easy familiarity with the tribes that Muhammad, Khaled and Abdullah did, nor did he seek to court them in the same way. By the late 1960s he was already noted for his frequent and prolonged absences abroad where he showed more abandonment than discretion. His exploits at the gaming table were a cause for comment in Riyadh. Spectacular losses suffered in casinos of the French Riviera in 1974 were well publicized. Feisal reproved Fahd for his conduct. Thereafter, the younger step-brother was more circumspect, unlike other members of the family, whose cheques bounced and who defaulted on gambling debts. Fahd did not have a taste for noxious fluids but enjoyed the pleasures of love afforded by his station and also kept a full complement of wives, changing junior partners as desire dictated.

Sultan, whose vigour on the couch was a cause for even more comment and respect, had proved a stern, tough and headstrong character. As General von Horn, the head of the U.N. observer mission during the Yemen war had observed, Sultan was volatile as well. Indeed, his temper could be explosive. His devotion and commitment to his defence fiefdom and the handling of foreign relations with the two Yemens was tenacious. In contrast to Fahd, he had gained a reputation for hard work, earning the nickname *bulbul,* or 'nightingale', through his un-princely ability and willingness on occasions to burn the midnight oil. Yet this dedication did not inhibit but rather encouraged him to profit from the Kingdom's finances, a great proportion of which was appropriated to his department. More devious than Fahd, he was correspondingly the object of greater mistrust amongst the more bluff and pious brothers. The extent of his ambition was obvious to everyone. It led to suspicions that he wanted to leapfrog over those senior to him, in particular Abdullah. With some good reason, he was suspected of wanting to bring the National Guard under the departmental authority of the Ministry of Defence and Aviation – an arrangement that would have made sense from the point of view of military efficiency but not self-defence.

Of the surviving sons of Abdul Aziz, Fahd and his six full brothers seemed to form a solid phalanx that, in the gloomy days of the late 1960s, gave some reassurance to concerned observers about the regime's chances of survival. The Al Fahd – frequently but erroneously referred to as the 'Sudairi Seven' – had been born over a period of sixteen years to one of the most prolific of the sire's favourite wives, Hassa bint Ahmed al Sudairi. They represented within the royal family a reservoir of common sense and knowledge of the world that was reflected in the trust placed in them by Feisal.

Abdul Rahman, the third of Hassa's sons, was sent to the Military High School, California, and then to the Military Academy there. He seemed destined for a career in the Armed Services. But he subsequently studied commerce at the University of California and opted for a business career that gave him substantial interests in building materials, cement, power generation, bottled gas, and farming. Naif, physically impressive like his father, was briefly Governor of Riyadh in 1953–4, when he was scarcely over twenty, and was subsequently appointed Deputy Minister of the Interior in 1970. Turki, who was named after Ibn Saud's first son, a victim of the 1919 influenza epidemic, was married to a daughter of Abdullah bin Abdul Rahman. He became Deputy Minister of Defence in 1969. Salman took over as Governor of Riyadh in 1962. Ahmed was made Deputy Governor of Mecca in 1971. Apart from Abdul Rahman, he was the only one of the seven to receive education abroad, studying political science at Redlands, California.

Despite their identification with each other, the brothers showed significant differences of chàracter. Most striking was the straightlaced puritanism of Naif and Ahmed. It did not inhibit their interest in availing themselves of the opportunities for enrichment provided by high office, however, even if their avidity was less spectacular than that of Fahd and Sultan.

In addition to Hassa bint Ahmed al Sudairi, Ibn Saud had taken two other Sudairi wives. Jauharah bint Saad al Sudairi had given birth to Saad, Musaid, and Abdul Mohsen. Haya bint Saad al Sudairi was the mother of Badr, Abdul Illah, and Abdul Majid. The clan provided for the Governors of all the border districts as well as the Deputy Minister of Information.

Overall, the weight of the Sudairis was a cohesive factor, even if their solidarity caused resentment in some quarters of the royal family. But as a whole its members had closed ranks. Any cloud of disgrace still hanging over Nawwaf, who had not gone into exile in 1962, was removed when he was enlisted by the King in 1968 as his Special Adviser on Gulf Affairs. Less than two years after Abdul Mohsen returned from Egypt, he was preferred with the important Governorship of Medina. In 1968 the undistinguished Badr was given the post of Deputy Commander of the National Guard. In 1971 Fawwaz, who earlier in the 1950s had had charge of Riyadh, was honoured with the Governorship of Mecca where he replaced the arrogant and reactionary Mishaal, eighteenth-born son of Ibn Saud.

Notwithstanding Feisal's tendency to harbour grievances, he was wise enough to realize that unity was better than alienation within the House of Saud, not the least because the Government needed the services of any prince of reasonable competence. Talal alone of the 'Free Princes' was not forgiven by Feisal. But he was at liberty to pursue his business ventures in the Kingdom and to amass more money – much of which he lost gambling. Only the exiled Saud and his sons were regarded as something like pariahs. In March 1968 the deposed ex-King had left Cairo after a stay of over a year and gone back to Athens. He returned to Egypt in August of that year declaring buoyantly that he would stay there until he returned to Saudi Arabia. In October he went to London to be treated for rheumatism and other ailments.

It was in Greece on 23 February 1969 that Saud finally expired. Feisal dispatched an aircraft to fetch his body. He recited verses from the Koran over the corpse before it was laid to rest in a grave near that of

Abdul Aziz, whose patrimony Saud had so nearly squandered. But the well-meaning, profligate King's name and remembrance have been expunged from official history. Feisal was in favour of disinheriting his dissolute sons but bowed to opposition from other members of the royal family, and so they returned to dispute the ownership of Saud's crumbling palaces, which stood on real estate of great value.

Feisal had been undisputed master of the House of Saud since 1964. But by 1969, greater political awareness brought about by higher standards of education and knowledge of a wider world was finding expression in rumbles of discontent against the strict application of Wahhabite precepts, the slow rate of progress in development, and the general inefficiency of the Government. An additional grievance was the concentration of wealth in the hands of relatively few people at the top of the ruling hierarchy. What in the West would normally be termed corruption had never had any stigma attached to it in Saudi Arabia. But the extent of graft was offensive to many young educated Saudis without the privilege of background that would have enabled them to take a share of the rising amount of the state's wealth being syphoned into the pockets of individuals. Feisal, undoubtedly, was aware of what was going on. But he would have appreciated that the concept of using influence in high places for personal gain was so ingrained in Saudi mentality and tradition that, even if he had felt able to eradicate the bribes and built-in 'commissions', any attempt to do so might have been dangerously disruptive to family unity. His main concern was that goods and services purchased were in the Kingdom's interest, even if the price was inflated by kickbacks.

Overall, the domination by the House of Saud over the Kingdom's wealth and its disproportionate share of it rankled with an identifiable middle class that was a significant element in society. The more sophisticated Hejazis resented the discrimination in official appointment in favour of Nejdis, who were using their superior influence to obtain a bigger stake in the country's commerce, as well as the spoils of office. Similar grievances were felt in Hasa where the Shiites remained a scorned minority.

General discontent amongst the less privileged was aggravated by the slowdown in the rate of economic growth experienced from 1968 to 1970 after five years of healthy expansion. As in the mid-1950s, oil production and revenues did not increase as they had done in previous years. In 1968 and 1969 income from petroleum, which made up 90 per cent of the Government's receipts, rose only marginally. The £50 million annual subvention paid to Egypt and Jordan amounted to rather more than one-tenth of oil revenue in 1969 and was a considerable burden. Defence expenditure had escalated and consumed about one-third of the budget. The Government had to dip into its reserves, ask Aramco for a loan of $150 million, and borrow almost as much from the commercial banks. State spending was responsible for generating the greater part of economic activity, which stagnated as a result of the squeeze on resources and the slow progress with civilian development. Credit was tight and some merchants were forced to turn to foreign suppliers to finance imports. The construction industry was at a low ebb. In 1969 Saudi Arabia was neither a cheerful nor an optimistic place.

Economic gloom only added to the uncertainty felt by contemporary observers about the future of the established order in Saudi Arabia. The somewhat amorphous, albeit often rabid, appeal of Nasser had declined with his prestige. A more potent threat came from the secretive Arab Baath Socialist Party and Marxists in the guise of the Arab Nationalist Movement. The representation of both was stronger in Hasa than in the Hejaz. During the 1960s and afterwards, the internal security service under the dour Omar Shams did not show scruples about arresting on mere suspicion and imprisoning without trial. It was wise not to grumble aloud about the royal family, the Government, or about state affairs within the Kingdom. Subversion and communism were loosely defined terms. If there was a principle about countering dissidence, it was 'better safe than sorry' – with the proviso that most of those who willingly repented would be freed. Saudi Arabia, it should be said, was little more inhumane than the 'progressive' Arab countries whose penetration it feared. Feisal took no risks. There had been a wave of arrests in the summer of 1964 as the Government denounced 'communists' who were seeking to spread a 'subversive creed originated by a vile Jew'.

Even so, the underground political opposition, though disquieting, appeared too fragmented and embryonic to mount a coup unless it received the fullest backing of the Armed Forces. They had been expanded considerably since 1962, with some misgivings on the part of Feisal and others, but more enthusiastically by those who stood to gain by the purchase of new equipment. The National Guard kept pace with the Army, whose units were stationed well away from the principal cities. The Navy had been formally founded in 1960, but plans to make it a meaningful force were shelved year after year to the disappointment of the British, who wanted to assist. In response to the Yemen war and the Egyptian raids, the biggest advance had been made in the development of the Air Force, which alone had the capability of covering the vast distances between population centres.

British withdrawal from Aden, begun in June 1967, was completed on 29 November that year when the last of the British troops left, effectively handing over Aden to the National Liberation Front, an act Feisal considered an unwittingly foolish or wickedly deliberate gift to communism. The N.L.F. – with a stronger indigenous base, the backing of the Arab Nationalist Movement, and far more ruthless gunmen – had easily overcome the Front for the Liberation of Occupied South Yemen, Cairo's petit bourgeois protégés. Salem Rubai Ali became President of the People's Democratic Republic of Yemen. But the real power was the ruthless, Moscow-orientated communist Abdul Fattah Ismail, Secretary-General of the N.L.F., and his Marxist colleagues. Feisal grumbled. But he had not proffered any assistance in consolidating the Federation of South Arabia that the British colonial authorities had tried to establish. Nor had he tried to build up tribal links to combat the N.L.F. In the event, the discredited, bickering sheikhs of South Arabia, who flattered themselves with the title of sultans, found that they had pathetically little grass-roots support. Nevertheless, tribal rivalries lived on. Not realizing the ideological form which they had taken, the Saudis sought from the beginning of 1968 to foment them as if they were dealing with North Yemen where their gold could be used to make the country ungovernable.

Sultan and Khaled Sudairi, Governor of Najran, directed the campaign. Their efforts certainly irritated the regime in Aden but never appeared to weaken the stranglehold of the N.L.F. In the course of 1968 a motley 'Liberation Army' was assembled at Najran. In December of that year the ungainly force moved into action. It was cut to pieces by South Yemeni Provosts bequeathed by the British colonial authorities. Differences between Sultan and Kamal Adham, who favoured the political approach, are said to have been one reason why the operation failed so lamentably. The King was furious over the fiasco. There were subsequent incursions on a smaller scale. But afterwards, efforts at creating unrest continued as a way of keeping the N.L.F. fully occupied within its own borders.

In North Yemen, Saudi Arabia could watch events unfold with relative, though not complete, equanimity. In the latter half of 1967, the Shah of Iran airlifted weaponry to Najran, including the heavier varieties which the Royalists had previously demanded in vain. Apart from allowing this limited traffic, Feisal fulfilled his part of the bargain struck at Khartoum by stopping aid to the Royalists. However, the Soviet Union intensified its material support for the Republicans, and Syria provided pilots as the Royalists laid siege to Sana early in 1968. The King resumed aid to them, but not for long. Despite the continued Soviet involvement, Feisal now felt confident that North Yemen could work out its own destiny and that there was no urgency in reaching a solution. The King's disillusionment with the Hamiduddin was total now, anyway. He wanted an all-faction ruling party established in North Yemen, but by the latter half of 1967 he seems to have been convinced that one – the Hamiduddin – would have to be excluded from it. Only in the early stages of the conflict had the principle of legitimacy been a real consideration. The motive in helping the Royalists had never been the restoration of the dynasty but ousting the Egyptians from the country and ensuring that a radical regime did not come to power there. For Saudi Arabia, one reassuring factor had been the rapid deterioration in relations between the Republican Government under the leadership, since Sallal's overthrow in November 1967, of Qadi Abdul Rahman Iryani and the N.L.F. in Aden, even if both camps still professed the ideal of unity of the two Yemens. The prospect of it materializing was abhorrent to the House of Saud which controlled a population so much smaller in a far greater expanse of territory.

By the spring of 1968 it was clear that the Royalists had lost for ever their big opportunity of taking Sana. The leadership disintegrated and one faction called for a ceasefire early in 1969. In March, Muhammad bin Hussein, the leading Royalist commander, gave up the struggle and retired to the Kingdom. During the *Hajj* that month, Republican representatives sought to make contact with the Saudi Government under the auspices of the Muslim World League, but received a cool response. The war spluttered on. In the early autumn Iryani complained that his Government's overtures were being brushed aside. In the winter of 1969 – 70 there was some friction over violations of Saudi air space by North Yemeni MiG 17s.

Finally the ice was broken when Feisal extended an invitation to Sana to send a delegation to the Islamic foreign ministers' conference held in Jeddah in March 1970. It was led by Mohsen Aini, Premier of the Yemen Arab Republic, and the way was paved for a reconciliation with some thirty leading Royalists, of which the most prominent was

Ahmed Shaami. Members of the Hamiduddin dynasty were not amongst them. In July, Saudi Arabia recognized the Republic. Feisal, Sultan and the others in the royal family power élite did not like Hassan Amri, the Chief of Staff and military strong-man behind the regime, but they were relieved that the traditionalist Zeidis were predominant. For although they were despised as accursed Shiites, they were preferable to the Sunnis of southern Y.A.R., known as Shafeis, who had provided the backbone of the Republican movement. Equally important, the regime, though reasonably broad based, was also sufficiently weak for the liking of the House of Saud. Keeping it that way would thereafter be a constant foreign policy objective.

Feisal was justified in regarding the Marxist regime of Salem Rubai Ali in the south as a threat to stability within the Kingdom itself even though the border was so remote from its main centres of population. In this situation he became even more cantankerously bitter when, early in 1968, the British Government made known its decision to withdraw from the Gulf by the end of 1971 and end the old treaty relationship with the nine Emirates there. Only two months previously it had confirmed its commitment to keep military forces in Bahrain and Sharjah.

Basic to the stability of the oil-rich region was co-operation between the two leading powers of the Gulf, Saudi Arabia and Iran. But notwithstanding the common interest they felt in confronting Nasser, Feisal looked with nervousness across the water at the Shah – 'King of Kings, Light of the Ayrians' and, moreover, a Shiite with clear pretensions to regional hegemony and vastly greater strength at his disposal. Already puffed up with illusions of grandeur and believing himself to be an extension of the God-head, the Shah took a supercilious view of the Saudi King. Worse, he claimed not only Bahrain as an ancient province of Persia but also the islands of Abu Musa and the Tunbs which had always been recognized by Britain as the possessions, respectively, of Sharjah and Ras al Khaimah.

Congenitally cautious and apprehensively preoccupied with the security of his own regime, Feisal had not the confidence to be a resolute champion of Arabism in the Gulf. In line with his territorial dispute with Abu Dhabi, the King had developed firm links with the crafty, one-eyed Sheikh Saqr bin Muhammad al Qawasim, Ruler of Ras al Khaimah, and in 1965 had begun giving financial aid to the astute Sheikh Rashid bin Said al Maktoum, Ruler of Dubai, who had to wait another five years for his own petroleum revenue. By way of reassurance, Feisal received Sheikh Issa bin Sulman al Khalifa, the Ruler of Bahrain, as his guest for three days in January 1968. In imperial pique the Shah cancelled a state visit that had been scheduled for the following month. Feisal was deeply offended. His reaction was uncharacteristically bold. He instructed Aramco to send a drilling rig into disputed territorial waters. It was promptly seized by Iranian naval vessels. The two monarchs had too much to risk from dissension, however. A rapprochement was achieved as a result of the Shah's initiative. In May he sent a conciliatory note to Feisal indicating some flexibility on the question of Bahrain and suggesting a meeting. On 3 June 1968, he stopped at Jeddah airport en route for Ethiopia and was greeted by the King. Their ninety-minute exchange was a cordial one and laid the basis for a somewhat fragile understanding based on a common interest in the stability of the Gulf. A state visit was rearranged for November. Agreement on the demarcation of an off-shore median line was reached. In other respects, especially over the question of Bahrain, the King's talks with the Shah in Riyadh were not a conspicuous success.

In the meantime, however, the British Government had launched its own highly sensitive and secret diplomatic initiative to solve the problem of Bahrain. The two key personalities involved in painstaking secret negotiations lasting over a year were Sir Geoffrey Arthur, the Assistant Under-Secretary at the Foreign Office then responsible for the Middle East, and Sir Denis Wright, U.K. Ambassador in Tehran. Their task was to devise a face-saving formula for the Shah and to persuade him to accept it. The Shah warned that the talks would be broken off if there was any leak about them. Publicity would make it impossible for him to recognize Bahrain's independence after British withdrawal from the Gulf. He was encumbered by his own magisterial pronouncements on the subject and also Persian claims long pre-dating his parvenu dynasty's rule. Secrecy was not only kept but maintained until after the deposition and death of the Shah. Apart from the Ruler of Bahrain and his Government, only Kuwait was consulted and kept informed of the fourteen-month-long exchanges between the Foreign Office and the Shah. The compromise, involving the Shah's acquiescence in Bahrain's independence, was reached just before the end of 1969. Only at that point was Feisal informed. Reconciled to what the outcome would be, the Shah, it was agreed, should publicly call a referendum to settle the future of Bahrain, as if it were his own magnanimous and statesman-like proposal. He duly did so in January 1970. Imperial pride was, thus, preserved. The subsequent involvement of Dr Ralph Bunche gave 'cover' to the collusion. In May, Pier P. Spinelli, the special representative of the U.N. Secretary-General U Thant, undertook an 'ascertainment of opinion', not a referendum. The Shah accepted his findings that a clear majority of Bahrain's inhabitants were in favour of independence. Although Feisal contributed nothing to the Bahrain settlement, later he liked to claim some credit for it.

The King's grim apprehensions about the revolutionary daggers pointing at him from South Yemen and Iraq were understandable, as was his concern about the vacuum that U.K. withdrawal might leave in the Gulf. But his ability to pursue a positive diplomacy that might help ensure stability there was impaired by his obsession over Buraimi. He could neither forgive nor forget this issue. For a while after the outbreak of the Yemen civil war it was shelved. Feisal raised it again in 1964 and rejected new proposals put to him by Britain. The deposition of Sheikh Shakhbut in 1966 and the accession of Sheikh Zaid brought no improvement in relations with Abu Dhabi. Feisal submitted to him fresh proposals on the frontier that represented a retreat from the claim of 1949 but embraced oil discoveries – the Zarrara field – made by the Abu Dhabi Petroleum Company within the *de facto* boundary recognized by Britain since 1935. The King's antagonism was intensified by Sheikh Zaid's failure to show the kind of humility that he expected.

After London had announced the plan for withdrawal from the Gulf, the British diplomatic machine was set in motion to create a union of all nine Emirates, including not only the homogenous Trucial States, but also Qatar and Bahrain. The King supported the impractical idea for the purely negative reason that a larger entity would not be dominated by Abu Dhabi. Because of his absorption with the territorial dispute and dislike of Sheikh Zaid, Feisal gave only half-hearted backing to the painstaking efforts of Sir William Luce, the U.K. special envoy, to formulate a union.

In May 1970, Sheikh Zaid went on his own initiative and against the advice of his British mentors to Riyadh in an attempt to win Saudi co-operation in countering subversion by the Aden-sponsored Popular Front for the Liberation of the Occupied Arab Gulf, which had extended its ambitions beyond Oman itself, where it was tying down the Sultan's Armed Forces in the southern province of Dhofar. The King's response was to demand as the price for his co-operation the settlement of the frontier question on the basis of a map which he thrust at Sheikh Zaid. It was again a modification of the 1949 claim but included the corridor to the Khor al Daid and a boundary along the twenty-third parallel running into what the Sultan of Oman considered to be his territory. Feisal proposed that a plebiscite should be held in the Buraimi Zone after the return of the tribesmen who had taken refuge in the Kingdom after the expulsion of Turki bin Utaishan's force from the oasis in 1955, to determine whether its inhabitants wished to come under the rule of Saudi Arabia, Abu Dhabi, or Oman. Feisal accompanied presentation of the map with a dictat that all oil development south of the parallel should cease and threatened physical intervention if it did not. Sheikh Zaid complied and sent a message two weeks later saying that work had stopped. It so happened that Aramco had discovered oil in commercial quantities just south of the 1935 *de facto* boundary that was later found – as had originally been suspected – to be part of the same structure as Zarrara. Zaid did not respond to the proposal for a plebiscite.

Prestige was involved for both Feisal and the Sheikh, who was alternately sullen and slippery in asserting shadowy historic rights. By 1970, the King's irritation with the Ruler of Abu Dhabi had intensified because of the latter's flirtation with Egypt. Yet, as a leader of international stature Feisal showed in this dynastic wrangle a lack of statesmanship that was all the more inexplicable given his obsessive concern with the communist peril and the paramount need for stability in the Gulf.

Worst fears about the revolutionary potential, or, at least, the subversive inclinations of the Saudi Armed Forces appeared to be confirmed by the discovery early in the summer of 1969 of a conspiracy involving both officers and civilians. The belief was widespread in the senior ranks of the military that the interception of a letter from a student in the U.S. after a tip-off from the C.I.A. led to the uncovering of the plot, and it persisted in the Kingdom a decade after the affair. That, perhaps, is a reflection of an almost paranoid Arab preoccupation about the supposed omniscience of the Agency. In fact, the C.I.A. was caught by surprise and subsequently spent six months trying to find out what had occurred, according to one former operative. It had no part, nor did M.I.6, in the subsequent crackdown that was an over-reaction to what, in substance, did not constitute an immediate or a serious threat. Nevertheless, the nervousness shown by the ruling hierarchy did reflect widespread doubt about its chances of survival, shared even by the Kingdom's friends. At the time, the possibility of revolution was discussed by both Saudis and expatriates.

Symptomatic of the uncertainty was the wave of arrests in May 1969 of people from the Hadramaut in South Arabia who were resident in the country, and Ghamidi tribesmen in the Asir. Then, on the night of 5 – 6 June, the security forces pounced. The immediate concentration was on officers of the Air Force, based in Jeddah, Dhahran, and Riyadh. The biggest proportion of those arrested were from the squadron of Lockheed C-130 transport aircraft that had their headquarters at Jeddah. Amongst them was the King's personal pilot. Also incarcerated were a number of key personnel at Dhahran, including Dawood Romahi, the Commander of the air base there. In June and the following month or so at least sixty to seventy members of the Air Force, Army and the police were detained, and a similar number of civilians.

There was a half-baked conspiracy, but in so far as any plot existed it was at an early stage of development. Central to the affair was Yusuf Taweel, a prosperous merchant. His father had served Ibn Saud after the conquest of the Hejaz. But in the minds of Nejdis, his family was looked upon with suspicion as being synonymous with the region's aspirations towards autonomy or secession or dominance. A cache of explosives was found in Jeddah. It did not indicate a sophisticated approach to the overthrow of the House of Saud – an operation that would have required secrecy, meticulous planning and the involvement of a significant number of military officers in key positions. As far as the explosives were concerned, the intention of the malcontents may have been no more than to create confusion and uncertainty about the future of the existing order.

There is a consensus amongst informed sources (none of which have a vested interest in minimizing whatever danger was posed to the regime) that probably no more than a dozen of the 130 or so men rounded up belonged to the circle of conspirators. They might, perhaps, have been more accurately described as a secret society. As such, it was a very indiscreet one. A few weeks before the arrests, an executive involved in the British air defence contract recalls speaking to one senior Air Force officer at the Dhahran base. He said with a flamboyant wave towards the capital: 'In a few weeks you will see me in Riyadh.' That was not the only instance of loose talk and bravado. It was indicative of a lack of organization and purpose on the part of the disaffected. There appears to have been no coherent ideology behind the conspicuous disloyalty. It hardly merited the term subversion beyond dislike of the royal family and vague aspirations of an Arab nationalist nature.

The number of those arrested convinced many – not the least the British personnel training the Air Force – that a coup d'état was imminent. It tended to give substance to the belief that the House of Saud escaped by only a whisker the fate that befell the moderate bourgeois democracy in the Sudan and King Idris of Libya in the course of the same summer. So, too, did the fact that military aircraft were grounded for several weeks. The gravity of the affair was further exaggerated by the silence of the Government. It never acknowledged that anything untoward had happened. In this atmosphere, bizarre rumours spread, including one that when distilled over a period of time took on the status of accepted fact: that Air Force officers had been dropped from Hercules C-130 transport aircraft into the sands of the Rub al Khali. The stories of brutal interrogation were better based in fact. One or possibly two of those arrested are reliably believed to have died in prison.

The Government did have reason to be worried about the extent of disaffection, especially in the Air Force. It was an élite body in which the royal family's representation was relatively strong. As many as a quarter of its officers were implicated, some justifiably but others not. Faced with evidence of dissidence, the regime lost its head rounding up wholesale anyone who might remotely be implicated in a plot, including all those known or suspected to have had Nasserite sympathies in the early 1960s. Some, it was known, had had contacts with the Egyptians in Beirut and Cairo. But that era had long passed and the disloyalty of those was based for the most part on opposition to Saud. Characteristically, the authorities recalled one military attaché and incarcerated him for no better reason than the fact that he was the brother of Abdul Aziz Muammer, who had been in prison since 1965 on suspicion of being a communist. According to his friends, even General Hashim S. Hashim, the R.A.F.-trained and Anglophile Commander of the Air Force who had been responsible for the recruitment of most of the men detained from that service but was in no way tainted by Nasserism, felt afraid that he might be arrested, not the least because he felt strongly resented as a Hejazi.

According to informed accounts, intensive interrogation revealed nothing to suggest that a successful coup d'état had been about to take place. Nevertheless, for the next few years Saudi Arabia experienced its own equivalent of America's McCarthyite era. In the following year, 1970, the Acting Commandant of the Staff College was put away. Other officers subsequently disappeared for a while, including Muhammad Duwaini, a military attaché posted to the Saudi Embassy in Tokyo, who had previously been head of Sultan's office and had obediently come home when recalled. More conspicuous by his sudden disappearance was Saleh Amba, the popular and liberal Dean of the College of Petroleum and Minerals, whose two brothers-in-law had radical views and were thought to have been involved in the bomb incidents of 1966. At the end of the same year, several hundred Saudis of the Shiite sect residing in Hasa were arrested on the vaguest suspicion of having Baathist links. Asked by a Western intelligence officer why they had been herded into the dreaded gaol at Al Khobar, Omar Shams, the Chief of the General Directorate of Intelligence, was at a loss to give any explanation other than the fact that they were Shias.

The House of Saud was as jittery as it had been at the time of the Republican coup d'état in Sana and the Egyptian intervention in Yemen, though it had much less reason to be. But the scale of arrests in itself was sufficient to undermine the confidence of the politically conscious and articulate in society. As a result, to many observers the prospective life of the regime looked as short as it had in 1958 and 1962.

Having over-reacted, the Government was confused and embarrassed. The problem was what to do with the alleged recalcitrants. Feisal, Fahd and Sultan had the sense to realize that they had gone too far and that any further draconian action would only compound the damage done. There were no executions, let alone parachute drops into the Rub al Khali. None of the apprehended had, anyway, committed offences under the *sharia* law by any conceivable interpretation of its letter. Another, more secular, consideration was that executions might have engendered blood feuds against the royal family. Moreover, its traditional style was one of conciliation, whether in the marital bed or through forgiveness fortified by generosity.

The Saudis were not sophisticated in the art of eliciting information. They were restrained from using bestial methods by religious, social and political inhibitions. But the ugly Omar Shams could claim credit for one original technique. It is known that as early as 1962 Bahrain had enlisted his assistance with the questioning of some suspected Baathists. He elicited confessions by leaving them in large refrigerators until they screamed for mercy. It was an ingenious method of avoiding Koranic prohibitions on causing bodily harm, apart from those punishments ordained by the *sharia*. No rumours suggested that those taken into custody in 1969 were subjected to this frosty treatment. But members of the Armed Forces imprisoned in the early round-up that summer were subjected to brutality. Omar Shams, Saudi citizens point out, was not restrained by fears of vengeance. A Muslim of Indian origin – and a fanatically orthodox one, too, who shared the House of Saud's loathing of Shias – he had no family ties.

After the initial alarm, the authorities made every effort to ensure that the detainees were well treated. They were housed in comfortable villas, well fed, and provided with television. Their families were paid half their salaries. From the beginning of 1972 they were progressively released – about half of them before Feisal's death in 1975. Saleh Amba was amongst the first. General Hashim, who left the Air Force in 1972 and went into business, employed several of his former subordinates who had been arrested. Yusuf Taweel eventually emerged to prosper more than ever in commerce. But not for several years was confidence in the reliability of the regular Armed Services restored. The Government's immediate reaction was to increase the strength of the National Guard.

It is possible that the trauma in the summer of 1969 could have encouraged South Yemen to launch the attack on the frontier post of El Wadieh later in the year on 26 November. The Aden regime might have calculated as well that such an incursion would cause more trouble and instability in the Kingdom. Without doubt the attack was also in response to the somewhat half-hearted attempts, backed and financed by Saudi Arabia, to stir up trouble in South Yemen. More specifically, the adventure could well have been prompted by the discovery by the N.L.F. regime of a plan involving units of the former Hadraumi Bedouin Legion, which had not yet been disbanded, to restore to power the young ex-Sultan Ghalib Qaiti. Coming just one month before the Arab summit conference scheduled to take place in Rabat, the aim may also have been to set the scene to air differences within the pan-Arab forum, to obtain the backing of the 'progressive' states, and perhaps also to create a *fait accompli* along the ill-defined frontier.

The ten bedouin members of the Saudi Frontier Guard manning the fort could put up only token opposition to the thrust by infantry and armoured cars. Riyadh reacted with uncharacteristic decisiveness. Saudi air power was mobilized within forty-eight hours and proved not only trustworthy but efficient. The first strike was made by a flight of Lightning F-53s (supplied as part of the main defence package) led by Wing-Commander Tony Winship, a British pilot. The fact was still denied in the Kingdom eleven years later and, indeed, behind the scenes in Whitehall there was something of a furore at the time about his participation in the counter-attack, even though he was a contract, rather than a seconded, officer. Pakistanis were also flying Sabres for the

Royal Saudi Air Force in an operational role. There was thus some substance to the Yemeni claim that 'mercenaries' had been involved.

Avoiding any aerial engagements, Yemeni Provosts bequeathed by the British and one MiG 17 supplied by the Soviet Union strafed the Saudi ground troops who went into action with supplies of pre-packaged, cellophane-wrapped airline meals provided courtesy of the Abela Catering Company. They occupied the fort ten days after its capture. Some Lightnings stayed on at Khamis Mushayt in the mountains of Asir, then undergoing development as a base, and a few months later forced down a military aircraft carrying the Iraqi Chief of Staff who was flying to Aden on a goodwill mission.

Despite the understanding reached at the Khartoum summit conference in 1967 that Egypt would cease to undermine the traditional regimes in Saudi Arabia's sphere of influence, Feisal's deep suspicion and distrust of Nasser had been by no means alleviated. Saudi leaders were even inclined to believe that Nasser had prompted the South Yemen attack on El Wadieh. The obsession with El Rais blinded them to the real nature of the Moscow-orientated regime in Aden which was far too independently minded and ideologically distinct to have taken even advice from Cairo. The King also suspected Nasser of plotting the conspiracy that summer. According to Muhammad Heikal, Feisal raised the subject with Nasser when they met just before the fourth Arab summit held in Rabat in December 1969, their first confrontation since the Khartoum conference in 1967. Feisal asserted uncompromisingly that the arrested officers had connections with Egypt and, in particular, with Sami Sharaf, Nasser's private secretary. Heikal wrote in his version of the exchange: 'I admit that before nineteen sixty-seven we may have been working against you, but after the Khartum conference I gave orders that all such operations should cease.'[7] The Egyptian leader was said to have added that people often claimed to be acting in his name without his authority.

Heikal's account is, at least, an indication of Feisal's lasting estrangement from Nasser and his feeling of besieged isolation at the time. After the overthrow of King Idris of Libya by Lieutenant Muammer Qaddafi and his brother officers on 1 September 1969, Saudi Arabia's only real allies in the Arab world were Morocco, Tunisia, and Jordan. Feisal showed his solidarity with the Hashemite King by continuing to keep a brigade stationed there. It was a symbolic gesture only. The Saudi force remained impassive in the face of punitive Israeli air raids on the East Ghor Canal in the winter of 1969 – 70, when it suffered some minor casualties.

The House of Saud took good care not to alienate the Palestinian resistance movement. In reality, Feisal's grief over the displaced people was great and as genuine as his hatred of the Zionists who had seized their land. Formally, Saudi Arabia's position on the Arab-Israeli conflict was impeccable. Having mobilized so ineffectually to aid Jordan in the 1967 war, the Kingdom had never subscribed to the U.N. ceasefire resolution and technically could be said to have been at war still. Feisal privately welcomed Security Council Resolution 242 of 22 November 1967 that in essence proposed Israeli withdrawal from occupied territories in exchange for Arab acceptance of its right to live in peace within secure and recognized boundaries, as well as calling for a settlement of the problem of refugees (without describing them as Palestinian). But the Kingdom did not vote, as Egypt and Jordan did, for what was to become the basis for all future diplomatic efforts to bring about a Middle East settlement. At the same time, Feisal feared the revolutionary proclivity of the dispossessed. Thus, the Kingdom allowed public fund-raising on behalf of Al Fatah, the moderate mainstream group of the Palestinian resistance movement founded by Yassir Arafat, Chairman of the P.L.O., in 1965, and the Government sanctioned the support of the Saudi press for it. But it watched with alarm the growing strength of the extremist Marxist fringe, especially the Popular Front for the Liberation of Palestine which was held responsible for the blowing up of the Trans-Arabian Pipeline in 1969. It was out of operation for nearly four months. While maintaining a line of communication with Yassir Arafat, Feisal suppressed political activity by the Palestinian community in Saudi Arabia and kept a close eye on it.

Because of the strength of Feisal's emotions about Jerusalem he could not support the idea of a settlement involving an international status for the Old City proposed by William Rogers, U.S. Secretary of State, at the end of 1969 as part of his peace plan. The King realized, however, that the impasse, by encouraging the radicalization of the Arab world and furthering Soviet influence, was endangering the House of Saud's hold on power. As it was, the Kingdom was on the sidelines of the conflict, as yet without any leverage despite its burdensome financial subventions to Egypt and Jordan. Just as the Kingdom's relations with the U.S. had been strained earlier by Kennedy's policy of cultivating Egypt, so they had become afflicted from 1966 by the pronounced bias of the Administrations of Lyndon Johnson and then Richard Nixon in favour of Israel.

In this situation, as a counter to Arab radicalisms, Feisal not only continued his pursuit of Islamic solidarity, but also tried to make it the framework for collective action against Israel. The long-awaited religious summit that Feisal and the Shah had called for in December 1965 eventually convened in Rabat on 22 September 1969. The original aim had been to confront Egypt and revolutionary Arab socialism. The actual catalyst for the gathering nearly four years later was the world-wide fury and indignation over the fire that had badly damaged the Al Aqsa Mosque in the Israeli-occupied Old City of Jerusalem on 21 August. King Hussein was the first to propose it. Convening the summit at all, and so quickly, was something of a triumph for Hussein, Feisal, and King Hassan of Morocco. But it was not an unqualified success. Representatives of twenty-five countries attended but amongst eleven invited that did not were Syria and Iraq. Nasser pleaded the state of his health in staying at home and sending instead Anwar Sadat, his specialist in the politics of Islam. Ritualistic resolutions were passed calling for the restoration of Jerusalem's status and Israeli withdrawal from the occupied territories. Turkey and Iran stoutly resisted efforts to drag non-Arab Islamic states into the Middle East conflict. In addition, they, together with five black African countries, refused the demand that they rupture all diplomatic and economic ties with the Jewish state. Feisal's own performance was dignified and impressive.

Aware of the extra financial demands that might be made upon Saudi Arabia, the King had resisted growing pressures for an Arab summit. With the failure of efforts by Dr Gunnar Jarring, U.N. special envoy, to bring an Arab-Israeli settlement into focus in line with Resolution 242, he could withstand them no longer. Eventually he agreed to attend the one that was held in Rabat from 21 to 23 December 1969. On the way there Feisal stopped in Cairo for talks with Nasser. They were meant to prepare the way for a smooth conference but did not save it from being a fiasco – except that it established the P.L.O. as a participant in pan-Arab counsels at the highest level. Nasser, whose war of attrition had cost Egypt dear, demanded an extra £40 million to bolster his confrontation with Israel. The King repeated in public what he had said earlier in Cairo. His country's budgetary commitments meant that he could not increase Saudi Arabia's subvention of £50 million a year. The Ruler of Kuwait said that he was prepared to make an extra once-and-for-all payment of £10 million, subject to approval by his National Assembly. Prior to the conference Saudi Arabia and

Kuwait had been goaded by the offer of Qaddafi to put all his country's resources at the disposal of the confrontation states.

The first meeting between the Guardian of the Holy Places and the young revolutionary firebrand of the Arab world proved to be a strain. Colonel Qaddafi – he had promoted himself immediately after the Libyan revolution – called the King 'Brother Feisal' without any honorific or royal embellishment. Although Feisal accepted unquestioningly such an address from his own bedouin, he did not receive it kindly from the brash young man. Yet this was a minor irritation compared with his anger when Salem Rubai Ali persevered with his attempts to raise the subject of the border dispute between Saudi Arabia and South Yemen. Then Nasser left in dramatic dudgeon, which may have been feigned, with Yassir Arafat and Abdul Khaliq Hassouna, Secretary-General of the Arab League, following in his wake. The session was boycotted by Syria, Iraq, South Yemen, and the Yemen Arab Republic. Nasser may not have been unhappy with the outcome. The lack of unity shown and the unwillingness or inability of the Arab states collectively to raise money for a meaningfully aggressive stance against Israel left the way open for him to explore seriously the Rogers initiative.

Whatever Nasser's intentions, Feisal was spared accusations by Cairo that he had betrayed the cause and emerged no weaker from the summit. With South Yemen having been chastised, the main aggravation came from Syria. The Trans-Arabian Pipeline broke down again early in May 1970. The reason given was that a bulldozer installing telephone lines had accidentally ruptured it. The radical and volatile Baathist regime's refusal to allow it to be repaired because the undertaking would be 'too dangerous' suggested a malevolent intent. Stung by the loss of some $200,000 a day, Saudi Arabia barred the entry of Syrian goods and vehicles. Syria retaliated by banning the over-flying of its territory by Saudi aircraft and the passage across its land of cargoes destined for the Kingdom.

Feisal continued in his campaign to rally the forces of Islam beyond the Arab world. A tour in June 1970 took him to Malaysia, Indonesia and Afghanistan. His itinerary concluded with his first visit to Houari Boumedienne's revolutionary Algeria. Feisal found to his surprise and delight that both the regime and the people were good Muslims even if they were socialists. His talks were a success. They might have seemed a portent of changes that would make the Arab world a more comfortable and congenial one for Saudi Arabia. At that time, however, the tension in Jordan between King Hussein and the Palestinian guerrillas was building up.

Feisal watched with impotent concern the hijack on 6 September of four aircraft by the Popular Front for the Liberation of Palestine and the forcible diversion of three of them to the remote Dawson's Field in Jordan where they were subsequently blown up on 12 September. Somewhat lamely, he is said to have proposed to Qaddafi that they should freeze their financial contributions to the guerrilla movement until it brought extremist elements under its control. It was a gauche suggestion. As it happened, the Libyan leader had suspended his subvention to Jordan just before the Popular Front's spectacular operation. Saudi Arabia kept a discreet silence throughout September and its brigade of troops, originally dispatched there at the end of the June War, stood impassively by as the Hashemite King's bedouin troops crushed the commandos. The sense of relief of the House of Saud over King Hussein's bloody victory was almost palpable.

At the summit called to discuss the crisis, Feisal maintained a low profile. The King withstood the impertinence of Qaddafi, though at one heated session restricted to heads of state he suggested, in a pointed barb directed at the Libyan leader, that, as all present seemed to be mad, perhaps a psychiatrist should be called in. At the summit's conclusion on 28 September he departed for Jeddah. Later that day Nasser was struck down by a massive heart attack. For the House of Saud it was the end of another era.

17 Friendship Across the Red Sea

1970–2

What finally got me involved in the execution of Middle East diplomacy was that Nixon did not believe he could risk recurrent crises in the Middle East in an election year. He therefore asked me to step in, if only to keep things quiet.[1] – HENRY KISSINGER

SUDDENLY the colossus who had dominated the Arab scene for fourteen years and seemed likely to do so for as long again was silent and inert. His presence on the global stage was such that the world at large, with the exception of his Zionist foe and die-hard enemies in the West, felt the loss of a figure who had provided a reference point on the shifting sands of international affairs. For many of his own people and millions of other Arabs to whom he had given pride and hope, even if of a most illusory kind, Nasser's departure caused anguish and grief.

It was surprising how little his image had been tarnished by Egypt's humiliation at the hands of Israel in 1967. Ten years later the charisma of the most outstanding Arab leader since Salahieddin, the hammer of the Crusades, still shone for the masses who had been inspired. But in death his dominion was gone. At the Saudi royal court news of his death was received with some unabashed jubilation. Egypt's defeat three years earlier had eased the predicament of a traditional and wealthy regime that by the year seemed to be more of a political anomaly amidst increasing Arab turbulence. It was one factor that in the course of the next two years would make possible a new regional conjuncture profoundly altering and fortifying. A second, equally important but related one was the sudden shift in the balance of power away from the oil consumers and international oil companies towards the producers. By coincidence, on the very day that Nasser's heart cracked, Esso and British Petroleum agreed reluctantly to Libya's demand for higher revenues. The solid façade maintained by the industry in the face of O.P.E.C.'s attempt to improve the real value of oil receipts had been broken. It was a development that would transform the fortunes of the Kingdom.

Amongst those who accompanied Fahd to the Egyptian leader's funeral was Muhammad Alireza, the former Saudi Ambassador to Egypt, who had been recalled from his post prematurely in 1968 because his life was threatened by Nasser. Feisal went instead to Geneva for medical attention, where, amidst much speculation to the contrary, he had no more than stringent medical check-ups and a short rest. He wanted an excuse not to attend the obsequies of a man, the 'Arab Castro' as Feisal privately called him, who had striven so strenuously to bring down the House of Saud.

Egypt's constitution provided for the Vice-President to become acting Head of State. Having been one of four Vice-Presidents in the period from 1964 to 1967, Anwar Sadat became the sole one in December 1969. At first it was by no means clear whether Sadat would take over on a permanent basis. Indeed, the assumption tended to be that he would only hold the post as a caretaker. Initially, there was speculation that had some basis in fact that the hard-headed Zakariah Mohieddin was Nasser's choice and would be the chief contender against Sadat. Regarded as the most conservative, pro-Western and anti-Soviet of the members of Egypt's original Revolutionary Command Council, he would have been more than acceptable to the Saudis. As it was, within a fortnight of Nasser's funeral the nomination of Sadat was decided upon by the Higher Executive Council of the Arab Socialist Union, Egypt's sole political party, approved unanimously by the National Assembly, and endorsed overwhelmingly by a national referendum.

Sadat was less objectionable to the Soviet Union than Mohieddin, a consideration of some importance both because of Egypt's massive dependence on the Soviet Union, not the least for weaponry, and because of the involvement in the selection process of prominent left-wingers like Ali Sabri, Sharawi Gomaa and Sami Sharaf. As ardent an Egyptian nationalist as Mohieddin, Sadat had also forcibly expressed in private his opposition to the spread of Soviet influence in the Middle East – a fact of which the Saudis were well aware. But because of

Sadat's major role in Egypt's conduct of the Yemen war, Feisal regarded him with a suspicion that almost equalled his hatred of Nasser. His advisers took a different, less paranoid view, in particular the King's Chief of Intelligence and brother-in-law Kamal Adham. They knew Sadat to have been a member of the Muslim Brotherhood, which was founded in the mid-1920s by Hassan Banna, an Egyptian. Known as the *Majallat al Ikhwan al Muslimin,* the movement set out to transform political, economic and social life on religious lines throughout the *umma,* or worldwide Islamic community. It had nothing to do with the primitive tribal warriors whose zeal Ibn Saud had finally been forced to suppress. Nor did the Brotherhood seek to dismantle national frontiers in the pursuit of an all-embracing Islamic state. Despite its religious fundamentalism, it was abhorrent to the Saudis because members believed in using violence to achieve its aims. Sadat's association with it arose mainly, it seems, from his passionate desire to free Egypt from the neo-colonial yoke of Britain. At least his connection with the Brothers indicated a right-wing temperament. The new incumbent of the Abdin Palace had also shown an interest in Islamic affairs, a field in which he had advised Nasser. Last but not least, the dapper Sadat, with his taste for Ted Lapidus clothes – apparel beyond the reach of any Egyptian without special access to scarce foreign exchange – was looked upon as 'accessible'. In the words of one long-serving Arab Ambassador, 'He liked the good things of life the Saudis could offer.' Adham, evidently, had helped provide them. Before Nasser's death he had become involved in business ventures with Jihan Sadat, the new Egyptian President's wife. It was Adham who convinced Feisal that he was a man with whom the House of Saud could have dealings and, at least, establish a *modus vivendi.*

Yet at this point there was no certainty – indeed, many doubts – over how long Sadat would remain at the helm of the Arab world's leading country. As well as being regarded as a compromise choice between Mohieddin and Sabri, many observers believed that he would be only a short-term incumbent in the Abdin Palace. He had inherited a Government that included a number of key political figures with their hands on critical levers of power who also showed a marked inclination towards Moscow. As events in the following year would show, they had views diametrically opposed to Sadat's and would be only too willing for a showdown.

The Soviet Union was more deeply entrenched in Egypt than ever. The installation of the surface-to-air missile system on the west bank of the Suez Canal had brought an end to Israel's deep-penetration bombing raids and had been an important factor leading to acceptance of the agreement on a six-month ceasefire arranged by the U.S. Secretary of State, William Rogers, though not, as far as Israel was concerned, the peace plan which he had put forward the previous December. However, the new defensive weaponry also increased the number of Soviet military advisers in Egypt to well over 15,000.

Feisal had become more demented than ever about the 'Zionist-Bolshevist' conspiracy. The causal connection between the two became even more of a travesty as the issue of the Jewish dissidents in the Soviet Union grew and received fuller publicity. But, whatever the nature of his logic on that score, the King had evolved one novel concept earlier than any other Arab leader including, it seems, Sadat. Feisal argued that the removal of the heavy Soviet presence in Egypt might induce a more even-handed U.S. policy and serious pressure by Washington on its client state to withdraw from Arab territories occupied in 1967. Such a move would also give the lie to the specious Israeli thesis, but one nevertheless readily accepted by American public opinion, that the Jewish state was not only a bastion of Western democracy but also one helping to contain what seemed to be a remorseless Soviet expansion in the Middle East. There was, indeed, substance in Feisal's *idée fixe*. The existence of Israel and its ever closer identity with the U.S. had furthered the Soviet advance in the region by forcing Egypt and Syria into the arms of Moscow. In this respect the Saudi King had a far sharper appreciation of the balance of power in the region than the U.S. Administration. Under sustained diplomatic pressure from Israel, and intensive, associated pressure from the Jewish lobby, it had wilted and failed to pursue the Rogers plan, which Jerusalem and its supporters had rejected even before the Arabs had had time to consider it. Instead, following a call by no less than seventy-four Senators, President Nixon had acceded to Israel's request for Phantom aircraft without exacting any flexibility in return, except the reluctant agreement of Mrs Golda Meir's Government to the ceasefire along the Suez Canal that effectively brought to an end the costly war of attrition.

At this point neither Feisal nor anyone else foresaw the enormity of the future escalation of petroleum prices and the financial inducements that the Kingdom could offer as a result. However, advised by Sheikh Ahmed Zaki Yamani, the Saudi Minister of Oil, the King was aware how demand was rapidly overtaking supply and that the prospect of the U.S. becoming a substantial importer of oil in the coming few years would make vital the role that Saudi Arabia could play in fulfilling its future requirements. The idea of partnership and co-operation with the U.S. based on the Kingdom's oil resources had taken root.

In his later years, Nasser had begun to appreciate the key role that the U.S. would have to play in any Middle East settlement satisfactory to the Arabs. Mindful of how President Eisenhower had pressurized Israel into withdrawing from Sinai in 1957, Nasser had always realized the theoretical advantages of taking a middle course between the two super-powers. If the circumstances had been right and such a strategy had looked as if it could regain the lost territories, he might possibly have played them off against each other. By 1969–70 he was thinking in terms of the super-powers co-operating in imposing a settlement that would avoid any direct negotiations with Israel. His reservations about the Rogers plan related to the fact that it was a U.S. initiative and a successful outcome would have given Washington the predominant role in bringing about a solution of the Arab-Israeli conflict. For precisely the same reason Moscow had originally rejected it. Following the visit of Joseph Sisco, U.S. Assistant Secretary of State, to Cairo in April 1970, Nasser slowly came round to the idea of acceptance after obtaining grudging approval from Leonid Brezhnev, the Soviet leader. Given the vacillations of the U.S. Administration in the face of Israel's intransigence, its rooted objection to any settlement not directly negotiated, and its reluctance to renounce Sinai in its totality, Nasser's positive response on 23 July of the same year, though significant, would probably have led nowhere even if he had lived.

The widely held view was that Nasser alone could lead the Arabs to recognize, if only *de facto,* the existence of Israel. However, his position had always been that the Jewish state would never be induced to an accommodation satisfactory to the Arabs unless they – and above all Egypt – maintained a plausible military option. Thus, dependence on the Soviet Union, with the concomitant large presence of military advisers, was necessarily greater than ever – especially with the missiles only recently installed, the lengthy period of time required to train Egyptians to operate them, and the need for other sophisticated weaponry. Sadat could hardly contemplate the situation in any other way as he viewed the legacy bequeathed him. In the winter of 1970–1 such a volte-face as a request, let alone an order, for the departure of the Soviet advisers was almost inconceivable.

Feisal was optimistic that the détente with Egypt could be transformed into a rapprochement under the new regime. He had kept in close touch with events in Cairo mainly through Kamal Adham. In November 1970, little more than a month after Nasser's death, he dispatched Adham to Cairo to put forward his proposition that the withdrawal of Soviet military advisers might encourage the U.S. to adopt a more forceful and, from the Arab point of view, more positive policy in the Middle East, a prospect that evidently had not yet occurred to Sadat. He responded by emphasizing Egypt's dire need for Soviet assistance to counter the weaponry being supplied by the U.S. to Israel, particularly the Phantoms. But Adham was able to report back a counter-suggestion by Sadat. Its gist was that he would ask the Soviets to leave Egypt if Israel withdrew its forces from the east bank of the Suez Canal and allowed it to be cleared for shipping. It so happened that Moshe Dayan, the Israeli Minister of Defence, had already aired the idea that Israeli forces should withdraw a distance of thirty kilometres from the waterway, vacating a zone that could be demilitarized. For Israel the rationale would have been that the tension along the Canal could be defused and the chances of a resumption of the costly war of attrition be reduced. The implied object was also to bring about the removal of the missile sites, a major Israeli preoccupation at the time. Dayan's proposal was opposed by others in the Israeli Cabinet, but the military establishment believed Sinai could best be defended by a mobile force on the hill ridges rather than by the static fortifications of the Bar-Lev line.

In the circumstances, Feisal's initiative was nevertheless a bold one. He never claimed credit nor did Sadat subsequently give it to him for planting the radical notion of evicting the Soviet advisers. But there is good reason to believe that the decisive move may have germinated from that proposal. Crucial contact had been made that was later to change the balance of power in the Middle East. In the coming months Adham was to pay frequent and regular visits to the Abdin Palace. 'This was not something to reassure the Russians,' Muhammad Heikal was later to comment.[2] It is doubtful whether Israeli intelligence was aware of Adham's calls.

There was a related and equally important development. Feisal had asked Sadat if he could convey to the U.S. Administration his proposal about getting rid of the Soviet presence in return for a limited pull-back by the Israelis. The Egyptian leader gave his assent and the message was passed on by Adham. It was at this point, early in 1971, that the King's brother-in-law established himself as a contact, approved by Sadat, between him and the White House. In his memoirs Dr Henry Kissinger, then Special Assistant to the President for National Security

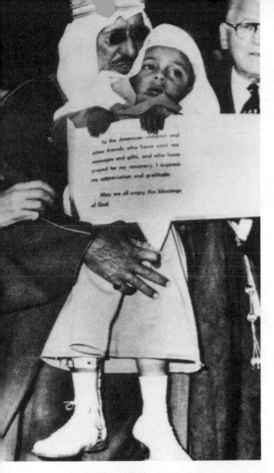

Saud holds up his twenty-fourth son, Mashur, after the boy's treatment for polio in the Walter Reed Hospital, Washington, 4 February 1957. 'Even the most cynical newspaperman was unable to discourage the response of the American people to the paralysed little boy'

Feisal with the architect of the Eisenhower Doctrine, the U.S. Secretary of State John Foster Dulles, in September 1957. Feisal spent half that year having treatment in the U.S., brooding about Saud's mismanagement of affairs

Above left: Talal, one of the 'Free Princes', at a press conference in Cairo on 13 September 1962, a month after his expulsion from Saudi Arabia

Above centre: Nasser and Saud at the Arab summit in Cairo, January 1964. Saud's presence there as leader of the Saudi delegation brought the power struggle with Feisal to a head

Above right: Reconciliation. Feisal and King Hussein of Hashemite Jordan at the Alexandria summit of September 1964

Right: Sultan, the Defence Minister, inspects the Saudi troops stationed in southern Jordan, February 1970. The brigade was originally dispatched at the time of the 1967 war but was to take no part in the bitter fighting between Hussein's army and the Palestinians later in 1970

End of an era: Feisal, Qaddafi, Abdul Rahman Iryani of Yemen, and Nasser at the Cairo summit called in late September 1970 to seek a solution to the Jordan crisis. A few day late El Rais was dead of a heart attack

Brothers in arms: Khaled, Feisal, Fahd, and Abdullah (left to right) take part in the *ardh*, t traditional Nejdi sword dance, in Riyadh

Affairs, is misleading. He says that it was in the first week of April 1972 that Egypt 'opened a secret channel to the White House'. Kissinger wrote: 'On April 5, a high Egyptian officer told an American official in Cairo that Egypt was dissatisfied with existing channels to the United States. In his government's view it was essential we communicate at the Presidential level, bypassing both foreign ministries.'[3] Sadat's chosen go-between was, in fact, Adham and his activity started fifteen months before the date given by Kissinger.

Sadat was still feeling his way, exploring possible options for recovering lost territories, and trying to establish his power base at home. From his lips and the Egyptian media, ritualistic tributes to the Soviet Union and denunciations of the U.S. continued. He declared that 1971 would be the 'year of decision'. Egypt's strategy would aim at preventing the ceasefire from becoming permanent – as a means of 'ensuring seriousness on the part of the other side'.[4] Preparations for war and a crossing of the Suez Canal that had started under Nasser went ahead. At the same time, Egypt engaged itself increasingly in the intensified diplomatic manoeuvres in the Middle East. The first half of 1971 proved to be totally inconclusive, but it did foreshadow and dictate the pattern of events in the region over the next six years, not the least the exaltation of the House of Saud to power and stature, the nature of which Ibn Saud could not have dreamt.

The somewhat desultory Big Four power talks – involving the U.S., the Soviet Union, Britain and France – on ways of guaranteeing a settlement took on a new urgency for a while. The mission of the U.N. envoy Dr Gunnar Jarring came to a climax as he strove to find common ground between the two conflicting parties. Most important of all, the U.S., appreciating the opportunities arising from the change of regime in Cairo and the dangers of new full-scale conflict, concentrated harder than ever on trying to find a solution. Washington was particularly encouraged by Sadat's public proposal on 4 February 1971 that Egypt would clear and re-open the Suez Canal if the Israelis were to withdraw an unspecified distance from it. With alacrity Rogers took up his proposal and American diplomacy concentrated on it as an interim arrangement. Despite left-wing opposition, Sadat extended the ceasefire which was due to expire the following day, but only for a month. Egypt responded more positively than Israel to the letter drafted by Jarring laying out the basic principles of a peace agreement and Sadat stated his willingness to sign a peace treaty if Resolution 242 was carried out in its entirety. Early in March he refused to extend further the ceasefire but stressed that 'this does not mean that political action will stop and that guns will start shooting'.[5]

In May 1971 the Rogers' initiative effectively ground to a halt and Sadat's turning towards the West was checked, though not reversed. That month and the ensuing one were both confusing and critical for Egypt, Saudi Arabia, and the growing rapprochement between the two. Significantly, the Secretary of State's Middle East tour in early May started in Riyadh where he assured Feisal of his Government's interest in obtaining Israeli withdrawal from the occupied territories. After proceeding through Jordan and Lebanon, the only Arab countries at the heart of the conflict that the U.S. could in any way regard as friends, he arrived in Egypt for a two-day visit on 4 May. Mahmoud Riad, the Foreign Minister, presented William Rogers with Cairo's plans for an interim arrangement involving Israeli withdrawal to a line from El Arish on the Mediterranean coast to the tip of the Sinai peninsula, the establishment of an Egyptian military foothold on the east bank and the re-opening of the Suez Canal. Exuding optimism, the Secretary of State spoke of Mrs Golda Meir's commitment to him that if any Arab leader would publicly declare his willingness to conclude a peace agreement, she would be 'prepared to put all her cards on the table'. But the visit left a sense of disappointment because Rogers had presented no American ideas or indication of the Administration's readiness to put pressure on Israel. Scepticism throughout the Arab world on the latter question was reinforced by reports of the Secretary of State's acrimonious exchanges in Jerusalem with Mrs Meir and her colleagues when he presented the Egyptian proposals and stuck to the outline of his original peace formula.

Meanwhile, Feisal watched with a mixture of bewilderment and concern the sequence of events in Cairo. Riyadh was now well informed about the internal situation in Egypt and fully aware of the struggle for survival and supremacy in which Sadat was engaged against the group within the ruling hierarchy opposed to any divergence from the already well established solid alignment with the Soviet Union. Through his own diplomatic intermediaries Feisal had conveyed to President Nixon Sadat's intention of reducing his dependence on the Soviet Union as soon as possible. The issue that forced the showdown was the U.A.R.'s agreement – very much at the behest of Sadat, with Sabri showing open dissent – to form a federation with Syria and Libya.

For his own reasons, rather than to reassure Washington, Sadat had dismissed Sabri from his post as one of the country's two Vice-Presidents two days before Rogers' visit. Two weeks later, on 16 May, he was arrested together with the rest of what Sadat then termed Sabri's 'clique' and years later 'the Soviet agents'[6] in the leadership. Foremost amongst them were Sharawi Gomaa, the Deputy Premier and Minister of the Interior, General Muhammad Fawzi, the Minister of War, and Sami Sharaf, Minister for Presidential Affairs.

On the morning of 16 May, Feisal left for a tour of the Far East in a mood of optimism. The outcome of the power struggle in Egypt had become clear enough three days earlier when on the night of 13 May Cairo Radio had announced Sadat's acceptance of the 'resignation' of the ring-leaders of the left-wing conspiracy against him. News of the dawn round-up in Cairo had broken by the time he reached Tehran for a ninety-minute luncheon stop-over with the Shah. The main purpose of the meeting was to discuss the future of the Gulf in the wake of British withdrawal scheduled for the end of 1971. The two monarchs, however, expressed pleasure at developments in Cairo.

After three days in Taiwan, Feisal flew to Tokyo for a state visit to Japan. While he was there, the King received an invitation from President Nixon to visit Washington. Feisal had let it be known that he would return to Saudi Arabia via the U.S. The way had been prepared for a meeting between the two leaders. Washington seemed suddenly concerned over Saudi discontent with U.S. attempts to bring about a comprehensive Middle East settlement and its supply to Israel of advanced combat aircraft. The State Department, at least, was aware of Saudi Arabia's growing rapprochement with Egypt, and keen to learn more about Sadat's real thinking.

Just hours before the ritual welcome on the White House lawn on 27 May, Cairo Radio announced that Egypt and the Soviet Union had signed a fifteen-year Treaty of Friendship and Co-operation. Heikal claimed that Sadat had put forward the idea in a message delivered to Moscow by Sharaf at the end of April.[7] Such a proposal had been put forward earlier by the Soviet leadership but rejected by Nasser. The Egyptian President's account is that President Nikolai Podgorny thrust the proposal upon him when he arrived in Cairo towards the end of May,[8] concerned about Sadat's crack-down on Moscow's friends and the Soviet Union's position in Egypt. Heikal's version should be regarded as the correct one. Feisal had not been informed in advance, but he was neither surprised nor angered by the development. He understood that Sadat had improved his chances of obtaining the arms required for a plausible military option with which to confront Israel, while at the same time pursuing a diplomatic settlement.

The White House talks centred around the treaty which, in practice, only helped to destroy the Rogers plan by making it easier for the Jewish lobby to rally opposition to it. Feisal assured Nixon that, whatever its links with the Soviet Union, Egypt would never be transformed into a communist state. Coinciding with the treaty, his talks provided the perfect opportunity for him to expound the theory that American bias towards Zionism had only served to further the interests of communism. To the King, it seemed a triumphant vindication of his thesis that the two were inextricably linked. He also hammered home the message that Saudi Arabia could not endorse any Middle East settlement if it failed to restore Jerusalem to Arab sovereignty. He made no mention of possible dangers to American interests in the region. Nixon's main preoccupations then were disengagement from Vietnam, the negotiations with the Soviet Union on strategic arms limitation, and the American opening to China that was being charted by Kissinger. Later, Nixon did not see fit to mention in his memoirs Feisal's visit to Washington. The President was over half way through the third year of his first term in office and concentrating on other foreign policy goals to help his re-election. Feisal departed more sceptical than before about the chances of bringing the U.S. effectively into the Middle East arena on the Arabs' behalf in the immediate future. But Saudi Arabia had formally entered it as an active participant.

On 19 June Feisal visited Cairo. Saudi Arabia's reconciliation with Egypt became an entente. The Kingdom had established itself in the mainstream of Arab affairs on the basis of partnership with it. The world recognized the joint communiqué on the discussions between Feisal and Sadat as signifying as much, even if there was no public appreciation of the intensive secret diplomacy preceding it. Contemporary commentators were surprised how – after every allowance had been made for Arab diplomatic parlance – the two leaders could have reached such an understanding so soon after the conclusion of the Egyptian-Soviet treaty. However, few doubts were expressed about its credibility. That could justifiably be seen as a reflection of the reputation for total integrity established by the King. A degree of solidarity was, in practice, reached that was not exaggerated by the communiqué's rhetorical reference to the two leaders' 'belief in the seriousness of the current stage in Arab history and the ferocity of the battle the Arab nation will wage in defence of its dignity and to recover territory against a usurping enemy who is defying international treaties and seeking to undermine all human values'.[9] Even Western commentators spoke in terms of the restoration of the Saudi-Egyptian axis that had been one of the main features of the Arab political scene in the early 1950s. The sober and independent Beirut daily newspaper *An Nahar* spoke of an alliance between the 'two Arab super-powers'.

The communiqué threw little light on substantive issues, in particular Sadat's strategy. But on 22 June he had outlined its shape in a speech to naval officers at Alexandria. In it he reasserted his aim of achieving a diplomatic solution based on the return of Egyptian forces to the east bank of the Suez Canal and his agreement to a six-month ceasefire during which Israeli withdrawal to Egypt's international border would be arranged by the U.N. envoy Jarring. The speech contained no attack on the U.S. Justifying his contacts with Nixon's Administration, Sadat declared: 'A person who does not consider the U.S. a basic element in the battle is like an ostrich which buries its head in the sand and sees nothing.'[10] The Soviet treaty, meanwhile, would ensure that Egypt would obtain all the weapons needed. If there was no diplomatic solution, then Egypt would feel at liberty to resort to force.

One question that was later to cause grave dissension in the Arab world was not tackled in detail: how the recovery of Egyptian territory would be linked with the liberation of all Arab land occupied in 1967. However, the proposals put forward by both Rogers and Jarring aimed at a comprehensive solution. Feisal also gave a vague and unspecified assurance of material aid. In immediate and practical terms the entente was marked by the King's agreement to facilitate the travel by Saudi citizens to Egypt, whose delights had long been denied them – to the profit of the Lebanon. Feisal, in turn, asked for Egyptian teachers who had been absent from Saudi Arabia since the days of the Yemen war. The request indicated the King's confidence in the close relationship developed within nine months of Nasser's death. The residue of suspicions about Egypt was still such, it is interesting to note in retrospect, that in Western circles the prospects of a return of Egyptian professional expertise was viewed with some misgivings. Several months later Saudi Arabia made its first offer of project aid to Egypt by making available $20 million of the finance required for the Suez-Mediterranean pipeline that was being negotiated with an international consortium, and also undertook to pump 5 million tons of oil through it each year.

Feeling more secure than at any point since 1956, Saudi Arabia was on the centre of the Arab stage on a basis of equality with Egypt. The gap left by Nasser had suddenly been filled by a duumvirate capable of unifying Arab ranks. Only Iraq and South Yemen were openly hostile to the new alliance. Feisal set about exploiting the situation with a diplomatic activism that few believed him to be capable of. His objectives were to unify Arab ranks, with the related aims of liberating occupied Arab territories and countering communist expansion. For the King, both were ends in themselves. However, progress towards the achievement of either would help consolidate the House of Saud, the overriding and underlying preoccupation of Saudi foreign policy.

The central Arab issue was the continuing conflict between King Hussein of Jordan and the Palestinian resistance movement. The ceasefire negotiated so painfully in September 1970 had broken down. In 1971, during 'Black June', as the Palestinians later called it, Hussein's bedouin troops had virtually eliminated the *fedayeen* from his territory. It presented an opportunity to Feisal who, Sadat advised Arafat, was the Arab leader best able to influence the Hashemite King. Omar Saqqaf now found himself charged, together with Hassan Sabri Kholy, Sadat's adviser on Arabian affairs, with trying to overcome the differences between King Hussein and the guerrilla movement. In the autumn, peace talks between the two sides were convened in Jeddah. No compromise was reached and Jordan was not accepted back into the pan-Arab fold, but Feisal, at least, drew him a little closer to it and prevented him from remaining a total outcast.

In that formative year, 1971, Saudi Arabia also achieved a rapprochement with Syria. As with Egypt, political changes made it possible for Feisal to contemplate something more than a *modus vivendi* with the regime in Damascus. In the previous autumn Hafez Assad had wrested control of power from the left-wing, civilian ideologues of the ruling Arab Baath Socialist Party and subsequently assumed the presidency. In doing so he confounded those who said that a member of Syria's minority Alawite sect – numbering some 450,000 out of a population of 7 million people – could not hold power. In so far as they existed, relations between Riyadh and Damascus had been considerably worsened by Syria's refusal to repair the Trans-Arabian Pipeline after its rupture in the spring of 1970. At the beginning of the following year Assad gave the go-ahead for its repair and the flow to Sidon was resumed.

The King took a sceptically realistic view of the Baath Party governing Syria, regarding it primarily as a vehicle whereby one clique held power. The ideological posturing of the ruling hierarchy and the extremely radical proclivities of some of its leaders had made it difficult to overcome a legacy of suspicion that had been sharpened on the Saudi side by the suspicions about Baathist cells in Hasa. The King also believed Syria to be inherently unstable because of its sectarian and regional differences. He appreciated the significance of the ousting of Ibrahim Makhous, Salah Jadid, and the other 'ideologues', but was by no means happy with Assad. Firstly, he distrusted him as a member of the Alawite sect which he regarded as heretical to the point of apostasy. Secondly, the King could not forget and found it hard to forgive the fact that Assad had been Minister of Defence during the débâcle in June 1967. In his opinion, he as much as Jadid and Makhous was responsible for the loss of the Golan Heights.

In 1971 Sadat played an important role in convincing Feisal that Assad was a reasonable man who should be helped. A start towards a better understanding was made when in May Saqqaf and Abdul Khalim Khaddam, the Syrian Foreign Minister, talked privately in Kuwait. They met privately on subsequent occasions to discuss the Palestinian-Jordanian question before a visit by Khaddam to Jeddah paved the way for one by Sultan to Damascus at the end of the year which effectively concluded the rapprochement. In the meantime, however, Feisal, who was well aware of the financial difficulties being suffered by Syria, had twice secretly dispatched an emissary with cheques of $100 million to help ease the strains.

To the west, across the Red Sea, Saudi Arabia watched with relief towards the end of July 1971 as President Jaafar Nimairi of Sudan regained power only days after being overthrown by hard-line communists with whom he had co-existed uneasily over the previous year. The King continued to think of Colonel Qaddafi of Libya as *majnoun* – 'mad' – and an undisciplined young man whose only virtue was his dedication to Islam and, at that time, aversion to Marxism. Yet he, like Sadat, was grateful to the impetuous Libyan leader for forcing down on 22 July (with the bizarre connivance of M.I.6) the British Airways VC 10 that was carrying Colonel Babakir Nour and Major Farouq Osman Hamadallah on their way back to Khartoum after what proved to be an unsuccessful coup d'état against President Nimairi. Later that year the King felt forced to congratulate Qaddafi for imposing the *sharia* as Libya's official law, though he regarded him more than ever as *majnoun*.

There remained two outstanding hostile elements externally in the form of the Iraqi Baathists in Baghdad and the increasingly Moscoworientated National Liberation Front in South Yemen. Of the two, the relatively silent Baathists seemed the most sinister. No one had any doubt how the most efficiently ruthless and determined of any Arab regime was eyeing and appraising the revolutionary potential in the Emirates from which Britain was withdrawing its protection. Indeed, for the indefinite future the subject would remain one tactfully not to be mentioned in the Kingdom. If any confirmation of Baghdad's malevolent intentions were needed, it came in February 1972, just after Bahrain had achieved its independence. The suitcase of an Iraqi diplomat was opened on arrival there to reveal a Kalashnikov machine-gun. The rapid build-up of the accredited staff at the Iraqi Embassy in Abu Dhabi, the temporary capital of the fledgling United Arab Emirates, to a number larger than that employed by any other diplomatic mission, was watched with some alarm and became the object of tight security surveillance.

For the protection of its northern and eastern approaches, Saudi Arabia had to rely upon the Shah of Iran, who had already made clear his intention of intervening if Iraq were to resurrect and press militarily its claim to Kuwait or any revolutionary change occurred along the south-western littoral of the Gulf. Relations between Riyadh and Tehran had become much more cordial than before but were still very reserved. On the Saudi side they were inhibited by the resentment felt over the presence of a stronger, more populous neighbour under an imperious leader more easily able to impress the West. In the event, the Kingdom, like Iraq, had no choice but to watch impotently as Iranian commandos seized the Tunbs and Abu Musa on 30 November, the islands regarded by Britain and the Emirates concerned as belonging respectively to Ras al Khaimah and Sharjah. Feisal himself had done nothing to forestall the Arab humiliation which, not without justification, he blamed on British weakness and cynicism.

The Shah purported to believe that there was a 'gentleman's agreement' whereby Britain would accept his sovereignty over the islands in return for his recognition of Bahrain's independence. In reality, the U.K. rejected his claim. At the same time, the Foreign Office told Sheikh Khaled bin Muhammad al Qawasim, Ruler of Sharjah, and his cousin Sheikh Saqr bin Muhammad al Qawasim, Ruler of Ras al Khaimah, that Britain could do nothing for them after the achievement of the Emirates' independence. Sheikh Saqr refused to cede his islands to the Shah, saying that he would rather die than do so. Shortly before U.K. withdrawal, Sheikh Khaled concluded a deal with the Shah under which he agreed to an Iranian military presence on Abu Musa in exchange for a grant of $3 million annually and a half share of the oil revenue from a promising oil strike in its off-shore waters. He was reputed to have commented that it would be 'the death of me'. It was. His cousin Saqr bin Sultan, whom Khaled had deposed in 1965, returned from exile in Cairo to regain power in January 1972. In the shoot-out at the *emiri diwan* Khaled was killed – to be succeeded, however, by his brother Sultan bin Muhammad.

In contrast to Feisal's deft diplomatic manoeuvres in the heartland of the Middle East, his foreign policy in his back yard was contorted. In the first half of 1971, Saudi Arabia and Kuwait had jointly tried to induce the nine Emirates, including Bahrain and Qatar, into a Federation. The Kingdom's efforts, however, were largely vitiated by the dogged obduracy with which Feisal stuck to his country's claim to the large slices of what the U.K. recognized as Abu Dhabi's territory. Talks in London in December 1970 between the British Government and Fahd had ended in total deadlock. In the following month compromise proposals were put forward by Sir William Luce, the U.K. special envoy charged with leaving the Gulf protectorates in an orderly state. They would have involved concessions by Sheikh Zaid, the Ruler of Abu Dhabi, but did not even elicit a response from the King. He became reconciled to the fact that Bahrain and Qatar, both homogeneous and viable, would not join the union. He was quick to recognize them and confident that he would bring both the newly independent states under his hegemony or, at least, influence.

Though Feisal had an outstanding territorial dispute with Oman arising from the Saudi claim to Buraimi, he extended recognition to it also. He did so after the young Sultan Qaboos, who had succeeded to the title after his father had been ousted in a British-inspired coup d'état on 23 July 1970, visited Riyadh in December 1971. The successor to the tyrannical old Said made humble obeisance of a kind that Sheikh Zaid had not been prepared to do. Even if he had, it would have made no difference in the absence of substantive concessions by Abu Dhabi. Nevertheless, Ahmed Khalifa Suweidi, Foreign Minister of the United Arab Emirates, arrived in the Wahhabite capital less than two weeks after the proclamation of the Federation on 2 December 1971, saying that he wanted to 'seek King Feisal's advice on means of creating brotherly and neighbourly relations'¹¹ between the two states. He was courteously, but inscrutably, received. Although the United Arab Emirates had already been accepted as a member of the Arab League and the United Nations, no Saudi acknowledgement of the official existence of the new entity followed.

The failure of Britain to resolve the dispute or of Saudi Arabia to renounce its territorial claims – depending on which way one looked at it – left a source of potential instability in the Gulf of a kind that was not in anyone's interest, least of all Saudi Arabia's. In 1974 the issue would be resolved in favour of the Kingdom. In strict terms of *realpolitik*, Britain should, some officials in London began to argue, have cultivated the Kingdom years before at the expense of the Emirates and have pummelled Abu Dhabi into submission to Feisal's demands before military withdrawal. Indeed, the U.K. had put mounting pressure on Sheikh Zaid to compromise. The final objective was, after all, to leave Saudi Arabia and Iran as the main pillars of stability in the Gulf. Those considerations apart, however, the affair revealed in Feisal a streak of perverse obstinacy. In coming to terms with Sadat's Egypt and mending the fences with Assad's Syria, Feisal had shown great flexibility and opportunism. Yet, more encumbered with the barnacles of history than ever, the King psychologically could not contemplate renouncing a notion received and fostered two decades before by self-interested senior employees of Aramco and self-seeking expatriate advisers.

Oman and Yemen remained outside the scope of Wahhabite influence. The Nejdis, who were increasingly dominating the Kingdom, regarded the people there as alien. As for the tribes of the Trucial Coast, the Saudis considered them as their legitimate subjects who would have been under their rule but for the neo-colonial tutelage and protection imposed by Britain. In name, though hardly in practice, Ras al Khaimah and Sharjah were Wahhabite. The Saudis were both aware and reluctant to concede that the Emirates had taken on a peculiar and altogether more tolerant character as a result of British protection. In a gesture totally foreign to Saudi mentality, Sheikh Shakhbut, for instance, had given land and permission for the construction of a Roman Catholic Church. Subsequently Sheikh Zaid did the same for the Protestant expatriate community. The hotels openly sold liquor, notwithstanding the fact that the only two sons of Sheikh Shakhbut had died of alcoholism and Sheikh Zaid was a devout abstainer. Free-wheeling Dubai was even more liberal. Ras al Khaimah, whose ruling family was open to Saudi influence and money, would soon open a casino.

Saudi incomprehension was almost wilful and perhaps best summed up by a remark of Saqqaf's shortly after the United Arab Emirates gained independence. He was asked by a Western envoy why, if the Kingdom felt unable to recognize the new entity, could it not at least establish a mission there to help create better understanding. He replied with a negative wag of his head that no Wahhabite Saudi could be expected to mix in a society where manners were so different and self-indulgent. The answer came oddly from the Minister of State for Foreign Affairs. He had been educated at the American University of Beirut and had served in the Saudi Embassy there. World weary and cynical, he preferred that his wife and family should live in her home country, Lebanon. He was partial to strong drink, card playing and other pleasures proscribed by the Koran.

Saudi Arabia's repulse of the South Yemeni incursion at the end of 1969 had put in due perspective the direct military threat from the National Liberation Front. Any such foray would be vulnerable to superior air power, which for the time being the Kingdom enjoyed. Nevertheless, nervousness about the danger from this direction largely accounted for Saudi interest in purchasing Phantom aircraft from the United States, as well as Sultan's increasing aggravation over the poor record of serviceability of the British-supplied Lightnings. In 1970, the communist regime in Aden was intensifying its support for the Popular Front for the Liberation of the Occupied Arab Gulf and the rebels had gained the upper hand in Dhofar. South Yemen also had the capacity and inclination to upset the delicate balance achieved in the Yemen Arab Republic or might have pressed for the union to which the Governments of both territories paid lip-service. Thus, towards the end of 1970, Sultan stepped up Saudi attempts to make trouble for the National Liberation Front by fomenting tribal unrest and encouraging raids by exiles armed by Saudi Arabia into the territory of what had become in November of that year the People's Democratic Republic of Yemen.

The 'National Deliverance Army', numbering some 5,000 men, was certainly more than an irritant to the regime in the south, which by 1971 was dominated by the hardline Marxist Abdul Fattah Ismail, the Secretary-General of the National Liberation Front, and was working more closely than ever with the Soviet Union. Valuable manpower and resources were diverted away from development to deal with the intruders and discontented tribesmen chafing under the centralized authority being imposed upon them. But there was no clear commitment by Saudi Arabia or proper co-ordination between the dissidents subsidized by it. The Saudis did not have the same power enjoyed in the mountains of North Yemen where they could ensure, through their gold, that no one ruled. The Saudi-sponsored irregulars never managed to get a guerrilla war under way. The Kingdom's policy did succeed, however, in so far as the P.D.R.Y. was kept on the defensive.

While hopes of undermining the National Liberation Front faded, Saudi Arabia kept a wary eye on the Yemen Arab Republic, which was in severe economic difficulties. The Kingdom, enjoying for the first time a healthy financial surplus, started a modest aid programme. It sought to bring the Government in Sana under a modicum of control and to thwart any meaningful move towards it uniting with the south. Nevertheless, moves in this direction started as early as the autumn of 1970 after Qaddafi had arranged a meeting between President Qadi Abdul Rahman Iryani and President Salem Rubai Ali during the Cairo summit at the end of September. The adoption by the Aden regime of the title People's Democratic Republic of Yemen had caused grave offence to Sana, but the flirtation went on, adding to the tensions in the north where a coup attempt in July 1971 and four changes of government in the course of the year took place. Sultan gave Iryani his customary ambivalent assent to efforts by his Prime Minister Mohsen Aini to pursue a policy of appeasements. That necessarily meant talking about union. But Feisal and Sultan were realistic enough to realize that such a marriage was unlikely to be consummated and they felt able to prevent it, anyway, through their connections with the powerful tribes in the mountainous north which continued to receive subventions. As it was, tension along the border between the two Yemens built up and fighting broke out at the end of September 1972. Pan-Arab intervention brought about one of those extraordinary turnarounds and an agreement on 'unity' was signed on 28 October by both Presidents. Feisal looked on sceptically. The price of his grudging acquiescence was to exact a renunciation by the Y.A.R. of any claim to the provinces of Asir, Jizan and Najran, which the Imam Yahya had ceded under Saudi military dictat in 1933 – 4. In practice, the agreement on unity between the two Yemens amounted to nothing more than a détente.

By the mid-summer of 1971 the efforts of Rogers and the State Department to bring about an Egyptian-Israeli interim agreement were flagging and becoming less plausible by the month. As long as Israel's own vessels would be able to transit the Suez Canal the Jewish state was not prepared to pull back from the east bank and allow the Canal to be cleared for shipping. But it would not contemplate an Egyptian military presence in Sinai nor consider a pact tied to further evacuation from the peninsula. Egypt wanted its forces, even if in limited numbers, back again on liberated territory and insisted that agreement should be linked to a phased withdrawal to the pre-June 1967 lines.

The main preoccupation of Dr Henry Kissinger, President Nixon's Adviser on National Security, continued to be extrication of the U.S. from its Vietnamese imbroglio and détente with the Soviet Union, but he began to turn his attention to the Middle East. He could hardly do otherwise since the conflict between the American and Soviet clients in the region provided the most dangerous ground for friction and confrontation at a super-power level. In so far as Kissinger had thought about the region in the context of his grand design, he had appeared to assume that the stalemate could be contained indefinitely because of the military superiority of Israel ensured by the U.S. In contrast, the Kremlin, distrustful of Sadat and patiently awaiting his overthrow, did not give reciprocal backing to Cairo despite the Treaty of Friendship and Co-operation. In his frustration, the Egyptian leader began to talk more and more of the 'year of decision', with the implied threat that Egypt would resort to arms if there was no political solution agreed before the end of 1971.

Kissinger was not unduly concerned. Contemptuously he dismissed the State Department's 'single-minded pursuit of unattainable goals',[12] meaning a partial settlement leading to a comprehensive one. His pessimism arose from the Soviet Union's refusal to budge from a 'total Arab position', or the recovery of all territory occupied by Israel in the 1967 war, and Sadat's inability to break his uneasy dependence on Moscow. Egypt was growing disillusioned about the chances of conventional diplomatic contacts producing anything positive. Happy to see the stalemate continue rather than let Israel be faced with American pressure for painful concessions, the Israeli Premier Golda Meir also preferred, as it happened, to deal with the White House directly through Yitzhak Rabin, her ambassador in Washington, who had developed a close relationship with Kissinger. Ali Hamdi Gammal, a director of Al Ahram, had approached Kissinger with a view to arranging a secret meeting between Nixon's National Security Adviser and Muhammad Heikal, the editor of the leading Egyptian newspaper, but the latter was not willing to act as an intermediary. At the end of 1971 Golda Meir reached an understanding with Nixon that the quest for a comprehensive settlement should be abandoned for the time being. Kissinger recounts that it was 'a step that began to establish my operational control of Middle East diplomacy'.[13] Kissinger calculated that any advance would be the fruit of his secret parallel diplomacy.

Sadat's 'year of decision' had passed – largely unnoticed because of the Indo-Pakistani war that provided him with an excuse for his failure to take action. It was not one that convinced students who surged through the streets of Cairo in January 1972 in the first riots since 1968. The spectacle was an alarming one to Feisal, who was banking heavily on Sadat's survival. Early in February 1972 Sadat visited Moscow, but came back with only hollow pledges.

Adham, as has been observed, had been active from the beginning of 1971. In early April 1972, the date given by Kissinger for the opening of the American 'secret channel' to the Abdin Palace, Egypt proposed through Adham that Kissinger should visit Cairo for talks with his opposite number Dr Hafez Ismail, or alternatively that Sadat's National Security Adviser should go to Washington. But Kissinger preferred to bide his time.

Towards the end of April Sadat went to Moscow again, returning more frustrated than ever. Following his own logic and instinct, as well as being encouraged by Feisal and his assurances of financial assistance, Sadat was now thinking more actively about breaking the 'no war, no peace' deadlock through the eviction of the Soviet advisers. Since Moscow would not deliver the weapons required by Egypt and apparently was either uninterested or incapable of bringing about progress towards a diplomatic solution, there would be nothing to lose. Not the least through his communications with Adham, Kissinger was aware that 'things were not going smoothly between Egypt and the Soviet Union'.[14] After Sadat's second visit to Moscow in 1972 he replied via the 'secret channel' that a high-level representative of Sadat would be welcome in Washington, but not until after the Moscow summit scheduled for June. The bland communiqué agreed on by Nixon and Brezhnev came as a 'shock' to Sadat. It made only a lukewarm call for a peaceful settlement and for military relaxation in the region. The Egyptian President's patience snapped when early in July he received the Kremlin's analysis of the situation, requested three months earlier. It did not even mention the question of the advanced weapons demanded by the Egyptian President. On 13 July, Adham passed on to Kissinger a

'Delphic missive'[15] from Sadat saying that he would be prepared to send a senior representative to Washington if the U.S. had anything to propose. Five days later Sadat announced that he was dispensing with the services of the Soviet experts. He gave them a week to leave the country.

According to Kissinger, the news came as a 'complete surprise' to the U.S. Administration.[16] Nevertheless, Nixon's National Security Adviser expected, or so he claimed later, that Sadat would sooner or later bargain over the expulsion of the Soviet military presence in return for U.S.-inspired movement towards a settlement. Probably Adham's suggestions would have been in no small measure responsible for any such anticipation on the part of Kissinger.

The King and his brother-in-law could take a large part of the credit for what seemed an impetuous action by the Egyptian President. The departure of the Soviets in no way strengthened his option to go to war. The consequences of his order looked as if they might weaken it considerably. Arguably the Egyptian President had thrown away his strongest card for nothing. More demented than ever by the 'communist-Zionist conspiracy', Feisal wanted them out anyway, regardless of his desire for the return of Jerusalem to Arab sovereignty and a just, honourable peace settlement that would make the world a more comfortable place for Saudi Arabia and its privileged royal family to live in.

At this point oil entered the somewhat confused Middle East equation. The Kingdom, it was emerging, had the capacity to raise its output by more than all the other members of O.P.E.C. Saudi Arabia was reckoned to possess up to 30 per cent of the world's oil reserves outside the Comecon and China. The U.S., it was realized, might have to import as much as 10 million barrels of oil a day as early as 1980. The Kingdom might have the leverage over Washington to counter Israel's manipulation of American political processes. A strategy was slowly evolving in the crabbed old King's mind.

18 The Jewel in Hand
1970-2

We do not believe in the use of oil as a political weapon in a negative manner. We do not believe that is the best way for true co-operation with the West, notably with the United States. In this way very strong economic ties are established which will ultimately reflect on our political relations.[1]

SHEIKH AHMED ZAKI YAMANI

BY the summer of 1972 the prospects of Saudi Arabia and the other oil producers had been transformed. The oil companies operating in Libya had put up fairly firm resistance to the demands of Colonel Qaddafi's regime, knowing the possible consequences. It was a fateful coincidence that Esso and British Petroleum should have capitulated on the same day as Nasser died, by agreeing to an increase in their tax rate from 50 to 55 per cent and a rise in the posted price of 30 cents per barrel. Starting with the American independents (as opposed to the international majors), other operators had given way to Libyan insistence on better terms as Arab leaders gathered in Cairo had grappled with the problem of ending the conflict in Jordan in September 1970. Shell refused to submit at first, but only held out for a few weeks.

The revolutionary regime in Tripoli had exploited the rising demand for Libya's light, sulphur-free, short-haul crude varieties. Its claim had been strengthened by the closure of both the Suez Canal and, early in 1970, the Trans-Arabian Pipeline carrying Saudi oil to the east Mediterranean terminal on the Lebanese coast. Advised and encouraged to take advantage of the situation by Algeria, Libya had built up pressure on the smaller companies, and in particular Occidental, who were heavily dependent on the country for supplies. It backed up its demands ruthlessly with orders to cut production. The dykes had been broken.

Libya had succeeded where the other members of O.P.E.C. had

failed, collectively and unilaterally, over the previous decade not only by obtaining better revenue terms but by creating a common front. There was a certain irony in this fact. Both Saudi Arabia and Iran had watched with jealousy and resentment the rapid growth of the production of a country that had only started exporting in 1959 and was not a founding member of O.P.E.C., having joined its ranks only in 1962.

As early as 1954, when Iranian production was restored following the fall of Dr Muhammad Mossadeq, the Saudi Government had made clear to the four partners in Aramco that resumption of Iranian oil production should not be at the Kingdom's expense. From 1965 the rivalry had become intense as the Shah sought to maximize Iranian output, claiming that a greater share of the market was justified by his country's population. Saudi Arabia's rejection of the argument and the competition between the two producers was a fundamental reason why Venezuela failed between 1965 and 1967 to persuade other members to implement a production programme. Such a policy would have been the only effective way to maintain realized values of crude oil in the conditions of surplus persisting throughout the 1960s. Iran had been in favour of 'prorationing' on the assumption that it could claim the biggest share of whatever cake might be agreed – in effect as much as it could contribute – on the basis of demographic criteria. Acutely sensitive about its own minuscule population, the Saudi Government exaggerated estimates of the number of inhabitants of the Kingdom – from six to fifteen million – and resolutely opposed any agreement that would mean curbing or reducing its own output.

Even if there had been a stronger collective will, the time was not ripe for a production programme because of the amount of spare capacity in existence and the psychological dominance of the oil companies. Although generally – and in this respect Libya was an exception – the gap between realized and posted prices had not narrowed, by 1970 market conditions were changing in favour of the producers. Overall, surplus capacity was still reckoned to be about 4 million barrels a day, but it was all in the hands of O.P.E.C. In the same year, by meaningful coincidence, U.S. production began to fall. Looking ahead, it was becoming apparent that, at the current rates of expansion, supply and demand would over the next five years come into something like balance. Nevertheless, the market did not seem ready for a significant breakthrough on the prices front. Yet Libya, with Algeria secretly aiding and abetting it, had driven a wedge between the competing oil companies. For the principal members of O.P.E.C., Libya's lead in militancy was an 'embarrassment', as a senior official of the organization was later to acknowledge. 'It rendered further silence almost impossible.'[2]

In the new situation created by Libya, however, Saudi Arabia was quick to assume the position in the confrontation with the industry that its possession of a quarter to one-third of the world's oil reserves warranted. In the climactic negotiations in Tehran which effectively broke the global power of the oil industry in January–February 1971, the Kingdom had as its protagonist for the historic moment Sheikh Yamani. Overnight, the Saudi Minister, with his soft but penetrating gaze, beguilingly gentle demeanour, and disarming smile, came to personify the Arab 'oil sheikh'. Presiding over O.P.E.C.'s camp was the haughty and arrogant figure of the Shah. But it was Yamani, wrapped in his gold-threaded black mishla, who caught the imagination of the world. Perhaps the contrast between his traditional dress and the massive transfer of wealth, at the expense of the industrialized countries, that was at stake accounted for the focus placed upon him. The attention was justified. Together with Dr Jamshid Amouzegar, the Iranian Minister of Finance, and Saadoun Hammadi, the Iraqi Minister of Oil, he made up the committee negotiating on behalf of the Gulf producers.

Of the triumvirate, the Saudi Minister of Oil was the most subtle and persuasive tactician. Based on his long experience of the shareholders in Aramco and his good personal relations with their leading executives, Yamani also had a greater understanding of the oil industry. The Tehran negotiations were confrontational, but also involved a measure of diplomacy. In this respect he excelled, not the least in the unofficial contacts that helped bring about the final compromise. Like Amouzegar, who would report several times daily to his master, Yamani was very much the servant of his King. Unlike the chief Iranian delegate, he was given a relatively free hand. The Shah, host of the cliff-hanging talks, did his best to give the appearance of dictating their outcome. But Yamani played an essential role in lubricating the way to the hard-wrought agreement. He emerged from the negotiations as the leading strategist of O.P.E.C.

Within the ranks of the oil producers, Saudi Arabia was not in 1971 the voice of qualified moderation that was heard after 1974. Nor did its solidarity with other producers have anything to do with the political prudence which would have been incumbent, anyway, on a conservative regime having to co-exist with a radical Arab neighbour like Baathist Iraq. Having seen a steady erosion in the purchasing power of its oil revenues since the unilateral cuts by the oil companies, the Kingdom was no less determined than other members of O.P.E.C. to obtain improved revenue terms. It was as tough as Iraq in its intransigent resistance to the oil companies' stipulation that the agreement should include the crude pumped from the two countries' fields to Mediterranean terminals.

The importance attached to this issue by the oil industry arose from anxiety that disproportionately higher rates might be obtained by producers outside the Gulf, particularly Libya, and be used to justify a revision of the five-year agreement under negotiation. For the same reason, the twenty companies represented in the London Policy Group, whose negotiating team was headed by Lord Strathalmond, Managing Director of British Petroleum, had initially insisted that they would bargain only with O.P.E.C. collectively. On both questions they were forced to give way. Far from being able to isolate the rogue elephant Libya in any way, the industry was confronted with a resolution by the producers supporting any measure that Libya might take against the oil companies. O.P.E.C. had never shown such purposeful solidarity, and Saudi Arabia was in the vanguard.

The negotiations lasted from 12 January to 14 February. Final agreement within hours of the second deadline set by the producers was greatly facilitated by two factors. First was the threat by O.P.E.C. that members would unilaterally impose their demands for a basic tax rate of 55 per cent, higher posted prices, and the elimination of certain discounts enjoyed by the companies. Second was the possibility, which the producers were seriously contemplating, of a cut-off in supplies. 'George, you know the supply situation better than I. You know you cannot take a shutdown,'[3] Yamani told Piercy, the Director of Esso, who headed the industry's team in the subsequent negotiations with Libya. The two used to consult privately.

Three years later, the revenue gains conceded by the industry would appear very modest, though at the time they seemed like something of a capitulation. More important than the extra cents per barrel involved was the way in which the industry, albeit not without a struggle, had given ground in the face of the producers' solidarity. One can only speculate what, if any, action O.P.E.C. might have taken to enforce its demands by cutting or curbing production. Some members, like Iran, which was particularly desperate for foreign exchange, could hardly have been able to afford to go without as little as one month's revenue.

The tax rate was established at 55 per cent. There was a basic increase in posted prices of 35 cents for all the varieties of crude oil produced in the Gulf, and a new system of differentials was adopted. In effect, this raised actual Government revenue for each barrel of the Kingdom's Arabian Light by 38 per cent from the 91 cents received before the Libyan breakthrough to $1.26. As a cover against inflation and in recognition of the increased demand for oil then anticipated, the agreement provided for four phased increments over the five years of its duration. Their implementation would have meant an increase in per barrel revenues of some 20 per cent by the beginning of 1975. The industrialized countries of the West, and especially Europe whose growth had been fuelled by cheap oil from the Middle East over the past dozen years, gasped at the addition to its oil imports bill in prospect. In practice, the immediate increment did not restore the real value of revenues to their pre-1960 level, but the oil company men departed looking forlorn, knowing that their complacent world would never be the same.

In a private briefing after the last marathon session of negotiations, Lord Strathalmond expressed his confidence that the agreement was 'watertight'.[4] Yamani, however, soon intimated his doubts to Piercy whether the oil companies would obtain the stability over a five-year period for which they had struggled so strenuously.[5] It was not a matter of unabashed bad faith on the part of the Saudi Oil Minister, rather a judgement based on the mood of the producers after what had clearly been a victory for them and a growing realization of how the supply and demand equation was changing.

There followed on 24 February negotiations in Tripoli with Libya that were to set rates for the Algerian, Saudi and Iraqi short-haul crudes shipped from the Mediterranean terminals. Again the Kingdom was no laggard in its militancy, twice joining with the other three producers in threatening to cut off oil supplies unless their demands were met. The outcome after six fairly agonizing weeks was a posted price for Libya's premium crudes far higher at $3.45 than the companies felt to be justified but lower than the $3.70 demanded. Yamani played an important role in bringing about the compromise, not the least through his understanding with Piercy. The other exporters from Mediterranean terminals, including Saudi Arabia, profited proportionately.

The high differential did not 'leap-frog' back to the Gulf in the form of higher demands. Yet within a year the five-year pricing agreement so painfully negotiated in Tehran was broken. Throughout 1971 the value of the dollar, in which oil prices were denominated and mainly paid, declined while pressure from the producers for compensation mounted. At the end of the year the U.S. currency was formally devalued and realigned against others. When they met the producers for another round of negotiations in Geneva in January 1972, the industry acquiesced in a further increase in the tax reference and indexation of posted prices in future according to the weighted average of a basket of currencies, even though the companies regarded the deal, known as the 'Geneva I' formula, as a breach of both the letter and the spirit of the Tehran pact. The immediate effect was to raise the actual receipts enjoyed by the Kingdom for each barrel of Arabian Light by 14 per cent to $1.45. Yet the meeting was of greater significance in another, more far-reaching, respect.

No sooner had the producers obtained improved revenue terms than they presented formally their demand for 'participation' in existing oil concessions and gave Yamani responsibility for negotiating a share of not less than 20 per cent. Now, with a growing consciousness of its oil muscle, Saudi Arabia became the front runner and led a pack which, with its appetite whetted, was hungry for more. The principle had been included in the original concession of the Iraq Petroleum Company. It had been enshrined in O.P.E.C.'s Oil Policy Statement adopted in June 1968. Finally, in the late summer of 1971, members had resolved to take immediate action towards implementation of participation.

In origin the concept was a peculiarly Saudi one. Frustrated by Aramco's refusal to divulge its accounts, thus making it impossible for the Government to know if it was receiving its 50 per cent of profits, Abdullah Tariki had been the first to voice the demand. In his quieter and more philosophical way, Yamani was also exasperated by being excluded from the inner counsels and calculations of the operating company. He took up and developed the idea of participation that was formally presented to Aramco. At a seminar in Beirut in the summer of 1968 he stated publicly the Kingdom's demand for a 50 per cent share in the concession, suggesting that compliance on the part of the share-holders would be the best way to guarantee their interests in the long term. The statement broke like a minor explosion, but its immediate effect was no more than that of a firework.

In 1968 the four companies felt too secure in their control of the concession really to take Yamani seriously. The Saudi regime felt no sense of urgency. In March 1969 Yamani told me that it might be a matter of one decade or several before the aspiration was fulfilled. But he was confident, saying: 'Time is on our side. We don't fight. We believe that our case is so strong that we can get what we want eventually.' In the longer term, he argued, it was in the interest of the major oil companies to concede participation because the natural alliance was between producing and consuming countries. National oil companies, like E.N.I. of Italy and Petrofina of Belgium, or the American independents, would be only too happy to step in as partners if the majors were not willing to accept a new arrangement. Yamani envisaged participation as 'an indissoluble relationship – like a Catholic marriage'.

It was obvious that he believed the revenue negotiations in Tehran had shown the limits of the companies' collective will to resist. The Saudi Minister of Oil was also well aware of what Aramco's concession meant to the U.S. and the West. Accustomed to abundant, cheap supplies of oil, neither perhaps fully appreciated just how vital an interest it constituted. The partners in the consortium were more conscious of what was at stake. Shortly after his retirement from Socal, Mr Otto Miller was to tell the U.S. Senate Foreign Relations Subcommittee on Multinational Corporations: 'Aramco is, by far, the most important and valuable foreign economic interest ever developed by U.S. citizens. When company officials first recognized the significance of the Aramco oil finds, it was immediately appreciated that its significance went beyond mere commercial implications. The finds were recognized as being of tremendous national importance to our country.'[6] Thus, Saudi Arabia's campaign for an equal or, if possible, a majority share in the exploitation of the country's great natural resource would encounter the enormous commercial interest of the four companies and a strategic concern of the world's leading power.

When the Kingdom came to grips with the partners in the consortium early in 1972, neither Yamani nor his Government appreciated fully the strength of Saudi Arabia's position as a supplier. One big concern in their minds was guaranteeing a market for a higher level of Saudi oil output. They continued throughout the year to talk of participation 'downstream' – meaning refining and distribution in the U.S. itself – as well as at the producing end. The aim was to obtain a secure outlet rather than an opportunity for investing the surplus revenues that were beginning to accumulate. In the circumstances – with the supply situation showing signs of tightening much sooner than anyone had expected and American oil imports beginning to rise – the preoccupation, one could say in the luxury of hindsight, could only be explained as a hang-over from the 1960s, when the main anxiety had been increasing the volume of crude exports, and as a sympton of congenital Saudi caution. Meanwhile, the notion of a special relationship with the United States based on a mutual inter-dependence was beginning to take a coherent form.

As for participation, Saudi Arabia took the initiative and pursued the goal on behalf of the other Gulf members of O.P.E.C. with a toughness and decisiveness that surprised the oil companies. In February 1971, Algeria had unilaterally asserted control over the French companies operating on its territory, effectively giving itself 80 per cent majority control. In December, Libya had seized the assets of British Petroleum in retaliation against the meek and abject failure of the United Kingdom to resist or protest against Iran's seizure of the Gulf islands. Saudi Arabia, as has been noted, was also shown a woeful, but perhaps more forgivable, diplomatic ineptitude in not protecting Arab interests. For a number of reasons, apart from the principle and the substance, participation was an ideal issue on which the Kingdom should take a stand.

Initially, the four partners of Aramco also took a hard, defensive line against the proposition. Their first response was to offer 50 per cent participation in the exploitation of proved but undeveloped oil structures, which was regarded as an attempt to make a special deal with the Kingdom. The oil companies' impassivity concealed apprehension rather than confidence. In contravention of the oft repeated assertion by them that they were totally independent of governments, they appealed to Nixon to intervene. His message urging restraint was received coldly, indeed indignantly, by both the King and Yamani. Feisal not only sent a brusque riposte but also had it publicly distributed on 18 February by the official Saudi Press Agency. It said: 'Gentlemen, the implementation of effective participation is imperative, and we expect the companies to co-operate with us with a view to reaching a satisfactory settlement. They should not oblige us to take measures to put into effect the implementation of participation.' Once again members of O.P.E.C. closed ranks. An extraordinary conference was summoned to take place in Beirut on 11 – 12 March 1972. On 10 March, just before the conference was due to convene, the Saudi Government received a letter from Aramco accepting the principle of 20 per cent participation. It acknowledged that the King's statement had been the occasion of the partners' decision. It was more like a shotgun wedding than any marriage, let alone a Catholic one.

Effectively, four of the seven leading international oil companies, together possessing assets of $50 billion, had capitulated both to the principle of state control and to another breach of the Tehran revenue pact. The six Gulf states had already made it clear that they wanted the initial 20 per cent share in the operators' concessions to rise to 51 per cent and that any of their entitlement purchased by the oil companies should be at a higher price than the basic revenue rate for what would henceforth be termed equity oil.

In yielding to the King, the four partners' primary consideration was to ensure access to as much oil as possible. The board had already approved an expansion of production capacity to 13.4 million barrels a day. The four majors concluded that the operation 'would so dominate the world of oil that their market shares would not only be protected but would increase as well – as long as they could buy back the increasing Saudi participation share of the Aramco oil'.[7] That became their main preoccupation.

On this issue the Kingdom proved to be amenable. Despite the Saudi Government's stated wish to invest 'downstream', it had no immediate ambitions to be involved directly and in a substantial way in the marketing of crude. Looked at cynically, its willingness to contemplate the four majors disposing of the bulk of its entitlement might have been regarded as a device to win their assent to the Gulf states obtaining a majority share in operations as soon as possible. In practice, however, the producers were uncertain how strong demand for oil would be in the future. O.P.E.C. as a whole wanted guaranteed 'buy-back' arrangements. Up until 1978 the state agency Petromin sold only a small proportion of Aramco's output. To a greater or lesser extent all members saw participation also as a means of winching up revenues again.

The obligation to purchase the Saudi Government's share of production might not worry its four American partners. They had the firm prospect of remaining shareholders in, or at least being privileged customers of, the oil supermarket with greatest growth potential. It did, however, cause misgivings to Gulf, British Petroleum and, to a lesser extent, Shell. These three majors did not have a stake in the Kingdom but had important investments elsewhere that would be affected by the outcome of the talks. As with the revenue negotiations early in 1971, the need for inter-industry consultation slowed up progress. The tough bargaining was over the demand made by Saudi Arabia on behalf of the other Gulf producers that the 'buy-back' price should be half way between tax-paid cost and posted prices, that compensation should be on the basis of net book value, and that the date for achievement of majority control should be as soon as possible. Then suddenly the pressures on Saudi Arabia and, particularly, Yamani mounted.

First, there was the abrupt and arbitrary nationalization on 1 June 1972 by the Baathist regime in Baghdad of the Iraq Petroleum Company whose fields accounted for the bulk of the country's production. The immediate cause was the concessionaires' failure to maintain a full flow of oil through the pipelines from the Kirkuk fields to the Mediterranean terminals of Banias in Syria and Tripoli in Lebanon because it believed that the price differential was too high in the market conditions then prevailing. The nationalization brought to an end the long and bitter disputes that had crippled Iraq's development as an oil power for over a decade. Yamani issued a statement in support of Iraq's 'legitimate demands' before the other members of O.P.E.C. endorsed its action and undertook not to benefit at its expense.

The second shock came from Iran and was far more disconcerting for Saudi Arabia. Early in July the Shah of Iran summoned a press conference in London. In the lavishly carpeted, Louis Quinze splendour of the Iranian Embassy, the 'King of Kings, Light of the Ayrians' blandly stated that the oil industry in his country had been under full state ownership since the reconstruction in 1954 of the consortium operating his country's main producing fields. His purpose was to announce agreement on a 'new kind of relationship with the oil companies'[8] and his rejection of the concept of participation. He told his audience that he was prepared to honour the 1954 deal with the consortium. In return, the companies had undertaken to maximize production over a twenty-five-year period, possibly to as much as 8 million barrels a day at its peak, and make available to the National Iranian Oil Company, the state oil corporation, any oil that it would be able to market profitably. The glee of the petroleum industry was ill disguised by the poker faces of the executives involved. They believed that they had ensured guaranteed supplies from the world's second most important producer. An unstated aim of the Shah and the industry had been to pull the rug out from the feet of the Arab producers of the Gulf, as well as to flatter the vanity of the Shah. The announcement was made only a day before delegates convened for an O.P.E.C. conference in Vienna where the main item for discussion was Yamani's progress with the participation negotiations.

Nominally, at least, Yamani had been negotiating on behalf of Iran. Neither Saudi Arabia nor any other member of O.P.E.C. had been informed of the six months of secret talks on its new relationship with the group of companies that included all the shareholders of Aramco. The announcement seemed like a calculated snub. The Arab delegates who assembled in Vienna were indignant and angry. Tempers cooled somewhat when Amouzegar explained that Iran had chosen its own alternative to participation. As a sovereign state it was entitled to do so. O.P.E.C. had never claimed to be a supranational body. However, the pressures on Yamani to bring about a new relationship between the Arab producers of the Gulf and the industry, giving financial terms at least as good as those obtained by Iran, increased.

The talks with the companies became badly bogged down on the complex issues of lifting arrangements, the price to be paid for different categories of the state's share of production, and the timetable for the achievement of participation. To strengthen Yamani's efforts, the King issued a solemn proclamation in August stressing Saudi Arabia's grave determination to bring about participation but at the same time giving a reassurance that any action taken to enforce it would be directed against the four companies and not harm the consumers. Saud al Feisal, the King's fourth son, who had been appointed Deputy Minister of Oil in July 1971 at the age of thirty, went to Washington to deliver a message to the U.S. Administration to this effect. The young prince threatened a cut in production. Having dangled the carrot, he 'waived' the stick, as one American official jokingly put it at the time. But Yamani warned the shareholders that any deal could be undermined by a more preferential one that might be obtained by other producers. His frankness reflected Saudi Arabia's desire not to be trumped. By then the market was beginning to turn decisively in favour of O.P.E.C.

Eventually, on 5 October 1972, the four majors concluded the General Agreement on Participation that gave Saudi Arabia, Kuwait, Abu Dhabi, and Qatar an immediate 25 per cent stake in their concessionaires' operations that was to rise progressively to 51 per cent by 1981. Its complex provisions had the effect of adding 9 cents to the average price of a barrel of Saudi crude. Historically, it should be seen as the second major milestone in the assertion of the producers' power and the emergence of O.P.E.C. as a unique oligopoly. However, Iraq, which had been represented in the negotiations with the status of an observer, was not a signatory to it. Libya was openly critical of the deal. Little more than two months later, on 21 December, at the first meeting called to discuss the implementation of the accord, the four Gulf states demanded a high price for their 'buy-back' oil, the crude that the producing states could require the companies to market on their behalf. In the event, its implementation hung fire as Saudi Arabia watched to see what other producers, especially Libya, would exact.

*   *   *

When the Kingdom embarked on its First Five-Year Plan in the late summer of 1970 it was undergoing the worst financial squeeze experienced since the near-bankruptcy of 1957 – 8. There were grave doubts within the Kingdom whether the 41,300 million riyals, about $8,000 million under the exchange rate then prevailing, projected for expenditure would be available. In 1970 petroleum income had just exceeded, for the first time, $1,000 million and accounted for no less than 90 per cent of state revenue. The modest assumption of the plan was that oil production would rise by 9.5 per cent annually.

There was speculation that, if the plan's target was to be achieved, then the Kingdom might have to borrow. Conversely, it was questioned whether, even if the money was available, it could be disbursed in full. The plan, drawn up with the help of experts from the Stanford Research Institute, amounted to little more than a loose framework and a list of projects. Some 44 per cent of the spending envisaged was for projects. In this respect, as in others, there was something notional about it. Hitherto, the Government's accounting had made no clear distinction between current and capital budgets. In practice, over the past three years the actual outlay on civilian projects had fallen short of allocations by about one-third. In general, Hisham Nazer, who had been chief of the Central Planning Organization since 1968 and was appointed Minister of State in July 1971, was under no illusions about how this performance might be due to the inadequacies of the bureaucracy, the rudimentary nature of the country's infra-structure, and the shortage of skilled manpower.

Caution and pessimism about finance were confounded by the sharp rise in oil production and per barrel receipts. Output increased from 3.79 million barrels a day in 1970 to 4.76 million barrels a day in 1971, and a little more than 6 million barrels a day over 1972, though by September of that year it had surged ahead to over 7 million barrels a day. Per barrel revenue was up from 84 – 99 cents in the autumn of 1970 to $1.36 – 1.50 (according to the crude variant) at the beginning of

1973, apart from the average increment of 9 cents per barrel resulting from the participation agreement. Oil revenue leapt from $1,149.7 million in 1970 to $1,994.9 million in 1972.

Already Saudi Arabia was earning more money than it could absorb. The sum allocated for Government expenditure in 1971 – 2 was 10,782 million riyals, nearly 70 per cent more than in the previous financial year. In August 1972 the *jihad* tax' introduced in the budget for the previous fiscal year was abolished and in the course of the year customs duties were reduced. The budget was about one-third under-spent. The total set aside for 1972 – 3 was 13,200 million riyals. In the financial year the Government greatly improved its disbursement rate to something like 75 per cent of the target, but revenues were running at about twice that rate.

Already the constraints on rapid development and absorption of the mounting revenue had become painfully apparent, not the least from the growing number of expatriate workers flocking into the Kingdom to do work that Saudis were either unable or unwilling to do. Gold and foreign exchange reserves that had been $662 million at the end of 1970 grew to $1,444 million a year later and stood at $2,500 million at the close of 1972. The world was reconciled to the fact that the Kingdom was going to accumulate a very large volume of foreign assets, especially if it wanted to expand its production capacity. It was estimated that by the early 1980s it might accumulate up to $30,000 million. Unlike Kuwait, which had long been building up funds to provide an alternative source of income, Saudi Arabia showed neither taste not aptitude for state investment, although the Government was not averse to earning interest on the money deposited short term with a handful of international banks. Moreover, the leadership had begun to ask itself whether it would not be preferable to let oil appreciate in the ground rather than in paper assets whose value, it was becoming clear, could only be eroded by inflation and depreciation of currencies.

The hierarchy was committed to the principle that the economy should be diversified and dependence on oil thereby reduced. It was fully appreciated that the exploitation of the Kingdom's hydrocarbon resources, particularly the gas associated with oil production, most of which was being flared and wasted, would be the basis for capital-intensive and export orientated industries. In addition, the Government sought in the longer term to process within the Kingdom as big a proportion as possible of its oil so that it could obtain the added value from it.

During the period of the First Plan, however, Petromin, the state agency chosen as the vanguard for economic diversification, was preoccupied with building up refining and other capacity to meet domestic requirements. It had sponsored the construction of the Saudi Arabian Fertilizer Company's plant by Occidental Petroleum. It took a 51 per cent participation while the rest of the shares were sold to the Saudi public. However, the technical difficulties encountered by the plant were such as to make the Government move slowly in industrial development and were largely responsible for the decision that in future projects foreign companies with the requisite expertise should take a 50 per cent stake in joint ventures. Such a vested interest, it was argued, would help ensure optimum efficiency, competitiveness and sales outlets.

Meanwhile, plans for heavy industrialization were postponed. In the summer of 1972 the regime had not worked out with any precision the scope and scale of the heavy industrial development that it wanted. Yamani had convinced Feisal of the advantages of downstream investment in the industrialized countries as a means of increasing the return from a depleting resource and securing a stable market for the Kingdom's crude oil. It had been an integral part of his concept of participation and now seemed a way of justifying the increase in Saudi oil production that the West was looking to Saudi Arabia for. The world had suddenly become aware of its future dependence on the producer possessing 28 per cent of the world's reserves.

On 30 September 1972, a week before the conclusion of the participation agreement, Yamani addressed the Middle East institute in Georgetown. He told a fairly startled audience that by 1980 the Kingdom would produce something like 20 million barrels a day, the equivalent of nearly three-quarters of total output by O.P.E.C. at its 1972 level. In return for satisfying the import requirements of the U.S., which were expected to reach 8 million barrels a day by the end of the decade, he proposed duty-free entry for his country's crude oil (exemption from import levies imposed under the quota system) and a privileged status for Saudi investment, especially downstream in the petroleum industry. He only hinted obliquely that Saudi Arabia would also look to the U.S. for a more even-handed approach to the Arab-Israeli problem. There was no implied threat or, indeed, suggestion that Saudi Arabia was thinking of using oil as a political weapon. Feisal had categorically ruled out any such policy.

In an interview published in *Newsweek* on 20 November Yamani once again deftly emphasized the Kingdom's privileged position. 'Don't forget that Saudi Arabia has the jewel in its hand,' he said. Through its huge potential capacity to pump more oil, Saudi Arabia would be able to ensure stability on the market, Yamani suggested. He was asked about the increasing volume of talk in the Arab world about using control of oil resources to influence the policies of Western countries towards Israel. He answered circumspectly: 'Yes and no. I should say that we do not believe in using oil as a political weapon in a negative manner. We think that the best way for the Arabs to use their oil is as a foundation for real co-operation with the West, mainly the U.S. We think of using Arab oil positively rather than negatively.' A similar concern was voiced by Saud al Feisal.

In a youthful way the King's son almost exaggerated his father's aristocratic bearing, manner and demeanour, though with a face leavened with a disarming smile. Educated at Princeton University, where he had received a B.A. degree in economics, Saud was well equipped to make a serious impression on Western ministers, officials and public opinion. He had done so in the summer when he delivered Feisal's warning about the need to reach agreement on the Saudi demand for participation. In an interview published on 20 November 1972 he told *Petroleum Intelligence Weekly:* 'You always hear you can't separate oil from politics. I simply do not see why not.' He pointed out that any arrangement about long-term supplies to the U.S. and preferential access to the American market should not be made at the expense of West Europe and Japan. He also observed that the Kingdom would be selling an increasing proportion of its crude oil production in competition with other members of O.P.E.C.

Sheikh Yamani had certainly made an impact. In an article accompanying the interview, *Newsweek,* a fairly well-balanced weather vane of American opinion, waxed lyrically, without conscious patronage and without saying what its criteria were, stating that Saudi Arabia was 'now ready to assume first-class citizenship in the Arab world'. The magazine singled out Yamani as the man responsible. 'Saudi Arabia is far more important to the West than Egypt, Syria or even Libya,' *Newsweek* went on. 'These nations have the potential to make plenty of trouble but it is Saudi Arabia that sits on top of the world's greatest known source of oil.'

The U.S. Administration was more reluctant to recognize the realities of the energy shortage. In Washington, Yamani elicited only one immediate and positive response to his proposal. It came from James

E. Akins, Director of the Office of Fuels and Energy at the State Department, who described the proposal as an important one and ventured to say it should be welcome even if it would be opposed in many quarters. The Administration's reaction was ambivalent to the point of being negative. When Yamani delivered a formal letter to the White House on behalf of the King strongly endorsing the proposals made in his Georgetown speech, he was told that any decision would have to await the formulation of Nixon's energy policy. There were other unstated considerations in the Administration's mind. One related to the prickly susceptibilities of the Shah and the interests of Iran, which had embarked on a programme to maximize production over a twenty-five-year period, though the prospect of an upper level from its main fields of 9 million barrels a day at the most was paltry compared with hopes about Saudi output raised by Yamani's Georgetown speech. The other concerned the requirements of West Europe and Japan which, relatively speaking, were far more dependent on imports of Arab oil. Not until the beginning of February, three months later, did Nixon reply to Feisal's letter. The delay in the President's response must have seemed to the King a calculated insult.

The explosion of wealth experienced by the Kingdom and its newly exalted status had not been matched by a comparable political evolution at home. In the autumn of 1970, Feisal at last established a Ministry of Justice as promised in the reform programme enunciated eight years previously. Not until 1971 did he sign the Civil Service Law pledged at the same time. The local representative councils foreshadowed also in the ten-point declaration did not materialize at all. His reluctant attitude to change even extended to bringing in new talent to the Council of Ministers. Over ten years he replaced only the Ministers of Health and Information. In 1971 he had appointed Fawwaz as Governor of Mecca in place of the unpopular and reactionary Mishaal, the only notable switch he made. But a number of promising technocrats had been given positions of responsibility.

Because of Feisal's advancing age and frequent trips abroad, which were very much the corollary of Saudi Arabia's enhanced position in the world, Fahd and Sultan emerged even more prominently as the two most influential princes. Crown Prince Khaled's physical capacity and his longevity became a matter of increasing concern after he was taken ill in January 1972 while on a private visit to London and flown to Cleveland, Ohio, where he underwent open heart surgery and had a pace-maker inserted. Already the assumption was growing amongst close observers that Fahd would exercise effective leadership and be *de facto* chief executive even if Khaled, as the next King, would sign royal decrees. In many respects the Second Deputy Premier seemed best fitted to lead the Kingdom into a new era of affluence and to preside over the country's development. But he was becoming overweight and had developed back trouble. Increasingly, he would absent himself for periods of rest. His chosen forms of relaxation were not necessarily best suited to alleviate the ailments. His bouts of indolence were becoming a subject of comment. There was a serious question mark over his ability to take a firm and sustained grip over the important affairs of Saudi Arabia.

By contrast, Sultan, the second of Hassa bint Ahmed al Sudairi's sons, continued to be a hard worker, but concentrated on the expansion and improvement of the Kingdom's military capacity. He, too, had a good understanding of the outside world and was aware of trends in technology that were of particular relevance to him as Minister of Defence. His interest in weaponry was heightened by the enormous commissions that could be made at the expense of the public treasury. In the early 1970s Fahd was cut in on the deals, but the majority share of transactions is said to have gone to Sultan in a ratio of about 60:40. Still in a key position and active in government was Feisal's uncle, Musaid bin Abdul Rahman, who had been Minister of Finance since 1962 and in a more affluent era as tightfisted as ever. The charmless and suspicious Nawwaf, still Special Adviser on Gulf Affairs, maintained some prominence in foreign policy matters. Muhammad, Khaled, and Abdullah constituted the conservative counterweight.

By the end of 1971 Saudi Arabia had emerged from the economic stagnation of the previous year. Increased disbursement of Government revenues directly stimulated business activity and profited the merchant class, as well as creating opportunities for new entrants to join it. At this point the marked increase in the volume of funds spent by the Government was a balm soothing the discernible frustration of the middle classes, especially in the Hejaz but also in Riyadh and Hasa, about the lack of any institutional change and slow social transformation. The system of government may have seemed more of an anachronism than ever, but the only potential threat to the established order looked as if it might come from the Armed Forces. However, the progressive release of men rounded up in 1969 from the beginning of 1972 reflected growing confidence on Feisal's part. Saudi Arabia was still a very backward place. Those – and there were now many thousands of them – who had received education or training in the U.S. or Europe, could not help but feel the debilitating boredom and cultural oppression. Yet the graduates from the West not only came back to the Kingdom, but stayed there. While there was resentment against the dominance of the royal family and the accumulation of enormous wealth in the hands of the few, it was mitigated by the involvement of an ever-increasing number of Saudis at senior levels of the Administration and the prospects of a share in the spoils of office, quite apart from widening business opportunities. A new pride and pleasure had been engendered by the country's wealth and status.

19 Oil Power Unleashed

1973

If the situation deteriorates, the day will come when oil will be used as the ultimate weapon in the battle. If the Arab governments do not do this the Arab peoples will take the initiative and do it.[1] -COLONEL MUAMMER QADDAFI

FEISAL had wanted to use oil to persuade rather than to pressurize the U.S. into changing its attitude towards the Middle East problem. He knew that the process could only be gradual, but was led to believe that it could be accelerated by the drastic shortfall in crude oil supplies that was looming. In the autumn of 1972 he was still adamant that oil and international politics should be kept separate, notwithstanding the hint in Yamani's Georgetown speech that progress towards solving the Palestinian issue, as well as special investment opportunities for Saudi Arabia in the U.S., could affect plans for raising oil production capacity. During the winter of 1972 – 3, however, the King's cautious, almost crab-like policy was overtaken by events.

In itself the U.S. Government's response, such as it was, to the proposals delivered by Yamani would have been enough to bring about the shift of emphasis from investing funds downstream in the U.S. to industrial development at home. However, the grand design came under criticism both at home and elsewhere in the Arab world because of the way in which it could heavily bind the Kingdom to the U.S. Saudi Arabia was very exposed to the radical taunts of Libya and Iraq. More important, however, was the growing Arab frustration over the lack of progress towards the liberation of territories occupied by Israel. By the spring of 1973 Feisal had decided that oil policy, both as regards pricing and production, could no longer be insulated from the Middle East deadlock.

The warm entente between Egypt and Saudi Arabia was not so close that Sadat confided vital secrets to Feisal. But as astute and experienced a statesman as the King, advised by a shrewd operator like Kamal Adham, would have been aware of the drift of the Egyptian leader's mind. Sadat had explored the possibilities of more decisive American diplomatic intervention through the 'secret channel'. As early as April 1972 the White House had responded to the suggestion that there should be a top-level secret meeting outside normal diplomatic avenues. One between Henry Kissinger and Hafez Ismail had been arranged to take place in October but was postponed because the former had other preoccupations. Indeed, the normalization of relations with China, the resolution of the Vietnamese conflict, and the search for détente with the Soviet Union – in the context of which Kissinger viewed the Arab-Israeli conflict – seemed to blind not only him but the U.S. Government and the West as a whole to rising tension in the Arab world.

Finally, in February 1973, Kissinger did meet Ismail. The message conveyed from the White House was at the same time bleak and challenging. Kissinger told Ismail, with an almost brutal realism, that as long as the Israelis were entrenched along the Canal there was nothing he could do for Egypt. 'How can you get them off it?' he asked. The Secretary of State was, no doubt, trying to buy time as he preoccupied himself with other global problems. But the secret meeting in Washington had a profound impact on Sadat. To him, Kissinger's words amounted to nothing less than an invitation to launch an offensive that would alter the status quo in the Middle East. Although almost certainly President Nixon had not intended it, the exchanges hardened Sadat's resolve to go to war. At the time, the Administration did not believe that he had any military option. It became more plausible later that month when the Soviet Union at last agreed to supply him with weapons that he had been seeking for nearly two years. That can only have fortified his resolution.

According to his own testimony, Sadat's patience had run out as early as February 1972.[2] Four months later he had expelled 15,000 Soviet military advisers to free Egypt's hand so that it could exercise the war option. In August he went on holiday to Alexandria and began to 'prepare for battle'.[3] He gave orders for the Armed Forces to be ready to fight by 7 November, the date of the American presidential election. The best dates were drawn up for an attack on Israel in 1973 – a series of days in May, August – September, or

October. In April, President Hafez Assad of Syria secretly visited Egypt and committed himself to a joint attack. The October date was tentatively chosen. In 1973 the drift towards military confrontation was remorseless.

By virtue of his country's wealth and his firm stance on the question of Palestinian rights, Feisal could not have stood aside even if he had wanted to. In April he dispatched Yamani and his Deputy, Saud al Feisal, to Washington with the message that Saudi Arabia could no longer regard oil and politics as inseparable. First they had a meeting with William Rogers, Secretary of State, and George Shultz, Secretary of the Treasury. In future, the Saudis said, the Kingdom's production and pricing policies would depend on progress towards a Middle East peace settlement. They then saw Kissinger, who was disturbed that they had spoken to Rogers and Shultz. He requested Yamani to speak to no one else on the subject. The Oil Minister suspected that Kissinger wanted to suppress the information in the hope of nullifying the new Saudi policy orientation. Consequently, Yamani told the *Washington Post* of his exchanges saying that the raising of the Kingdom's oil production from 7.2 million barrels a day to 20 million barrels a day was still a 'good possibility' provided that the U.S. 'creates the right political atmosphere'. He pointed out that Saudi Arabia was already receiving more petroleum revenue than its small economy could absorb. 'If we consider only local interests, then we shouldn't produce more, maybe even less.'⁴ Yamani added that Saudi Arabia could set a much higher price per barrel simply by cutting its output by 25 per cent.

The warning delivered by the Oil Minister was oblique and expressed in the politest possible terms because he was well aware of the sensitivity with which Feisal, as well as the Administration, regarded any linkage between the question of oil supplies and the Middle East problem. At this point the King envisaged no more than freezing Saudi output at 7 million barrels a day, the rate reached early in 1973. But the message was now clear enough and understood by the Administration. However, it chose not to acknowledge the gravity of the situation posed by the lack of progress towards a Middle East settlement.

Two weeks later the King and Yamani were angered to read senior officials quoted as denying that any such warning had been formally passed on. Equally aggravating were suggestions emanating from the White House that the linkage between the raising of Saudi Arabia's producing capacity to progress towards a resolution of the Middle East conflict was Yamani's policy, not Saudi Arabia's. Thus, the King made another attempt to convince the U.S. Government, indirectly and in the politest possible way, that he was serious. Early in May he spoke to Frank Jungers, the President of Aramco, describing the pressures that he was under and explaining that he could 'not stand alone much longer' in the Middle East as a friend of the U.S. Jungers reported him as saying that 'it was up to those Americans and American enterprises who were friends of the Arabs and had interests in the area urgently to do something to change the posture of the U.S. Government'. Even a simple disavowal of Israeli policies would do something to change the anti-American climate in the region, the King advised.

Feisal went to Cairo on 12 May for two days of talks with Sadat. In advance of his visit the Egyptian leader had praised the role that Saudi Arabia, together with Kuwait and Algeria, were playing in the 'battle'. Meanwhile, Libya's state-controlled media charged that the King was going to Egypt 'in the service of colonialism' and any hopes of lighting a lingering spark of humanity in him were misplaced because 'a bent stick cannot cast a straight shadow'. *Al Thawra,* the mouthpiece of Iraq's ruling Arab Baath Socialist Party, bluntly charged Saudi Arabia of 'playing a basic and important role in implementing and ensuring the U.S.-Zionist scheme to dominate the Red Sea'.

Feisal strengthened his commitment to bring what pressure he could to bear on the U.S. and to give financial support to Egypt. In a way the embarrassment of the two leaders was complementary. Nearly eighteen months had passed since Sadat's 'year of decision' had expired with hardly a whimper and he seemed no nearer to breaking the 'no war, no peace' deadlock. Feisal had failed to obtain any results using oil as an economic threat.

On 15 May, the twenty-fifth anniversary of the foundation of Israel, a one-hour stoppage of oil production was ordered by Algeria, Iraq, and Kuwait. Libya suspended operations for a whole day. Two days later Colonel Qaddafi said: 'If the situation deteriorates, the day will come when oil will be used as the ultimate weapon in the battle. If the Arab governments do not do this the Arab peoples will take the initiative and do it.'[5] Feisal proceeded to Paris for a state visit. Negotiations began on the purchase of Mirage Vs on behalf of Sadat to supplement those being passed on to Qaddafi who at the time was strenuously trying to bring about a merger between his own country and Egypt.

In the third week of May, after a tour of five countries, the King conferred in Geneva with senior executives of the four U.S. partners in Aramco. Again he hammered home the nature of his predicament. The word was passed to the U.S. State Department where, according to the Saudi Minister of Oil himself, it was again dismissed as 'Yamani's line'. The wishful thinking and evasiveness shown by the U.S. Government was all the more gauche for ignoring the fact that Yamani proved to be amongst the most moderate of those advising the Saudi hierarchy. As it was, Omar Saqqaf, Minister of State for Foreign Affairs, delivered the most stern warning yet at a press conference in Brasilia on 28 May. He did not mention the U.S. or Israel by name, but clearly alluded to them. He said that Saudi Arabia felt a responsibility to sell oil to whoever wanted it 'but when it comes to threatening our cause and to helping our enemy, then we think there is a limit'.[6]

Qaddafi, meanwhile, did not feel confident enough to expropriate the main American oil interests in his country. He did, however, lash out at the most easy and vulnerable target in sight. On 11 June, at a ceremony marking the third anniversary of the U.S. evacuation of the Wheelus air base near Tripoli, he announced nationalization of the 50 per cent stake in a concession in the east of the country owned by Bunker Hunt, the Texan oil billionaire. Previously, he had shared it with British Petroleum, whose assets had been summarily expropriated by Qaddafi at the end of 1971 in retaliation against what the Libyan leader regarded as U.K. connivance over the appropriations by Iran of the Gulf islands that year. Standing at Qaddafi's side as he delivered his 'slap in the face' to the U.S. were Sadat and Idi Amin of Uganda.

These pressures only bound Feisal closer to Sadat and fortified the King's determination to prevent the merger that Libya was pressing upon Egypt. In the summer of 1973, the Kingdom's positive diplomatic achievement was bringing Jordan into the pan-Arab fold. Egypt had broken off diplomatic relations after Hussein had proposed early in 1972 the establishment of a 'United Arab Kingdom', a federal arrangement that would have given Jordan and the West Bank equal status under his rule. Amidst talk about the reactivation of the Arab 'Eastern Front', Sadat, Assad and Hussein eventually met in Cairo on 9 September 1973. Egypt announced a restoration of diplomatic relations after the talks. Restoration of formal links between Syria and Jordan was postponed pending settlement of the demand of the Palestinian guerrillas evicted from the Hashemite Kingdom in June 1971 that they should be allowed to return to Jordan.

Washington's obtuseness knew no bounds, as far as Feisal was concerned. On 1 June he felt forced to send a personal cable to Nixon emphasizing that Saudi Arabia was serious in its warnings about a production freeze. The Administration's reaction was described in a telex from Joe Johnston, an Aramco executive based in the U.S., to Frank Jungers in Dhahran. It read: 'The general atmosphere encountered was attentiveness to the message and acknowledgement by all that a problem did exist but a large degree of disbelief that any drastic action was imminent or that any measures other than those already underway were needed to prevent such from happening. It was pointed out by several from the government that Saudi Arabia had faced much greater pressures from Nasser than they apparently face now and had handled such successfully then and should be equally successful now.

'The impression was given that some believe H.M. [King Feisal] is calling wolf when no wolf exists except in his imagination. Also, there is little or nothing the U.S. Government can do or will do on an urgent basis to affect the Arab/Israeli issue.'[7]

On 4 July, Feisal pronounced on the subject of relations with the U.S. and the situation in the region. He told two visiting correspondents from the *Washington Post* and the *Christian Science Monitor* that the future strength of links would 'depend on the United States having a more even-handed and just policy' in the Middle East. Saudi Arabia would find it 'difficult' to continue close co-operation with the U.S. if American support for Israel continued at such a high level, he said. In Jeddah, the U.S. Embassy staff sought to play down the warnings. So, too, did senior officials in Washington. It was easy for them to do so because the King was 'making an entreaty rather than a threat',[8] as one contemporary Western Ambassador to Saudi Arabia put it. Moreover, there were differences amongst the senior princes and their commoner advisers.

Finally, the U.S. did acknowledge publicly that oil was an element in the Middle East equation. In an interview with Israeli television on 1 August, Joseph Sisco, the U.S. Assistant Secretary of State, acknowledged: 'There is increasing concern in our country . . . over the energy question and I think that it is foolhardy to believe that it is not a factor in the situation.' Belatedly recognizing the essence of the Middle East problem, he even conceded that any peace agreement would have to take into account 'the legitimate interests and aspirations of the Palestinian people'.[9] A few days later Nixon was pressed upon the question of the use of oil as a 'club'. He admitted that it was 'a subject of major concern'.[10] But West Europe was more dependent on the Arab world for petroleum supplies than the U.S., he also pointed out.

Their remarks more or less coincided with what turned out to be the last two appeals either made or authorized by Feisal. In a programme screened on 31 August he told the National Broadcasting Corporation: 'We do not wish to place any restrictions on our oil exports to the United States, but America's complete support for Zionism against Arabs makes it extremely difficult for us to continue to supply the United States' petroleum needs and even to maintain our friendly relations with the United States.'[11] In an interview with *Newsweek* published on 10 September he spoke with equally guarded deliberation. Logic, he said, required that Saudi Arabia should not produce oil at a rate surpassing its own financial requirements. It would do so only if the West helped with the country's industrialization plans, on the one hand, and if a 'suitable political atmosphere, hitherto disturbed by the Middle East crisis and Zionist expansionism, were created', on the other. The King said that he was still against the use of oil for political ends. The only question was whether a rise in production could be justified by the realization of the two conditions. 'The U.S. would be responding adequately if it abstained from adopting biased attitudes and from giving unlimited aid to Israel.' Expressing his desire for friendship with the U.S., West Europe and Japan, he stressed, 'We do not think with a punitive mentality.' A few days previously, Yamani had felt forced to correct a report in the normally reliable Beirut weekly *Al Hawadess*. Written in the wake of Sadat's visit to Saudi Arabia – for an Arab readership – it quoted Feisal as telling him that oil should be used to buy arms rather than withheld as a means of making the U.S. put pressure on Israel. The King had been misrepresented, Yamani said.

At least some principles had been established by August. First, Saudi Arabia should decide itself when and how the oil should be utilized to influence the U.S. Second, any policy should distinguish the U.S., Israel's benefactor, from West Europe and Japan. Third, any leverage would have to be applied gradually. Beyond that, the ruling hierarchy was far from reaching a consensus.

'Oil is an economic weapon and like all economic weapons it needs study and time to produce results,'[12] said Saud al Feisal. Fahd was most loath to contemplate hurting the U.S. in any way or to risk antagonizing 'our American friends' whom, he suggested, should be allowed more time. 'I am Chairman of the Supreme Petroleum Council and I will make the decisions. I have decided to give our American friends an opportunity to change their policy,' he confidently told one visitor at the time. He instructed his protégé Hisham Nazer, the head of the Central Planning Organization who was going to the U.S. in September, to say nothing 'foolish' that might be interpreted as a threat, and he was insistent that the question should not be discussed at the Fourth Conference of Non-Aligned Nations that was to begin in Algiers on 2 September. Sadat, as it happened, had already agreed with Feisal not to press the issue. Fahd even went so far as to say that all the Arabs were ready to make peace with Israel and recognize the Jewish state, though the King had yet to accept U.N. Security Council Resolution 242. U.S. optimism appears to have arisen in no small measure from the contact that Nicholas G. Thacher, the U.S. Ambassador in Jeddah, had with Fahd. Sultan, who was primarily responsible for the negotiations to purchase Phantom F-4 aircraft from the U.S., seemed to take the same accommodating line as his brother.

At the other extreme was Saqqaf, who favoured a complete freeze. He was more exposed than anyone in the Council of Ministers to the full brunt of Arab opinion through his meetings with foreign ministers. Saqqaf was also always likely to take a different view from Yamani because of his personal dislike for him, if for no other reason. Muhammad Abal Khail, Deputy Finance Minister, took a similar view to Saqqaf arguing that it was pointless for Saudi Arabia to increase production further when it was only able to spend 60 per cent of the state budget. Also steering clear of weightier considerations of foreign policy, Hisham Nazer proposed that the expansion of production should be limited to 10 per cent annually in line with the projections of the Second Five-Year Plan that was being drafted under his supervision.

The position of Yamani appeared to be more flexible and non-committal. In public, though, he generally expressed himself in favour of Saudi Arabia fulfilling its obligations as a supplier. The Oil Minister was probably closer to the King's thinking on the subject than Fahd, who was very much less than punctilious in his attendance at Feisal's *majlis*. Yamani questioned whether Thacher was conveying the right information to Washington. Moreover, in the last analysis the decision would be made by the King and not Fahd. The King's view was that if there was no meaningful progress towards a Middle East peace settlement within six months, then oil sanctions would have to be imposed. The time was significant for him because he wanted to give Kissinger such a period to apply himself to the problem and believed Sadat would also wait as long.

In the last week of August Sadat had completed his diplomatic preparations with a quick round tour that took in Saudi Arabia, Qatar, and Syria before he returned to Cairo for talks with the Ruler of Kuwait. Many Saudis insist that at this point the King was made party to the Egyptian President's secrets about the opening of hostilities. The fact is that Sadat went no further than discussing contingency planning, without revealing how imminent war was. But he succeeded in exacting firmer – but unspecified – pledges from Feisal of support through the use of Saudi oil resources and increased financial aid, if and when the battle were to commence. Similar assurances were obtained from Qatar and Kuwait. Yamani pondered deeply how the Arab producers should use the oil weapon.

As the summer dragged on, *The Times* of London splashed its edition of 26 September with a story, attributed to Kuwaiti sources, that Henry Kissinger, just elevated to the position of Secretary of State in place of William Rogers, had drawn up a peace plan and it had been accepted as a basis of negotiations by Feisal. On the face of it, that seemed improbable since one of the provisions of the supposed framework was the inclusion of Jerusalem in Israel. The occasion of the newspaper's gaffe was the meeting of the new Secretary of State with Arab Ambassadors to the U.N. All he had said was that for the time being he would confine himself 'to hearing and only hearing'.[13] In response to *The Times'* report, the garrulous Jamil Baroody, Saudi representative at the U.N., said with unusual brevity: 'There is no truth whatsoever in the report; we do not know of any Kissinger plan.'[14]

*   *   *

Pressures, meanwhile, had been building up within O.P.E.C. towards total rupture of the Tehran Agreement that was to coincide, fatefully for the West, with the explosive unfolding of President Sadat's military plans. The five-year pact had already undergone one important revision early in 1972 when members had reached agreement with the oil companies on the so-called 'Geneva I' formula under which prices were adjusted monthly according to changes in a basket of currencies. First, the 10 per cent devaluation of the dollar in February 1973 led to an immediate demand for compensation for the loss in purchasing power suffered by the oil producers, as well as a new basket of currencies for calculating adjustments that would give less weight to the dollar, and, therefore, bigger increments. Second, the eleven members of O.P.E.C. (Ecuador and Gabon had yet to join) were becoming increasingly dissatisfied about the rising rate of inflation of the cost of goods purchased from the industrialized countries that was by no means covered by the 2.5 per cent plus 50 cents annual increment allowed for in the Tehran Agreement. Last, but not least, as demand began to squeeze available supply O.P.E.C. became acutely conscious and resentful of the fact that prices ultimately realized on the market were far higher than their basic revenue rates, that the gap was widening, and that most of the difference was swelling the oil companies' profits. Three rounds of negotiations in the spring of 1973 in Cairo, Vienna, and Tripoli failed to resolve the issue. A fourth began in Geneva at the end of June

At this point both Libya and Iraq proposed that O.P.E.C. should take unilateral action by raising the tax reference substantially. But Saudi Arabia, backed by Iran, said that the Tehran Agreement should be modified rather than breached to take account of the changed circumstances. Finally, under the threat of an extraordinary ministerial conference that would give the go-ahead to members to enact unilateral legislation setting their own terms, the oil companies capitulated after five days of cliff-hanging talks. Final agreement was held up by the difficulties encountered by Izzedin Mabrouk, the Libyan Minister of Oil, in contacting Major Abdul Salem Jalloud, Colonel Qaddafi's right-hand man, for approval of the terms. Finally his assent was obtained. The deal gave an increase in the posted price of 11.9 per cent and a revised currency index excluding the dollar under what became known as the 'Geneva II' agreement. The outcome was a somewhat abject capitulation by the industry negotiators who now seemed to be playing for time, blithely unaware how far the balance of power had tilted against them.

In the early hours of the morning of 3 June, following conclusion of the agreement, Yamani took three journalists, including myself, for a vigorous walk – sometimes breaking out into a jog – in the deserted suburban streets behind the Intercontinental Hotel. He predicted in confidence that the year 1973 would see the end of all negotiations between O.P.E.C. and the companies. By then, it was clear, he was convinced that the time had come to sweep away the deal so painfully negotiated a year and a half before. This conviction had probably hardened as a result of the sale held early in May 1973 of Saudi Arabia's first entitlement of 'participation crude' for supplies to customers over a three-year period. It received the prices that it asked for, a record $2.55 per barrel of Arabian Light, compared with the $1.70 rate then applying to the same crude lifted by the partners in Aramco as part of their equity entitlement. In addition, allowance had been made for a rise in line with the expected compensation for the dollar's devaluation. By the end of August, Yamani and Saud al Feisal had convinced the King of the need to do away with the strait-jacket that the Tehran Agreement constituted on the heightened expectations and ambitions of the producers. That much became clear when Yamani was quoted in the *Middle East Economic Survey*, the authoritative oil journal, in its edition of 7 September 1973: 'The Tehran Agreement is either dead or dying and in need of revision.' Revival or resuscitation were obviously the words that he was searching for. But the message was clear. The companies would have to submit to O.P.E.C.'s demands.

Saudi Arabia's final conversion to the idea of discarding the pact altogether followed a visit to Riyadh of an Iraqi delegation towards the end of August. The last obstacle in discarding it was Iran. Its consent was obtained after Yamani flew to Tehran for talks with Amouzegar. It was decided that there should be a straight increase in posted prices to bring the tax reference into line with the realities of the market, a mechanism should be devised to maintain a suitable differential between them in future, and the allowance for inflation should be amended to reflect more accurately the actual rate. An offer put forward by the representatives of the companies early in September was dismissed out of hand. Subsequently, an extraordinary conference held on 15 – 16 September in Vienna approved the proposals of the Gulf committee handling the negotiations for a wholesale revision of the Tehran Agreement. The proposals were largely the work of Yamani. The scene was thus set for O.P.E.C. to set its own prices unilaterally. The conjuncture was one that would shake the foundations of the world's economy. Even if there had been no Middle East crisis the outcome would almost certainly have been the same.

In the meantime, the leading militant Libya had forced Occidental on 11 August and the Oasis group, mainly U.S.-owned but including Shell amongst the shareholders, on 16 August to agree to 51 per cent participation. Having been frustrated in his negotiations with other producers, Major Jalloud announced on 1 September a similar takeover of majority control of all other producing operations with the exception of two minor ones (Mobil and Exxon) to commemorate the fourth anniversary of the revolution. (It was not until April 1974 that Mobil and Exxon finally accepted Libya's unilateral assertion of a 51 per cent majority stake in their operations.) The following month the Kuwait Government reached an agreement with British Petroleum and Gulf whereby the Government would take a 60 per cent controlling interest in the companies' concession. But the National Assembly did not immediately accept it.

The Saudi Government felt that it had been trumped and immediately showed signs of dissatisfaction with the arrangement that Yamani had negotiated on behalf of it and the other Arab producing states of the Gulf. But the Oil Minister was not under pressure to react; his immediate preoccupation was the next meeting with oil company representatives prior to the O.P.E.C. ministerial conference scheduled for 10 October in Vienna.

*   *   *

The combined Egyptian-Syrian assault on Israel in the early afternoon of 6 October 1973 came as a surprise and a shock for the Saudi royal court, the Government and Armed Forces. Just how ill informed they were can be seen from the fact that General Zuhair, the Commander of the Royal Saudi Air Force, spent several hours after hostilities started on the Suez Canal and Golan Heights, unaware of them, arguing with the British Aircraft Corporation about the interpretation of a contract.

When the scale of the military operation against Israel was appreciated, there was a feeling of desperation in Riyadh. Feisal had counselled Sadat against resorting to arms and still hoped, somewhat naively, that he could bring about the evacuation of the occupied Arab territories through the threat of a squeeze on oil supplies. The King took a pessimistic view of the outcome of any military venture. A grave setback or a catastrophe such as the one suffered in 1967 could shame and even bring down the fabric of his realm and the other conservative Arab regimes of the Arabian peninsula, he feared. It was difficult to decide what would be more dangerous – entering the war or maintaining a cool detachment. With its territory situated only twenty miles from Eilat and even less from Sinai towards its southern extremities, Saudi Arabia was regarded as a confrontation state by Israel and its American lobbyists, as the Congressional opposition to sale of Phantom F-4s to the Kingdom had shown in the summer. However, because of its dependence on the U.S. for weapons and finance, Israel, unless it was on the verge of being overwhelmed and felt forced to bring down the roof of the Temple on the West, could be relied upon not to strike at the oil fields of Saudi Arabia and Kuwait.

Iraq and Jordan could make meaningful military contributions and did so. Assistance was also forthcoming from Morocco, Algeria, and Sudan, which sent contingents. Saudi Arabia and Kuwait salvaged their honour by sending, respectively, a motorized infantry brigade equipped with armoured cars and artillery units, which arrived on the Golan Heights just before 14 October. The Saudi contingent joined the Arab Foreign Legion that was deployed to the south of the main fighting to stem the advance. The Saudis saw one day's action on 19 October. Two days later a military spokesman said that they had destroyed five Israeli tanks and damaged three others, killing a number of men. Their losses were four armoured vehicles, four 106 mm. self-propelled guns, and nine men killed, wounded or missing. About this time the Israelis reported the capture of a number of Panhard armoured cars, presumed to be Saudi because none of the other Arab forces possessed them.

On the diplomatic front, Feisal was involved from the beginning. On 8 October Kissinger followed up his messages to the Israeli and Egyptian Foreign Ministers, who were in New York for the U.N. General Assembly, with one to the King. It asked him to use his influence to bring about a ceasefire. Feisal sent in reply a warning to the effect that 'if the U.S. does not act to deter Israel from continuing her aggression...the volcano which so far has been inactive in the Middle East will be active again and its holocaust will not be confined to this region alone...The U.S. should force Israel to withdraw from Arab territories and to grant the Palestinian people their rights in their territory and homeland.'

On the second day of the war Libya offered finance and oil supplies to Egypt and Syria. Iraq seized the 23.75 per cent share in the Basrah Petroleum Company (untouched by the 1972 nationalization of I.P.C. which was under the same ownership) that was shared between Exxon and Mobil. The other Arab oil producers were inert for the first few days of the war. Their options for wielding the oil weapon were limited. First, there was expropriation of the assets of U.S. oil companies. That was generally ruled out as a blunt and inflexible instrument that could only be used once. It could also be counter-productive if, as seemed almost certain, retaliation through the withdrawal of expatriate personnel adversely affected producing operations and oil movements. Iraq could nationalize the American stake in the Basrah Petroleum Company without fear of any such repercussions because it was the only concession left in the country. Second, embargos could be imposed on specific destinations, in particular the U.S. and other countries judged to be helping or friendly to Israel. Here again there were misgivings about the Arab oil states' ability to monitor the complex global distribution of crude, not the least because the bulk of it was controlled by the major international companies with their integrated systems. Third, overall Arab output could be cut by varying degrees or completely. In general this method commended itself most because it looked simple to administer and at the same time was an adjustable lever. But while an across-the-board cut in production could squeeze the West into putting pressure on Israel it would hurt friends as well as benefactors of the Jewish state. Several leading Saudis, including Saud al Feisal and Abdullah, had expressed concern on this score.

Sadat had persuaded the conservative Arab states to give their support when the hour for battle came. But he had neither agreed with them a plan of action nor worked out how the oil weapon should be deployed. Later, Muhammad Heikal claimed credit for devising the tactics for its use. He recounted[15] how he had commissioned Dr Mustapha Khalil, a former Egyptian minister and 'a brilliant administrator', to make a study of the energy crisis in the U.S. and its implications for the Arab world. Heikal said that on its completion on 1 October he showed it to Sadat. From his account, the essence of the document appears to have been the proposition that the availability of oil to West European countries and Japan should be related to their attitude to the struggle to liberate occupied Arab territory and their willingness to exert pressure on the U.S. It was decided that Dr Said Marei, whom Sadat had charged with co-ordinating Egyptian policy with other Arab governments, and Khalil should take the report unaltered as a basis for discussion. On 8 October they flew to Riyadh for talks with Feisal who, Heikal says, 'accepted completely the policy outlined'.[16] The King made fulsome tribute to the sacrifices being made by the Egyptians and promised an immediate grant of $200 million, telling the delegation also that Saudi troops were to be dispatched to the Syrian front.

Marei and Khalil proceeded to Kuwait. After a hastily convened meeting of the Council of Ministers, the Kuwaiti Government called for the convening of a conference of the Organization of Arab Petroleum Exporting Countries on 17 October. This body had been formed in the wake of the Khartoum summit conference in the late summer of 1967 by Saudi Arabia, Kuwait, and Libya – all then producing states governed by conservative regimes – after they had agreed to provide financial aid to Egypt, Jordan, and Syria. The original purpose of the founder members had been to limit the interference of the Arab League in their own petroleum affairs. Following the Libyan revolution, membership had expanded to include also Iraq, Algeria, Egypt, Syria, Abu Dhabi, Bahrain, and Qatar. O.A.P.E.C. had become the recognized Arab forum for consultations relating to oil.

Marei and Khalil may have helped strengthen the resolve of the ten Arab oil producing states to unsheath the oil weapon. In truth, however, the shape of it – albeit with some roughness at the edges – was fashioned, and the manner in which it would be wielded was provisionally decided, by Yamani and the ministers of the other four Arab states making up O.P.E.C.'s Gulf committee. The crucial discussions took place during their meeting with oil company representatives in Vienna on 8 – 9 October. Earlier in the summer a consensus had formed that graduated across-the-board reductions in output would sooner or later be imposed, whether or not an Arab-Israeli war broke out, if the diplomatic impasse continued. In the Austrian capital Yamani and his colleagues reached a fairly specific agreement on the scale of the initial percentage cut in output and the phasing of subsequent ones.

More immediately, the gathering in Vienna was of historic significance. It marked the end of negotiations between O.P.E.C. and the international oil industry that Yamani had predicted four months earlier. Representatives of the companies put forward an offer of an immediate 15 per cent increase in posted prices and the introduction of an inflation index. The producers' committee proposed a 100 per cent rise in the tax reference. It was not prepared to allow for a two-week postponement that the companies wanted for consultation.

The six chiefs of the Gulf delegation assembled in Kuwait on 16 October. They raised posted prices by 70 per cent so that the rate for Arabian Light rose from $3.01 per barrel to $5.11. Actual Saudi Government revenue for the equity crude lifted and marketed by the U.S. partners in Aramco went up slightly more, from $1.77 per barrel to $3.04. As the battle raged in Sinai and on the Golan Heights, the enormity of the increase in prices went largely unnoticed except by those whose business was oil. So, too, did the fact that producers had set prices unilaterally for the first time.

In the meantime, Saqqaf had been instructed by Feisal to see the U.S. President. His first duty was to enquire about the truth of Egyptian press reports to the effect that arms replenishments were being flown to Israel. The second purpose of this mission was to convey with greater foreboding than before the consequences that could arise – and now seemed highly probable – from the biased policy of the U.S. towards the Middle East. Saqqaf delivered a letter, covering both points, to the White House on 12 October. The massive airlift of weaponry to beleaguered Israel began on 13 October, Nixon subsequently revealed.[17]

After Amouzegar, the Iranian chief delegate at the Kuwait meeting, had departed for home, the five Arab ministers were joined by their counterparts from Algeria, Libya, Egypt, Syria, and Bahrain to take part in the O.A.P.E.C. conference. The majority favoured imposing an immediate embargo on supplies to the U.S. On behalf of Saudi Arabia, Yamani resisted, arguing that Washington should be given more time to formulate a fair policy towards the Arabs. It was accepted by all that each producer had a right individually to halt exports to the U.S. Yamani was successful in getting any reference to an American embargo excluded, but he aroused Iraqi bitterness and hostility in the process. His firm stand strengthened the belief amongst some delegates that Saudi Arabia had given assurances to Washington about the maintenance of supplies. By contrast, agreement was reached rapidly – but not without difficulty – on an across-the-board cut in production, immediately by a minimum of 5 per cent and thereafter by a similar amount each month 'until such time as total evacuation of Israeli forces from all Arab territory occupied during the June 1967 war is completed, and the legitimate rights of the

Palestinian people are restored'. The flow of oil, it was decided, should remain unaffected for 'any friendly state which has extended or may in the future extend effective concrete assistance to the Arabs'.[18] They were not specified.

As the Arab oil ministers were ending their meeting in Kuwait, Saqqaf called upon Nixon on 17 October, together with the Algerian, Jordanian, and Kuwaiti foreign ministers. After his second White House encounter, the Saudi Minister of State for Foreign Affairs optimistically told reporters: 'We think the man who could solve the Vietnam war, the man who could have settled the peace all over the world, can easily play a good role in settling and having peace in our area of the Middle East,' a remark that was smugly recorded by the former President in his memoirs.[19] That afternoon Nixon sent his request for $2,200 million in emergency aid for Israel, which had gained the upper hand militarily by this time. Sheikh Zaid of Abu Dhabi was the first to declare an embargo against the U.S. on the following day, just beating Libya and Qatar. Other Arab oil producing states quickly followed suit. Saudi Arabia announced that it would reduce its output by 10 per cent and try to persuade the U.S. to modify its stand on the war. Two days later, on 20 October, it also declared its decision to boycott the U.S.

Having been converted to the use of the Arab oil weapon, Feisal showed little restraint in its use. At the meeting in Kuwait the assumption had been that oil affected by destination embargos would be covered by the straight percentage cutbacks. If they exceeded the across-the-board reductions, the difference would be available for distribution to other consuming countries. The hurried compromise, it materialized, had not been carefully worked out and for the next six weeks, as the winter approached and the weapon began to penetrate, the world watched with pained bemusement the Arab producers work out its implementation. Saudi Arabia, like Kuwait, added the volumes that would have been exported to the U.S. to the overall 10 per cent slashed from their output. Aramco was thus instructed to bring down output by nearly 22 per cent from the 8.3 million barrels a day produced in September to 6.5 million barrels a day. The level was no less than 29 per cent down on the one that had been planned for November. The greater part of the difference, nearly one million barrels a day, was attributed to crude and refined products that would have been shipped either to the U.S. directly or to destinations, like the Caribbean, known to be refining for the American market.

There followed the embargo of Holland that had repercussions extending beyond it because of Rotterdam's importance as a refining centre. No further reduction in the rate of output was made in respect of the Netherlands, however. In pursuit of a policy of reward and punishment, a number of consuming countries were placed on an 'exempt' or 'most favoured' basis, including France, Spain, and (provisionally) the U.K.

Iraq chose not to subscribe to the Arab programme for co-ordinated reductions in oil production adopted by the other members of O.A.P.E.C., largely in protest against Feisal's success in obtaining exclusion from the agreement of any mention of an embargo against the U.S. In an official statement issued on 21 October, the Baathist regime in Baghdad argued that 'it was essential to differentiate between hostile countries, which should be boycotted, and friendly countries, which should be assured of normal oil supplies, and that this would necessitate an embargo on oil exports to the United States'.[20] Iraq called instead for the nationalization of all American commercial interests in Arab countries, the withdrawal of funds invested by them in the U.S., and a complete rupture of diplomatic relations with Washington. The statement was a deliberate challenge to Saudi Arabia's leadership of the Arab oil producers and intended to embarrass it. It marked the start of a period of strained relations between the two countries in the sphere of oil politics.

On 22 October, the U.N. Security Council passed Resolution 338 that called on the belligerents to end the fighting. On the following day, a ceasefire came into effect and the Egyptian Second Army was saved. Technically the Kingdom had been at war with Israel since 1967. However, it subscribed to Resolution 338, which also called upon the parties immediately to start implementation of Resolution 242. Thus, Saudi Arabia implicitly accepted the latter in what was regarded by contemporary diplomats as a considerable concession given the Kingdom's fundamentalist stand on the Palestine question and Feisal's virulent anti-Zionism. The engagement of the Saudi unit on the Golan Heights may have been peripheral. But in 1973, in contrast to 1967, the Kingdom could congratulate itself on the role that it had played, through forging and applying the oil weapon, in the outcome. The balance of power in the Middle East had changed and Saudi Arabia's stature in the world was greater than it had ever been.

20 Beyond the Dreams of Avarice

1973–4

For two years we have tried to make America understand the realities of the situation, and we have warned U.S. officials of the damage their interests in the Arab world will suffer. And this was before we spoke of our just rights, the exhaustion of the Arabs' patience and their inability to continue to suffer injustice and oppression from the Zionists. But it was to no avail.[1]

KING FEISAL

SADAT would later say, only half jokingly it seems, that he had never expected Feisal to use the oil weapon. But the tide of emotion running through the Arab world, the suddenness of the outbreak of war, and the apprehensions of the Kingdom generally proved stronger than fine calculations about ends and means. In the heat of the war, Saudi Arabia had gone further than the options discussed in the summer of 1973. A profound shock had to be administered to the West as a whole in the short term if the status quo was to be changed, and in the long term if a lasting impression was to be left. From the point of view of co-existence within the Arab world and the safety of the realm, the ruling hierarchy could be said to have manoeuvred with some skill.

Despite the embargo, American imports rose sharply in 1973, especially those from the Arab world which now accounted for about 10 per cent of the super-power's overall consumption and 28 per cent of total requirements from abroad. But the U.S. still had greater self-sufficiency and ability to switch sources of supply than West Europe and Japan had. It was clear from the start that they would be hardest hit by the across-the-board cuts. The prospect about which Abdullah and Saud al Feisal had voiced misgivings in August was soon becoming a reality.

There was an arbitrary element in the classification of countries. There seemed no outstanding reason why Britain, which had been accorded 'exempt' status, should be considered more friendly than West Germany and Italy, apart from its refusal to send Israel spare parts for Centurion tanks. But with supplies so much reduced, 'exempt' status did not mean much, as the U.K. soon discovered. Its imports had fallen 20 per cent below requirements by December, chiefly because international companies, not the least British Petroleum, were spreading available supplies amongst their customers. Following a visit to Riyadh by Lord Aldington, Chairman of Grindlays Bank and a friend of British Premier Edward Heath, Saudi Arabia agreed to alleviate the problems of the U.K. by undertaking to sell it an additional 200,000 barrels a day temporarily. Holland was victimized for the pro-Israeli statements made by ministers and officials early in the October War. The situation was not changed when The Hague subscribed to the joint European Community statement declaring Israeli occupation of Arab lands illegal, or even when the Dutch Foreign Minister went slightly further in recognizing the need to fulfil Palestinian aspirations.

Saudi Arabia was in potential danger of exposing itself dangerously as a result of the firm and fundamentalist stand that it had taken. As a Western-orientated oil power, responsible for over 40 per cent of Arab production before the war and with the scope for increasing it considerably, it was bound to be the focus of U.S. pressure. From the Arab point of view, the onus was on the Kingdom to deliver a satisfactory settlement. At this stage its position was most definitively spelt out by the King in an interview with the Egyptian newspaper *Al Gomhouriya* on 22 November. 'The effectiveness of this important weapon cannot be relaxed nor the determination to use it weakened under any transitory circumstances,'[2] he said. Its withdrawal would depend on three conditions: total Israeli disengagement from all the occupied territory; the confirmation of the right of the Palestinian people to determine their own destiny; and the affirmation of the Arab character of Jerusalem. The phrasing of the last condition seemed to hold open the possibility of the Kingdom accepting something less than full Arab sovereignty over the Holy City. But Feisal would not concede that Jerusalem had any sacred significance to Jews. Asked about the remains of Solomon's Temple, he replied with evident contempt: 'The Wailing Wall is a structure they weep against, and they have no historic right to it. Another wall can be built for them to weep against. We will not go back on our position and determination.' It is easy to imagine his thin lips curling with derision.

Jerusalem apart, senior diplomats posted to Jeddah at the time believed that Feisal would settle for anything that the confrontation states would. But now only wishful thinking could envisage the Palestine Liberation Organization not being party to any settlement sanctified by the blessing of the pan-Arab consensus. A high-ranking member of the P.L.O. visited Riyadh in the wake of the war. A few weeks later, Al Fatah, the mainstream group under its umbrella, opened an office in Riyadh.

The Arab summit held in Algiers on 26–28 October took a hard line, deciding that the implementation of the progressive cuts should continue. But it was agreed that individual producers' income would not fall below 75 per cent of their monthly average revenue in 1972. With the rises in per barrel revenue of the previous five months, this point would not be reached for the best part of a year. In addition, the Arab leaders charged their foreign and oil ministers to draw up a common system for classifying the consuming countries by the three categories of exempt, neutral, and embargoed. Japan was excused the 5 per cent cutback scheduled for December following a policy declaration in favour of the Arabs.

On the question of withdrawal from the occupied territories, Arab oil muscle had succeeded in detaching West Europe and Japan from the U.S., to the indignation of Kissinger, who in private was contemptuous of them. But Feisal would not have dissented from Sadat's view that 'the U.S. holds 99 per cent of the cards'. So began in earnest the bargaining – it was no less – over the resumption of oil supplies in return for American pressure on Israel of a kind that could show clear signs of progress towards the achievement of Arab goals. The exercise was a difficult and painful one, not the least because Feisal looked to the U.S. as a friend and the power best able to provide his country with its requirements for development and defence. Diplomatic communication was improved by the appointment of James E. Akins as U.S. Ambassador in Jeddah in place of Thacher. Akins arrived shortly after the October War.

Kissinger did not appear to appreciate fully the nature of the Kingdom's predicament or its thinking. However, well aware of Feisal's obsession about Jews and his almost total identification of them with communism, the normally ebullient Secretary of State is said to have betrayed some uncharacteristic nervousness as he headed for Saudi Arabia, his fourth destination on a tour that had already taken him to

Morocco, Egypt and Jordan. In Cairo he had drawn up with Sadat six points as the basis for an agreement on implementing the ceasefire through the separation of forces, the beginning of his 'step-by-step' peace-making efforts.

The first engagement between the King and the Secretary of State on 8 November was fraught with tension. The Secretary of State, perhaps uneasy about Akins' rapport with his Saudi hosts, sought to exclude him from his first encounter in the King's study. Akins told me he threatened: 'If you go in there without me you'll have my resignation when you get out.' Kissinger realized that, as a representative of the U.S. President, the Ambassador was entitled to insist on being present and gave way grudgingly. The exchanges within were sharp, frosty, and to the point. Kissinger tried to create the right atmosphere by paying tribute to the King's perspicacity in his warnings about the dangers of the Middle East situation. He presented the untimely announcement on 17 October about financial aid to Israel as an act aimed at stopping communist expansion. Feisal cut in to assert the categorical need for the U.S. to force Israeli withdrawal from the occupied territories and reminded him of Eisenhower's firm stand after the Suez War in 1956. On the contrary, he emphasized, fulfilment of Soviet aims could only be helped by a continuation of the status quo. Kissinger replied that the U.S. was committed to achieving a settlement, but had to take into account domestic political considerations. Its task would be made less difficult if the oil embargo was lifted. The King explained that Saudi Arabia, too, was in a predicament. It could not afford to be accused of bowing to American pressures. He also stressed that the U.S. must pay attention to Syria and assured him that Assad would be prepared to receive a visit from him. At dinner afterwards Kissinger was spared nothing in Feisal's monologue about the iniquity of Jews and his theories about Zionist-communist conspiracy. As he left the banqueting chamber, Kissinger muttered to Akins, 'That was your idea of light dinner conversation?'[3]

The experience was not a comfortable one for Kissinger who was far more conscious of his background as a Jewish refugee than was generally appreciated and, therefore, emotionally committed to Israel as well. A month later, at a closed meeting of Jewish intellectuals, most of them former Harvard colleagues, he described the King as a 'religious fanatic, concerned mostly about Jerusalem with little interest in the Sinai or the Palestinians'.[4]

Despite the mutual distaste that they felt for each other at the first meeting, Kissinger was said to have been impressed by the King's forthright consistency and patent honesty. He was one of the few Saudis for whom he had respect or much time. The American Metternich who used bonhomie as a diplomatic tool was neither intrigued nor had much patience with the solemn inscrutability and polite evasiveness characteristic of Saudis generally. The lifting of the Arab oil embargo and ensuring of adequate oil supplies to the West at a reasonable price were important aspects of his grand global design, but he was not convinced that Saudi Arabia was a central part of the Middle East equation and, it seems, less than certain what weight should be given to it. In this respect the Secretary of State appraised correctly the essentially defensive outlook of the Kingdom and the limits to its influence amongst Arab states. It quickly became apparent that Saudi Arabia could not now sheath the oil weapon without Egyptian and Syrian approval. That necessarily involved Israeli withdrawal.

But Kissinger was slow to appreciate the extent to which the Saudi regime politically had to align itself with the pan-Arab community on the question of Palestine and how members of the ruling hierarchy as well as citizens of the Kingdom identified with a cause that was integral to their religiously impregnated view of the world. His remark about the King's 'little interest in the Sinai or the Palestinians' indicated an ignorance about the basis of inter-Arab relations. Kissinger had difficulty in understanding why, out of self-interest, Feisal could not or would not align more fully with the U.S. and thus fall with more tidiness into his conceptual framework of a worldwide balance of power between the two super-powers. His attitude underestimated the ability of the King and his advisers to make their own judgement as to how the fate of the West and, conversely, the stability of the Kingdom might be affected by the price of oil. For his part, it is fair to assume that the King would have regarded Kissinger from the start with grave suspicion because he was a Jew. He would have recognized early that the Secretary of State was the kind of man who, in the words of Kissinger's own doctoral dissertation, did 'not shy away from duplicity, cynicism or unscrupulousness, all of which are acceptable tools of statecraft'.[5]

It is uncertain whether or not Kissinger was following consciously any such Machiavellian principles when on 21 November he spoke of the need for the U.S. 'to consider what counter-measures it may have to take'[6] if the oil embargo was not lifted. He was answering a question at a press conference during which the tenor of his language was generally persuasive. He stressed, as he had done in his conversation with Feisal, that the Arab oil producers should consider 'whether it is appropriate to engage in such steps while peace negotiations are being prepared, and even more while negotiations are being conducted'.[7] His reply took no account of Saudi hyper-sensitivity. On the following day Yamani reacted sharply. Saudi Arabia's oil installations would be blown up if the U.S took any military measure to counter Arab policy, denying oil to West Europe and Japan in the process, he said in an interview in Copenhagen.[8] This was an over-reaction: he had not seen the full text of Kissinger's remarks, which showed that the Secretary of State had been talking in terms of economic counter-measures and the use of force had not been discussed. Yamani's riposte was nevertheless symptomatic of the tension between the U.S. and Saudi Arabia.

On 27 November talks on the implementation of the ceasefire agreement were broken off by Egypt because of Israel's 'elusiveness' on the question of the disengagement of the opposing forces. Kissinger set off for the Middle East again on 13 December. In Cairo he discussed the agenda of the forthcoming peace talks in Geneva with Sadat before going on to Riyadh amidst speculation that Feisal might be prepared to relax the squeeze on oil supplies. When they met on 14 December, the King told Kissinger uncompromisingly that there could be no easing of the oil embargo until Israel made a clear commitment to withdraw from the occupied Arab territories. The U.S. alone was responsible for the implementation of U.N. Security Council Resolution 242, he asserted.

On 21 December the Geneva Conference was convened under the co-chairmanship of the U.S. and the Soviet Union, with Egypt, Jordan, and Israel represented but Syria absent. The session consisted of little more than a formal statement of the positions of the three states that merely underlined the enormous problems to be resolved. It was agreed between Egypt and Israel that the so far abortive talks on the separation of forces in Sinai should be continued in Geneva after Christmas.

In the meantime, the oil producers were moving to capitalize on the drastic shortage in supplies caused by the war and the progressive reductions in Arab output. A full ministerial conference of O.P.E.C. was convened in Vienna on 18 November. The majority of members, though not Saudi Arabia, were in favour of using direct Government sales of certain key Gulf crudes as a basis for setting new prices. In the autumn of 1973 these accounted for only marginal volumes of oil but one for which there was an artificially high demand because of cutbacks in Arab production. In their anxiety to obtain any supplies, buyers were prepared to pay what a few months earlier would have seemed impossible prices. In the second week in December, Iran obtained no less than $17 per barrel, compared with the post-16 October posted price of $5.34 and actual Government receipts of $3.17.

O.P.E.C. met again in Tehran on 23–24 December. Shortly before the actual meeting, Yamani gave notice that the prices achieved in the auction of crude by the National Iranian Oil Company should not be taken as the determinant for basic official selling prices. They reflected, he said, 'to a large extent the effects of the oil embargo and cutback measures taken by the Arab oil producing countries and since these measures are of a political nature they should not have an economic effect'. To take the auction realizations as the guideline 'would ruin the existing economic structure of the industrialized countries, as well as of the developing countries, and very soon the entire amount of money available for financing international trade would not be enough to pay for our oil'.[9] It was a significant statement heralding struggles within O.P.E.C. which were to strain Saudi Arabia gravely. Never before on a matter of vital issue had Yamani struck in public a moderate line of a kind likely to provoke other members.

The Tehran conference was essentially one of the Gulf committee. Four other members sent delegations, and Venezuela, preoccupied with Christmas and recalling a dry Yuletide O.P.E.C. gathering in Riyadh in 1963, just one observer, who was a teetotaller. Imperiously, the Shah of Iran had over the previous weeks admonished the Arabs for continuing its restrictions on oil production. He wanted prices directly related to those obtained in the N.I.O.C.'s auction which, he said, showed the 'real value' of oil. On his own soil, he wanted to dictate the outcome. He pushed for a fourfold increase in the producers' per barrel receipts to $12–14 with a tax reference of $17–20. Feisal and Yamani knew that another increase was inevitable and also believed it to be justifiable. As early as the summer, Saudi Arabia had been aiming at a basic rate for revenue of $5 per barrel that would have meant a posted price of $7–8 per barrel.

Never again would Yamani feel so lonely and vulnerable at an O.P.E.C. conference. In trying to curb the Iranian demand, however, he received some support from Iraq. In advance of the meeting both Vice-President Saddam Hussein and the Iraqi Oil Minister Saadoun Hammadi had called for caution. The Iraqi delegation also resented the manner in which the Shah, bloated with illusions of regal authority, attempted to direct the proceedings. That was true of other Arab delegations but they did not object to the maximalist demands of their host. Yamani fought bravely, to the point that Amouzegar eventually was authorized by the Shah to come down to, but insist on no less than, a posted price of $13.33 which would yield a Government 'take' of $8. Yamani tried to call the King in Riyadh but could not reach him. Instead, he talked to Fahd, who merely told him not to risk splitting O.P.E.C. Earlier he had even sought advice about the economic repercussions for the world of the kind of rise in prospect with a senior American oil executive and an independent authority on the oil industry who were observing the conference.

Yamani gave ground, though not without forcing Iran to lower its final demand. The result was a new posted price of $11.65 per barrel of Arabian Light 'marker' crude, an increase of 127 per cent over the level set in mid-October. Per barrel revenue went up from just over $3 to rather more than $7, an increase of about 130 per cent. Yamani's predicament had been a painful one. Looking pale after the meeting and not a little shaken, he said: 'In our opinion a lower posted price would have been more equitable and reasonable. However, we went along with the majority.'[10] It was a frank admission. He had gone beyond the very limited scope of his negotiating brief, however. The King was very angry when he heard the news but did not blame Yamani.

The Shah again ruffled Arab susceptibilities by calling a press conference to announce the new price level even before the ministers had finished completing other business or had drawn up the communique. He aroused more resentment by the way in which he announced a plan for talks between producers and consumers on the relation between the price of oil and the cost of manufactured goods imported from industrialized countries as if he was launching an initiative on behalf of O.P.E.C. The proposal was aired at the meeting in Tehran and was one of the main reasons for calling a fully-fledged ministerial conference two weeks later in Geneva. There, other delegations made it clear that they were not prepared to allow the Shah to make the running in this way. They also objected to the restricted framework envisaged by him for such a dialogue. Algeria, in particular, foresaw talks embracing a whole range of subjects of interest to members.

But the truth was that O.P.E.C did not know where to proceed to next. It had discovered and asserted its power but had hardly started to think how it should be used in global affairs, except for the protection and furtherance of the producers' interests. With tripling of revenue in a period of less than three months the question of compensation for inflation and also the formula for adjusting prices according to fluctuations in exchange rates lapsed. The dollar, as it happened, had been recovering its strength since the autumn. But there were rumbles of discontent over the increased 'windfall' profits being made by the major oil companies as a result of higher prices. Pressure was building up for a change in the taxation system to reduce them.

Other Arab oil producers, apart from the Kingdom, were now concerned about the wounds inflicted on the economies of West Europe and Japan. Meeting in Kuwait on 24–25 December, the oil ministers of ten of the Arab producers decided to restore output that had been down to 25 per cent below the September level to a rate 15 per cent below. Iraq, who had opted out of the across-the-board reduction in output and profited by considerably increasing production, did not attend. It was decided that Japan would be permitted aggregate supplies at September rates. Belgian imports via Rotterdam were allowed provided that none of the oil was directed to other destinations. Most favoured states could receive all their requirements.

It was obvious that the U.S. would be assisted by the increased amount of oil available, but public opinion was aggressive now. Early in January 1974, James Schlesinger, Secretary of State for Defense, warned the Arabs of the danger of provoking force against them if they carried the use of the oil weapon too far. In public, Kissinger was far more cautious, saying only, 'It becomes increasingly less appropriate for Arab governments to pursue discriminatory measures against the U.S. when the United States has publicly declared its support for Resolution 242 and has been the principal country promoting a settlement in the area.'[11] In December, prior to the meeting with the Arab oil ministers in Kuwait, Sadat had given him an undertaking that he would obtain a lifting of the embargo.

The first agreement between Egypt and Israel on the disengagement of forces had finally been reached on 17 January. The day after it was signed, Sadat set off for Saudi Arabia on a tour that subsequently took in Syria, Kuwait, Bahrain, Qatar, Abu Dhabi, Algeria and Morocco. The U.S. had done enough to earn a reciprocal favour in the form of a raising of the boycott, the Egyptian leader pleaded. The Saudi King had just been briefed by Khaddam, the Syrian Foreign Minister, and would concede nothing. Assad told Sadat in Damascus that he wanted at least an equivalent disengagement before the oil weapon was put back into its scabbard.

The focus of both Egyptian and American attention was heavily on the Kingdom, not only as the largest Arab producer but as the one that might be most amenable to pressure. Through diplomatic channels, Kissinger sought to undermine Feisal's resistance. At first he argued that a continuation of the embargo could only be counter-productive. By further eroding the position of Nixon, who was beginning to buckle under the Watergate scandal and was now faced with an Arab-induced recession, it would only make it harder for him to pursue efforts towards a comprehensive peace settlement. Then a letter signed by Nixon but written by the U.S. Secretary of State was sent to the American Embassy for Akins to pass on to the King. Akins was shocked by its threatening nature and called Saqqaf. The Minister of State for Foreign Affairs agreed that it should not be delivered. It could lead Feisal to break off diplomatic relations. The letter indicated that the U.S. would abandon attempts to bring about a peace settlement if oil sanctions were not lifted. It is believed also to have emphasized the degree of hostility in Congress towards Saudi Arabia on account of the steep rise in oil prices decided upon by O.P.E.C. the previous month. It was said to include a threat to reveal certain embarrassing correspondence detailing military assistance discussed or agreed in the winter of 1962–3 at the time of the outbreak of the Yemen civil war. Akins sent back a message to the White House warning of the vengeful reaction if the letter was handed to the King. He proposed that it should be rewritten. It was.

The period was one of intense strain in Saudi-U.S. relations at a time when the producers were watching with great suspicion U.S. plans for an energy conference of the industrialized countries in Washington. Then, at a press conference in Washington on 22 January, Kissinger went further than he had in public before by saying that failure to end the embargo 'would raise serious questions of confidence in our minds with respect to the Arab nations with whom we have dealt on this issue'.[12] He expected an end to sanctions before the forty-day period for the completion of the forces' disengagement in Sinai. It rested on nothing more than the fact that the Arab oil ministers had scheduled another meeting for 14 February in Tripoli. The meeting never took place because all the Arab states involved, with the exception of Egypt, believed that it would be premature. Instead, the burning issue was discussed at a restricted summit in Algiers by Sadat, Assad, Feisal, and Boumedienne. They agreed in principle that the boycott of the U.S. should be dropped if the U.S. showed a demonstrable will to make a 'constructive effort' to bring about a military disengagement on the

Golan Heights. It remained unclear precisely what Syria itself expected. A few days later Kissinger was informed of the outcome of the Algiers meeting.

The message galvanized the Secretary of State into his third visit to the region and a shuttle between Damascus and Jerusalem. Having presented disengagement proposals to Assad on 1 March, he went on to Riyadh. There he gave assurances to Feisal which, backed by Syrian indications that progress was being achieved, persuaded the King that the embargo could be provisionally lifted. His meeting with Kissinger was more relaxed than the previous two. The Secretary of State, it seems, expressed confidence that he could deliver Israeli withdrawal from the River Jordan and the dismantling of Jewish settlements in territory evacuated. The King was pleased with what he had to say, but nonetheless sceptical.

Amidst uncertainty, the Arab oil ministers eventually assembled in Tripoli on 13 March in a raging storm that bent the palm trees on the city's waterfront nearly double. The conference was also tempestuous. Saudi Arabia was in the majority, backing Egypt and calling for a lifting of the embargo with as few specific conditions as possible. Syria, it turned out, believed that the U.S. had not done enough to warrant the removal of oil sanctions. Algeria supported that view. Libya uncompromisingly insisted that the embargo should remain in force. Iraq, which was absent, was equally adamant that the U.S. should still be punished.

A provisional compromise was reached. The ban would be raised for a limited period and with an extension conditional on further progress towards a settlement. But Libya was not a party to it, arguing that the disengagement agreement only benefited Israel and, on the contrary, measures against the U.S. should be strengthened. Syria also dissented from the agreement to lift the embargo. Algeria saw its suspension as 'provisional in nature and limited to the period expiring 1 June 1974'.[13] Unanimous agreement had been made harder by Nixon who, in response to the positive signals reported from Tripoli, warned that any conditions relating to the lifting of the embargo would not be encouraging for peace efforts. The moderate majority – Egypt, Saudi Arabia, Kuwait, the United Arab Emirates, Qatar, and Bahrain – announced no restrictions. But they agreed to meet in Cairo early in June to review the situation. However, Yamani let it be known that Egypt and Saudi Arabia expected by then not only a disengagement agreement on the Syrian front but one between Israel and Jordan, further progress towards the implementation of U.N. Security Council Resolution 242, and a just settlement of the Palestinian problem. Feisal, it seemed, had been led to believe that such was possible. On 28 May, after a final frenzy of shuttle diplomacy in which Kissinger visited Damascus and Jerusalem no less than eleven times each in a period of a fortnight, Syria finally agreed to the proposals for the disengagement of forces on the Golan Heights accepted by Israel. The agreement was signed on 31 May.

Immediately after the Arab ministers' meeting, O.P.E.C. called a ministerial conference in Vienna to review the oil prices set for the next quarter. Yamani was briefed by the King to resist all pressures for another increase and, indeed, to argue for a reduction. That was a thankless task because the other producers, fresh from their climactic triumphs in 1973, would never again be able to contemplate a lowering of prices in nominal terms and were discussing ways to ensure that they should be held steady in real terms. The King's position was based on nothing more complicated than his statesman-like view that the world could not afford to pay the prices set at Tehran. He had kept separate in his mind the issue of the embargo and the level of prices. In practice, the two were very related, as all members knew very well. The production cutbacks had made possible the tripling of prices in the last quarter of 1973.

O.P.E.C.'s Economic Commission recommended that market conditions in the first quarter justified a rise in the tax reference. It suggested an increment that would give actual revenues of 12–24 per cent. Inside the closed meeting there were some fierce exchanges. Abu Dhabi and Kuwait accused Saudi Arabia of sabotaging their oil auctions, whose yield had been disappointing, through its call for lower prices. Yamani, however, stood his ground and even threatened that the Kingdom might unilaterally bring down its prices unless the others withdrew their demand for an increase. The weight of Saudi Arabia as the world's most important exporter was enough to bring about an uneasy compromise on a three-month price freeze. As if in compensation, Yamani agreed that the Economic Commission should embark on a study on the co-ordination of output but without acquiescing in the principle of a production programme. He was also genuinely a party to an understanding reached in Vienna in March 1974 that the tax structure should be changed to cut back the enormous windfall profits enjoyed by the oil companies as a result of the price increases of the previous year.

On a wider front, Saudi Arabia then took an important initiative. At the Sixth Special Session of the U.N. General Assembly held in April 1974, which was devoted to raw materials and the problems of producers in obtaining a fair return for them, Yamani first aired the idea of a dialogue at a high level between countries representing the industrialized and developing world to discuss a global economic system. 'We in Saudi Arabia view the world's economic situation as an indivisible whole,' he stated on 16 April. So, too, was the international community, which he described, quoting the words of the Prophet, as 'similar to the human body: when one part is hurt, all the other parts share its pain'. The Kingdom, he claimed with justification, was making a major sacrifice by maintaining oil output at a level twice the one required for its own financial needs and in the process depleting an irreplaceable asset. 'In spite of the fact that our immediate interests and those of our future generations would warrant that we reduce production and raise prices or, at least, permit them to find their own levels, it is our sense of responsibility towards the world we live in that charts our course for us.'[14]

In the course of 1974 the idea of an international conference aimed at redressing the injustices suffered by the developing countries and the creation of a new economic order took root. Conscious of its power and success in dictating the price for the vital raw material produced by its members, O.P.E.C. took up cudgels on behalf of the Third World as a whole. Saudi Arabia's involvement in the great design tended to mitigate the anger behind charges from other members that it was being too accommodating to the West in its attempts to restrain price rises. By the summer, however, Saudi Arabia would find itself at odds with other members of O.P.E.C. and bound ever more closely to the U.S.

21 The Arabian Janus
1974–5

The Saudis are caught in a dilemma. It is exceedingly difficult if not impossible, for them to accommodate the United States while the United States provides the money and arms which enable Israel to occupy Arab lands.[1] – SENATOR JOHN W. FULBRIGHT

ON 4 June 1974, Fahd arrived in Washington with a full galaxy of Saudi talent including Yamani, Muhammad Abal Khail, the Minister of State for Finance and National Economy, Hisham Nazer, Minister of State and Director-General of the Central Planning Organization, Muhammad Masaud, Under-Secretary at the Ministry of Foreign Affairs, and senior Army officers. Their purpose was to conclude a far-reaching agreement on economic and technical collaboration with the U.S. that had been under preparation since Kissinger's visit to Riyadh in March. The arrangement envisaged was a modified version of the design outlined in Yamani's speech delivered at the Middle East Institute, Georgetown, in the autumn of 1972. It was also one that, the ruling hierarchy hoped, would be less open to criticism at home and within the Arab world.

The U.S. Government, for its part, wanted nothing less than to bind the Kingdom in a 'special relationship', a phrase used by John Sawhill, former head of the Federal Energy Administration, in his testimony to the U.S. Foreign Relations Committee Subcommittee on Multinational Corporations.[2] The concept of close and formalized collaboration fitted in with Kissinger's two-pronged policy of containing the Arab-Israeli conflict, on the one hand, and the massive sale of American technology on the other. The agreement would be a device for reducing the American balance of payments deficit. It was also an important part of the U.S. strategy for meeting the energy crisis. America was looking for Saudi efforts to restrain oil price rises and willingness to maintain output at a rate higher than its financial requirements might justify. A senior U.S. diplomat in the Middle East said at the time: 'The purpose of the operation is to show it's worth producing the oil rather than keep it in the ground.'

The U.S. Government assumed that existing institutions could absorb the enormous surpluses that looked as if they would accrue mainly to the Kingdom. It did not take into account the *quid pro quo*, in the form of a comprehensive settlement of the Middle East conflict involving total withdrawal from the territories conquered in 1967 and the satisfaction of Palestinian grievances that Saudi Arabia expected from the U.S., and other Arab states would look to Riyadh for in view of its close relationship with Washington. Nor, it seems, did Kissinger and his colleagues appreciate fully Saudi Arabia's commitment to the reform of the international economic order and the interests of the Third World in general that had been spelt out in some detail by Yamani and Boumedienne at the Sixth Special Session of the U.N. General Assembly in April. The Administration's evident assumption continued to be that Saudi Arabia, because of the endemic insecurity felt by both its regime and now its citizenry, who had emerged as a peculiarly privileged one by virtue of the Kingdom's oil wealth, might collapse with the failure of a peace initiative.

The Saudi delegation met Nixon as well as Kissinger, Schlesinger, William Simon, Secretary of the Treasury (who had succeeded George Shultz in the previous month), and Frederick Dent, Secretary of Commerce. Within the context of global peace and security, the situation in the Arabian peninsula was discussed. As a framework for American assistance with the Kingdom's development programme, the Joint Commission on Economic Co-operation was established on 8 June under the U.S. Secretary of the Treasury and the Saudi Minister of State for Finance. Working groups covering industrialization, manpower and education, technology and agriculture, were established. On the American side, its members included representatives from the Departments of State, the Treasury and Commerce, and the National Science Foundation. On the Saudi side were officials from the Ministries of Foreign Affairs, Finance and National Economy, Commerce and Industry, and the Central Planning Organization. A parallel joint commission under Sultan and the U.S. Assistant Secretary of Defense for International Security Affairs, then Robert Ellsworth, was set up 'to review programs already under way for modernizing Saudi Arabia's armed forces in the light of the Kingdom's defense requirements especially as they relate to training'.[3]

It was the most far-reaching agreement of its kind ever concluded by the U.S. with a developing country, almost similar in scope to a Soviet treaty of friendship and co-operation. It had the potential to entrench the U.S. deeply in the Kingdom, fortifying the concept of mutual inter-dependence.

The new entente was marked by Nixon's two-day visit to Saudi Arabia on 14–15 June 1974 as part of his tour celebrating Kissinger's achievement in negotiating the agreement on military disengagement in Sinai and on the Golan Heights. He noted that Feisal looked older than his years, adding that according to American intelligence reports he was seventy-two years old. (The date of his birth was, however, generally accepted as being 1904 or 1905.) Nixon wrote: 'Feisal saw Zionist and Communist conspiracies everywhere around him. He even put forward what must be the ultimate conspiratorialist notion: that the Zionists were behind the Palestinian terrorists.'⁴ For the President the dinner given by the King on the day of his arrival was evidently a strain. Feisal, however, seemed determined to make the most of the occasion. Thinking the meal was over, Nixon and his party were anxious to leave and about to rise from the table. At this point, according to one American official present, the King reached out a long bony hand and grasped an apple, which he then took over five minutes to consume.

The accord on military collaboration was to provide a framework for rationalizing co-operation. American investment in Saudi Arabia's defence programme was already deep, if somewhat haphazard, not the least as the result of the efforts of Adnan Khashoggi, who was the well-established agent for Lockheed, Northrop, and Raytheon. In the American fiscal year 1973/4 the value of U.S. military sales agreements with Saudi Arabia totalled $459 million. In the following year they were valued at $1,993 million, more than any other country with the exception of Iran.

The Shah's grandiose spending and pretensions were both a source of concern to Saudi Arabia and also a stimulant to its own ambitions. Iraq was another preoccupation, and also one that the U.S. could cite in putting the case for supplying modern weapons systems to the Kingdom. The cause of guarding the world's greatest repository of oil still required heavy pleading in the face of the pro-Israeli and Zionist orchestration of Congress.

Genuine paranoia about Israeli intentions and the feeling of obligation that it should lend its weight to confronting the Jewish state (even if it would strenuously avoid becoming a confrontation state) was another factor governing Saudi Arabia's attitude towards arms procurement and the development of its limited martial potential. A fourth, undoubtedly, was the opportunity presented to some of those who made the decisions to either profit directly from them or benefit their relatives.

Deliveries of the Lockheed C-130 Hercules transport aircraft, originally ordered in 1965, had continued. In 1971 the company was given the contract to provide systems management, organization and manpower for the radar and associated control centres installed under the original air defence deal concluded in 1965. In the late summer of 1971, an order was placed for twenty Northrop F-5E Tiger fighter-bombers. With his eye still on Iran, Iraq, and Israel, Sultan soon set his heart on the superior and more sophisticated McDonnell F-4 Phantom that was the main war-horse in Israel's aerial armoury. In 1973, Saudi Arabia seriously pursued negotiations on the purchase of Phantoms, but talks on their acquisition were effectively ended by the October War. In 1974, thirty more F-5Es were ordered, followed by another batch early in 1975, bringing the full complement to 100. In April 1974, two months before the joint commission was established, Saudi Arabia ordered advanced Hawk surface-to-air missiles from Raytheon at a cost then set at $270 million. In its final form the contract, signed early in 1977, was worth over $1,000 million.

As early as 1972 Abdullah had agreed, though apparently with some reluctance, to a proposal by the King that American expertise should be called upon for the reorganization, training and equipment of the National Guard. In April 1975 a government-to-government agreement was reached with a provisional figure of $330 million set for the deal. Subsequently, the U.S. Government decided that the package was too large for any one company to carry out. One award, with an initial duration of three years and a value of $77 million, was made to the Vinnell Corporation. It undertook to train four of the National Guard's twenty battalions in infantry tactics, the handling of armoured vehicles and firing of missiles. Nearly 1,000 American ex-servicemen (many of them Vietnam veterans), Jordanians and Pakistanis were hired. A by-product was the sale of Cadillac Gage V-150 armoured cars and Vulcan anti-aircraft guns. It was speculated that the arrangement had greatly furthered the C.I.A.'s penetration of the Kingdom. The Bendix Corporation was contracted to organize logistic support for the Saudi Army.

When in 1974 the U.S. had signed an agreement to supervise the

expansion of the embryonic Royal Saudi Navy, the main outlay was to
be on the construction of base facilities at Jubail and near Jeddah. In
train came orders for fast patrol boats, corvettes and minesweepers. The
programme, which was costed at $150 million, escalated beyond $2,000
million in less than three years. Meanwhile, the U.S. Corps of
Engineers, originally engaged in 1951 to build the Dhahran air base,
was more heavily involved than ever. Under a subsequent agreement
concluded in 1964 it had been contracted by the Ministry of Defence
and Aviation to supervise the construction of military facilities. By the
spring of 1975 it was handling projects worth over $10,000 million.
There were well over 1,500 American citizens in the Kingdom working
on defence projects, including nearly 300 seconded military personnel.
The number increased as Lockheed and Northrop became more
entrenched. Using its advisory contract as a springboard, Lockheed
obtained in 1976 a contract, initially worth $611 million, to construct
an air traffic control system for the whole country and catering for the
needs of civil aviation as well. In March of the same year Northrop won
a $1,500 million contract for support services and training.

Feisal had no desire to be wholly dependent on the U.S. and a
continuing role was found for Britain, despite Sultan's dissatisfaction
with the Lightnings. The original deal negotiated by Edwards in 1965
was superseded in the spring of 1973 by a new government-to-
government agreement. The British Aircraft Corporation became the
main contractor for training and maintenance as well as building work
and the operation of hospitals. About 2,000 instructors, engineers,
fitters, and administrators were involved in work on various projects in
Riyadh, Dhahran, Jeddah, Khamis Mushayt, Tabuk, and Taif. At the
same time, the small British military mission remained with the
National Guard.

More dramatically, Sultan announced in Paris at the end of December
1974 that Saudi Arabia had ordered weapons from the French at a
cost of about 4,000 million francs, or about $900 million. Amongst
them were thirty-eight Mirage V fighter-bombers that the French
Government had undertaken to supply after Feisal's visit to Paris in May
1973. The aircraft were destined for Egypt and some of them had
already been delivered there. Saudi Arabia's purchase of the aircraft on
behalf of Sadat was a means of circumventing President Georges
Pompidou's commitment not to supply arms to any of the parties to the
Middle East conflict (amongst which the Kingdom was not numbered
despite the engagement of its forces on the Golan Heights). Also

included in the deal, but for Saudi Arabia's own use, were tanks, armoured cars, and combat vehicles. Two hundred AMX-30 tanks had, in fact, been ordered in the previous year and the number was subsequently raised to 250. Also bought were 200 AML-60/90 armoured cars, 500 AMX-10P mechanized infantry carriers, and AMX-30SP anti-aircraft guns.

The Saudi Government agreed to finance the development costs and also purchase a more sophisticated and flexible version of the Crotale low-level surface-to-air missile, with a longer range, to be named the Shahine. In addition, France obtained a special oil allocation of 200,000 barrels a day. It was a notable triumph for President Valéry Giscard d'Estaing, who had been elected head of state in May 1974 following the death of Pompidou. The preferential supply agreement helped the Elysée to resist American pressures to join the International Energy Agency. All the other leading industrialized countries joined the institution, which was regarded with suspicion by O.P.E.C. as a vehicle designed to unite them in confrontation with it. The agreement was an extension and a very considerable enlargement of collaboration in the military field dating back to the purchase of 220 Panhard armoured cars at a price inflated by built-in commissions said to amount to as much as 25 per cent of the total cost. For members of the Saudi ruling hierarchy, their princely protégés and the middle-men who served them, West European suppliers were attractive. They did not raise ethical objections of the kind that surfaced suddenly in the U.S.

The hearings of the Senate Foreign Relations Subcommittee on Multinational Corporations,[5] together with the Northrop and Lockheed papers released to it, revealed in 1975 a fragment of the truth about the award of public contracts in the Kingdom. The findings focused on Adnan Khashoggi's remarkable feat in gaining a near monopoly in the handling of military sales as a result of his relationship with Sultan, but gave only a hint of how the country's Exchequer paid inflated prices to profit a limited number of its sons. Testimony and evidence drew attention to the large fees paid to agents, other intermediaries, and officials, but did not cast any light upon the practice of padding contracts with large sums that would accrue to those with the power to decide whether or not a project should be given the go-ahead. Hidden commissions could easily be disguised or absorbed within a large package, described as 'promotion charges' or such like. The beneficiaries were senior princes who regarded the state's wealth as belonging to their family anyway. By the early 1970s, some who had been in key positions for many years, like Sultan, had either grown extremely rich or could more or less draw money directly from the Ministry of Finance. Others of second degree had to peddle their influence more surreptitiously to divert public funds to their own bank accounts.

By hard graft and dedicated application Khashoggi, the son of one of the first Saudi doctors to qualify who became a personal physician to King Abdul Aziz, managed to make himself indispensable to those wishing for success in selling to the Ministry of Defence, at least. In the early 1960s, however, he showed little of the flamboyance of the commercial empire-builder of a decade later. In the foyer of the Yamama Hotel in Riyadh he would seek out Western businessmen and diplomats in his dogged search for new agencies. He would invite them home to his modest flat in the round building at the end of Airport Road. Acquaintances remember him when young as already rotund and somewhat unprepossessing, but an agreeable host who, untypically of Saudis, served fine wines. He hardly seemed like a man who would achieve international renown and notoriety.

Undoubtedly, however, the key to his outstanding success was his close friendship with Sultan and the Minister of Defence's enormous power to bestow or withhold. Khashoggi's critical move towards armaments is said to have been winning the contract for the running of the Dhahran air base on behalf of Commonwealth Services, an American concern. It led to a moderate $2,000 per month retainer from Lockheed in 1964 to investigate market opportunities. In the following year the company had appointed him its fully-fledged agent as it competed with its F-104 Starfighter to win a part of the air defence deal. Given Khashoggi's vital connection with the Minister of Defence and the fact that the Pentagon recommended this aircraft at the time, it is surprising that Lockheed did not clinch the contract. Yet the order for Hercules aircraft and the 1971 agreement were enough to earn Khashoggi $106 million in 'agent's fees' from 1970 to 1975 (before the deal for the air traffic control system), according to the company.

Khashoggi took over responsibility for Raytheon's business in the Kingdom in 1967 through purchasing its local marketing company. The Northrop representation came to him in 1970. M. S. Gonzales, a Vice-President of Northrop, said that Sultan recommended him indirectly through Kermit Roosevelt, the former C.I.A. agent largely responsible for bringing down Mossadeq in Iran, who had become a consultant to the aircraft manufacturer. Khashoggi was to claim that Lockheed had put his name forward. But no one has denied the intimacy of their relationship. Khashoggi was also the agent for Sofma, the French arms export agency, and received a $45 million commission for the sale to Saudi Arabia of the AMX-30 tanks, 7.5 per cent of the $600 million they cost. He also acknowledged receiving $5.8 million from a British helicopter manufacturer and $4.5 million from a Belgian company in respect of ordinance.

Feisal was content to allow Sultan a free hand within the domain of the Ministry of Defence and Aviation. But he had come to resent Khashoggi's role as an intermediary for the bulk of Saudi arms contracts. The philosophy of the royal family, to which Feisal would have subscribed, was that they should be distributed more evenly. The King also disapproved of the agent on account of his increasingly conspicuous lifestyle.

Khashoggi's influence evidently did not extend to the National Guard, the preserve of Abdullah, who would have disapproved of his lifestyle. When the U.S. Government was approached with the request for assistance with the reorganization and training of the security force, it informed Riyadh that the work would have to be done by private contractors and that the only one it could fully guarantee was Raytheon. The official agent named in its bid for the contract was Abdullah al Feisal, the King's eldest son, who may have received up to $75 million for his services. In September 1975, Jim Hoagland of the *Washington Post* reported: 'Word of a business arrangement between Khashoggi and Prince Abdullah Faisal reached the King, according to a well informed source. Reacting in anger, the King and his brother rejected Raytheon.'[6] As a die-hard conservative, Abdullah had probably acquired an even more acute distaste for Khashoggi than the King himself. The contract was awarded instead to the Vinnell Corporation. The intermediaries who profited from it were believed to be Kamal Adham and Ghassan Shaker, a Saudi-born banker of cosmopolitan disposition who had been a contemporary of Khashoggi's at Victoria College, Alexandria, the 'Egyptian Harrow', run on British public school lines where many prominent Saudis were educated, including Hisham Nazer who rose to be head boy there. King Hussein was another alumnus of the institution, which was subsequently nationalized by Nasser and named after him. Shaker had also developed close links with Sultan Qaboos of Oman. He became a part of the Sultan's tight-knit coterie and was greatly enriched in the process.

Adham was preferentially placed. In 1965, when the quadrilateral struggle for the aircraft order was reaching its climax, Kermit Roosevelt sent a cable to Northrop in which he referred to a meeting with the King's brother-in-law 'who, as I understand it, already has a piece of the Lightning deal, the Mirage deal and the Lockheed deal and is trying to complete the square by an arrangement with Northrop'. He suggested that 'without him we are going to be weakly represented'.[7] The sophisticated Adham, who maintained a residence in Paris, had good French connections. He had also developed a close working relationship with Ashraf Marwan, Nasser's son-in-law, who was his regular political liaison in Cairo and had become a business associate, and was a favourite of Jihan Sadat. Adham was the main intermediary in the deals whereby Saudi Arabia bought French weapons on behalf of Egypt. The commissions from these sales, including the Mirage aircraft, were understood to have been channelled through his company Arcan. In turn, Feisal's Chief of Intelligence started investing heavily in Egypt as Sadat instituted his economic 'open door' policy. In contrast to Khashoggi, Adham today keeps a discreetly low profile, but his international operations, though less spectacular, are probably as substantial and widespread as Khashoggi's.

The 'Lockheed deal' referred to by Roosevelt in 1965 did not materialize. Clearly, at that point Adham and Khashoggi were not incompatible. More interesting was the clear indication that, if Lockheed had been successful, Adham would have stood to gain, which points to how others could benefit from state contracts quite apart from the agent himself working on a percentage commission. In the absence of any public accounting system it proved easy, with the co-operation of companies supplying goods and services, to pad contract prices – even after they had been signed or sealed.

Khashoggi and Northrop have given differing accounts of what prompted the company to pay $450,000 to General Hashim S. Hashim, Commander of the Royal Saudi Air Force, who retired 'at his own request' in September 1972 and was succeeded by General Zuhair, a Nejdi. Northrop's own documents showed that Khashoggi had recommended that the two senior officers should be paid to ensure conclusion of the first contract for the Northrop F-5E Tigers and facilitate future sales. Khashoggi protested that the company had taken the initiative. He later said: 'I pocketed it myself to save Northrop making a dreadful mistake. I told them that it was not necessary to do this. And I was disappointed and very unhappy that Northrop decided to pay the money.'[8] It was a small amount compared with the

$45 million which Northrop doled out to Khashoggi in the early 1970s or, indeed, the contract sales that ensued. It can only be a matter of speculation what proportion of the payments from Lockheed or Northrop were passed on to others in the form of sweeteners.

In this respect the most revealing evidence that emerged from the Senate subcommittee related to the successful attempts by Khaled bin Abdullah bin Abdul Rahman, son of the revered elder statesman of the royal family and a nephew of Abdul Aziz, to obtain a share of the commission for the first batch of Northrop aircraft. From the papers released by the company's accountants Ernst and Ernst it emerges that the prince, then only twenty-five years old, was to have had a portion of the 12 per cent fee of the deal but was squeezed out when it was reduced to 4 per cent because the transaction was to be made through the Pentagon. Northrop had already had contacts with the young Khaled who, despite his denial, they believed represented Turki, Deputy Minister of Defence and also Sultan's full brother – not the least because Khashoggi said as much. The company was puzzled as to what amount of influence Khaled did wield, but Khashoggi was anxious that he should be rewarded. When the appointed date for concluding the deal came in July 1971 the signing of the contract was postponed by Sultan for a day without explanation. Khaled alighted on the bewildered Northrop team in the evening and told them bluntly: 'If I get nothing, then I will make sure that Adnan gets nothing. If he gets one dollar, then I want 50 cents.'[9] He pointed out meaningfully that he had been in touch with Turki about the matter. On the following morning Sultan did sign, but not before pointing out that such contracts usually took much longer to complete. Later, General Hashim requested Gonzales, one of the Northrop men involved, to settle the 'disagreement with Prince Khaled',[10] who received a slice of the commission. Some $3 million was paid by Northrop to Khaled's Contina Company.

*   *   *

Towards the end of 1974, Dr Anwar Ali, Feisal's trusted adviser who had been appointed Governor of S.A.M.A. fifteen years before, died and was replaced by Abdul Aziz Quraishi who contracted the services of Baring Brothers and White Weld to help with the formulation of loan and investment policies. An important part of their role was to fend off the droves of bankers who poured into Jeddah in greater number than ever from the beginning of 1974. It was a problem also for the Finance Minister Musaid bin Abdul Rahman. To cope with the influx he decided at one point to see them in groups of up to a dozen, much to the distress of the men offering their expertise.

Up until 1974 nearly all the Kingdom's reserves had been held on deposit with a handful of international banks, the most significant of which were Morgan Guaranty and Chase Manhattan. With the growth of the surplus, the number of approved institutions was quickly increased to about thirty from all leading Western countries and Japan. The bulk of Saudi Arabia's greatly increased foreign assets tended to be held short term but, though theoretically liquid, the money could in no way be called volatile. Asked about investment policy towards the end of 1974, Yamani replied: 'We never withdraw what we deposit at short notice. Short-term investment, as far as we are concerned, is the same as long-term investment.' He emphasized the Kingdom's anxiety not to weaken the stability of Western economies. The Saudi Arabian Monetary Agency became more active with its purchases of U.K. and U.S. bills and other prime government and quasi-state bonds. Partly for political reasons, it deposited $1,000 million with the Bank of Tokyo. Unlike Kuwait, it had no interest in equity portfolios, although downstream investment was not ruled out. In the same interview Yamani commented: 'We don't think it represents the answer to what we should do and what you have to do. I think the other type of investment is the answer, whether it is a new investment related to what we have back home or buying into existing enterprises.' From the point of view of maintaining real values, property and land were more attractive propositions, he suggested. But Saudi Arabia never did diversify. Already the official line was that in the long term Saudi Arabia would not be a country generating enormous surpluses and would absorb all its oil revenues. There was little sign of it in 1974.

At the end of that year another unpublicized accord with the U.S. was reached as a result of a visit to Saudi Arabia by Simon. An understanding was arrived at whereby S.A.M.A. would 'purchase new issues of marketable U.S. Treasury obligations with a maturity of one year or more through a special arrangement involving the Federal Reserve Bank of New York as agent',[11] according to a memorandum written in February 1975 by Jack Bennett, an Assistant Secretary to the Treasury and revealed under the Freedom of Information Act. The expectation was that S.A.M.A. would purchase $2,500 million worth over the coming six months, quite apart from any other acquisitions of

U.S. Government paper that it might make, he said. The assumption was that the Saudis would continue to buy short-term Treasury securities as well through normal channels.

The issue was subsequently raised in 1979 by a sub-committee of the House of Representatives Committee on Government Operations chaired by Benjamin Rosenthal, the Republican Congressman from New York who was a leading opponent of the 'special relationship'. Back in private life as a lawyer, Simon testified in March 1979 that 'a separate facility, or add-on, was set up to handle the sales'. Explaining the advantages of the deal to the U.S., the former Treasury Secretary said: 'The fear was that the Saudis would have bought securities from the United States when they felt like it. In order to bring more order to this purchasing system and because the United States wanted the investment, the Saudis were asked to purchase large amounts of securities.' In return the Kingdom was guaranteed confidentiality, although, as Rosenthal's questioning showed, it was doubtful whether any such understanding conformed with the Security and Exchange Commission's rules of disclosure. In his memorandum Bennett had said: 'The principal advantage to the Saudis of the arrangement is that it will avoid the disruption to the market occasioned by large security purchases or sales on their part. It should be emphasized that the purchases are at the auction average. Thus we are giving the Saudis no interest rate advantage.'[12]

The arrangement was well geared to the cautious and secretive attitude of the Saudi Government towards its rapidly accumulating surplus. It was cautious because of the still vivid remembrance of the near bankruptcy of 1957–8, and secretive because of the envy, as well as demands, that its wealth might inspire. For this reason the regime did not even claim credit for its massive financial backing, almost wholly in the form of grants, of the front-line states. From the time of the outbreak of the October War of 1973 until the spring of 1975 it had committed and mainly disbursed about $2,500 million to Egypt, Syria and Jordan.

Characteristically, it responded to the West's call for money to be recycled through international institutions. It contributed $1,000 million to the International Monetary Fund's oil facility designed to assist oil importers to carry the burden of higher prices (the I.M.F. was unprepared, however, for the Kingdom's hard bargaining for a higher rate of interest than was being offered). It lent $750 million to the World Bank. There were other subventions, including $30 million to the U.N. Special Fund. In 1974 Feisal announced the formation of the

Saudi Fund for Development, with a capital the equivalent of $865 million. The Kingdom was one of the main financers of the new Islamic Bank which was headquartered in Jeddah. Saudi Arabia decided finally to participate in the Arab Fund for Economic and Social Development formed in 1968 and it took a share in the Arab Investment Company, in which thirteen other Arab governments were shareholders. Yet its reserves as published by the I.M.F. leapt in the course of 1974 from $3,780 million to $14,280 million, a figure that included the money lent to the I.M.F. and the World Bank.

Of all the members of O.P.E.C., Saudi Arabia could best afford to stand for moderation on oil pricing. But Feisal's approach was based mainly on an objective view of global economic realities. Given the Kingdom's nervous sensitivity, he showed courage in fighting for restraint and Yamani great skill in his manoeuvres. At the same time, Saudi Arabia fully supported other objectives of fellow producers and was, indeed, in the forefront of the concerted drive to achieve them.

Together with other oil producers, the Kingdom had also resolved to slash back the profits of the international oil companies that had swollen as a result of the price increases in the last quarter of 1973, though by no means as much as the producers' revenues. In the twelve months beginning 1 October 1973, those of Exxon, Gulf, Mobil, Socal, and Texaco were up 56 per cent on the previous year. For the five leading American independents the increase was 92 per cent. As early as the O.P.E.C. conference in Geneva in January 1974, members were fully aware of what was happening, and at their meeting in Vienna in March of that year had vowed to take action when the oil ministers convened again at Quito in June. By then, however, the issue had lost relevance as far as Yamani was concerned and also had become embroiled in the debate on the basic price level.

In the rarefied atmosphere 10,000 feet up in the Andes the O.P.E.C. meeting proved to be a breathtaking battle. Yamani was as isolated as he had been at Tehran six months earlier but determined to concede as little as possible. The ministers had before them recommendations from O.P.E.C.'s economic experts that would have raised the basic price for a barrel of oil by $4 from $7 to $11. A small part, yielding about 9 cents, would have come from a straight increase in the posted price. In addition, it was proposed that the tax rate should go up from 55 per cent to 87 per cent of posted prices involving extra revenue of rather more than $3.20 per barrel. Yamani's response was to say that posted prices should be reduced by $2 per barrel. He argued that, out of

nothing else but self-interest, other members could not detach themselves from the burden that the proposed increases would place upon the
consuming countries and the Western economy. Furthermore, there
was no point in changing the tax structure to cut the profits on the
cheap equity crude lifted by the companies because the Kingdom was
negotiating the total takeover of Aramco. If they went ahead with the
increases, then Saudi Arabia would lower its prices and expand its
output by up to 3 million barrels a day. The militants said that they
would bring down their production to support higher revenues,
regardless of what Saudi Arabia did.

Finally, under fierce pressure, Saudi Arabia compromised on a 2 per
cent increase in the royalty rate, from 12.5 to 14.5 per cent. That meant
a nominal effective increase in posted prices of 10–11 cents per barrel, or
1.5 per cent. Yamani had called the bluff of the other producers. For
him and the Kingdom it had been an exhausting and strained meeting.

For a while the Kingdom did not impose the new rate. It intended to
hold an auction of its participation crude in the summer. The sale was
planned to support the thesis that the O.P.E.C. price was not justified
by the state of the market. Simon visited Saudi Arabia in July and
Yamani confirmed to him that the auction would be held in September.
However, the event on which the U.S. Government pinned so much
hope did not take place. Feisal did not feel strong enough to provoke
other members of O.P.E.C., particularly Iran and Iraq, at a time when
the Arab world was growing restive at the failure of Kissinger to follow
up the Syrian disengagement by exacting further Israeli withdrawal.

In Vienna in September Saudi Arabia resisted proposals to raise the
posted price and argued again for a reduction, but agreed to higher tax
and royalty rates as a means of lowering the oil companies' profit
margins. It was a contorted compromise that meant an extra 28–35
cents, or 3.5 per cent, for each barrel. As at Quito, Saudi Arabia
disassociated itself from the new increment, disgruntled that Yamani
had not been given any concession in return for the Kingdom dropping
plans for the auction and not expanding its own output.

On 13 October, Feisal and Kissinger met in Riyadh. In contrast to
Sadat, the King neither liked nor trusted the U.S. Secretary of State.
His misgivings were increased following their talks. In an attempt to
win Feisal's support for his efforts to bring about a second
disengagement agreement in Sinai, Kissinger invited him to present him
with a list of armaments that Saudi Arabia wanted to buy on behalf of
Egypt. Sadat's foremost priority was the Northrop F-5E aircraft. The

Secretary of State told the King that its purchase could be arranged, assuring him that neither the U.S. Congress nor the Foreign Military Sales Act should present any problem. It was a foolish ploy. The Saudis knew enough about American democratic processes and legislation to be sceptical. Nevertheless, the King felt betrayed when Kissinger failed to deliver on what he regarded as a pledge concerning the F-5Es for Egypt.

Tension in the region was rising. On the day before the meeting between Feisal and Kissinger, 7,000 Israelis had demonstrated in front of the office of Golda Meir, the Prime Minister, protesting against any curbs on Jewish settlements in 'Judea and Samaria'. From the radical flanks of Araby there was talk about unsheathing the oil weapon again. It took on a more serious tone when the Arab summit convened at Rabat on 26 October. The role of oil in Arab strategy was discussed. Omar Saqqaf warned that any use of it as a weapon at this stage could have disastrous consequences. The heads of state decided that more effective economic, as well as military and political, co-ordination should be achieved. The oil producers pledged $2,500 million in grant aid for the coming year for Egypt, Syria, Jordan, the Palestine Liberation Organization, and the People's Democratic Republic of Yemen (in return for the Arab League taking a ninety-nine-year lease of Perim Island, strategically placed at the narrow mouth of the Red Sea). The Kingdom's share was put at $400 million and the money was promptly paid up. But the most significant aspect of the summit was the resounding endorsement given to the leadership of the P.L.O. by the conferring on it of responsibility for any territory recovered by war or negotiation, and the rejection of any partial settlement – a term that was understood generally to include any more interim military disengagement agreements.

There were gloomy predictions about Arab-Israeli hostilities breaking out again. Even President Ford recognized the possibility. Israel warned that it would not stand idly by if Arab oil-producing states sent troops to the front line. Saudi Arabia still had brigades in Syria and Jordan and the expectation was that the latter would be drawn fully into a new conflict. Israel stepped up the pressure on the Palestinians in Lebanon with aerial strikes on guerrilla camps in the environs of Beirut and blasted the movement's offices in the city with car bombs in November, fatally exacerbating the internal tensions in the country, which were to break into serious intercommunal violence in the spring of the following year.

In this situation it was becoming more difficult politically for Saudi

Arabia to resist pressure from other members of O.P.E.C. to obtain
better revenue terms. Kissinger visited Riyadh again on 6 November.
On his departure, Omar Saqqaf was reported as saying that Saudi Arabia
would endeavour 'to keep oil prices down' and possibly lower them
slightly. He added that 'if we could lower them more than symbolic-
ally, we would'.[13] Yet within three days Saudi Arabia took an initiative,
after consultations with the United Arab Emirates and Qatar which
followed its lead, that seemed to contradict its stated intention. After a
meeting in Abu Dhabi on 9–10 November, the three producing states
announced a change in the method of assessing the fiscal obligations of
their operating companies. They lowered posted prices but raised both
tax and royalty rates in a complicated manoeuvre designed to limit
profit margins.

   Yamani's aims were to strengthen the position of the state oil
corporations and thereby help the independents which had suffered in
the market from the majors' access to cheaper equity crude; to expand
the producers' direct dealings with consuming countries, thereby
reducing the role of the big companies – the so-called 'Seven Sisters' –
still further; and to prepare the financial basis for the 100 per cent
takeover of Aramco. In the process, Saudi Arabia incorporated in its
price structure the increments decided upon by other members of
O.P.E.C. in Quito in June and Vienna in September, from which it had
originally disassociated itself. The move by the three Gulf producers also
effectively raised average per barrel revenue by 40 cents. From America
there were cries of 'perfidy'. On the face of it, Yamani appeared to be
disingenuous in his talk of the new system lowering the price of oil to
consumers. Apart from building the Quito and Vienna increases into the
new structure, his intention certainly had not been to raise prices. The
simple, somewhat embarrassing, fact was that he and his colleagues
made an arithmetical error in their calculations, according to one well-
informed diplomat. At the next full O.P.E.C. ministerial conference
held in Vienna on 12 December, members adopted and refined the new
tax system agreed upon by Saudi Arabia, the United Arab Emirates and
Qatar, so that the posted price, though it remained in being, ceased to
have much relevance. Yamani duly negotiated and reached agreement
on a 100 per cent takeover of Aramco.

   Various revenue gains achieved in 1974, including the 'hidden
escalation' resulting from the imposition of 60 per cent participation,
raised the average price of a barrel of oil by about one-quarter over the
level set in Tehran at the end of the previous year, despite the

Kingdom's policy of restraint. The limits to what it could do alone to protect oil consumers had been made abundantly clear. They were set in no small measure by Kissinger's failure to make further progress towards a comprehensive settlement of the Arab-Israeli conflict. The point was put forcibly by Senator John W. Fulbright, the Chairman of the U.S. Senate Committee on Foreign Relations, in a speech entitled 'The Clear and Present Danger' given on 2 November at Westminster College, Fulton, Missouri. 'The Saudis are caught in a dilemma,' he said. 'It is exceedingly difficult, if not impossible, for them to accommodate the United States while the United States provides the money and arms which enable Israel to occupy Arab lands. Further – and this is the heart of the matter – King Feisal feels a special responsibility – indeed a stewardship – for the holy places of Islam.'[14]

By the turn of the year Kissinger recognized publicly the limits of the power and independence of Saudi Arabia in the sphere of oil politics, notwithstanding its predominance as a producer within O.P.E.C., even if he did not seem properly to appreciate how the diplomatic impasse in the Middle East inhibited it. In an interview published in mid-January 1975 the Secretary of State said: 'The Saudi Government has performed the enormously skilful act of surviving in a leadership position in an increasingly radical Arab world. It is doing that by carefully balancing itself among the various factions and acting as a resultant of a relation of forces, and never getting too far out ahead. Thereafter, I never for a moment believed, nor do I believe today, that the lead in cutting prices will be taken by Saudi Arabia. On the other hand, the Saudis will happily support a cut in prices proposed by others.'[15]

At the same time Kissinger heightened the growing tension in the Arab world by suggesting in the same interview that the U.S. would contemplate using military force if faced with 'some actual strangulation of the industrialized world'[16] as a result of oil supplies being withheld. Speaking at a press conference in Washington on 14 January, James Schlesinger, Secretary of State for Defense, declined to define what circumstances would justify military intervention, but asserted that armed intervention would be possible.[17] Their statements provoked a furore in the Arab world. Saudi Arabia was the only member of O.P.E.C. not to react officially. But units of the National Guard were ordered to take up positions around the oil fields.

Thus, the atmosphere was fraught when Feisal had his last meeting with Kissinger in mid-February. The F-5Es which Saudi Arabia wanted to purchase would not be forthcoming, it was now clear. The King regarded with concern Kissinger's renewed attempt to negotiate a second military disengagement agreement in Sinai. On the one hand, Feisal understood the meaning placed upon the resolution of the Rabat summit rejecting partial solutions to the Arab-Israeli conflict and feared the Egyptian-Syrian rift that might result from one. On the other hand, the King understood Sadat's political need to recover more occupied territory and therefore was generally in favour of another interim accord. Feisal grimaced as the Secretary of State sought to explain away his failure to achieve one. Kissinger complained that he had been a victim of the Zionist lobby. It was hardly an excuse calculated to enhance his prestige in the eyes of the aged King.

Serious moves were now being made to establish a framework for the dialogue between industrialized and developing countries first suggested by Yamani in April 1974. The necessary momentum was given, through intensive diplomatic consultations, by France which sought to win credit from the oil producers. It also took a more flexible line than the U.S. and other industrialized countries, which wanted any exchanges to be limited to the oil producers and confined as much as possible to questions relating directly to energy. Thus, at a press conference in Paris on 21 October, Giscard d'Estaing proposed the convening of an international economic conference. Then on 1 November, President Houari Boumedienne of Algeria called for an O.P.E.C. summit so that members could agree upon and proclaim a common position on questions relating to the Third World as a whole. For the oil producers much of the impetus came from what was perceived as an aggressive attitude of confrontation on the part of the industrialized countries, particularly the U.S., in the face of O.P.E.C.'s demands.

Saudi Arabia vied with Algeria in espousing the cause. A prominent role for the Kingdom in the fight for a better economic deal not only proved a good defence against allegations about it being too sympathetic to the West. Saudi Arabia also had its own special concerns about guarantees for the safety of, and the return from, accumulated financial surplus assets invested abroad, and the reform of the international monetary system, including active participation in its management by developing countries – most specifically itself.

Towards the end of January 1975 oil, finance and foreign ministers of O.P.E.C. member states met in Vienna to draw up an agenda for a summit. Though a protagonist of the proposed dialogue, Feisal at first opposed the idea of a gathering of O.P.E.C. heads of state because he feared the radicals would dominate one and sober calculation would be overwhelmed by rhetoric. Impressed, however, by the reasonable nature of a working paper presented by Algeria at the tri-ministerial meeting he agreed to the convening of a summit. Included amongst the proposals in the Algerian document was one, favoured by Saudi Arabia, stating that O.P.E.C. should give assurances about satisfying consuming countries' requirements of oil. But because the event coincided with the visit of President Sergio Leone of Italy to Riyadh, Feisal did not attend the first O.P.E.C. summit that took place in Algiers from 4 to 6 March.

Its prime purpose was to formulate a Solemn Declaration that would set out the collective objectives that O.P.E.C. members wished to achieve in a dialogue. As the meeting debated the content of the policy statement, Fahd and Yamani were confronted by demands, most vigorously aired by Algeria, for an oil production programme. Fahd resisted them, arguing that the slack in the oil market apparent from the summer of 1974 was a response to high prices rather than the result of a co-ordinated attempt to curb consumption. He emphasized at the same time the obligation felt by the Kingdom to satisfy the needs of its customers and refused to entertain any suggestion that other members of O.P.E.C. should have any control over its level of output. In line with its belief that the price of oil was too high, it rejected proposals for a system of indexation. It had the backing of other members of O.P.E.C. in dismissing the plan put forward by Boumedienne for a fund with a capital of $10–15,000 million to provide aid for developing countries. In drafting the section that related to threats of 'confrontation' by the industrialized countries, Saudi Arabia also withstood pressures from Iraq and Libya for a specific mention of the U.S. in this context.

The document that emerged was a moderate one. It gave an assurance about members' willingness to meet the 'essential needs' of the industrialized countries and to negotiate the stabilization of oil prices. But the offer was made uncompromisingly dependent on the readiness of the industrialized rich to assist with the development of the Third World. O.P.E.C. gave notice that the discussion of energy questions should be made conditional on equal precedence being given to the whole range of raw materials and the problems of the Third World generally. The main objective of the Solemn Declaration was the formation of a binding alliance between the producers and developing countries. The scene was set for the preparatory talks in Paris in April

1975 to discuss the format of a full-scale dialogue bringing together the industrialized countries, the oil producers and the developing countries.

Saudi Arabia was at the centre of the world stage, deporting itself with dignity and aplomb as never before, but also still with some ill-disguised uncertainty. In its own immediate environment the Kingdom remained anxious about subversive movements and radical threats, but Feisal was still hesitant and indecisive in dealing with them, despite his greatly increased power.

The People's Democratic Republic of Yemen on the Kingdom's south-east border remained an anxious preoccupation as its links with Moscow grew closer. Early in 1973, the P.D.R.Y. and the Yemen Arab Republic to the north agreed to merge following a period of border conflict. Predictably, the union was never consummated. Nevertheless, the Saudi hierarchy had to take any such development seriously because it remained deeply apprehensive about a fulfilment of Yemeni aspirations towards unity. More disconcerting was the Iraqi seizure of a Kuwaiti border post on 20 March 1973 and Baghdad's assertion of a claim to the islands of Bubiyan and Warba. Nothing could have been better calculated to make the Saudi royal court shudder. It was a measure of an almost paranoid suspicion felt about the Baathists of Baghdad that Iraqi pilots were believed to have manned the P.D.R.Y. MiG 17s that violated Saudi air space two days later. The Iraqi-Kuwaiti crisis was fairly rapidly defused, though the dormant tension remained. For tactical or other reasons, by the beginning of 1975 Baghdad had begun to show itself less interested in spreading subversion in the Gulf. Then, at the end of Feisal's reign, an unexpected event occurred that made possible much better relations: on 5 March, during the O.P.E.C. summit in Algiers, Iran and Iraq became reconciled.

The most positive achievement in the Arabian peninsula was the agreement reached in the summer of 1974 with the United Arab Emirates, or more precisely with Abu Dhabi, on the common frontier with that state. It had generally been assumed that because of deep-seated dynastic animosities there would be no solution to the problem in the lifetime of Feisal. Fahd's influence was crucial and it was he who, together with Sheikh Zaid, initialled a rough map in Abu Dhabi on 29 July. In so far as it was a compromise, nearly all the concessions were made by Zaid. It was his realization of the expansion of Saudi power and his willingness to surrender considerable tracts of land in consequence that made possible the agreement.

In outline it was not unlike the one that Sir William Luce had proposed four years earlier. Sheikh Zaid ceded the corridor to the Gulf, giving Saudi Arabia a stretch of its coast to the east of Qatar known as the Khor al Bdaid, and to the south a long slice of what he had considered his territory, including the Zarrara oil field. Under the accord, the Liwa settlements and Buraimi Oasis remained a part of Abu Dhabi. The new line was roughly drawn and it was left to a technical committee to define more precisely. It transpired that the two maps were not identical. That did not really matter given the resolve of both sides to settle the dispute. Later, however, it was discovered that Abu Dhabi had surrendered to Saudi Arabia a part of what Oman considered to be its territory. Henceforth, Saudi Arabia would regard the United Arab Emirates and also Qatar as being within its sphere of influence, an assumption that Kuwait did not happily accept.

Apprehensive about the Shah's ambitions and the chances of him intervening on the southern littoral of the Gulf – an act that could only humiliate the Kingdom – Saudi Arabia looked ambiguously at Iranian military support for the Sultan of Oman's Armed Forces in their struggle against the guerrillas of the Popular Front for the Liberation of the Occupied Arab Gulf. Feisal and his brothers did not deny Sultan Qaboos's right to call for help, but it was galling that it should come from a non-Arab power seeking hegemony over the region. Riyadh was relieved when the British-officered Omani troops and the Iranian expeditionary force scored successes against the communist dissidents and insurgents in 1974 and the early part of 1975. The Saudi regime was wary, however, when the Shah proposed a Gulf co-operation agreement. Undoubtedly, the King preferred British military support for Sultan Qaboos to a conspicuous Iranian presence. Saudi Arabian confusion over political collaboration with Iran, whose relations with O.P.E.C. had become very strained, was reflected in Sultan's comment on 29 April 1974: 'We are happy that Iran is strengthening itself in order to safeguard its security, but we do not believe that there is a vacuum in the Gulf nor do we recognize any country's right to impose its domination. Security in the Gulf is primarily the responsibility of the Arab nation and the Gulf peoples.'[18]

Feisal was at the height of his stature as a leader. To a great extent he had been exalted to it by Saudi Arabia's good fortune in finding itself the repository of a large part of the oil reserves available to an energy-hungry world. As the financial wealth of the Kingdom had grown, so too had the King's influence. Yet he managed to keep his head above the currents of world affairs that for the most part seemed to flow in favour of a political entity that was still far less than a nation or a state. Great riches faced his realm with proportionately large responsibilities and difficult decisions, particularly in the wider pan-Arab domain. Feisal proved equal to both. In his latter years his outstanding achievement was to keep in the Arab mainstream while preserving a close, if sometimes strained, relationship with the U.S., and as chief custodian of the oil weapon handling it with a statesman-like care despite the radical pressures on him.

Many members of the royal family and Saudi citizens, with wealth thrust upon the country, saw themselves as peculiarly blessed by Allah – almost a chosen people. This helped to bind together a variegated, heterogeneous society but also fortified the natural arrogance of a people who assumed that as inhabitants of the birth place of Islam they were somehow special. Though fully aware of his position of Guardian of the Holy Places, Feisal preserved a simple and modest detachment in keeping with the virtues of his religion. In February 1974 he had turned down the proposal of the second Islamic summit meeting held in Lahore that he should be called the 'Commander of the Faithful'. Fundamentally, he believed that policy should be based on right rather than political expedient.

After eleven years on the throne, Feisal was as much as ever an absolute monarch even if his power was conditioned by the respect paid to the *sharia* law and the traditional forms of direct democracy which allowed citizens the right of access or appeal to the monarchy. He was a feudal patriarch who would respond to the bedouin cry, 'Ya, Feisal!' When questioned about the pace of change in the Kingdom, he replied: 'Revolutions can come from thrones as well as from conspirators' cellars.'[19] It was incumbent on his regime to disburse oil revenues and spread wealth as fast as the economy could accommodate them – if only for the sake of its survival. Feisal was committed to the expansion and improvement of the educational system, health services and social welfare as well as the build-up of its infrastructure. But the slowness of the decision-making and his own reluctance to delegate would have ensured that development was a measured, even halting, affair even if the surge of income had occurred earlier. For him, stability was the first priority. There was sound common sense in his comment, 'We cannot make miracles overnight.'[20]

By 1975 he looked old beyond his generally accepted age of 69–70. He was also saddened by the death of Anwar Ali on 5 November 1974 and Omar Saqqaf on 14 November (from a heart attack suffered in

New York), who, together with Rashad Pharaoun and Yamani, were the non-royal advisers closest to him. At this point the health of Pharaoun and Yamani was poor as well. The King was feeling the strains of the past dozen years. He found it increasingly hard to concentrate in the afternoons and evenings. The old authoritarian was slowing up. But his disposition towards dealing with as many affairs as possible himself was in no way weakened. As late as 1975 applications for visas from Jews and journalists were still being referred to him. At the same time, the increase and weight of governmental business meant that the Council of Ministers had asserted a great deal more independence. The King's veto would be respected, but it was less certain that a positive order would be quickly implemented. Fahd was assuming a leadership of a technocratic nature, notwithstanding his bouts of indolence and frequent absences abroad that were often prolonged. There was speculation that Crown Prince Khaled might yet decline the throne and that Fahd might succeed Feisal.

The King believed that death was preordained to the exact second of the day appointed by Allah. That partly explained his contempt for security arrangements in general. He was irritated by guards whom he looked upon as an unnecessary encumbrance. They were, too, at variance with the ease of access that was an essential part of the Saudi system of government, indeed the basis of the claim that it was in some way 'democratic'. The trappings of safety around his person were therefore minimal compared with those protecting most heads of state.

It was always incumbent on the King to grant an audience to a member of the royal family seeking one. But on the morning of 25 March, which happened to be the anniversary of the Prophet's birth, he was unaware that Feisal bin Musaid, a nephew whose father was the fifteenth-born son of Ibn Saud, had gained entry to the Al Raisa ('Headquarters of the Chief') where Feisal was receiving visitors. If he had been, it is unlikely that he would have been pleased to see the 28-year-old prince. Feisal bin Musaid had pleaded guilty to a charge of conspiracy to sell the drug L.S.D. when he was a student at the University of Colorado in 1970. The King had refused to intervene on his nephew's behalf, according to one of his courtiers. The prince was released on probation and continued his sojourn on the campus until 1971. On his return home, Feisal insisted that he stay at home. His attitude may not have been unrelated to the incident in 1965 when television was introduced to the Kingdom. It was his brother, Khaled bin Musaid, who had been shot dead by a policeman after leading the abortive attempt to destroy the new transmitter in Riyadh. Musaid, the father, claimed subsequently that the King had instigated the killing and he remained resentful.

At Colorado University, Feisal bin Musaid was said to have absorbed some half-baked radical ideas from his American girlfriend and he had been exposed to decadent influences also in Beirut, where he had been heard to express his intention of killing the King. He had also been close to the family of the deposed Saud. His father's palace had been adjacent to the former King's at Nasariya and as a child he was said to have become closely acquainted with the children of the prodigal monarch. Recently the errant prince had become engaged to one of Saud's daughters, Sitta bint Saud. Only the previous week his father and mother had flown to Beirut to make final preparations for the marriage. There was another dynastic feature of the young prince to which importance would later be attached. His mother was from the Al Rashid clan which Ibn Saud had fought for possession of Nejd and Hasa in the early years of the century.

Feisal bin Musaid knew that Abdul Mutaleb Kazimi, the Kuwaiti Minister of Oil, was about to be presented at court by Yamani. The prince and Kazimi had known each other as students at Colorado University. Feisal bin Musaid pushed himself into an embrace with Kazimi. Ahmed Abdul Wahhab, Chief of Royal Protocol, did not recognize him. On learning the young prince's name, Abdul Wahhab went to consult the King. But the doors were prematurely flung open for Kazimi. Thus, Feisal bin Musaid was able to follow the Kuwaiti Minister into the audience room, obscured from the King's view by Kazimi's bulky frame. As the Kuwaiti Oil Minister was being introduced by his Saudi counterpart, Feisal bin Musaid peered round his shoulder and fired three shots from a .38 calibre pistol. The first and fatal one struck the King in the throat, the others only grazing his head and shattering an ear. The Chief of Royal Protocol lunged at the Kuwaiti Minister, pushing him back. He then saw the miscreant and wrestled him onto a sofa where he was overpowered by security guards. The dying King was rushed to the Central Hospital in Riyadh but the blood transfusions and heart massage given him there did not revive him.

News of Feisal's death spread through the capital and beyond to Jeddah like wildfire before the official announcement of his death was broadcast at 11.12 G.M.T. There was shock throughout Saudi Arabia and instant alarm abroad over the stability of the Kingdom on which the West had become so dependent. The senior princes acted quickly in obtaining confirmation of approval from the *ulema* for Khaled to succeed as King and Fahd to be nominated Crown Prince. The proclamation naming them as such was made at 12.20 G.M.T., little more than three hours after the assassination. The Kingdom relapsed into a tomb-like silence, but there was no outward show of grief.

Feisal was buried just over twenty-four hours later, a little past the time for the internment of a Muslim but in time for the arrival of foreign dignitaries. Never had Riyadh seen such a gathering. The convergence on the Saudi capital was in contrast to the hurried and informal laying to rest of Ibn Saud by Feisal in 1953. That perhaps was a measure of the stature gained by the Kingdom by virtue of its oil resources and Feisal's authoritative influence in pan-Arab and Islamic affairs. The Arab heads of state present symbolized the achievement of the King in establishing Saudi Arabia in a secure position in the diversified and divergent Arab family of states. Prominent amongst them were Anwar Sadat of Egypt, Houari Boumedienne of Algeria, King Hussein of Jordan, Ahmed Hassan al Bakr of Iraq, and Jaafar Nimairi of Sudan. Muhammad Daoud of Afghanistan came, while Pakistan was represented by Zulfiqar Bhutto, its Premier. Others who drew attention were Prince Juan Carlos, shortly to assume the Spanish throne, and Idi Amin of Uganda, who was wearing his Field Marshal's uniform but with a Black Watch kilt, accompanied by one of his sons, also dressed in a kilt. Muammer Qaddafi of Libya was notable for his absence. The notice was evidently too short for any Indonesian representative to arrive in time, an absence that caused offence to the Saudi regime and something of a diplomatic frisson.

Fahd and Abdullah, the latter implicitly acknowledged as third man in the hierarchy, led the procession following the bier from the Al Raisa to a small mosque nearby around which a crowd estimated to number 100,000 had gathered. Muslim funerals are not recited in the graveyard itself, which is reckoned to be too polluted a place for so sacred a ceremony. Representatives and ambassadors of Christian countries were permitted to enter the mosque for the traditional prayers without taking off their shoes. Ending the short ceremony the congregation intoned, 'It is the decree of God.' As chief member of the family, Khaled replied: 'I am pleased with the will of God.' The body was taken in a white ambulance to the stony and bleak cemetery outside Riyadh where Feisal was laid to rest with his head to the north and feet to the south, his face turned towards Mecca. Amin rushed forward to the grave, threw himself to the ground, seemingly in a paroxysm of grief, a display which was embarrassing to stoical Wahhabite onlookers.

Feisal's unmarked grave lies near that of Ibn Saud. Afterwards Khaled and Fahd received men of religion and tribal elders, ministers, military officers, and senior civil servants. Eventually men of the National Guard and the police were unable to hold back the throng anxious to offer their condolences and show their respect. The new King and Crown Prince were almost engulfed. It was a show of love and loyalty that impressed even foreign observers. As one young prince of the second generation told me: 'We had a hundred nervous breakdowns. For a week everyone went to bed at eight o'clock in the evening.'

In a statement read on Khaled's behalf by Fahd and broadcast on 1 April, the King declared with the fatalism of his religion: 'It was the will of God that the world as a whole and the Arabic and Islamic peoples in particular should lose a leader who believed in the Lord and strove for his country, who defended the principles of right and justice, and who worked for international cooperation and world peace.'[21] Nevertheless, the assassination led to speculation, some of it mischievously inspired, about international conspiracies. Arab attention focused on the C.I.A. – inevitably, given the general preoccupation with the Agency. It could be and was tortuously argued that it was in the interests of the U.S. to remove a King who was so intransigent on the question of Jerusalem and replace him with a more amenable leadership. Such conjecture could only be reinforced by language such as the *Washington Post* used in its editorial commenting on Feisal's death. It said: 'Feisal probably did more damage to the West than any other single man since Adolf Hitler.' The fact that many Westerners referred to him as a 'friend' or moderate was a 'revealing measure of his power'.[22] No credence, however, was given by the Saudis to widespread rumours that the C.I.A. had been at work on Kissinger's instructions following his last frosty interview with the late King.

A more plausible thesis was that Feisal was a target of international communism. No one could have any illusions that Moscow was pleased to see him depart from the world stage. A senior Soviet diplomat, a K.G.B. officer, was quoted as saying: 'Is it not remarkable that John Kennedy was killed by a "madman" two years after brutally removing Soviet units from Cuba; and that King Faysal was killed by a "madman" two years after having been instrumental in the equally brutal removal of Soviet units from Egypt.' He referred to the 'Hamlet syndrome'.[23] It was far more likely, if conspiracy there was, that Feisal would have been a victim of Arab radicalism. It was feared that the young prince might have had contacts with the Libyan regime or Palestinian extremists.

The initial announcement of the King's death had described Feisal bin Musaid as 'mentally deranged'. A few days later, Khaled, in an interview with the *Daily Star* of Beirut, said that the killing had been 'an isolated act by a deranged person without any foreign scheming'. Doctors in Beirut were quoted as saying that the prince had been under treatment there for drug addiction and alcoholism. Under the *sharia* code, if insanity was proved, then the assassin would be spared the death penalty.

Feisal bin Musaid was agitated and nervous during the first two weeks of interrogation by the official investigators. They did not tell him of Feisal's death and left him with the impression that he was still alive, recovering from his wounds. Questioning elicited little from the prince other than a confession, a contempt for all his uncle had stood for, not the least the Islamic faith, and also an implacable resolve to kill him. Then about a fortnight after the assassination he was shown a film of the funeral. On seeing it he became completely serene and at peace with himself. The Foreign Liaison Bureau believed that any chance of extracting information about a plot – if there had been one – was lost thereafter. Eventually it was concluded that the prince was guilty of 'wilful and premeditated murder'. His beheading in Riyadh on 18 June, eighty-five days after the murder, was witnessed by a crowd of 20,000. The formal announcement spoke of his rejection of 'True Faith' and opposition to it 'because it encouraged passivity and hindered progress'. For good measure the Saudi Press Agency added the following day that the investigation had shown 'no external motive for the crime'. Within the royal family and on its fringes, meanwhile, there was muttering about 'bad blood', a reference to Feisal bin Musaid's maternal ancestry.

22 Mammon Rampant

1975–7

'My Kingdom will survive only in so far as it remains a country of difficult access, where the foreigner will have no other aim with his task fulfilled, but to get out.' [1] – IBN SAUD

'MY God, without the House of Saud, what would happen to this country?' said one minister – a commoner – soon after Feisal's death. With Khaled safely on the throne after a smooth, undisputed succession, the vast clan looked a force for cohesion, the cement for a political entity that, otherwise, would have been very fragmented.

The understanding reached in the autumn of 1963 when, some months before the final deposition of Saud, Muhammad had waived his claim and the consensus decreed Khaled should be heir apparent to Feisal, was respected. In the event, Ibn Saud's eldest surviving son was consulted again in conformity with family practice. The time that elapsed between the death of Feisal and the announcement confirming the succession is reliably said to have been due to having to fetch Muhammad from the desert near Riyadh where he happened to be.

On the face of it, though, there seemed some justification for questioning whether a consensus formed nearly a dozen years ago was still valid. The leader of the tribe had to be the man best qualified to lead. In this respect there were obvious doubts about Khaled, though his mind was by no means as limited as many foreign observers supposed. But there appeared to be a danger that, once he had mounted the throne, he would be as vulnerable as Saud to intrigue. Moreover, his health was poor in the extreme. His heart was weak and his expectancy of life seemed to be short.

Khaled had travelled abroad, visiting Germany, Britain and the U.S. as a young man. In 1938 he had dined with Hitler on the night that

Czechoslovakia ceded his claim to the Sudetenland. Khaled joined in the toast and was evidently impressed by his host. Nearly six years after his accession, when Saudi Arabia was seeking to purchase Leopard tanks from Bonn, he surprised a foreign diplomat by saying he believed the Führer to be a maligned man. It could be said that Khaled's tact and judgement in matters relating to foreign affairs seemed to be lacking. Perhaps he regarded the Sudetenland as the European equivalent to the Buraimi Oasis.

Viewed as even more conservative than Abdullah, Khaled did not seem – to all appearances – the best man to hold the balance between the traditionalists and the more 'progressive' faction led by Fahd. Yet there were sound reasons for choosing Khaled. He commanded affection, as well as a great measure of esteem, and was the prince least likely to become involved in factional disputes. According to most accounts, he was the most agreeable of the senior princes and the best loved. His presence was regal but without overbearing authority. His face reflected a genuine sense of humour uncommon in his solemnly self-important clan. He shared the charm of Muhammad but lacked his violent temper and, since an open heart operation in 1972, his taste for strong alcohol. Khaled's warmth and sympathy to the most humble of supplicants, immediately obvious to any foreigner at his *majlis,* was in contrast to the austere and forbidding demeanour of Feisal. He was more in the style of Saud. Hundreds would come to his weekly *majlis,* held from 10 o'clock to 12 o'clock each Monday. He would entertain up to 1,000 for dinner every night after sunset prayers. Unlike others in the close inner circle and their princely progeny, there was no evidence that he had tried to enrich himself from the public exchequer by muscling in on the award of Government contracts. Nor did he display the haughty arrogance characteristic of a clan that regarded the Kingdom as its own vast estate and its bountiful revenues as the royal family's own income.

Khaled's long absences in the desert with his falcons had also given him one very important qualification. A bedouin at heart, he had developed an instinctive understanding of the tribes, still the essential human fibre of the country, whether settled or leading a nomadic, pastoral existence. In cultivating a close relationship with them, Khaled, together with Abdullah and Muhammad, had performed an essential service for Feisal, who in his preoccupation with central Government and the handling of the Kingdom's foreign affairs, had neglected the travels round his realm that, arguably, were still an important function of a Saudi King.

The conclusive argument in favour of Khaled was the opposition of Abdullah, Muhammad, and influential elders of the royal family to Fahd's succession. Feisal had trusted Fahd's intelligence and ability, despite his half-brother's perfunctory attendance at the *diwan*. That, however, was symptomatic of Fahd's independence, at which other more clannish or sycophantic princes looked askance. He was fond of gambling. This in itself was no great matter, but his huge losses in the South of France in 1974 had been well publicized and this the House of Saud, which valued appearances before all things, could not ignore. Reports of his indiscretions were attributed to Zionist malice. Nevertheless, it was difficult to dispute that by experience, capacity and stature Fahd was the best man to become chief executive under the new King who, it was assumed, would be little more than a figurehead. As early as 1953 Fahd had become the Kingdom's first Minister of Education, responsible for starting the state system. He had presided over the drafting of the First Five-Year Development Plan and since the end of 1973 had taken a close interest in the drawing up of the Second Development Plan, which was about to be unveiled. Though Feisal had kept tight control of foreign policy, Fahd had been more actively involved in it than any other prince. And he was conversant with oil and its wider ramifications. His hooded eyes, which blinked independently of each other, seemed to convey languor and boredom at times, especially when presiding over his *majlis* and coping with petitioners, but his ambition was greater than his desultory manner indicated.

Following Feisal's example, Khaled kept the title of Prime Minister. Fahd became First Deputy Prime Minister and Abdullah Second Deputy Prime Minister. Khaled quickly declared his willingness to delegate. Shortly after the succession he issued a royal decree empowering Fahd 'to handle matters which are addressed to us'. It went on to say: 'We would also like you to take measures, hand down directives, both foreign and domestic, and issue administrative decisions in accordance with policies which are in effect, and after consulting with us.' In practice, the King's endorsement proved to be far more than a formality. Before giving his imprimatur, he would consult with others individually and, more often, collectively. The senior hierarchy, particularly Muhammad and Abdullah, would not allow him to be a simple cypher. When a group of American politicians, who happened to be in Riyadh at the time of the assassination, went to pay their condolences to the senior princes, their leader, Nelson A. Rockefeller, went straight for the Crown Prince who, it was assumed, would be the effective ruler. He was peremptorily reproved for his lack of respect by the Chief of Royal Protocol, Ahmed Abdul Wahhab. 'You are not standing for office here, Governor,' he said pointedly.

From the start, Khaled and Fahd conferred a kind of sanctity upon the course set by the late King. The concern about continuity was illustrated by the custom whereby ministerial and official walls were hung almost invariably from the spring of 1975 onwards with a kind of triptych of portraits. The composition had two standard varieties showing either Feisal, Khaled and Fahd, or Ibn Saud, Khaled and Feisal. To the world at large, however, Fahd soon emerged with greatest prominence. It seemed significant that on 1 April, a week after Feisal's death, Fahd should have delivered a policy statement on Khaled's behalf. In summary it amounted to no more than a commitment to follow the lines in domestic and external affairs pursued with such deliberate caution by the lamented King. It reiterated the Kingdom's commitment to 'restoring the rights of the Palestinian people and liberating the occupied territories'. But no mention was made in the statement about Saudi Arabia's disillusionment with Kissinger's step-by-step approach towards a peace settlement in the Middle East following the U.S. Secretary of State's failure to bring about a second Egyptian-Israeli disengagement. In an interview with the Beirut daily newspaper *Al Anwar* on the same day, however, Fahd recalled how he had told Kissinger on his last visit to Riyadh in February that 'an agreement on one front or a settlement that did not satisfy the Palestinians was unacceptable'. The Crown Prince gave notice of his intention not only to don the mantle of Feisal in running the country's foreign affairs but also to conduct a vigorous policy. Two of the most salient features of his interview were a reference to 'my brother Saddam Hussein' of Iraq and an explicit overture of friendship towards the People's Democratic Republic of Yemen.

Not until October 1975 was a new Council of Ministers appointed and the Government expanded to provide the wider institutional framework demanded for carrying out the Second Development Plan. An infusion of expertise was a prerequisite. In the event, six new departments were created and three Ministers of State nominated to give added strength to a twenty-six-member Council of Ministers. But the changes did not signify any delegation of authority by the royal family. As expected, and entirely in keeping with the Saudi system, all the key, sensitive positions were kept firmly in princely hands. The Ministry of Defence and Aviation remained under Sultan, who had held it for the past thirteen years. Abdullah kept his National Guard. On Fahd's elevation, Naif, formerly his deputy, succeeded him as Minister of the Interior and their full brother, Ahmed, took the second position. The department appeared to have become a fief of the Al Fahd. Saud al Feisal, who had been made Minister of State responsible for external affairs shortly after the succession, was handed the Foreign Affairs portfolio that his father had held for over thirty years.

Two princes were given charge over two of the new departments. Mitab, born in 1928, the twentieth of Ibn Saud's sons, became the first Minister of Public Works and Housing, and Majid, thirty-third son, born in 1937, was appointed Minister of Municipalities and Rural Affairs. Both were obscure figures – whom, it is reliably said, Feisal had held in some contempt. The Central Planning Organization was upgraded to full ministerial status under the ebullient Hisham Nazer. A Ministry of Posts, Telegraphs and Telephones was carved out of the Ministry of Communications and placed under Dr Alawi Kayyal.

In the shuffle five ministers were tranferred to other posts. Apart from Sultan, only two kept their old portfolios and one was Ahmed Zaki Yamani. The belief in circles close to the ruling hierarchy, one apparently shared by the long-serving Minister of Oil himself, was that he would be removed from office – probably with the tactful politeness of being made a Royal Adviser. Yamani's relations with Fahd had varied, but the Crown Prince's attitude to him was generally one of animosity. It was shared by some other princes who were jealous of the dash that Yamani cut on the international scene and the fact that a Hejazi could make such an impact. Khaled and Fahd evidently decided that he was too valuable an asset to lose. But at home Yamani's province was attenuated when, shortly after the formation of the new Council of Ministers, the decision was made to take responsibility for hydrocarbon-based industrial development from him as overlord of Petromin and give it to the new Ministry of Industry and Electricity, where it was put under a newly created state agency, the Saudi Arabian Basic Industries Corporation. That would hold the state's share in the prospective joint ventures with foreign companies. Petromin kept control over the refining of crude oil. The new ministry was placed under the care of Ghazi Qusaibi, a 35-year-old who had been educated at the Universities of Cairo, Southern California, and London where he was awarded his doctorate in international relations.

Two more members of the Al al Sheikh family were brought into the Government, making a total of three. Ibrahim bin Muhammad al Sheikh became Minister of Justice and Abdul Rahman bin Abdul Aziz al Sheikh, who had a Ph.D. from the University of Arizona, was appointed Minister of Agriculture and Water. Sheikh Hassan bin Abdullah al Sheikh, who had been Minister of Education since 1962, was nominated as the first Minister of Higher Education.

Two prominent members of the Sudairi family were widely expected to be given portfolios because of their ability and experience. But neither were because, it was reliably said, the senior princes at the heart of the regime were embarrassed by the extent of Sudairi intrigue and concerned about the resentment within the Kingdom, not the least within the royal family itself. Touchiness about the clan's power was such that at about this time foreign publications were censored if they carried any reference to the 'Sudairi Seven', as the Al Fahd were known.

After thirteen years as Minister of Finance and National Economy, Musaid bin Abdul Rahman gave way to Muhammed Abal Khail, then forty-four years of age. An engaging and unpretentious man of Nejdi origin, he did not boast a doctorate like most of the newcomers, merely a Bachelor of Commerce degree from Cairo University. He had been Director-General of the Public Administration Institute before joining the Ministry of Finance and National Economy in 1964 and becoming Musaid's deputy in 1970. The new Minister of Health, Hussein Juzairi, aged forty-one, was chosen for his specialist qualifications. Having graduated as a surgeon from Cairo University and become a Fellow of the Royal College of Surgeons, London, in 1966, he had worked as Dean of the Faculty of Medicine at Riyadh University since its foundation in 1967. Suleiman Suleim took over at the Ministry of Commerce at the age of thirty-four. He had obtained an M.A. at the University of Southern California and a Ph.D. at Johns Hopkins, both in international relations. Suleim had spent several years in the civil service and then been Professor of Political Science at Riyadh University before being appointed Deputy Minister of Commerce in 1974.

About a quarter of the sixty or so ministers and their deputies had doctorates, a proportion that probably no other country – let alone a developing country – could claim. The vastly expanded programme for expenditure and development greatly enhanced the position of the commoners. One of them could claim a few years later, 'We have the total confidence of the royal family and we can, therefore, influence decisions.' Essentially, however, the system of 'parallel government', under which the weightiest decisions – especially those relating to foreign policy – were made informally by a group of princes, remained in being.

In contrast to official fears expressed in the autumn of 1970, the revenue gains achieved in the three fiscal years ending in July 1973 were sufficient to give the Government a total budgetary surplus of 5,500 million riyals even before the exponential oil price increases at the end of that year. Unspent revenue swelled to 23,000 million riyals in 1973–4 and 65,000 million riyals ($18,413 million) in 1974–5. Much better developed countries could hardly have absorbed such a flood, regardless of any considerations about inflation. As it was, cumulative spending on projects from 1970 to 1975 was about 70 per cent of the total target.

Over the five-year period of the First Development Plan, receipts from oil had risen from rather less than $1 per barrel to nearly $10, while the volume of output rose by 131 per cent. The tenfold rise in prices and the weight that they reflected in the national accounts exaggerated the extent of economic growth. At current prices it was reckoned that, having increased from the equivalent of $4,900 million to $11,300 million over the 1970–3 period, Saudi Arabia's Gross Domestic Product rocketed to $42,200 million by the end of fiscal (mid-summer) 1975. That would have given a notional rise in per capita income from $910 million to $6,806 million. The calculation is based on an estimated population of 6.2 million implied by the Second Development Plan, compared with an official figure of 7 million given after its publication. In translating these calculations into constant – or real – increases, the Minister of Planning introduced a 'deflation factor' to minimize the contribution of oil. On this adjusted basis it was calculated that national income would have grown annually at a rate of about 44 per cent over this period in real terms. Entrepreneurs and their princely allies may have benefited on such or an even greater scale, but the average citizen would not have done so. For the latter, a more relevant calculation perhaps was that non-oil Gross Domestic Product grew by an average 13.3 per cent each year. But as Faiz Badr, a Sorbonne-educated Ph.D. in economics, who was then deputy head of the Central Planning Organization, put it in 1974: 'Growth increases as such mean very little. We are very realistic and apprehensive about our dependence on oil.'[2] The fact was that oil, gas and refining accounted for 87 per cent of total G.D.P. in 1974 while employing only about 3 per cent of the labour force.

The First Five-Year Development Plan had been very much a matter of learning and experiment. The main value of the exercise could be seen, without cynicism, as the collection of data and the more precise identification of obstacles standing in the way of development projects on the scale and in the numbers demanded by the country's sudden wealth. The most critical of them, the shortage of skilled and semi-skilled manpower, had been foreseen in 1970. As it was, only a few targets were met or exceeded, one of them the education of girls, who were allowed only a limited role in the labour force. In 1974, 70 per cent of Saudis over the age of ten were still illiterate. The expansion of the Kingdom's telecommunications system, road network and health services had fallen well behind schedule. The greatest shortcoming was the failure to anticipate the Kingdom's requirements for greatly increased port capacity.

Outside the purview of the Central Planning Organization and jealously guarding his own preserve under the benign eye of Yamani, Abdul Hadi Taher, Governor of Petromin, had nearly fifty projects under review. He had also commissioned a number of feasibility studies, and had signed letters of intent with two oil companies on the construction of fertilizer plants. On the ground, little progress had been made towards economic diversification centred upon hydro-carbon-based industries. Indeed, there could have been little because work had yet to start on the programme for harnessing the Kingdom's natural gas associated with oil production which, for the most part, was still being flared and going to waste.

Most disappointing and serious of all, however, was the per-formance of the agricultural sector, which included the country's declining but still substantial nomadic population. Between 1970 and 1975 growth was reckoned to have been at an annual rate of 3.6 per cent compared with the target of 4.6 per cent. By the summer of 1975, Saudi Arabia had not felt the full impact from the flood of oil revenues. Nevertheless, the infusion had been such as to attract a significant number of people away from the rural areas, especially Jizan and Asir, the most productive of them all, to the rapidly growing conurbations of Jeddah and Riyadh. The makeshift shacks made of metal containers with corrugated iron roofs and the breezeblock houses around the two cities bore witness to the exodus of people in search of more remunerative and less arduous employment. This was in spite of the lavish incentives and grants offered to maximize agricultural output.

Government attempts to stimulate output seemed haphazard, and figures relating to it were more notional than others. The sector defied precise definition and delineation not only because of the difficulties of counting a wandering community, estimated at anything from 15 to 50 per cent of the indigenous population, but because more and more bedouin were adopting a way of life both nomadic and settled, with the balance set by the collective inclination and the exigencies of sustaining a livelihood. Increasingly, the younger men would drift to urban centres in search of lucrative work and then return for a time to their tribes, seeking the comfort of kinship and tradition.

Aramco's early experience had shown that given the right incentive bedouin were happy enough to settle, a process aided by their remarkable adaptability and aptitude for the products of modern technology as well as taste for them. Men of immediate nomadic origin made up over half the established labour force that came into being after the initial stage of expansion and numbered 10,000 in 1974. But for the majority the transition was slower and often painfully uncomfortable, characterized by their state of 'semi-settlement'.

The settlement of the nomadic tribes in the Ikhwan communities, or *Hujar*, marked the first attempt by the state to exercise control over the fluid and amorphous bedouin; treaties and frontier agreements further limited the tribes' freedom of movement. In 1958, after several years of drought, the Government, under Feisal, again attempted to settle the nomads – on farmland retrieved from the desert by drilling new water wells at Sirhan, Haradh and Jabrin. Tribes had suffered by the loss of their livestock, and the projects made good humanitarian sense, but they also furthered what seemed to be a policy of detribalization. In both respects the schemes had mixed results. More significant had been Feisal's 1967–8 decree distributing parcels of land to the bedouin. By creating notions of individual ownership, it further eroded the basis of tribal life to which common rights to the *dira*, or grazing ground, were fundamental.

The bedouin's per capita income was comparatively low, in spite of the fact that by the mid-seventies they supplied about 50 per cent of the Kingdom's growing demand for meat – mainly sheep but also goats. Because of drought and the rapid deterioration of pastureland through over-grazing, Feisal's decree and the establishment of villages centred around wells exploiting underground aquifers lured an increasing proportion of nomads to put down roots of a kind. Other attractions were the few schools, clinics and other services that were provided. But the state did nothing to preserve the traditional way of life that was an essential sub-stratum of Saudi society. To ease the dislocation the Government used the time-honoured, somewhat arbitrary method of distributing occasional largesse, both to allay discontent and to strengthen loyalty on the basis of dependence. The nomadic, tribal community was caught in a limbo between a traditional past and an indeterminate future. The development of the Kingdom had largely passed the bedouin by, and in so far as they had been touched by it, the effect had been disrupting and disturbing.

The nomads were only one part of the economy where official attitudes were fogged in confusion. In attempting to identify where the accumulated billions could best be spent, the planners were hampered by the lack of reliable information on population and manpower. Yet attempts to redress the situation were complicated by the royal family's nervousness about the size of the population.

The first attempt at a census, carried out by the Central Statistics Department in 1962–3, came up with a figure for the indigenous population of 3.3 million – a figure so distressingly low that it was suppressed. A more serious effort was made in 1973 as a basis for the Central Planning Organization's development strategy. It probably covered only 70 per cent of the population, and, once again, the figures were disappointing. The C.P.O. reacted by overcompensating for the uncounted heads and underestimating the foreigners. The unpublished conclusion was that there were 5.9 million Saudis and 790,000 foreigners. The figure adopted as a guideline for the Second Development Plan was 6.2 million. After its publication towards the end of 1975, the Government announced formally that the census had shown a total population of 7,012,642, no more no less. Subsequently, two economists from Durham University, J. S. Birks and C. A. Sinclair, revised the figures on the basis of factors that could be ascertained with some precision, notably school enrolment. It was clear that large numbers of Yemenis, who until 1972 had not needed passports to work in the Kingdom, had been counted as Saudis, along with other Arabs. Even some contemporary advisers to the C.P.O. reckoned the Yemenis to number 500,000 or more. Birks and Sinclair estimated the population in 1974 as 4.3 million Saudis and some 1.5 million foreigners.

The two Durham economists calculated that in 1975 total employment was 1,799,900, of which 1,026,500 were Saudis and 773,400 non-nationals. That would have given the indigenous inhabitants a 57 per cent share of active manpower. But with some 530,700 of them believed to be engaged in agriculture and fisheries (together with 54,900 foreigners), Saudis would have accounted for only 40 per cent of people working in the so-called 'modern sector'. Over half of them, some 277,100, were employed in 'community and social services', in other words, for the most part the civil service, the Armed Forces and National Guard, and other security services.

Saudi Arabia abounded in entrepreneurial flair. The Ph.D. and other graduates emerging from places of higher education in the United States, West Europe, Cairo and Beirut, and the national institutions, made the Kingdom proportionately rich in talent. But defence and security apart, the majority of largely illiterate Saudis drifting in from the country had neither the qualification not the inclination to do anything much more than drive taxis and trucks and become tea-makers or doorkeepers, or find employment in other peripheral occupations that did nothing to reduce dependence on expatriates. The disdain for demeaning manual labour on the part of men of desert origins had become more manifest than ever. It seemed almost intensified by the wealth thrust upon the country and the perceptible feeling that Saudi citizens were God's chosen people. Religious and social taboos meant that the female contribution to the labour force was minimal and probably declining. Except in the south-west, where they continued to be noticeably active, the tendency in other areas was to keep indoors women who formerly had done most of the work in the fields, because the men felt that they should be able to afford to hire labour. The role of women was limited to teaching, health and social services. But their participation in these spheres was restricted to the cities and towns. Women teachers were badly needed in the rural areas but their families would not accept that their daughters could go away anywhere alone.

Against this background of genuine and wilful ignorance, the Kingdom embarked on its Second Development Plan and the disbursement of revenues greater than any other country had enjoyed. When the implications of the rise in oil prices of 1973–4 became clear, the plan under formulation was suddenly revised and, in the words of one expatriate adviser, 'hastily thrown together'. The same person commented: 'The ministers bungled in everything they did with no knowledge of the constraints involved.' Hisham Nazer almost certainly was in favour of a measure of restraint. He was not without influence with the Crown Prince, who had become increasingly interested in planning, but at the same time very much beholden to him to the extent of earning the sobriquet 'Hisham Abdul Fahd' – or 'Slave of Fahd'. Nevertheless, the urbane and charming planning chief, who had entrenched himself in a trusted position through dedication and hard work, strove to achieve the correct balance.

Review boards were set up under Musaid bin Abdul Rahman and attended by the various ministers, advisers and consultants concerned, as well as Hisham Nazer. The latter's role was very much one of trying to co-ordinate and reconcile the demands of different ministers, some of whom were too preoccupied to read properly the experts' reports submitted to them. After hearing the views expressed, it was the Minister of Finance who had the last word – which often was not what the planners hoped for. Faiz Badr, a hard worker and a man of ability and imagination but also of almost disruptive pugnacity, was the one Saudi to make his presence heavily felt amongst Hisham Nazer's stable of Ph.D.s and graduates in the preparation of the 579-page tome that eventually emerged.

Scheduled for implementation from 9 July 1975, the plan submitted to Fahd was an amalgam of aspirations that hardly amounted to a programme for development. In terms of strategy its biggest and most easily identifiable shortcoming was the vagueness relating to the plans of Petromin for a large expansion of the Kingdom's capacity for refining crude oil, constructing the gas-gathering system and erecting petrochemical plants. These were fundamental to the Kingdom's strategy of economic diversification. Plans had been only slowly formulated. They depended very much on feasibility studies and negotiations with foreign companies which, it had been decided, should be brought in as partners on a 50:50 basis as a means of guaranteeing the economic viability of projects involved. An equity stake would ensure the best management, a vested interest in the success of the venture by the companies providing technology, and markets for their produce. It would take almost another five years before the prospective partners could commit themselves, not the least because of the Saudi Government's difficulty in making up its mind as to how many barrels of crude oil for each million dollars invested – the so-called 'entitlement' arrangement – they should be allowed as an incentive to participate in the Kingdom's grand industrialization.

There was, however, another equally important reason why the Second Development Plan could not take proper account of the schemes of Petromin. This was the jealous determination of Abdul Hadi Taher to keep his preserve to himself, out of range of Hisham Nazer's planning department. The two men were known to be great rivals, to the point of being mutually hostile. Taher was immediately answerable to Yamani, who in turn was reputed to be not on the best of terms with Nazer. As it was, it seemed almost deliberately perverse when the Governor of Petromin revealed the agency's five-year programme about a week before the Council of Ministers formally approved the Second Development Plan. He gave far more detail than the long-awaited document. It was Nazer, however, who was mainly instrumental in persuading the ruling hierarchy and the Government to establish the Royal Commission for Jubail and Yanbo, the sites on the Gulf and the Red Sea coast designated for refining centres and the hydrocarbon-based industrial complexes. The decree founding the commission, which was given special powers to by-pass delays inherent in normal bureaucratic routine, was announced in December. Hisham Nazer's aim was to gain overall control of the hydrocarbon-based industries through his seat, as Minister of Planning, on the Royal Commission. This, in itself, was sensible if there was to be co-ordination between the various agencies.

On 21 May 1975, five months before the appointment of the new Council of Ministers, approval of the Second Development Plan was announced and its outline publicized. Expenditure was projected at the breathtaking figure of 498,230 million riyals, the equivalent of $141,000 million. The industrialized world, still recovering from the blow delivered by O.P.E.C. in 1973, gasped at the opportunities for recycling revenue and reducing payments deficits presented by the programme. The sum was nine times the total allocations of the First Development Plan. While Feisal had been open to the charge of moving too slowly, it seemed clear that Fahd in his enthusiasm might go too fast. The impression was that Hisham Nazer, as a known conservationist and if left to himself, might have preferred a lower rate of spending. Scepticism was expressed as to whether the Kingdom could disburse so much. That was understandable given its increasing inability to absorb its mounting revenues over the previous five years. But expression of such a notion caused resentment within Saudi Government circles. Such was the weight of Fahd's backing for the plan by now that total fulfilment became, officially, an article of faith. On the one hand, there was from the domestic point of view the need to modernize and develop the Kingdom as fast as possible, to spend the swollen revenues for the good of the common weal and distribute the riches. On the other hand, Saudi Arabia had to answer its critics and opponents within O.P.E.C., particularly its Arab members, about a high level of output that was, in reality, maintained largely to enforce a measure of price restraint. It was a combination of self-interest and idealism. Cynical Western commentators tended to ignore the genuine concern about the health of the Western industrialized countries. At the same time, as Yamani put it in one interview, 'We know that if your economy falls we fall with you.' It was the kind of frank and honest remark that did not necessarily endear him to some in the princely hierarchy.

In slight contradiction to the obligatory lip-service paid to the letter of the plan, the document itself was realistic and frank in a qualification contained in the preamble: 'It is anticipated that actual expenditure for a variety of reasons will fall short of appropriations. The development plans of individual ministries and agencies are not beyond accomplishment but in combination they represent a formidable task. Bottlenecks and other problems may be expected from time to time and the achievement of many may require extra time.' Nothing was said in the plan about revenue, but projections were based on the unstated and conservative assumption, inevitably arbitrary, that oil production would rise from 5 million barrels a day in 1975–6 to 7 million barrels a day in 1979–80. That compared with an actual rate of 8.48 million in 1974 and 7.07 million in 1975 (including Saudi Arabia's share of production from the Neutral Zone operations). In addition, it was reckoned that receipts from other sources, including investment income, would be in the order of 40,000 million riyals, or the equivalent of about $11,300 million. The budgeting was so prudent that the expenditure forecast could be met without any rise in oil prices over the five-year period and at what would certainly be a minimal level of production. Meanwhile, it was calculated that average annual growth over the 1975–80 period would be 10.2 per cent for the economy as a whole – with 13.3 per cent for the non-oil sector and 9.7 per cent for the petroleum sector. The goal seemed modest enough given the resources likely to be available, but realistic in view of the impediments that would certainly be encountered.

Seven 'fundamental values and principles' guiding Saudi Arabia's development were spelt out. First was the maintenance of the religious and moral values of Islam. Second came the defence and internal security of the Kingdom. Third was placed the maintenance of growth by the development of economic resources and the maximizing of earnings from oil over the longer term while conserving the state's great but depleting asset. Fourth was the reduction in the Kingdom's dependence on the export of crude oil. Fifth in the list was the development of human resources by training, education and health. Sixth was the increase in the well-being of all groups within society and the fostering of social stability. Seventh was the development of infrastructure.

For the development of economic resources, on which expenditure was estimated at 92,135 million riyals, no breakdown was possible because of the confusion over what heavy industry projects would be implemented and how soon. In addition to the hydrocarbon-based projects, the heading included also water resources and electricity. The plan itself put industrial investment at 50,690 million riyals, a sum including the cost of the gas-gathering scheme and pipelines carrying crude oil and natural gas liquids (N.G.L. – propane, butane and natural gasoline) to Yanbo on the Red Sea coast where the plan envisaged the construction of an export refinery and a petrochemical complex. For Jubail on the Gulf coast it spoke more vaguely of 'the production of petrochemicals, the refining of products for export, fertilizer production and the manufacture of steel and aluminium products'. The plan gave a cautionary proviso about the completion of feasibility studies.

Taher was far more specific and ambitious. He talked of an industrial development programme costing about $13,000 million. Of this sum he foresaw expenditure of about $5,000 million on a comprehensive system for harnessing gas amounting to no less than 6,000 million cubic feet daily. Such a volume would be the result of an output of oil of some 12 million barrels a day, the total capacity then under installation by Aramco. The output of N.G.L. promised to enhance immeasurably Saudi Arabia's ambitions to maximize its earnings from oil, giving it about 10 per cent of the market for that family of by-products obtained from associated gas. After the extraction of this richer derivative, the 'dry gas' would be used as fuel and feedstock for the petrochemical and other industrial ventures, to fire the power stations providing electricity for an integrated network covering Hasa and Riyadh, and for Aramco's own operational purposes, not the least re-injection into the oil fields for the maintenance of pressure. Cost escalations eventually made nonsense of Petromin's estimates.

No one could doubt Taher's determination. Whether so much could be achieved over a period of five years was open to question. The foreign companies, in particular, were wary of committing themselves to what amounted to a massive transfer of technology from the industrialized countries. Increasingly, they wanted firm guarantees about access to supplies of Saudi crude oil in return for their investment.

Better grounds for optimism about rapid progress lay in the reliance placed on the private sector. A lively response was expected to the generous credit being extended by the Saudi Industrial Development Fund (S.I.D.F.) which was set up in 1974 to provide cheap long-term finance, to the facilities made available on state-financed industrial estates, and to the prospect of a fast expanding market that the Government was prepared to protect with limited tariffs in order to nurture local industries. In the event even the strict appraisal of requests for loans by the staff seconded from Chase Manhattan failed to inhibit the call upon the Fund's resources by merchants, to the extent that its capital soon had to be raised from 500 million riyals to 3,000 million riyals.

Saudi planners were under no illusions that success in the implementation of the Second Development Plan would in no small measure depend on the ability of the Kingdom's infrastructure to cope with the demands made upon it. Nearly a quarter of total appropriations were directed to it. Of critical importance were the ports of Jeddah and Dammam through which 90 per cent of the goods entering the country passed. With construction reckoned to absorb more than half of anticipated expenditure, the pressure that would be put on limited facilities by imports of building materials and related equipment was obvious. Congestion at the country's ports was building up even before the plan had been published. To cater for increased demand, something like a five-fold increase in capacity was planned. The allocation of 6,925 million riyals ($1,700 million) seemed meagre in terms of the challenge.

Another high priority was the unfinished telecommunications programme of the First Development Plan and then the expansion of direct-dial telephones from 200,000 to 600,000 to give a ration of twenty for each 100 inhabitants in the larger towns and five for every

100 elsewhere. The plan also aimed at the completion of no less than 13,066 kilometres of trunk, secondary, and other paved roads.

Financial provision and Government schemes for housing seemed inadequate given the scale of the shortage that had become fully apparent in 1974. The state's own projects did not extend beyond the construction of 52,000 dwellings for low-income families, a similar number of temporary units to accommodate imported labour, and the preparation of nearly 45,000 fully serviced building lots for families to construct their own homes.

With the prospect – based on those faulty assumptions about population and nationality – of the need to import 500,000 workers, education was a bigger priority than ever. A crash programme was drawn up to nearly double school enrolment from 760,000 to 1.3 million, an expansion requiring no less than 24,000 new classrooms. A new King Feisal University was planned for Dammam. But deterioration in quality seemed threatened by the kind of advance in quantity planned.

For health the Second Development Plan was almost as ambitious as for education, with the state aim, for instance, of increasing availability of hospital beds from 7,734 to 19,234 and doubling the number of dispensaries from 215 to 427. Preparation was at too early a stage to give any real hope of quick implementation. Only belatedly were consultants commissioned to produce designs for five standard hospitals. The intention was that seven major complexes should be built. No master plan existed. The University of Michigan had been asked to carry out a study on an all-embracing service but withdrew after the Saudi Government had declared unacceptable a Jewish member of the team designated. Subsequently, other American concerns which were approached showed no interest in taking up the contract. The programme, it seemed, would proceed on a piecemeal basis. The Ministry of Health, once again, faced the problem of attracting sufficient staff of quality.'

Compounding all the difficulties that the administration was to face by way of bottlenecks was the need to improve the facilities for the pilgrims making the *Hajj* and overcome the dislocation caused by the annual influx. The multitude bearing in upon the Kingdom had grown by the year from 107,632 in 1950 to 283,319 in 1960, 406,295 in 1970, and 918,777 in 1974. To handle the mounting tide through the provision of quarantine quarters, accommodation, road and airport extensions, and medical services – including a mobile hospital – the

Ministry of Pilgrimage and Religious Endowments was given a
budget of 5,000 million riyals ($1,416 million).

The bedouin were not completely forgotten. Although they were
dealt with under the heading of 'social development', it was promised
that policy for them would be drawn up on the basis of economic,
rather than welfare, considerations. None had been drawn up in the
previous five years and so the new plan had nothing very specific to
propose. It did, however, establish a Unit for Bedouin Affairs under
the Ministry of the Interior to co-ordinate with other departments in
evolving a strategy to overcome the grievances of this important part
of the community.

From the autumn of 1974, the pressures from wealth on a scale
unforeseen had begun to build up perceptibly. For the foreign visitor
the most nightmare recollections of those days of unbridled, demented
bonanza from 1974 to 1976 would probably be the struggle to obtain
seats on internal flights and to secure a hotel room. Businessmen
without the right connections could queue night after night for the
boarding pass that had to be obtained the night before departure on
the Arabian Express from Jeddah to Riyadh or vice versa. A lucrative
secondary market grew up in those precious red and green slips as
could be witnessed by the clusters of people around some citizens who
appeared to be able to lay their hands on fistfuls. But even expatriates
endowed with the most generous of expenses and prepared to abject
themselves for an assured seat seemed deliberately excluded. Not until
the aircraft was actually in the air could a passenger be sure of reaching
his destination because if a member of the royal family, who did not
need a boarding card, decided at the last minute to catch a flight he
might have to give up his seat for the royal latecomer. One British
banker recalls seeing in 1975 no less a person than Yamani, boarding
pass in hand, unable to take his first-class seat because an acned young
prince was occupying it. The impasse was only broken when a veiled
woman was persuaded to take the one vacant place in the tourist
section at the rear.

Such hotels as there were witnessed scenes of ill temper, jostling and
unabashed bribery as visitors vied with each other for hotel
accommodation that was scarcely adequate for half the hoard of hapless
businessmen who descended upon the Kingdom in the hope of making
fat profits for themselves or their companies. The desperation was all
the worse for the fact that it was a mad pursuit in which many were
called but – it soon became clear – relatively few were chosen as far as

riches were concerned. Even the most sympathetic hotel receptionists were reduced to giving world-weary shrugs of their shoulders when faced with the turmoil in the lobbies every night as international flights arrived. Second- and third-class establishments found that they could charge the equivalent of $100 a day for a room and even more if they were shared – as frequently they were. Some well-heeled foreigners were reduced to repairing to doss houses in downtown Riyadh, while others preferred the option of hiring a taxi for the night, which could cost $50. The rapacity of the drivers, in Riyadh especially, came to know no limits and only a few delightful exceptions would pay any regard to the rates laid down by the municipalities even if the khawaja,[3] or foreigner, could make himself understood in Arabic.

Driving became progressively more hazardous. Perversely, as traffic snarled up in Jeddah, Riyadh and Dammam the mad recklessness of Saudis at the wheel increased. So, too, did the rusting wrecks left by the roadside. In 1977 there were estimated to be nearly 80,000 in and around Jeddah alone. It was debatable whether the drivers' abandonment reflected Islamic fatalism. It certainly paid no regard to the prospect of insalubrious incarceration or the payment of 'blood money' to the unfortunate victims of fatal crashes, which in 1976 totalled 17,000. Passengers, Saudi and expatriate alike, had to face with equanimity the possibility of being held under detention indefinitely as witnesses of accidents. Despite the state airline Saudia's desperate inability to cope with growing demand, few foreigners took the 1,000-kilometre land route across the desert between Jeddah and Riyadh that Abdullah Suleiman had covered in a record eighteen hours in 1938. Thus they missed seeing perhaps the bloodiest stretch of macadamized road in the world. At first sight nothing seemed to symbolize the quickening, frenzied tempo of life and the concomitant increase in frayed nerves and irascibility better than traffic conditions. Veteran foreign residents bewailed the passing of more leisured days. They noted with concern signs of xenophobia that in the past had undoubtedly existed but had been masked by traditional politeness and hospitality.

In the three years before the impact of the oil price explosion had been felt the value of the Kingdom's imports had doubled to just over $2,000 million. They reached $6,592 million in 1975, $11,031 million in 1976 and $15,689 million in 1977.[4] Even allowing for inflation, the expansion was phenomenal and unprecedented. Growing congestion at Jeddah and Dammam was made worse by poor management and the archaic procedures of the customs authorities who seemed more interested in discovering illicit shipments of alcohol, pornography and pork than in facilitating the flow of imports. Their cumbersome activity certainly cost more than the revenue raised. Operations were further slowed down by what seemed an inordinate preoccupation with security and a preference – inevitably bound to further disorder – given to some importers by officials who received bribes in return for their favours. Another factor making for delays was the restriction on the employment of non-Saudis. As a result there was a shortage of fork-lift truck drivers. Tally clerks could not read English. And at Jeddah the new short-wave radio system could not be used because there were not enough trained nationals to operate it.

By the autumn of 1975, about 130 vessels were waiting at anchor at Jeddah and Dammam with an average waiting time of more than a month, some even experiencing delays of up to three months, at a cost of $3,000 to $5,000 a day in demurrage. In Jeddah the situation deteriorated as ships bringing pilgrims were given preference. By the end of December, the number of merchantmen lying at anchor numbered 200. Waiting times lengthened to between four and six months and surcharges charged by the shipping lines varied from 50 per cent to 250 per cent. Spared the extra burden of the *Hajj*, Dammam was in better shape, though waiting times there increased to three months with surcharges up to 90 per cent. In 1976, it was calculated, the congestion added almost 40 per cent to the cost of imports, upon which the Kingdom depended for most of its requirements. The effect on the price of building materials was considerable and made worse by profiteering merchants who held back supplies that were available to exact the maximum price.

In the first half of 1976 Jeddah seemed to illustrate something even worse than acute developmental dyspepsia, rather a country choking over its own riches. Foodstuffs rotted on the quayside where the city's rat population multiplied and flourished. Bags of flour piled up and burst, their contents turned into paste by the occasional rainstorm. Badly needed construction equipment and less vital luxurious automobiles waited for months in the humid, corrosive atmosphere. One incident seemed to sum up the squalor deriving from such opulence. Early on humans and livestock had been given priority. However, one of a shipload of camels from Sudan expired, presumably through lack of nutrition. Its body was thrown overboard but refused to sink and floated upon its distended stomach. A marauding shark zeroed in but failed to get a grip on the carcass, which bobbed balloon-like on the water. The carnivore tried again with similar results. On the next attempt the shark's fangs penetrated the stomach. It exploded, sending putrid pieces of flesh over the port area, and the shark went without its meal. The stench lasted for days.

Not without plenty of unsolicited, self-interested advice, the Saudi Government showed itself flexible in responding to the situation and adopting unorthodox measures. A General Ports Authority with powers to cut through the administrative tangles was set up. Gray McKenzie won a contract to assist bringing a semblance of order to the port of Jeddah. Over Christmas the company organized an airlift of over 100 Chicago stevedores, several of whom thought better of the enterprise at a refuelling stop in Athens and elected to spend the festive season there before returning home to the U.S. At Dammam, Mersey Docks and Harbour Company undertook a similar task to Gray McKenzie's. More dramatically, Carson Helicopters of the U.S. were hired to satisfy the construction boom's hunger for cement by lifting it from anchored vessels at the astronomical cost of $185 per ton. From dawn to dusk helicopters, usually trailing a plume of cement dust, became a common feature of Jeddah's landscape. The introduction of roll-on, roll-off facilities helped to alleviate the congestion. So, too, finally did the appointment in the autumn of Faiz Badr, extracted from his post of Deputy Minister of Planning to become President of the Saudi Ports Authority with ministerial status. He badgered merchants into clearing quays and warehouses under the draconian threat of auctioning off their goods if they were not cleared within a few days. Advertisements placed in local newspapers warned of the deadline for the discharge of cargoes. Shippers were told in no uncertain terms, often in lengthy telex messages, what was expected of them and, in particular, the savings that might be made through containerization. Miraculously, by January Dammam was decongested and there were even empty berths. For Jeddah, Faiz Badr banned stevedoring through ships' agents and contracted Filipino labour directly. Having installed special cement-handling facilities he terminated the spectacular but costly arrangement with Carson. He was deeply resentful of any suggestion that foreign expertise had helped him fulfil his objectives. The backlog was cleared in February 1977, a month before the deadline set.

Opening up this critical bottleneck was justifiably regarded as a triumph and a very vital one in the midst of a boom in which one constraint rubbed against another, sending the Kingdom into a vicious spiral of inflation. One of the most acute problems, with serious political and social implications, was the rising cost of housing that affected Saudis at both ends of the social spectrum. Even worse was the plight of poorer Arab expatriates working for the Government or private sector who did not enjoy the benefits of citizenship or the support of international companies. The shanty satellites around Jeddah, Riyadh and Dammam and the makeshift erections within their confines bore witness to the problem caused by the drift from rural to urban areas. But not until the summer of 1976 were contracts awarded for the construction of 41,000 units in Riyadh, Jeddah Dammam, Mecca, Medina and Al Khafji that had been put out to tender at the beginning of 1975. It was a classic example of the kind of bureaucracy that would delay achievement of the physical targets of the Second Development Plan.

The plan recognized the opportunities given to greedy, rack-renting landlords. From 1972 rents had begun to rise inexorably. In 1975, the Government limited rent increases to 5 per cent and then froze them for 1976. It gave security of tenure to tenants except when the landlord could claim his property was needed for his own family. Under this pretext and various unscrupulous pressures, rentiers succeeded in clearing premises. Landlords were able to double and triple rents. Such was the rancour and misery caused that in 1976 the Government established a committee to mediate in disputes and review rent legislation. At the end of 1973 a good three-bedroomed villa in Riyadh or Jeddah cost around 30,000 riyals (about $8,500) a year to rent. At the end of 1975 rates for such property had gone up to 80-90,000 riyals, and even more in the capital. At the luxury end of the market, rents of 500,000 a year were recorded. A good four-bedroomed villa in the better residential suburbs cost something in the region of 150,000 to 200,000 riyals annually.

Inevitably, the property boom was accompanied by wild speculation in land from which the princes stood most to gain. Because Saudi society, for the most part, had only emerged from tribalism over the past three or four decades, notions about land tenure were vague. Claims based on custom or oral tradition had little validity, and possession or use were the mainstays of a claim. Much land without clear title tended to become royal property and ownership to it was established after it had, in the first instance, been distributed as a gift, usually to princes but sometimes also as a reward to commoners. Land values in and around the main urban centres had multiplied several thousand times compared with what they had been in 1970, making Saudi real estate as expensive as that of Manhattan or the City of London.

The influx of foreigners required to implement the projects of the Second Development Plan was a major factor in the inflationary process. It was quickly identified as such by the Government and provided a ready scapegoat. Hisham Nazer commented: 'A company that can rent a villa which is worth 60,000 riyals for 400,000 riyals is a company that has really inflated cost estimates. Now I think that they have come to the point that the situation which they created was intolerable even for them.'[5] He had a point. The Government resorted in 1976 to the expedient of preventing companies with contracts worth 100,000 riyals or more from renting either offices or houses in the cities and instead making them build their own accommodation on rented land. But the root problem, the inadequacy of the basic stock of housing, remained.

The Government provided generous loans through the Real Estate Development Fund, set up in 1974, and looked to Saudi developers to remedy the housing shortage. The terms offered by the fund not only provided citizens with 70 per cent of necessary finance bearing an 'administration fee' of only 2.5 per cent on concessionary repayment terms, but also a 20 per cent rebate on building costs. It was also empowered to provide loans covering 50 per cent of the cost of commercial developments. Its capital had been steadily increased to 12,000 million riyals by November 1976. The private sector responded as expected. By the beginning of 1977, eighteen months after the start of the plan period, one-third of its housing target had been fulfilled. That year rents began to stabilize at a level about ten times on average that of 1972. Resigned to the fact that attempts to enforce a freeze of rents could only discourage the profit motive, the Council of Ministers allowed them to be increased by 15 per cent from January 1977.

In 1948 Ibn Saud was reported to have told a Western diplomat: 'My Kingdom will survive only in so far as it remains a country of difficult access, where the foreigner will have no other aim, with his task fulfilled, but to get out.' It was not possible, he said, for a country 'to emerge from the Middle Ages and enter the twentieth century on steady feet' without a long preparation. Much had changed. But it was still pertinent to ask how the Kingdom, and not the least its traditional values, would withstand the vast influx of 500,000 foreigners anticipated for the implementation of the Second Development Plan. The number would amount to more than one-tenth of the native population and a bigger proportion if their dependents were accounted for. In addition, there were those already in the country whose numbers exceeded the figures officially acknowledged. It remained to be seen what the reaction of the indigenous temperament, especially that of the Nejdi, would be to such an enormous increase in the presence of *khawajas*.

As Arabians, the ubiquitous Yemenis blended relatively well, though they were conspicuous in contrast to the Saudis not only in their dress but also in their energy and natural aptitude. But even they were regarded with some apprehension. Late in 1975 I remember a Saudi official in the Ministry of Planning commenting, without paying recognition to their labour, that by living so cheaply and saving their money the Yemenis contributed nothing to the economy – meaning presumably the profits of the merchants. Hadraumis of southern origin had established themselves with some success, mainly as small traders, being far more adept than Saudis at retailing. It seemed a particularly graceless protectionist act of the Government, albeit a response to antagonistic Saudi public opinion, when late in 1976 it banned 'foreigners' from engaging independently in commerce. As a result there was a decline in standards of service and an increase in the cost of goods sold.

Apart from the Yemenis, there had been a perceptible, unregistered rise in the expatriate population from an increasing number of pilgrims lingering on in search of gainful employment that they could not find back home in Somaliland, Oman, Sudan, India, Pakistan. More, however, came as immigrants legally, particularly Egyptians and other northern Arabs, as well as Muslims from the Asian sub-continent, sponsored by state departments or private employers. Faced with the invasion and worried by the prospect of uncontrolled mobility of labour, the Saudi Government laid down that employers should retain an immigrant's passport and the internal travel of expatriate workers not carrying passports was heavily restricted. The privilege of citizenship was extended rarely and only through the favour of high princes. The Kingdom was alarmed at the possibility of large-scale permanent settlement by Arabs and Muslims rather than Americans or Europeans whom Ibn Saud had had in mind in 1948 when he warned of the danger of alien influences. The majority of Westerners, who benefited from free housing and cost-of-living allowances, would be only too anxious to depart having saved what they could.

More immediate was the question of how a foreign work force on the scale envisaged and with the right variety of skills could be obtained. Early in the implementation of the Second Development Plan the Government made the contractors responsible for recruiting abroad in bulk. Inherent in the approach was the politically and socially reassuring certainty that the 'packaged' expatriate workers could be segregated in their camps and transported home when projects were complete. From 1975 there was the novel sight – that did not, however, seem to raise any Saudi eyebrows – of bronzed British workers, dressed only in shorts and desert boots, wielding pneumatic drills in the main streets of Jeddah as they carried out the sewage contract being undertaken by Streeter's of Godalming, England. The shortage of resident manpower was such that unskilled workers also had to be imported. Early in 1977, for instance, a joint venture between Waste Management of Chicago and the Pritchard Services Group of Britain, the latter in conjunction with a Saudi partner, won a unique five-year contract for disposing of the accumulating rubbish in the streets of Riyadh which had rapidly taken on the appearance of a cross between the garbage capital of the Middle East (not excluding Basrah) and the region's biggest building site. No sooner had the prize been secured than over 2,000 trainee refuse men had to be hired in bulk from Pakistan and a camp built for the task force. The massive mobilization programme also included the purchase of 400 specialized vehicles.

For the lower grades, the Indian sub-continent and then increasingly the Far East were the source of what became for the oil-rich Gulf region as a whole an international trade in indentured workers. Labour proved to account for anything from one-third to one-half of the cost of contracting a project. The economic logic of the situation was that in areas of development not requiring higher forms of technology, countries able to provide their own cheap labour were soon prominent in the list of contract winners. That meant business for construction companies of such countries as Turkey, Greece and Cyprus. But in the most advantageous position of all were those Asian states that could provide even cheaper labour. South Korea thrust itself to the forefront, winning a disproportionate share of the most spectacular contracts. In 1974 there were about 4,000 South Koreans

in the Kingdom. By 1978 there were 35-40,000. The presence of these
Asians first impinged heavily on the consciousness of bemused Saudis
when one concern began reclaiming land off Jeddah's shore. Saudis
watched energetic orientals of fierce demeanour with amazement and
not a little awe. Around the clock day and night lorries thundered up
and down the Medina Road carrying and fetching sand and ballast in
what seemed a display of martial discipline. South Korean operations
were in fact run along military lines, utilizing craftsmen and operatives
trained during their national service and given an early discharge as
tied labour assisting with their Government's export drive.

In 1976 South Korean companies won contracts worth $2,100
million, including one valued at over $1,000 million awarded to
Hyundai for the industrial port at Jubail. They were admired not only
for their efficiency and competitiveness but also for their discipline.
That broke down dramatically, however, in March 1977 when
Hyundai's indentured labour laid down their tools and protested
against harsh working conditions, spartan living accommodation, and
low wages that were 40 per cent below those paid by their European
and American competitors in the vicinity. Though the site was
isolated, the affair was enough to make Saudi authority, unused to
open demonstrations of any kind, twitch, especially as the disruption
had occurred in the region where the Kingdom's spurned and
disaffected Shiite subjects lived. Fortunately, the company's
management was able to quell the disorder without bringing in the
Saudi police. There was also trouble with Turkish workers about this
time at Tabuk in the north-west. There, the Government reacted fast
by sending C-130 Hercules aircraft of the Royal Saudi Air Force to
transport them home.

An internal study carried out by the Ministry of Planning calculated
that foreign manpower towards the end of 1978 totalled 1.3 million.
The native labour force was put at only 1 million and, therefore, was
significantly out-numbered. Saudi ministers and officials pointed out
that a large proportion of the expatriates were engaged on
infrastructure projects and would return home. But packaged labour
could not reduce the economy's dependence on the existing and
overstretched pool of casual labour. From 1975 to 1978 the pressure of
demand for it had led to a significant rise in daily wage rates. Amidst
the bedlam of dislocation caused by unco-ordinated expenditure and
planning, the regime was faced with the dilemma, even if it could
hardly rationalize it, that the wholesale expulsion of illegal immigrants
could only add to inflationary pressures.

In 1975, 1976, and the early part of 1977, inflation stretched the country's social fabric and threatened to tear it apart. From the middle of 1976 onwards it became a major preoccupation of both the ruling hierarchy and the Council of Ministers, though the unthinking ignorance of most of those constituting the former and the rivalries within the latter made it difficult to adopt a coherent policy to solve the problem. Official statistics relating to the question were published by S.A.M.A. long after the year concerned had passed. Privately, those in authority were forced to agree that they minimized greatly the extent of the phenomenon. Based on an absurdly out-of-date pattern of consumption well below the lifestyles and expectations of most Saudi urban dwellers, the cost-of-living index showed a rise of 31.5 per cent in the fiscal year 1975–6 and 11.2 per cent in 1976–7. But inflation had been building rapidly in the year before implementation of the Second Development Plan when it was running at a rate generally acknowledged to be in excess of 30 per cent. In 1975–6 bankers and diplomats reckoned that it had reached 40–50 per cent and as much as 70 per cent if rents were taken into account. In the second year it eased slightly, but by far less than the official figures suggested, and was still at a totally unacceptable level.

The Government reacted with piecemeal measures to bring the beast under some sort of control. In 1974 it had extended the list of subsidized goods (hitherto flour, rice and sugar) to include imported lamb carcasses, milk, milk products, vegetable oils, and medicines. It had eliminated customs duties on many goods, removed the excise on petroleum products and abolished road tax. Early in 1975 the wages of public employees had been raised by 30 per cent and the salaries of medium-to-high categories by 15–18 per cent. Another across-the-board award had been made in November of that year. Then, confused by the need to protect citizens from inflation and its realization that pay increases were a factor contributing towards it, the Government decided to restrain them. And so, squeezed unmercifully, civil servants sought employment with the private sector and reduced the capacity of an under-developed bureaucracy to cope with the pressures of unprecedented growth. But in the midst of this flirtation with monetarism, the Council of Ministers saw fit to double overnight the pay rates of the Armed Forces.

Through a series of measures by the Government, several *ad hoc* in nature but all of them fairly swift responses to the problem of breakneck growth and the injection into the economy of more money than it could contain, the country was eased out of the roaring inflation that undermined it. Fundamental to the recovery was the slower growth rate of Government expenditure. The rate of increase from 81,000 million riyals in 1975–6 to 128,000 million riyals in 1976–7 was 56 per cent but fell to 8 per cent in 1977–8 when the actual outlay amounted to 138,000 million riyals ($37,155 million). An important factor was the suspension of the operations of the Real Estate Development Fund in March 1977. By then it had committed over 20,000 million riyals in loans. The programme had made the major contribution to the Government's success in overcoming the shortage in housing and to that extent reduced the cost of living. But it was a wildly extravagant one that by disbursing an enormous sum of money over a short period was very inflationary. Not until February 1978 did it resume operations.

The greater part of Saudi Arabia's inflation from 1975 to 1977 had been generated domestically. But there was no doubt also that some foreign contractors, as well as suppliers, raised their bid prices to levels far higher than the average in the West. They did so, however, not so much in any reasonable hope of a fat profit margin, but as a prudent safeguard against the exceptional contingencies and high risks involved in undertaking work amidst the economic turmoil that only began to subside in 1977. In preparing bids for contracts put out to tender by the Kingdom, companies were faced with a number of variable factors. The country's domestic rate of inflation, the price of building materials and locally recruited labour, the commissions that would have to be paid, and the costs incurred as a result of bottlenecks in transportation could not be calculated with any precision. The Government, with some rare exceptions, insisted on fixed-price contracts and permitted no escalation clauses. Ministries would make changes in specifications without any warning or compensation. Contracts contained harsh penalty clauses that made no allowance for delays caused by factors beyond the control of companies implementing projects. The Government would not consent to any independent or international arbitration. It was not surprising that tenders included an element to cover a wide variety of contingencies which, though not unique to Saudi Arabia, were endemic there. Some companies did perhaps attempt to exploit the premium that Saudi Arabia appeared to be willing to pay for rapid development. Yet they were unlikely to be successful in a competitive market – unless they purchased or enjoyed the right princely patronage.

The extent to which the price of projects was swollen by commissions became fully apparent in February 1977 when the Ministry of Posts, Telegraphs and Telephones cancelled a prospective contract with Philips of the Netherlands that it was on the verge of concluding after a year of negotiations. The project in question involved a nationwide fully electronic switch-gear system designed to raise the number of automatic telephones operating in the Kingdom from 200,000 to 600,000. The contract, which promised to be the most valuable ever in the field of telecommunications worldwide, was never put out to open tender. The Saudi Government chose to single out for exclusive negotiations the Dutch company which, as it happened, was represented by Al Bilad, the company of Muhammad bin Fahd, the sixth son of the Crown Prince, who at the time was about twenty-five years old.

Inevitably, other giants of the industry became aware of the project, including the International Telephone and Telegraph Company (I.T.T.) and Western Electric of the U.S., Nippon, Hitachi and Mitsubishi of Japan, Siemens of West Germany, and Thompson C.S.F. of France. Their approaches were politely received by the Ministry of P.T.T. but the department did not seem to be exercising any autonomy. None of the companies was given the opportunity to make a formal presentation and no exchanges that could be remotely characterized as negotiations with them took place. William Porter, who had been appointed to succeed James Akins as U.S. Ambassador in August 1975, sought to intervene on behalf of the American companies but neither the King nor the Crown Prince would receive him. He protested vigorously in writing about the way in which the deal was being settled, nearly making himself *persona non grata* in the process. There were good reasons for his righteous indignation. Not only had the contract not gone out to competitive bidding but the two American companies could fairly claim at the time to be amongst the front-runners in developing the technology required. Moreover, the cost in prospect appeared to be inflated. Early in 1977 it had become known that the price virtually agreed upon with Philips was about 20,000 million riyals ($6,700 million). That was about five times the consultants' estimates and also, as it happens, more than the budget of the department for the whole of the 1975–80 period.

Muhammad bin Fahd was exacting commissions not only from the Dutch company but separately also from the sub-contractors lined up by it. 'Double dipping' was by no means uncommon as a growing proportion of state expenditure was recycled into princely bank accounts. If they wanted the business, contractors normally had no choice but to comply with the demands of influential middlemen who in turn could give assurances that the price of the contract could be stretched to absorb their commissions. But the enormity of the sum that would have accrued to Muhammad bin Fahd and his collaborators was a subject of astonishment amongst expatriate businessmen and envy amongst other Saudi agents. It would have amounted to no less than 20 per cent of the total contract price, or something in excess of $1,300 million, according to a former executive of one of the American contenders.

There was nothing illegal about the payment or receipt of commissions and no restrictions on their size in Saudi Arabia. However much practices differed from those prevailing in the West, foreign companies competing for contracts in the Kingdom had no choice but to accustom themselves with local ways of doing business. These did not always conform to U.S. domestic legislation, however, in particular the so-called Foreign Corrupt Practices Act. Thus, American companies were at some disadvantage. But presentation in accountancy could disguise commissions as items of expenditure.

I.T.T. had as its accredited agent none other than the formidable Muhammad, the senior member of the family, while amongst those advising Western Electric (though not formally their agent) was Kamal Adham. The situation was such as to strain severely one critical area of tension within the royal family – the relationship between Fahd and Muhammad. The Crown Prince was clearly determined that his own son, together with his coterie, should benefit. Fahd presided over economic development. Muhammad was held in greater awe and respect within the House of Saud but had no direct control over the administration and departments of state. However, the scandal over the manner in which the Ministry of P.T.T. was forced to conduct the negotiations and the cost involved led to the cancellation of the proposed deal.

With some changes in specifications, the contract was put out to tender formally in March 1977. The project was divided into three separate and distinct parts. It was made clear also that contenders were not obliged to bid for all of them. When the tenders were opened at the end of September, I.T.T.'s offer prices were lowest for part one, the heart of the project, at $1,242 million and part two at $198,758 million. The Philips consortium's bids stood at $1,488 million and

$234,572. Those of the Western Electric group were $1,517 million and $246,088. For part three there were wide variations. Here evidently some confusion arose. I.T.T.'s figure of $1,951 million was a comprehensive one. The Philips consortium's $467,620 provided only for management, not personnel. The Western Electric group's $1,179 million excluded the civil construction works. There were also muddle over the specifications. I.T.T. was accused of not having met properly those laid down by the Saudi Government. The American multinational, in turn, let it be known that it would like to meet at the 'negotiating table' and sort out discrepancies in the bids. The company felt that the contract had only been put out to tender in the formal sense and that the outcome had been effectively decided in advance. I.T.T. had spent over $22 million in preparing its bid and felt very bitter.

Towards the end of December 1977 the contract was awarded to the Philips consortium. Muhammad bin Fahd did not go unrewarded. According to reliable sources, his commission amounted to about $500 million which was accommodated comfortably in the price finally agreed.

Three weeks after the award the Government announced a 5 per cent ceiling on agents' commissions. Draft legislation on the question had existed for the previous eight months and originally included a clause laying down a maximum commission of 3 per cent. It was open to question whether the regulation would be observed by princes able to bend the rules. But Muhammad bin Fahd was launched on a truly spectacular business career.

Most newly formed Saudi registered companies from about 1975 would contain a princely name or two. A typical one, taken at random, was the Contractor Company. Prominent amongst its shareholders, each with 29 per cent, were Saud bin Naif and Abdul Aziz bin Salman, both the offspring of prominent sons of Ibn Saud. They would, no doubt, have had no difficulty in contributing their share capital. But dark tales were told of enterprising commoners embarking on new ventures being approached by more obscure and penurious princes, enjoying little or no influence in the upper realms of the hierarchy, demanding a stake and obtaining it without contributing a riyal. It was only natural that even the less well-connected members of the family should want to enjoy the dividends from what they still regarded as their feudal estate to supplement their stipends.

23 Lords of the Ascendant

1975–7

I feel ashamed for those who declare that their only concern in the event of a fresh outbreak of hostilities is the continued flow of oil, without regard for the bloodshed that would result.[1] – CROWN PRINCE FAHD

FROM Feisal the House of Saud inherited fairly well-defined guidelines for the survival of the established order and the preservation of the country's territorial integrity. He had set an example in the way that he followed them with consistency and flexibility. The conjuncture in the heartland of the Arab world, for which he was largely responsible, could hardly have been better when he died. Only in one vital respect – relations with the states of the Gulf and the Arabian peninsula – had there been a disturbing, even alarming, lack of Saudi clarity and deftness at a time when the stability of the region was becoming a global preoccupation. Here the Kingdom's suspicions and rivalries, ideological and dynastic, died hard.

In contrast with their domestic confusion in 1975–6, Feisal's heirs moved with increasing assurance and aplomb in their conduct of external affairs. Although Fahd immediately moved to the forefront, it was soon apparent that leadership in this field would be collegiate. Sultan kept the Yemens as his preserve. Abdullah was to develop a special interest in Syria and Iraq. Saud al Feisal, though having to defer to his elders as was the custom in the family, soon showed himself to be far more than a golden messenger boy. Both for his advice and diplomatic skills he was to prove a priceless asset to the regime. But the Foreign Minister's influence is said to have been more limited than it might have been because of Fahd's less than friendly attitude towards him. Kamal Adham still held the title of Royal Adviser and kept his line to Sadat open. The wealthy Saudi-born businessman Ghassan Shaker was used as a link with Oman. Rashad Pharaoun's views were still sought and listened to. He was probably no better counsellor than he was physician but he represented a strand of continuity going back to the days of Ibn Saud.

Relations with Egypt and Syria were solid and friendly. Overall, Sadat's success in consolidating his position and the reversal of Arab fortunes as a result of his military adventure in 1973 had altered the balance of power in the region in favour of the conservative regimes. By the same criterion, however, the weight of the responsibilities and burdens placed upon the Kingdom grew. Its enhanced position did not necessarily involve a corresponding assurance of serenity or security. A greater premium than ever was placed upon the promotion and preservation of Arab unity, an objective that Saudi Arabia was better able and more psychologically inclined to pursue thanks to the accumulating billions of petrodollars and as a result of Feisal's demise. Yet two factors, almost coinciding, immediately threatened the harmony so keenly sought.

Having failed in the previous summer, Kissinger had embarked on a renewed and intensified attempt to bring about a second Egyptian-Israeli disengagement agreement. Achievement of one was almost bound to antagonize Syria and in the process disrupt Cairo's alliance with Damascus. The fact was that Kissinger clearly saw another interim accord as a means of easing Egypt fully out of the confrontation with Israel. There would be no guarantee that Syria in return would regain more occupied territory. Thus, U.S. diplomatic moves were watched apprehensively by the Saudi ruling hierarchy.

In Lebanon, meanwhile, violence had flared in the south in and around Sidon where clashes between the security forces and left-wing elements in February and March 1975 led to accusations of involvement being levelled against the Palestinian guerrillas and a call by Pierre Gemayel, the right-wing Christian Maronite leader, for a revision of the accords regulating their activity. On 13 April at Ain al Rummaneh near Beirut, as Gemayel was attending the consecration of a church, members of his Phalangist paramilitary organization fired on a bus filled mainly with Palestinians from the Tal Zaatar refugee camp, killing twenty-seven passengers and wounding nineteen. The Lebanese civil war had begun, although at the time no one could predict the consequences of the atrocity. Even harder to foresee was the manner in which the conflict would become the focus of inter-Arab disputes and present the Kingdom with a formidable challenge.

On the international front, Saudi Arabia's preoccupation following Khaled's accession was with the call for a dialogue between industrialized countries and the developing world. The issue was one amongst several complicating and exacerbating the nexus with the United States. The Kingdom was one of the four oil producers represented at the tedious preparatory conference convened by President Valéry Giscard d'Estaing and held in Paris early in April 1975. Concerned wholly with procedural affairs, the abortive gathering failed to agree upon a format or agenda for a full-scale conference on substantive issues. The producers and other Third World delegations were clearly irritated and angered by a seemingly inflexible insistence on the part of the U.S., reflected in almost equal measure by the European Economic Community and Japan, that any dialogue should concentrate mainly on the supply and price of energy. The industrialized representatives would not agree to placing on the agenda unequivocally raw materials other than oil, the reform of the international monetary system and protection of the real value of the financial assets of developing countries. The last of those three subjects was obviously of special interest to Saudi Arabia. Even the accommodating Fahd was aggravated by what the Saudis saw as the obdurate stance taken by the United States.

On 9 May there appeared in the *Middle East Economic Survey* a report that had about it the smack of authority and was clearly designed as a warning to the U.S. It quoted an authoritative Saudi source, believed to be Yamani, as saying: 'Saudi Arabia's future oil policy in the crucial matter of production levels and prices will depend very largely on what sort of attitude is adopted by the industrialized oil-consuming countries towards O.P.E.C. and the Third World.' He gave three possible scenarios. If the industrialized countries entered into a meaningful dialogue, Saudi Arabia would continue its traditional policies of moderation within O.P.E.C. If they did not show a serious interest, then the Kingdom would agree with other members of O.P.E.C. to prices being raised 'by some reasonably modest amount'. The third scenario – open confrontation between the two sides – would result in Saudi Arabia completely abandoning its moderate stand.

The response of the U.S. came towards the end of the month in the form of new proposals. They did not meet the demands of the oil producers and other developing countries but did give some hope for the convening of a conference that might grapple seriously with the whole gamut of problems covered by the O.P.E.C. Solemn Declaration. A brief flurry was caused by what seemed to be a veiled threat that the United States would be prepared to use force to secure oil supplies in the event of another Arab oil embargo. James Schlesinger, the U.S. Defense Secretary, said no more than 'we might not remain entirely passive to the imposition of such an embargo'.[2] The normally mute Abdullah snapped: 'Is Mr Schlesinger so angry from Vietnam that he would explode his anger here? We pray to God that his comment is not the first sign of the reassessment of American policy in the Middle East.'[3] President Ford attempted to soothe ruffled feathers by stressing that American policy was one of co-operation, not confrontation. This was duly welcomed by Fahd, who added, 'I feel ashamed for whose who declare that their only concern in the event of a fresh outbreak of hostilities is the continued flow of oil, without regard for the bloodshed that would result.'[4]

In the course of 1975 American democracy began to strain the special relationship. In the aftermath of the October War, Israel and its lobbyists had been quick to appreciate the diplomatic weight acquired by Saudi Arabia as a result of its emergence as the predominant oil power. Congress responded and the Kindgom became in 1975 a special target. An early irritant in the summer of that year was the beginning of the campaign to prevent compliance by American companies with Saudi requirements discriminating against the employment of Jews on contracts undertaken in the Kingdom and of the Arab Boycott of Israel generally. Saudi Arabia's arms requests and United States involvement in its defence programmes came under increasing fire on Capitol Hill. As Arab impatience and frustration over the lack of further progress towards a Middle East peace settlement grew, Riyadh could point to no gains from its policy of moderation in oil pricing and its special relationship.

Towards the end of May, Khaled made a tentative but, in the Saudi context, bold suggestion that the Kingdom might be prepared to concede Israel's right to exist within the *de facto* borders prior to the June War of 1967 in return for withdrawal from the Arab territories occupied then and the establishment of a Palestinian state upon them. There was a certain ambiguity about his statement. But it indicated a greater flexibility than Feisal could have shown, a straw in the wind perhaps but one that reflected a genuine anxiety about stability within the region. Undoubtedly, too, it was cast in the direction of the U.S. Government and American public opinion. It went unnoticed.

When Kissinger's reactivated peace mission gained momentum in the early summer, it was apparent that the price for Israeli co-operation would be a high one in terms of American military aid and supplies of sophisticated weaponry. But the Arabs appeared to have scored a success when in April the Administration approved the sale of Hawk anti-aircraft missile batteries to Jordan that were to be financed by Saudi Arabia. Then, in the face of Congressional opposition, Ford climbed down and at the end of July withdrew the offer. An injured feeling that the Kingdom was being taken for granted might have been one reason for the guarded and cautious hints that it might be contemplating dealings of some kind with the Eastern bloc. 'The Kingdom of Saudi Arabia pursues an open-door policy towards all countries of the world, whether Eastern or Western, and has economic relations with all states,'[5] Fahd was reported as saying early in July. On the following day he was quoted as saying, 'We want good relations with both East and West on the same footing, and according to their policies.'[6] But the context of his statements indicated that concern about the Muslim minority in the Soviet Union and a recognition of Moscow as an important part of the Middle East equation were more important factors in these typically oblique utterances. In practice, Marxism remained anathema and the threat posed by it to traditional regimes a continuing preoccupation.

In August, the Saudis were disturbed by Kissinger's abrupt and cynical dismissal of James E. Akins, who had won the friendship and trust of the ruling hierarchy. The Secretary of State leaked his decision to the New York Times from which Akins first learnt of his recall. His outspoken honesty perhaps was too much for Kissinger. More important may have been the Secretary of State's resentment of Akins' consolidation of his position as accredited representative of the President of the United States. That, together with the distrust Kissinger himself had engendered in Feisal, made it impossible for him to bypass normal diplomatic channels in Saudi Arabia or to establish the same kind of link that he had forged with Sadat, who was wont to call him 'Dear Henry'. The incident in 1973 when Akins had refused to be excluded from the meeting with Feisal may have been significant. Akins was replaced by William Porter, who was nearing the end of his career and – in the eyes of the Saudis at least – seemed more concerned with his impending retirement than their own particular problems. In the same month, Riyadh was embarrassed by the revelations from the Senate Foreign Relations Committee Subcommittee on Multinational Corporations about the 'irregular' payments made by Lockheed and Northrop. A theoretical understanding of the workings of American democracy and media was not sufficient to make the Saudi leaders forgive such leakages of confidential information. At the same time, Saudi appreciation of the way in which both were subject to Zionist influences was objectively correct.

Feisal's heirs were quickly convinced of the need to strengthen and develop ties with other Western countries, the only possible alternative for Saudi Arabia as far as both armaments and technology were concerned. Hence the significance of Fahd's official visit to France from 21 to 24 July 1975. It resulted in an agreement on economic collaboration, including the setting up of a joint commission modelled on the Saudi-U.S. one. An understanding was reached whereby Saudi Arabia would invest part of its financial surplus in France which, reciprocally, would participate in development projects in the Kingdom. Discussions on a long-term oil supply deal were inconclusive.

Saddled with the 1973–4 oil price increases, all the industrialized countries were courting Saudi Arabia. In the clamour to share the business opportunities provided by the Second Development Plan, West Germany, Italy, and Japan also signed accords with Saudi Arabia on joint collaboration. Britain began to cultivate the Kingdom more seriously than before. On a private visit to London in the summer of 1975, Fahd enjoyed a discreet evening gaming at the Clermont Club after dinner at the exclusive club Annabel's. He returned to London on an official visit in October. In the following month James Callaghan, the Foreign Secretary, flew out on the first leg of a tour of the Gulf states aboard an R.A.F. Comet, exuding bonhomie and well briefed with proposals for assisting the Saudi Government to overcome some of the problems already being confronted in the implementation of the grand economic design.

With the death of Feisal, the climate of relations with Britain much improved. The resentment that hung so heavily on Feisal's mind because of Palestine, Buraimi and the Abu Dhabi border issue, the scuttle from Aden, and the apparent connivance over the Iranian seizure of the Gulf islands, quickly evaporated after his death. On a government-to-government basis the air defence contract that had

Meeting of opposites: Feisal's last meeting with Kissinger, Riyadh, February 1975.
Fahd hovers in his brother's shadow

Below left: Naif, Interior Minister, an upright if unimaginative servant of the
Kingdom. Below right: A chip off the old block: Saud al Feisal addresses the U.N.
General Assembly as Foreign Minister

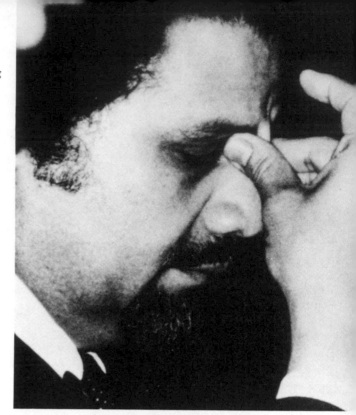

Ahmed Zaki Yamani showing
the strain of representing the
world's greatest oil power

Feisal's assassin, Fe
Musaid, and his A
girlfriend Christina
The couple met at
University of Colo
where the young p
absorbed some radi

Feisal's bier is carried from the funeral mosque to his burial place outside Riyadh, 26 March 1975

The family's choice: although Khaled had neither the skills nor the experience of Fahd, only he could unite the family's different factions

The Grand Mosque in Mecca, scene of bloody revolt in November 1979

been renewed in 1973 was being smoothly carried out. Britain was trying to interest Saudi Arabia in the Jaguar fighter-bomber, an Anglo-French product, one of which arrived at Riyadh shortly after Callaghan's departure to show its paces. The Saudis showed only polite interest. Despite the desire to diversify, they were in reality heavily committed to a U.S. system for the future development of the Air Force. Though the Lightning that had given so much trouble in its early days remained the Kingdom's front-line interceptor, the British role would be limited to basic, if essential, functions, of which the most important were maintenance and training. As it was, the Foreign Secretary left having signed a Memorandum of Understanding on Economic, Technical and Industrial Co-operation. In Riyadh, the attitude, in the words of one British diplomat, was 'suck-it-and-see' as the Saudis contemplated what the U.K. could offer compared with its more dynamic American rival. In subsequent years the Saudi Arabian-United States Joint Commission on Economic Co-operation would generate a multiplicity of technical aid programmes, but its British equivalent – and, indeed, those of the other industrialized countries – produced very modest results by comparison. In the meantime, the United States Corps of Engineers was more deeply entrenched in Saudi Arabia than ever. By 1975 it was supervising the construction of projects, mainly in the field of military infrastructure, that had a value approaching $15,000 million.

The maintenance of the best possible relations with Egypt and the fortification of Sadat's moderate, pro-Western regime remained a cardinal aim of foreign policy. To that end, Saudi Arabia was paying a price in aid that was not insignificant, even in terms of its own bulging oil revenues, which many in the ruling hierarchy were beginning to begrudge. In addition, it had agreed in May 1975 with Kuwait, the United Arab Emirates and Qatar to contribute $1,000 million as the initial capital of a joint enterprise to manufacture armaments that subsequently became known as the Arab Industries Organization.

Well aware of the schism in the heart of the Arab world that could result from Kissinger's renewed attempt during the summer of 1975 to bring about a second Egyptian-Israeli disengagement agreement, the rulers of Saudi Arabia watched with cautious reserve, reluctant to commit themselves to it. Sadat saw another interim accord as a step towards a final settlement involving the recovery of all lost territory. At his meeting in Salzburg on 1–2 June with President Ford, the

Egyptian leader had also pressed for any second disengagement agreement to be linked firmly to a subsequent one with Syria.

The Saudis appreciated that he felt the need, for domestic political reasons, for a second agreement giving Egypt a more substantial slice of Sinai than the first one had. Even though the balance of power in the Middle East had been changed by the October War, Sadat had nothing much in the way of gains from it to show his people, who had yet to feel any economic benefits. The disengagement agreement concluded early in 1974 had enabled Egypt to reopen the Suez Canal on 5 June 1975. The second one being negotiated by Kissinger held the promise of Egypt getting back the oil fields on the eastern side of the Gulf of Suez that Israel had exploited since 1967, the opportunity to implement plans for enlarging the waterway, and the chance to rebuild the wrecked towns along it. Conclusion of another agreement would also enhance the prospect of increased aid from the West. Sadat argued persuasively in favour of another interim accord when the Saudi King visited Cairo from 16 to 20 July. He sought to reassure Khaled that Egypt would not betray the Arab cause in contravention of the Rabat summit's resolutions rejecting partial solutions. Their talks began only two days after Sadat had informed the U.N. Security Council that he would not agree to a renewal of the mandate of the U.N. Expeditionary Force that supervised the ceasefire in Sinai. In this atmosphere of deliberately heightened tension, Khaled's presence in Egypt appeared a demonstration of Saudi Arabia's solidarity with it and the axis between the two countries. The King further cemented the alliance by granting Egypt another $600 million in budgetary aid.

Preservation of friendship with a moderate Egypt was the foremost axiom of the Kingdom's pan-Arab policy. Yet it was also desperately concerned to have the best possible links with Damascus. But the Saudi ruling hierarchy felt powerless to take an open and clear-cut position on the widening rift between Egypt and Syria. At this critical juncture the only relevant public utterance from the leadership came from Fahd and this was related to the unlikely possibility of war breaking out. 'If we are forced to go to war in order to preserve our rights and liberate our occupied lands, and Jerusalem in particular, then it is our right to fight with all our legitimate weapons in order to guarantee our victory,'[7] he said in an interview published on 2 August. Just over a week later the Crown Prince visited Damascus. There, it can be assumed, he assured Assad of Saudi support for any Syrian demand for a reciprocal interim accord with Israel. In practice, the geography of the Golan Heights and the limited room for manoeuvre ruled out any chance of Israel agreeing to one. As far as Syria was concerned, the only diplomatic progress could be towards a comprehensive settlement that also satisfied Palestinian aspirations.

Tension built up as Sadat despatched his adviser Ashraf Marwan and Vice-President Hosni Mubarak to Riyadh in a bid to win Saudi endorsement for the deal with Israel which had virtually been finalized. He succeeded. The Kingdom, guardedly but quite positively, approved the second disengagement agreement that was initialled on 1 September in Cairo and Jerusalem, then signed in Geneva on 4 September. Saud al Feisal described it as an important step in the right direction deserving the appreciation of all Arab states. The accord gave Egypt back 1,000 square kilometres of Sinai and the Abu Rudeis oil fields. That in itself was acceptable to the Arab consensus. But other provisions of the agreement confirmed Syria's fears. The two signatories to it pledged themselves not to resort to force in future. Israel was to withdraw from the strategic Mitla and Giddi passes facing the Suez Canal. They were to be demilitarized and 200 U.S. civilian personnel were to be stationed there to monitor observances of the accord with electronic surveillance equipment. Egypt agreed to allow non-military cargoes destined for Israel to pass through the Suez Canal.

Damascus attacked the arrangement and the P.L.O. condemned it. Saudi Arabia was placed in a quandary. The ink on the contentious document was hardly dry before Abdul Khalim Khaddam, Syrian Foreign Minister, and Yassir Arafat, Chairman of the P.L.O., arrived in Riyadh determined to sway the Saudi regime to Damascus's view. Khaled, Fahd and Saud al Feisal explained that they had only supported the move because of Egyptian promises that it would not relinquish Syrian and Palestinian claims. They felt that Sadat had a right to recover what Egyptian territory he could, but they were nearly convinced by Assad's argument and charge that the Egyptian leader had betrayed the Arab cause. The Kingdom felt pulled between Cairo and Damascus. It had failed to use its weight to ensure that the U.S. undertook to do something for Syria. For over a year it would be troubled by the deep division in Arab ranks resulting from the second Sinai disengagement agreement, which would be widened by the growing conflict in Lebanon.

In the Kingdom's efforts to assert itself in its own immediate environment and work towards stability there, Fahd's first priority was to initiate closer consultations with Iraq in an attempt to improve relations with it. The Baathist regime was still regarded as the most sinister threat to the Kingdom. The construction of the Al Batin military cantonment in the far north-east of the Kingdom was not coincidental in this respect. Apart from the Baathist commitment to the unification of the Arab world under its own woolly creed – an amalgam of socialism and nationalism – the Baghdad regime had not renounced its claim to Kuwaiti territory or jurisdiction over the channel linking it with the Iraqi port of Umm Qasr, including the islands of Warba and Bubiyan that had always been under the control of the neighbouring state. But in the spring of 1975 conditions were ripe for Fahd's initiative. The pact concluded between Iran and Iraq in March had not only eased what then was the biggest source of tension in the Gulf, but also seemed to indicate a new policy of moderation on the part of the Revolutionary Command Council. That impression appeared to be confirmed by the indications over the past year or so that Iraq had ceased trying to subvert the traditional Arab regimes of the Gulf. In 1972 it had entered into a fifteen-year Treaty of Friendship and Co-operation with the Soviet Union. By 1975 it was fully evident that President Hassan Bakr's regime was trying to reduce its dependence on and loosen its alignment with Moscow. The deal with the Shah had enabled it to set about crushing the Kurdish rebellion that had plagued successive regimes in Baghdad for over a decade. Riyadh was reasonably satisfied that the clique in power in Iraq now wanted to concentrate on consolidating its hold on the country and pressing ahead with economic development.

Fahd's trip to Baghdad in June for talks with the Iraqi Vice-President, 'our brother Siddam Hussein'[8] as the Crown Prince referred to him in an interview published on 1 April, had been planned before the death of Feisal. Nothing could have prepared the ground for it better than Yamani's efforts to mediate a settlement of the bitter dispute between Iraq and Syria over the utilization of the waters of the Euphrates River. This arose following the completion of the Tabqa Dam in Syria and the filling of the Lake Assad reservoir behind it, depriving Iraq of water, in the spring of 1975. As a result, Iraq was threatened with and, in the event, did suffer the loss of a large part of its rice crop. In May, Syria built up troop concentrations on the border. The Iraqis did not respond. Given its close political and economic links with both countries, the Soviet Union seemed to be in the best position to intervene. Saudi Arabia, however, seized the initiative. No precise agreement was reached, but at least the Saudi initiative defused the tension. On 3 June, a day after Khaddam had delivered a message from the Syrian President to Khaled in Riyadh, Damascus announced that it would make available extra supplies of water to Iraq.

Fahd also wished to ease the tension in the Gulf posed by Iraq's territorial dispute with Kuwait. This was the main subject of his talks in Kuwait prior to his arrival in Baghdad on 12 June. But the Iraqi regime was not disposed to entertain a third party intervening in what it saw as a bilateral matter and also one involving a militarily weak and exposed neighbour. The Revolutionary Command Council's clear preference was that the question should remain unresolved. Its existence afforded Iraq a means of putting pressure on Kuwait. Fahd made little progress. One immediate result of his meetings in Baghdad, however, was an agreement on the partitioning of the Neutral Zone between Saudi Arabia and Iraq on 2 July. Moreover, some basis of trust seemed to be established by the Crown Prince's visit.

Momentum towards filling the vacuum, for which Feisal's inflexibility, hesitancy and resentments had been largely responsible, was building up. In March 1975, the Shah had proposed a security pact grouping all the Gulf states. He had taken up the idea when he visited Riyadh on 28 April to make the acquaintance of the new Saudi King. The subject dominated the Crown Prince's talks with the Shah in Tehran in mid-summer. The communiqué issued on 3 July called for close co-operation amongst the littoral states of the Gulf. Saudi Arabia was opposed to a formal agreement of the kind that the Shah was pressing for. Such a pact, it was feared, would amount to an acknowledgement of Iran's leadership and dominance. The Saudi ruling hierarchy was as resentful as ever about the Iranian forces in Dhofar, the southern province of Oman, where the war against the Aden-backed insurgents was being won. It was not just a question of the slighted pride of a state that was the richest Arab power but one still weak in military terms. With good reason, Khaled and his colleagues were apprehensive that the involvement could serve as a precedent for Iranian interference or intervention in the smaller states of the Gulf. Bahrain and the United Arab Emirates, on the face of it the most unstable entities in the Gulf, looked the most likely areas where His Imperial Majesty might decide to exercise his pretensions as the policeman of the region. His high-handed seizure of Abu Musa and the Tunbs was still fresh in the minds of the Arabs and a source of festering indignation. For that reason alone the U.A.E., a strange hybrid confederation undergoing a crisis of identity, could hardly be expected to join an alliance with Iran.

Prior to Fahd's visit to Tehran, contacts had taken place amongst all the Gulf states. It had been agreed tentatively that a summit should be held in October. In the course of the Islamic meeting in Jeddah in mid-July, their foreign ministers met and established certain principles. The most important of them was the agreement to exclude the superpowers from the region and deny military bases to them. Other principles on the agenda that were approved were military co-operation to ensure freedom of navigation, the peaceful resolution of regional disputes, and collective guarantees concerning the territorial integrity of the states concerned.

Not until the end of September did Bahrain half-heartedly announce that it had requested the U.S. to withdraw by mid-1977 from the naval facility bequeathed by the British Government on its evacuation from the Gulf. Saudi activity in the region focused on the populous state whose benign, relatively liberal ruling family appeared to be the most vulnerable of all but also amenable to stern demands and financial inducements from the Wahhabite power across the narrow stretch of shallow water. In an unprecedented, albeit unconvincing show of muscle-flexing, Saudi paratroopers had participated in military manoeuvres with the Bahrain Defence Force on 4 June. The ruling hierarchy was intent on showing Iraq that it had the ability to intervene to preserve the established order. It had eyed with grave misgivings the experiment in parliamentary democracy initiated by Sheikh Issa bin Sulman al Khalifa, the Ruler of Bahrain. The Bahrain National Assembly had proved to be a rumbustious, dangerously radical body that thwarted Government legislation. Fahd in person insisted on its suppression. On 28 August, Sheikh Issa dissolved the National Assembly, suspending the constitutional provision that required two months' notice for such action.

With more decisiveness than in Feisal's time, the Kingdom was using its strength as the main paymaster of the Arab world to create conditions of stability. To this end it went in the summer of 1975 about as far as it was capable in collaborating and promoting moves towards greater co-ordination of the oil-producing states. Eliminating rifts that it deplored, rather than uneasily co-existing with them as Feisal had, had become the basic concern of its foreign policy. Yet in the Yemen Arab Republic, for which Sultan continued to take responsibility, Saudi Arabia still displayed an ambivalence born of more atavistic instincts.

Ibrahim Hamdi, who had become President of the Y.A.R. in mid-1974 following a military takeover, had asserted enough strength, moderation and popularity as to suggest that he was a man well suited to unite the divisive factions in the land. Towards the end of 1974 he had ended a flirtation with left-wing elements in Sana, which was bound to offend the Soviet Union, the Y.A.R.'s supplier of weapons. The extent of the opposition within the Army could be seen from the fact that the number of officers in the Command Council had been reduced in number from fifteen to five. In the summer of 1975 there was evidence that Muhammad bin Hussein, the most able of the Royalist commanders in the civil war, had been trying to stir up trouble with the Hashed and Bakhil, the great tribal confederations in the north. Hamdi looked to Saudi Arabia for the backing that it could give in the form of finance and also influence over the tribes. In August 1975, $100 million was given in the form of budgetary support, and $360 million in development aid. Sultan managed to exact a pledge that the Y.A.R. would sever its military connection with the Soviet Union. But its Armed Forces could hardly dispense immediately with Soviet military advisers. For its part, Saudi Arabia was ambivalent and hesitated about committing aid to finance armaments from other sources. Meanwhile, a cautious approach towards weaning the People's Democratic Republic of Yemen away from Moscow was under way. An opportunity for direct contact was made during the Islamic foreign ministers conference in Jeddah in July which was attended by Muhammad Saleh Mutia, the P.D.R.Y. Foreign Minister.

Militating against Saudi diplomacy in the Gulf were the growing stresses and strains within O.P.E.C. that centred largely on the Kingdom's determination to limit price increases. Other members had been growing restive because of the erosion in the purchasing power of their oil revenues due to inflation and the depreciation of the dollar. But as a result of the world-wide economic recession, for which O.P.E.C. was largely responsible, demand for oil had shown a marked decline. Saudi Arabia was more than happy to bear the brunt of the fall. Its revenues were far exceeding its expenditure and early in June Yamani had pointed out in a speech to the London Stock Exchange that, while the Kingdom was still prepared to produce at a rate of up to 8.5 million barrels a day, its maximum permitted ceiling, its financial requirements warranted no more than 3.5 million barrels daily. Thus, the agreement to freeze prices up until the end of the third quarter of 1975 did not come under serious question at the regular biannual ordinary conference held in Libreville, Gabon, in June. Neither Yamani, engaged in his mediation between Iraq and Syria, nor Jamshid Amouzegar, organizing elections at home, thought it necessary to attend in person.

By September, diplomatic moves following consultations between the U.S. and France were firmly under way towards the convening of a substantive dialogue between the industrialized and developing countries, thus satisfying Saudi Arabia's main stated condition for continued moderation within O.P.E.C. on the issue of oil prices. But just prior to the O.P.E.C. conference held in Vienna on 24 September the Shah of Iran gave his imperious, magisterial view that there should be an increase of no less than 15 per cent. That was enough to set the scene for a tense and dramatic meeting. Officially, the Kingdom was in favour of an extension of the freeze for an indefinite period – though that depended on the industrialized countries showing some honest endeavour about lowering export prices, an issue inextricably linked to the forthcoming dialogue.

When the meeting began, the battle lines were soon drawn up. Iraq emerged as Iran's ally, leading the campaign of the militant group that included also Libya, Nigeria, Qatar, Ecuador, and Gabon – several of them pressing for as much as a 20 per cent increase. Saudi Arabia obtained support from Algeria, hitherto regarded as a radical. As it happened, Yamani had visited Algiers before the conference. Their alignment was based upon a common interest and role in promoting the dialogue. Saudi Arabia was assured of the backing of the United Arab Emirates, while the position of Kuwait was ambiguous. Belaid Abdul Salem, Algerian Minister of Energy, took the initiative in proposing a compromise under which there would be an immediate rise of 5 per cent and a similar one at the beginning of 1976, followed by a freeze for the whole of that year. The maximalists kept their ranks, conceding only that a 15 per cent increment should be staggered, with 10 per cent from the third quarter and the balance of 5 per cent from the start of 1976 – with no commitment to any subsequent standstill. There was deadlock and an atmosphere of confrontation.

On 26 September Yamani flew to London without explanation. His sudden departure seemed like an act of brinkmanship. O.P.E.C. was more seriously split than ever before. The Saudi Oil Minister's apparent walk-out may have been calculated partly to cool tempers or jolt the majority into a more compromising frame of mind. His main purpose was to obtain better – and, it is believed, more secure – communications with Khaled and Fahd through the Saudi Embassy in London than were available in Vienna. His trip served to emphasize that skilled negotiator though Yamani was, he was constricted to guidelines laid down by his masters. On his return he found that the militants had not shifted their position. He then played the most powerful card possessed by the Kingdom – but one involving serious political risks – by threatening to raise Saudi oil production to full capacity. That was then officially said to be 11.2 million barrels a day, a rate that fifteen months later was shown to be much higher than the Kingdom could actually sustain technically or without causing damage to reservoir pressures, thereby endangering the life of important fields.

There may have been an element of bluff as well as courage in the tactics pursued. They worked, however. Faced with the prospect of over 4 million barrels a day of cheaper Saudi oil flooding the market, Iran and Iraq gave way. All members accepted a Kuwaiti compromise proposal that there should be a 10 per cent increase and price stability for a further period of nine months until mid-1976. Iran, however, did not accept gracefully what amounted to a defeat. The Shah's pride had been wounded. As a result of the 1973–4 escalation in oil prices his plans for economic development and the purchase of ultra-sophisticated weaponry had become more grandiose than ever. Speaking on Iranian television, Amouzegar referred to an 'unholy' alliance between Saudi Arabia and another unspecified country – Algeria. The unnamed partner had only followed the Kingdom's lead for fear of its leaving O.P.E.C., he said.

Tayeh Abdul Karim, the Iraqi Minister of Oil, put a better face on the whole affair. He pointed out that the agreement had strengthened O.P.E.C. unity and (in its moderation) concern for the world's economy. He also stressed the need for O.P.E.C. to agree on a system of differentials fairly reflecting the quality and transportation costs of various crudes. In sagging market conditions disparities had become marked. The heavier oils produced in the Gulf and the lighter one, including Iraq's, transported through the pipeline system and shipped from the Mediterranean, were over-priced in relation to Arabian Light, the Saudi staple that by then was firmly established as the 'marker' for others. O.P.E.C. economists had worked out a formula for co-ordinating differentials but it had not been properly discussed, let alone approved, at the Vienna conference because of the frenzied and fraught concentrations on the basic price increase. Kuwait called for a conference of Gulf producers to work out a structure. Saudi Arabia demurred. Yamani and his colleagues argued that it could only be solved in the wider O.P.E.C. context because the rates being charged not only by Libya but also Iraq for the crude pumped through the pipelines to the Mediterranean terminals were intrinsic to it. The meeting never took place.

Kuwait then adjusted and lowered the price of its heavy oil to a level that conformed with the recommendations of the O.P.E.C. experts. Iraq had a comparable crude that would be undercut in price. The Baathist regime, accustomed to brow-beating and bullying its small, rich neighbour, reacted with anger. The Iraqi Foreign Minister summoned the Kuwaiti Ambassador in Baghdad and delivered a stern protest. On 17 November, Al Thawra, organ of the ruling party, warned that if other oil producers insisted upon taking unilateral action without reference to O.P.E.C., then Iraq would have to 'take corresponding steps to protect its own interests and prevent any imbalance in the Arabian Gulf area'. The sinister threat was not lost upon any readers or listeners to Baghdad Radio, which duly transmitted such commentaries. The Kingdom was not directly attacked. But as the heavyweight oil producer most insistent on ironing out price disparities it was clearly a target. Only a few days earlier Yamani had stated that another increase in the price of oil would not be justified until the end of 1976 in the absence of a remarkable recovery of the worldwide economy. As the time approached for the next ordinary biannual O.P.E.C. conference, another fraught meeting seemed assured.

Since the three-fold escalation of prices at the end of 1973, for the world at large the stakes at any O.P.E.C. ministerial conference had been immeasurably raised. Yet the gatherings were still very informal affairs. The O.P.E.C. Secretariat in Vienna, where delegates met on the morning of 21 December, was hospitable, and anxious that member states should be allowed an opportunity to convey their views to the world. Thus, it has always allowed journalists and television crews to assemble in the very restricted area. They used to sit on the floor, covering it with their bodies and camera equipment and spilling out into the main corridor that led to the conference room. In December, for the first time, the presence of the media men there had been forbidden because of the overcrowding. If it had not, then the dramatic events that day could have had a different outcome.

O.P.E.C. was holding a reception in the evening to which some members of the press were invited and it was necessary to pick up a formal invitation from the Secretariat, which was then housed on the second floor of an office block on the Dr Karl Luger-Ringstrasse, the road encircling the medieval heart of Vienna. As I approached the building to collect my invitation I saw six figures trooping through the entrance. Half a minute later I passed through the same swing doors, heard shots from above, and was knocked to the ground by one of four journalists who had been lurking outside the conference room upstairs when Ilich Ramirez Sanchez – alias Carlos – and his companions burst in to hold the O.P.E.C. delegates hostage. The capture of the eleven ministers as they were engaged in a fierce debate over differentials, the aides accompanying them, and the secretarial staff was achieved with astonishing ease – at the cost of three lives. The first to die was Inspector Anton Tichler of the Austrian State Police Bureau. The other two fatal casualties were Ali Hassan Khalafi, one of the Iraqi Oil Minister Abdul Karim's bodyguards, and Yusuf Ismirli, a Libyan economist based in Vienna who was one of the organization's experts. The woman amongst the terrorists, later identified as Gabriele Kröcher-Tiederman of the Baader-Meinhof gang, gunned down Tichler and Khalifi. She later apologized for killing the Iraqi. Carlos killed Ismirli as the Libyan bravely grappled with him.

Carlos's importance as an operator for the Popular Front for the Liberation of Palestine and the extent of his terrorist activities in the past were not then understood. He was the playboy son of a Marxist millionaire who had enjoyed a sybaritic existence as an expatriate in London. It was not known at this point that he had attended the Patrice Lumumba University in Moscow and that he had trained in P.L.F.P. training camps in Lebanon. But he had earned himself considerable notoriety as the young man identified as having tried to murder, and very nearly succeeding, Lord Sieff, Vice-President of the Zionist Federation of Great Britain, at the Jewish businessman's home in St John's Wood, London, in December 1973. More dramatically he was known for having killed a Lebanese betrayer and two agents of the Direction de la Surveillance du Territoire when they had sought him out for interrogation at a party in a flat on the Rue Toullier, Paris, in June 1975. He was quick to introduce himself proudly as the

famous Carlos to the O.P.E.C. delegates and secretarial staff corralled
inside the conference chamber.

For the House of Saud and the world at large the motivation behind
the rude interruption of the O.P.E.C. meeting had become fully
apparent by the evening of 21 December. It was revealed all too clearly
by a statement broadcast over the Austrian state radio that evening.
Issued in the name of the 'Arm of the Arab Revolution', the
manifesto was indelibly imprinted with the jargon of the P.L.F.P.
Badly written in French and transmitted in the same language, it
condemned the 'treacherous agreements over the Sinai and the
reopening of the Suez Canal to Zionist trading'. The testament
asserted the 'principle of full sovereignty over "our" petroleum and
financial wealth through nationalization of petroleum monopolies and
the adoption of a national petroleum and financial policy which will
enable the Arab people to use its resources for its development, its
progress, the safeguard of its national interests, and the strengthening
of its sovereignty alongside the friendly people of the Third World so
that they can emerge from their economic stagnation, on condition
that priority be given to the financing of the confrontation countries
and the Palestinian resistance'. There was no explicit mention of what
the P.L.F.P. was wont to refer to as 'Arab reactionary circles', but it
was abundantly clear that the message was aimed at Saudi Arabia as
well as Iran.

Both the background to the affair and the development of the siege
have been well reconstructed and told.[9] Not the least of the sources was
Yamani, who gave his account on Saudi radio: I felt certain that I was
going to die and began to repeat from the Koran the following verse:
"To the righteous soul will be said: O thou soul, in complete rest and
satisfaction! Come back to thy Lord, well content thyself and well-
pleasing unto Him! Enter thou, then, among thy devotees! Yea, enter
thou my Heaven."' Yamani's initial instinct seemed to be correct. As
he crouched under the long green baize-covered table together with the
other oil ministers, their aides and secretarial staff, he heard one of the
Arabs in the gang ask, 'Have you found Yamani?' The Saudis, together
with their colleagues from Iran, the U.A.E. and Qatar, were separated
from the others into what the terrorists termed the 'criminal group'
and herded to the end of the room. The Iraqis, Libyans, Algerians, and
Kuwaitis were classified as 'liberals and semi-liberals', and the other
delegations as 'neutrals'. Carlos wrote a note addressed to the
Austrian Government and gave it to Grizelda Carey, the blonde

English secretary, who bravely took it to the police outside. It was a political statement, said Carlos, that he demanded should be broadcast over Austrian radio. If it was not, then he would kill a member of the United Arab Emirates' delegation first. An hour later a Saudi would be named and then an Iranian. Soon after, he enquired who was the deputy leader of the Saudi delegation. Abdul Aziz Turki, the bright-eyed and good humoured Deputy Minister of Oil, identified himself. He seemed to be the first Saudi victim singled out. But early in the afternoon, at about 3 p.m., delegates thought the hour had come for the Kingdom's most famous citizen when Carlos led Yamani away to an empty adjoining room. Carlos told him that if the Austrian Government had not, by 4.30 p.m., agreed to broadcast their statement, he would be killed at 6.30 p.m. 'Carlos then said that he hoped I would not feel bitterness against them because they intended to kill me, and that he expected a man of my intelligence and mind to understand their noble aims and intentions.' Yamani asked how, having been threatened with death, he could be expected to show no resentment. The Venezuelan then revealed that if and when the Austrians complied with his demand his intention was to go to Libya where Izzedin Mabrouk, the Libyan Oil Minister, would be released, and also Abdul Mutaleb Kazzimi if the terrorists decided not to stop off at Kuwait, his home state. The final destination would be Aden, Carlos told Yamani. Wadi Haddad, the master-mind of P.F.L.P. terrorist operations, including the multiple hijacking of aircraft to Dawson's Field, Jordan, in 1970, resided and had his headquarters in Aden. There, Yamani understood, he and Amouzegar would be killed. The 4.30 p.m. deadline passed and there had been no broadcast. Carlos, with a smile on his face, came to remind Yamani what would happen. At 6.20 the statement was transmitted in badly translated French, the language in which the message had been written.

At an early stage, Riyadh Azzawi, the Iraqi Chargé d'Affaires who had happened to be in the vicinity, had volunteered his services as an intermediary and acted as a go-between with the Austrian authorities. Within the O.P.E.C. camp, the Venezuelans had helped defuse the situation considerably by engaging Carlos in conversation, but the crucial role in handling the situation was played by the Algerian Minister of Energy, Belaid Abdul Salem, a formidable intellectual who ranked with Yamani and Amouzegar as one of the three heavyweights in the organization's debates and deliberations. Those present, not the least Yamani, later gave credit to him for his tact and diplomacy.

It was he who persuaded Carlos to agree that the captors and their hostages should fly first to Algiers rather than Tripoli, Libya. Of those representing the other 'liberal' regimes (Kuwait clearly was the 'semi-liberal'), Mabrouk of Libya seemed genuinely bemused and Abdul Karim of Iraq strangely silent.

Early the following morning the Austrian Airways DC9 made available to Carlos departed for Algiers. There, Abdul Salem and the Algerian authorities sought to persuade Carlos to release all the hostages 'in return for complying with his demands', as Yamani put it. By then, it was later revealed, they had come to include the payment of a large ransom from the Governments of Saudi Arabia and Iran, a condition that was conveyed to them by the Algerians. In the event all but fifteen of the hostages were released in Algiers, including Abdul Salem, but he insisted on staying with the remaining captives. Yamani recalled: 'The Algerians also made an attempt to force Carlos to promise that after going to Baghdad he would return to Algiers and make that his final stop. He agreed to this, unless he received orders while in Baghdad to continue his trip to Aden.' But he insisted on going first to Tripoli before proceeding eastwards to Iraq and the P.D.R.Y.

On the flight to Libya Carlos told Yamani how the Algerians had made entreaties to save his life and that of Amouzegar. 'He said that he was surprised to see their concern about two reactionaries, and then he spoke again about the plane that the Libyans would place at our disposal.' Manfred Pollack, the Captain of the DC9, had explained that the aircraft did not have the fuel capacity to reach Baghdad. Damascus would be possible. Carlos retorted that he did not regard it as revolutionary any longer. It was, indeed, under the influence and within the orbit of Saudi money.

It became clear that the terrorists were neither expected nor particularly welcome in Tripoli. For half an hour the DC9 had to circle the airport before Major Abdul Salem Jalloud, the Libyan Premier, was sought in the hope that he would make available an aircraft with a range capable of reaching Baghdad. He either could not or would not comply with the request. So the party returned to Algiers, although not before Carlos had tried a diversion to Tunis in an attempt to evade the firm moderating Algerian aegis.

Despite, or perhaps because of, the amphetamines that had sustained them, the terrorists were becoming jittery and indecisive. On the DC9's return to Algiers in the small hours of 23 December, Carlos and his colleagues discussed again the fate of their prime captives. The 'democratic decision' conveyed by Carlos to Yamani and Amouzegar was that they would be released at midday, though evidently this was displeasing to Kröcher-Tiederman and one of the Arab terrorists who was named Khaled. Carlos then told the ministers that their sentence was being deferred only. He warned them that if they escaped death this time, the terrorists would find them wherever they might be.

In Algiers a deal was finalized to the apparent satisfaction of the terrorists. The details have never been revealed but certainly involved a large financial transaction, as well as safe conduct. The hostages were freed. As they were relaxing for the first time in two days in a room adjoining the airport terminal, Khaled, it seems, was still determined to kill the Saudi Oil Minister. Having handed over a pistol to the Algerian security men, he was granted a last request to speak to Yamani and Amouzegar. In violent language he repeated the threat made by Carlos. A suspicious movement of his hand over his chest alerted Abdul Aziz Bouteflika, the Algerian Foreign Minister, who distracted Khaled by offering him a fruit juice. Security men then disarmed him of another gun in a shoulder holster concealed under his coat.

For the terrorists, the publicity for their cause and the extortion of a large ransom seem to have been the foremost reasons for their operation. On the final leg of the journey, from Tripoli to Algiers, Carlos himself revealed that he wished to protract the drama as long as possible. In practice, the actual limit was probably set by the gang's mounting fatigue. Although Yamani was convinced that the terrorists were set to assassinate him and Amouzegar, neither the ruthless war of nerves waged against them nor the pathological hatred shown towards them by Kröcher-Tiederman and Khaled proved conclusively that this was true. In the last analysis, their fate probably depended on the satisfaction of financial demands, The size of the payment has remained a secret, characteristically well kept by the Saudi Government. No details emerged from Iran when, after the revolution, the archives of the Shah were ransacked. The sum has variously been reported as anything from $5 million to $50 million, a portion of which is assumed to have gone into the private bank accounts of Carlos, Wadi Haddad and George Habbash, the political leader of the P.F.L.P. It would be surprising if the P.D.R.Y., perennially short of hard currency, did not also enjoy a share of the spoils.

For the Kingdom, a substantively more important question was which Arab regimes had been accomplices to the crime and to what extent. In the preamble to this account, Yamani explained: 'My freedom of speech is restricted by my feelings of responsibility, but there will come a day when the facts will be fully known, God willing, and the light of truth will unveil those who work in the dark.' In effect, he was constrained by political considerations about telling all he knew, deduced or surmised about what was no less than an indirect assault on the princely dynasty which he served. Thereafter, Yamani would be protected around the clock by a team of British security men provided under contract, all of them former members of the élite and covert Special Air Service regiment.

Algeria gave Carlos temporary refuge. His father was appreciative of how President Houari Boumedienne's Government treated him and even made a payment, according to the Marxist millionaire. He did not specify the amount. That has led some observers to believe that Algeria was a party to the conspiracy. Yamani's tribute to the manner in which 'our Algerian brothers' handled the situation indicates the contrary. For them to have been a party to the plot would not have been consistent with the country's general conduct at the time. But it was accustomed to giving sanctuary to a wide variety of revolutionaries, amongst whom were numbered the American black militant Eldridge Cleaver. The regime had given refuge to other hijackers acting on behalf of the Palestinian cause and could hardly have been expected to treat Carlos in a less courteous fashion. The probability also is that one of the Venezuelan's conditions for releasing the hostages was an invitation to stay in the country as the guest of the Government. Certainly, neither Saudi Arabia nor Iran raised any protest with Algiers.

Carlos referred to the Libyan regime as his 'bosses'. Qaddafi was a sworn enemy of Iran and, to a lesser extent, Saudi Arabia. He was known to finance handsomely, if capriciously, terrorist or revolutionary movements that might in any way discomfort his foes and, in particular, the trinity of evils that he saw as being represented by the U.S., imperialism and international capitalism. Qaddafi had funded, on a large scale it was believed, the P.F.L.P. Tripoli was Carlos's first chosen destination. Yet he was neither expected nor welcomed. One explanation for the reluctance of the Libyan regime to greet and harbour the terrorists could have been that it did not want to be seen to be associated with the holding to ransom of O.P.E.C. The Libyan behaviour betrayed an element of pique and embarrassment.

It is difficult not to conclude that these 'bosses' were not a party to the planning of the operation and probably had not been consulted about it. The fact that later, after his sojourn in Algeria, Carlos is generally assumed to have been given a home and protection in Libya, even if true, is not in itself evidence of active complicity.

The Arab country directly involved was Iraq. Almost exactly two years later Kröcher-Tiederman was arrested, together with another member of the Baader-Meinhof gang, after a shooting affray on the West German border. Her capture provided the first chance to interrogate a member of the Carlos group that had kidnapped O.P.E.C. According to two reliable informants, she revealed under intensive questioning that a senior member of the ruling Baath Party had provided detailed plans of O.P.E.C.'s headquarters. It was ironic that Abdul Karim's bodyguard should have been killed when the Secretariat was stormed and that the Iraqi Chargé d'Affaires should have acted as mediator between the terrorists and the Austrian police.

Five years later much still remained obscure about the kidnapping, the motivation behind it and the combination of forces conspiring against the two leading members of the producers' organization which, in a further irony, were at loggerheads over pricing policy. At least the affair highlighted the malevolence of revolutionary Arab elements working against a conservative oil power behaving responsibly towards the wider world. Saudi Arabia's use of the oil weapon in 1973 should have established its pan-Arab nationalist credentials. Yet, perversely, its enhanced status and wealth only increased the reliance placed upon it to use its power to give vital assistance in bringing about a Middle East settlement. Conversely, the Kingdom's patent desire for one, its general moderation and close links with the U.S., exposed it to the hostility of radical Arab states and ideology that were opposed to any compromise with Israel. Moreover, the holding to ransom of O.P.E.C. occurred at a time when the community of Arab states was in the process of being torn apart more seriously than ever by the second Israeli-Egyptian disengagement agreement and the related civil war in Lebanon.

The divisions over the regional issue and the more localized strife became inextricably entangled and the misery of that hapless country compounded as inter-Arab rivalries were fought out, by proxy, on its soil. By the end of the year the fighting between the right-wing Christian paramilitary groups and their predominantly left-wing opponents, who actually supported the Palestinian cause, had reached a new level of intensity. The guerrilla presence in Lebanon that had swollen since King Hussein of Jordan's showdown with them in 1970 – 1, the license granted to the guerrillas to mount operations against Israel from a constricted area of the southern border, and the resulting Jewish retaliation had been the underlying cause of the rift between the two communities. At the end of 1975 the only redeeming aspect of a grim situation had been Yassir Arafat's partly successful efforts to keep the mainstream guerrilla movement out of the conflict. Even so, despite the Syrian attempts to mediate, the manoeuvres of President Suleiman Franjieh and Rashid Karame, the Prime Minister, to achieve a 'national reconciliation' seemed to grow more hopeless by the week.

Franjieh's policies were generally blamed for preparing the ground for the conflict. A feudal chieftan from the north, he was a Catholic Maronite in accordance with Lebanon's unwritten 'National Covenant' on which the country's delicately balanced confessional system had been based since its independence in 1946. In 1957, after his own clan had perpetrated a bloody massacre of a rival one in the Zghorta area – at a funeral service where the two had met to put an end to a blood feud – he had sought refuge in Syria and there made friends with Assad. Franjieh scarcely seemed the right person to bring about national reconciliation. Rashid Karame held out far better hopes. He had been called upon to form a government in the early days of the conflict as the most acceptable candidate to the left and Muslims, as well as a politician with close links with Damascus.

Syria had been using its influence to bring about an accommodation. It had a very direct interest in bringing about stability in Lebanon for a number of reasons. Discord there could easily have disruptive repercussions in its own heterogeneous society. Damascus had been and still was the main champion of the Palestinian resistance movement, but sought to prevent it from provoking a war with Israel at a time not of Syria's choice. Looking further ahead, Assad wished to exercise a measure of control over Lebanon and extend a wider front against Israel.

For the House of Saud nothing could have been more reassuring than the triangular configuration established in 1972 – 3 with its points reaching Cairo, Damascus and Riyadh. The fragile geometrical pattern was shattered by the second Egyptian-Israeli disengagement agreement. The fragmentation was made even worse by the Lebanese civil war. Sadat set out to embarrass Assad as much as possible over his involvement in Lebanon. Both sought the support of the main paymaster of the Arab world. For the Saudi hierarchy the equation was complicated by the Iraqi Baathist regime's bitter antagonism towards its Syrian rival and the opportunities presented by the Lebanese imbroglio to embarrass it. A further dimension, both dangerous and challenging, was added by Libya. Qaddafi was determined to use his petrodollar surpluses as a means of influencing Syrian policy decisively in favour of the Palestinians and a Lebanon in open confrontation with Israel. Relieved at the complete suspension of guerrilla activity on the border, the Jewish state watched with pleasure the spectacle of the Arab world tearing itself to pieces and sought to exploit the situation by providing the Maronite Christians with weapons and the build-up of a buffer zone along the border defended by friendly Lebanese.

As the possibility of Lebanon becoming dismembered became real, Saudi Arabia came out firmly in favour of the maintenance of its territorial integrity. Ever fearful of the radicalization of the Arab world, it was anxious to preserve the traditional system and leadership, Christian as well as Muslim. For rich Saudis, not the least princes, Lebanon had become the most important summer playground, where decadent Western delights could be enjoyed in an Arab environment, especially after Nasser's nationalization measures and the puritan stamp of his regime had made Cairo and Alexandria less attractive. During the Yemen civil war they had become inaccessible and then unwelcoming. Since the late 1940s, Saudi princes and merchants had developed business relations with the prosperous bourgeoisie of Lebanon dominating the country's political structure and commerce, and had invested heavily there in real estate. The Lebanese, in turn, had profited from contracts carried out in the Kingdom. The royal family in particular had long cultivated links with the conservative Sunni Muslim establishment.

The most important nexus was with Saeb Salem, the conservative Sunni politician whose family power base was Beirut. As the situation deteriorated, Saeb Salem had visited Saudi Arabia in November 1975. It was significant that on his return Salem delivered a verbal onslaught on communism and other radical ideologies. He was believed to be the main Sunni conduit for Saudi subventions aimed at propping up the crumbling system.

The Salems and other leading Sunni families had only a limited armed following amounting to no more than bands of retainers kept to protect their households and assert control over their immediate sphere of influence. Organized Muslim strength under traditional control was concentrated in the forces of Kamal Jumblatt, the leader of the heterodox Druze sect, and his Progressive Socialist Party. Given the feudal nature of its chieftain, the P.S.P. seemed inaptly named, but Jumblatt and his fierce warriors were committed to the left and the Palestinians. Their armed potential was spread amongst no less than a dozen parties and groups financed by Libya, Iraq and, to a lesser extent, also Egypt, which was bent upon causing Syria maximum discomfiture over its Lebanese involvement.

In this situation the conservative Sunni establishment appreciated that its common interest with its Maronite counterpart overrode religious loyalties and differences. The latter had paramilitary forces organized and trained. Foremost amongst them were those of Pierre Gemayel's Phalangist Party and former President Camille Chamoun's National Liberal Party. The Saudi hierarchy was well acquainted with the Maronite leadership, regarding it with less distaste than they did the Shiites or members of heterodox sects such as the Druzes and the Alawites. As trouble brewed and tension rose in 1974, Gemayel had flown to Riyadh and opened the way for covert Saudi assistance, mainly financial but also material in the form of some military equipment. It was a measure of the complexity of the civil war and its regional ramifications that Saudi Arabia should have been in the same camp as Israel through the sustenance given by it to the main and most conservative Christian community. The paradox was that the Kingdom's aid was directed at the preservation of Lebanon as an entity while that of the Jewish state was aimed at prolonging the chaos and, in the long term, restoring the dominance of the Maronites or – if that proved impossible – bringing about the partition of the country. However, Saudi Arabia is believed to have stopped aid to the Christian Maronites at about the turn of 1975. In the meantime, the Kingdom continued granting subventions to the P.L.O. and Al Fatah, the mainstream group of the Palestinian movement, whose arch-enemies were those same right-wing Christians.

Sharing its concern about the maintenance of the unity of Lebanon, Saudi Arabia supported Syria's attempts to mediate a compromise amongst the various political factions and its creeping intervention. Cairo responded to Damascus's strident attacks on its second disengagement agreement by both criticizing its policy in Lebanon and trying to thwart it. Riyadh felt in danger of being caught in the cross fire of recrimination between them and being forced to take sides. The Saudi hierarchy was prepared to pay a heavy price in aid to keep on the best terms with both Egypt and Syria which, in turn, both wanted the patronage of the greatest Arab oil power. Nevertheless, the balance was a difficult one to keep. Sadat could only regard Khaled's visit to Damascus in December 1975 with jealousy and suspicion. A month previously Syria had gained prestige through its initiative in securing a debate in the U.N. Security Council on the Palestinian issue that was scheduled to begin on 12 January 1976. It ended in predictable anti-climax two weeks later with a U.S. veto of a hard-line resolution which had Saudi Arabia's backing.

By January the threat of partition had become very real as the Maronites consolidated their enclave based on Mount Lebanon, a process that was given added impetus by the influx of refugees from isolated Christian communities and most notably the town of Damour (the home of the Chamoun family). On 7 January, Khaddam, the Syrian Foreign Minister, had given a warning that Syria would not tolerate the dismemberment of Lebanon and hinted that it would absorb the country rather than allow it to be fragmented. The point was emphasized heavily when in the third week of the month Assad mobilized a brigade of the Palestine Liberation Army which crossed the Lebanese border and took up positions on the eastern side of the central mountain range. The deployment of the Syrian officered and controlled force, over which Arafat exercised no authority, amounted to the first direct military intervention, finely balanced to deter secessionist moves by the right-wing Christians and to reassure their left-wing Muslim opponents. Assad summoned Franjieh and Karame to Damascus. They formulated a seventeen-point programme based on Syrian proposals that promised a more equitable distribution of power between Christians and Muslims but reaffirmed the restrictions on Palestinian guerrilla activity dating back to 1969. There was a brief respite in the conflict.

By backing Syria in its Lebanese imbroglio, Saudi Arabia had perceptibly strained its relations with Egypt. It may have been a slip of the tongue by Sadat in an interview, but one bound to touch Riyadh's sensitivities, when having lashed out at Syria, he pointed out it was receiving support from the Kingdom. Even such a passing reference could not be ignored. Nevertheless, plans for a five-day state visit by Sadat in February at the start of a tour of the Arab oil states went ahead. Saudi Arabia knew its strengths, an abundance of surplus funds, and Egypt's weakness, a desperate lack of finance and a yawning balance of payments deficit. The Egyptian leader's main intent became clear when he arrived accompanied not only by his Foreign Minister but also his Ministers of the Economy and Planning. Sadat was given an immediate grant of $300 million, rather less than he probably hoped for. A statement in Khaled's name called upon all Arabs to render urgent assistance to Egypt. But Saudi Arabia, like Kuwait, was becoming restive at the prospect of having to alleviate indefinitely Egypt's hand-to-mouth, undisciplined attempts to overcome chronic short-term economic problems through large dollops of aid not tied in any way to development projects. On the pan-Arab front, the communiqué issued at the end of the five-day visit was non-committal.

From a position of some confidence Saudi Arabia set about in earnest, but with discretion, to reconciling the regimes of Sadat and Assad. For its initiative it found a ready collaborator in Kuwait which felt equally vulnerable to rifts in the Arab world. It was a mediation attempt that sought to avoid antagonizing Iraq and the Palestinians but risked provoking it because of the basic approval clearly given by the two oil powers to Assad as he became more immersed in the turbulent waters of the Lebanon. The potential hazards became apparent as a result of an extraordinary tactical change of direction by Damascus enforced upon it by events that it could not control. The uneasy and by no means total ceasefire imposed by Syrian persuasion and cajolery in February broke down in the second week of March. Suddenly, Assad was faced by the prospect of anarchy on a bigger scale than ever, placing in grave jeopardy his attempt to reconstruct a disintegrating entity. The Lebanese Army finally broke apart along confessional lines. Precisely because of this danger it had for the most part stood aside from the communal fighting and made no effort to intervene. The predominantly Muslim left-wing forces and Palestinian guerrilla groups, with the exception of the Syrian-sponsored Saiqa, turned upon their adversaries, driving them from the luxury hotels in central Beirut and the port area, threatening the Maronites altogether. Responding to the challenge, Syria asserted itself more strongly. Suddenly, the regime that had been the main protector of the Palestinian resistance found itself in confrontation with it. Even more ironically, its main instruments on the ground at this point were both Saiqa and, to a lesser extent, the Palestine Liberation Army, both of them nominally Palestinian.

Towards the end of March, Jumblatt went to Damascus to plead with Assad, but in vain. Relations with the Lebanese left were virtually ruptured. Even so, Assad had finally concluded that Franjieh would have to go if there was to be any chance of a national reconciliation. As he orchestrated Lebanon's bourgeois politicians, the Chamber of Deputies met on 10 April dangerously near the so-called 'green line' separating Christian and Muslim combatants. All ninety members present agreed and voted unanimously to amend the constitution so that new elections could be held six months before the expiry of Franjieh's term of office. In a second ballot on 10 May, Elias Sarkis, the Governor of the Central Bank, who was the Syrian nominee, received the sixty-six votes, a two-thirds majority, required to succeed.

In April, President Ford praised the Syrian role as being constructive and indicated that direct Syrian military intervention on a limited scale would not be unacceptable. The Israeli Government, relishing the sight of the Syrians coming to blows with the guerrillas, acquiesced in their increased involvement. Saudi Arabia was the party making it possible for Syria to bear the mounting cost of the adventure. Egypt was too far out in the cold, because of the second interim accord, and also too much in need of finance which only the Arab oil producers could give, to capitalize on the situation, especially as far as its main benefactor was concerned.

At daggers drawn with its Syrian rival, the Baathist regime of Iraq tried to bring pressure to bear on Saudi Arabia. In mid-April Siddam Hussein alighted upon Riyadh in a bid to undermine Saudi support. He was followed, hot foot, by Major-General Naji Jamil, Syrian Deputy Minister of Defence and Commander of the Air Force, bearing a letter from Assad for the King. Saudi Arabia and Kuwait stuck to their course. Saud al Feisal and Sheikh Sabah al Salem al Sabah, his Kuwaiti counterpart, obtained provisional agreement on a meeting at prime ministerial level that was to have taken place on 19 May. It did not materialize, primarily because of Syria's insistence on putting the question of the second Sinai accord on the agenda on a level of equal parity to the Lebanese issue. More important was Damascus's determination to impose its own will on Lebanon before having any dealings with Cairo.

At the end of May, Syria sent its own troops in for the first time to relieve beleaguered Christian villages in the far north of the country. Three days later a force of more than 4,000 and 250 tanks thrust into the Bekaa Valley to relieve the town of Zahle which was under siege by Palestinian guerrillas. A thrust to the south-west met stiff opposition from the commandos. The Syrian military presence in the country had become an established fact. From the beginning of June until the end of the conflict, moreover, Syria sided with the right-wing Christian militias against the Lebanese left and their Palestinian allies, a policy that created discontent amongst Syria's Sunni majority but caused no qualms to the Saudi leadership which continued to give its undemonstrative blessing to Assad.

Libya, in the person of Jalloud, made a vigorous attempt to persuade Syria to adopt Qaddafi's view of what Lebanon should be – a state with a predominantly Muslim outlook in open confrontation with Israel. From early June he spent nearly four weeks in pursuit of this objective, dangling offers of financial aid in front of the Syrian regime and tirelessly trying to arrange a ceasefire. Arab foreign ministers met in Cairo on 9 – 10 June and decided to establish a joint peace-keeping force, which would include Syrian troops. Mahmoud Riad, Secretary-General of the Arab League, attempted a mediation mission and Hassan Sabri Kholy was appointed as the organization's on-the-spot negotiator. Unflaggingly he tried to arrange ceasefires. Assad tolerated these efforts at mediation but, having committed himself, continued to tread doggedly a path that exposed him to the hostility of other Arab countries, the risk of Israeli retaliation, and the displeasure of the Soviet Union. Tension rose as Iraq massed troops on Syria's eastern border. There was a corresponding strain in relations between Riyadh and Baghdad.

The Kingdom had the first visible reward for its patient, somewhat stealthy, petrodollar diplomacy when on 23 June the two Prime Ministers, Mahmoud Ayyoubi of Syria and Mamdouh Salem of Egypt, accompanied by their Foreign Ministers, met in Riyadh in the presence of Saud al Feisal and his Kuwaiti colleague. On balance, Assad emerged the victor. There was a reciprocal understanding on a cessation of hostile propaganda. Though it did not recognize the legitimacy of Syria's role in Lebanon, Egypt agreed not to thwart Assad's design. But the lip-service in the communiqué meant nothing. The Christian militias were unimpeded in their siege of Tal Zaatar. The Palestinian refugee camp and guerrilla stronghold finally fell early in August. Assad was left to bend the political shape of Lebanon to his will and physically subdue the greater part of the country. The Christian Maronites, saved by Syria's direct intervention, were left in control of their mountainous heartland. The south remained a vacuum, where the civil war would continue to be fought out on a smaller scale, because Assad could not dare cross the 'red line' set by the Israelis, an undefined concept but one understood to approximate geographically to the River Litani, without incurring retaliation that he could not face militarily. Syrian forces encroached on Beirut but for the time being did not enter it.

By the middle of October Assad was ready for the grand reconciliation with Sadat that the Saudi leadership had invested so much effort and money to achieve. On 18 October they attended the restricted summit in Riyadh presided over by Khaled, together with President Elias Sarkis and Yassir Arafat. The main points of a settlement were quickly agreed: a ceasefire to take effect from 21 October; the creation of an enlarged 30,000-man Arab peace-keeping force under Lebanese command; a return of the combatants to positions held before 13 April 1975; and a cessation of hostile propaganda between Egypt and Syria. The full-scale meeting of Arab foreign ministers that convened in Cairo on 25 – 26 October endorsed the agreement and approved that a committee made up of representatives of Saudi Arabia, Kuwait, Egypt, and Syria should be responsible for implementing it. Protesting that decisions had been taken at a limited gathering of select states, in contravention of Arab League rules, Iraq voted against it. Libya would have voted against it but its delegation, which had not been expected anyway, arrived too late to do so, while Algeria expressed its dissent.

The Syrian presence in Lebanon had been legitimized. From the beginning it was obvious that the Lebanese command of the Arab Deterrent Force (A.D.F.), as the peace-keeping operation became known, would be only titular and that the A.D.F. would be predominantly Syrian. Indeed, the A.D.F. had not been properly constituted before Syrian armour and troops descended from the hills into Beirut and took up positions between the Muslim and Christian sectors. The contingents provided by Saudi Arabia, the United Arab Emirates and Sudan seemed almost superfluous to the 22,000 or so Syrian troops occupying the country.

The outcome constituted a triumph for Assad but it was one in which the House of Saud shared and for which it was largely responsible. For his part, Sadat was relieved to be swimming again in the mainstream of Arab life. The essence of the compromise was that Cairo would accept the primacy of Syria's role in Lebanon and reciprocally

Damascus would acknowledge as an accomplished fact the second Sinai disengagement agreement. The reconciliation was consummated at the tripartite summit held in Riyadh in December where Egypt and Syria agreed once again to put their armed forces under joint command. Thus the solidarity of the three Arab powers was recreated. The maintenance of close harmony was necessary if they were to exact an honourable peace settlement from the renewed Middle East peace initiative promised by the incoming U.S. Administration of Jimmy Carter.

The general détente in the Arab world and the re-establishment of the Saudi-Egyptian-Syrian axis was greatly reassuring to the Kingdom. But Iraq did not disguise its anger over the Lebanese settlement, venting its wrath on the vulnerable oil-producing states. Early in November a statement was issued by the Baghdad-based 'National Command' of the Arab Baath Socialist Party (the directorate concerned with pan-Arab unity, as opposed to the 'Regional Command' whose concern was internal affairs). It warned that the Arab 'masses' would settle accounts with the 'reactionary regimes' of Saudi Arabia and Kuwaiti – and seemed a sinister reminder of the holding to ransom of O.P.E.C. nearly a year previously.

Whilst the main concentration of Saudi diplomacy had been on ending the Lebanese civil war and mending the Egyptian-Syrian rift, apprehensions about Iraq's intentions only served to spur a more active policy in the immediate region aimed at consolidating the Kingdom's position in the Arabian peninsula and the Gulf. The contacts made with Aden through Mutia, the P.D.R.Y. Foreign Minister, bore fruit in March 1976 when diplomatic relations between it and Saudi Arabia were established. Thanks to Iranian, as well as British and Jordanian assistance, the Sultan of Oman had, as he claimed in December, effectively defeated the rebels backed by the P.D.R.Y.'s Marxist regime. But cross-border fighting had continued. Now Saudi Arabia was able to use its good offices to arrange a ceasefire. The exchange of ambassadors between Riyadh and Aden seemed to indicate that the more moderate faction in the ruling N.L.F. might have gained ground on the militants. Hopes were raised that the P.D.R.Y. might be weaned by plentiful aid.

To the north, President Hamdi of the Y.A.R. looked with some misgivings on the possibility of a genuine rapprochement between the Kingdom and the P.D.R.Y. He also evinced a desire to lessen his regime's dependence on the Kingdom by embarking on a tour of the other oil-producing states of the Gulf, and later in the year visited China, North Korea, and Pakistan. The Y.A.R., however, seemed more safely secured within the Western sphere of influence by President Ford's approval of $100 million in military aid in June.

With solemn purpose the Kingdom set about asserting its hegemony over the conservative states of the Gulf. Towards the end of March 1976 Khaled made a regal progression through Kuwait, Bahrain, Qatar, the United Arab Emirates, and Oman. The King and his entourage stressed the need for close co-operation to guarantee the security of the region. The possibility of a Gulf summit was discussed. But the Saudi leadership wanted it restricted to Kuwait, the Emirates, and the Sultanate with their common interest in survival. The Shah's proposal for a joint pact including Iran and Iraq was looked upon in Riyadh with as much suspicion as when it had been put forward a year earlier. This was made as clear as Saudi politeness and ambiguity would allow when Abbas Hoveida, the Iranian Premier, came to Riyadh for talks early in April.

Another issue cooling the atmosphere was the continued presence on Omani soil of the Shah's expeditionary force. With the end of the insurrection in Dhofar, their continued presence seemed a bigger violation of the essential Arabism of the peninsula and an embarrassing insult to the Kingdom. Hoveida was able to claim that they were there at the request of Sultan Qaboos. For his part, the Shah, his Persian pride as prickly as ever, had been infuriated by the plan being considered by a committee of the six traditional Arab regimes for an 'Arabian Gulf News Agency'. As always, he saw red when the Gulf was referred to as anything but the 'Persian Gulf'. His displeasure was conveyed by his faithful servant Hoveida. Mutual comprehension appears to have been slightly improved by Khaled's state visit to Tehran in May, but the latent strain caused by differences over oil pricing was not eased. Meanwhile, one can only speculate how the Shah reacted, if his sycophantic advisers who sifted the world's press for his viewing saw fit to show it to him, to an article by the American columnist Jack Anderson in which he quoted Yamani, in private conversation, as describing the Shah as 'highly unstable mentally'.[10]

Saudi Arabia's conservative shadow seemed to be cast beyond its own frontiers when at the end of August the Ruler of Kuwait suspended the National Assembly in the thirteenth year of its existence. For good measure he also stopped publication of two radical magazines. The ruling Al Sabah family would not have welcomed or, necessarily, followed any heavily-weighted Saudi advice on the matter, as the Al Khalifa of Bahrain evidently had a year previously. But it was frightened enough on its own account by the passions unleashed by the Lebanese civil war, especially amongst its Palestinian residents, who were denied participation in the restricted democratic exercise, and the sympathies evoked by them amongst liberal young Kuwaitis. The factious little parliament had, by any standards of good administration, proved to be an obstacle to progress through its refusal to pass much-needed legislation. Nevertheless, the Saudi regime silently nodded its approval of the measure and the mass deportation of illegal or dangerous Arab immigrants rounded up by the Kuwaiti authorities.

The United Arab Emirates was also the object of Saudi concern. The political and economic development of the federation was being stultified by the differences between the Rulers of Abu Dhabi and Dubai. Towards the end of 1976 it seemed to be drifting out of control. In these circumstances Naif embarked in October 1976 on a round tour, hardly noticed at the time but of some significance, for consultations with fellow Ministers of the Interior of Kuwait, the Emirates, and the Sultanate. For the six conservative Arab states, the somewhat nebulous concept of Gulf security had taken on a more coherent meaning. Specifically, it embraced the pooling of information about criminals, malcontents and dissidents, and some kind of scheme whereby they should collectively ensure the respective established orders of each other's states. It was a process, at an early stage of evolution, that none of them, for obvious reasons, wanted Iraq or Iran to be involved in. Yet some kind of obeisance had to be made to the two strongest political entities in the Gulf. Thus, they were invited to attend the meeting of the region's foreign ministers that took place in Muscat at the end of November 1976. Anxious in their different ways to extend their influence in the Arabian peninsula, as well as to exclude super-powers from the Gulf, they argued in favour of a pact. The six politely rejected the proposal, suggesting in response that no formal arrangement was necessary.

At the end of 1976 O.P.E.C. politics once again impinged upon Saudi Arabia's preoccupation with ensuring maximum stability and security within its own immediate environment. In the first few months of the year, however, it seemed as if the oil producers – even those which might have been a party to the kidnapping – had been numbed to the

point of concussion. It was as if Carlos, in contradiction to his stated
political aims, had stunned the argument over differentials and also
rendered impossible any further questioning of the nine-month freeze
in prices virtually dictated by Saudi Arabia at the O.P.E.C. conference
in Vienna in the previous autumn.

There were other reasons for the eerie lull, however, which
coincided with the build-up of pan-Arab tension over Lebanon.
Demand for oil was recovering fast, especially for the lighter, short-
haul crudes shipped from the Mediterranean. Thus, Algeria, Libya,
and Nigeria were content for the time being. In the spring of 1976
Iraq, as a result of a dispute with Syria over transit dues and the
Lebanese civil war, stopped pumping crude altogether through the
pipeline system to Banias and Tripoli. Pushing all its exports through
its Gulf outlets, which had undergone an expansion in capacity, Iraq
was content to let the issue of differentials rest. In addition, member
states of O.P.E.C. other than Saudi Arabia wished to see how the
dialogue progressed before deciding their stance. In December 1975,
under the title of the Conference on International Economic Co-
operation (C.I.E.C.), substantive discussion had begun in Paris
according to a format acceptable to the developing countries and oil
producers that gave equal weight to the questions of energy, raw
materials, development, and financial affairs. In the early part of 1976
Yamani had intensified his talks with the four U.S. partners in
Aramco on a full state takeover and was able in March to announce
that the principles of an agreement had been concluded, though no
details were released.

Well in advance of O.P.E.C.'s ordinary ministerial conference in
Bali in June 1976, the Saudi Oil Minister had given notice that the
Kingdom was in favour of a prolongation of the standstill in prices
until the end of the year. Another confrontation seemed to be building
up. In anticipation of it Yamani visited Kuwait, the U.A.E. and Qatar
in a bid to rally support. Other members sought an increase of at least
20 per cent, justifying the demand on the basis of the calculation by
O.P.E.C. economic experts that the cost of their imports from
October 1975 to mid-1976 would have risen by 26 per cent. Pressure
for a significant increase seemed irresistible.

Draconian security was imposed on the meeting on the idyllic island
leaving the media men gathered for the occasion with little to do
except surf, enjoy the fabled female company, and eat magic
mushrooms. The barricades erected around the meeting could not

conceal the intensity of the debate as Saudi Arabia found itself cornered alone with, in the final outcome, some sympathetic support from Kuwait. Iraq and Libya headed the pack baying for fullest compensation for the erosion in the purchasing power of their oil revenues. Venezuela and Algeria assumed the role of mediators. Iran stood back awaiting a chance to exploit the situation but avoiding any risk of humiliation. The majority eventually were willing to compromise at 5 per cent. Yamani stuck to his brief and held his ground.

Saudi Arabia emphasized that its oil pricing policy was conditional on progress in the C.I.E.C. No agreement was in sight. The prospect of a package deal heralding the bright new economic order envisaged by O.P.E.C. seemed remote. The Third World cried out for a solution to the problem of its indebtedness, protection of its export earnings, and a fairer deal generally. The participants in the C.I.E.C. set a deadline of mid-December for a final ministerial session. So far from agreement was the conference, however, that it was postponed indefinitely and the work of its four commissions suspended for three months. Then, shortly before O.P.E.C. convened for its conference in Doha, Qatar, in December 1976, a leaked cable from the U.S. State Department to the American delegation seemed to prove conclusively that Washington's tactics were mainly, as the Third World representatives suspected, aimed at driving a wedge between the oil-producing states and other developing countries.

Tension had been raised prior to the Doha conference by accusations levelled by Yamani against members who 'felt free to reduce prices as they wish' – a clear reference to Iraq which, with its Mediterranean outlets closed, was believed to be selling oil at a discount in an attempt to increase shipments from the Gulf. Nevertheless, the deadlock in the dialogue reinforced the impression that Saudi Arabia regarded a price increase as inevitable. At a secret meeting in Taif in September, Yamani had agreed with his Algerian and Venezuelan peers that there should be one of 10 per cent, if there was progress in the dialogue, and as much as 15 per cent if there was none. The Shah had given notice that he wanted an increase of 15 per cent. Iraq, it materialized, was aiming for nothing less than 26 per cent, the rate of inflation that O.P.E.C. economists calculated for the cost of goods imported by members since October 1975. On his arrival in Doha, Yamani said that the Kingdom was in favour of a continued price freeze. It seemed like an initial bargaining position. Just how little scope Yamani had been allowed by Fahd when the hour came emerged on the second day of the conference, 16 December, when he flew to Jeddah for consultations with the King and Crown Prince. On his return the following morning he told the other delegates that the maximum the Kingdom could offer was 5 per cent. Even the more conciliatory delegations – Venezuela, Indonesia, and Kuwait – were dumbfounded. Only the U.A.E., whose chief delegate Mana Otaiba was under strict instructions from Sheikh Zaid to follow the Saudi line, raised its voice in support of the Kingdom.

O.P.E.C. was hopelessly divided. The majority of eleven members decided to go their own way by setting a price increase of 10 per cent. Saudi Arabia and the U.A.E. stuck at 5 per cent. That in itself was an act of bold defiance exposing the Kingdom to the wrath of other members, not the least Iraq and Iran, and allegations of being too responsive to Western and, in particular, U.S. pressures. Even more provocatively, Yamani declared in the wake of the conference Saudi Arabia's intention of taking the battle to the market place. It would open the taps to the full extent of Aramco's capacity which, he said, amounted to 11.8 million barrels a day, in an attempt to force down the average O.P.E.C. price. The latter, Yamani assured, would not rise by more than 5 per cent and the production of the eleven members would fall by a quarter. As for the future of the organization, he said with realistic confidence: 'Saudi Arabia needs O.P.E.C. and O.P.E.C. needs Saudi Arabia.'[11] Some members, like Venezuela and Kuwait, expressed sorrow. Others, notably Iraq and Iran, were almost explosive in their anger.

On his return to Baghdad, Abdul Karim accused the Kingdom of acting 'in the service of imperialism and Zionism'. His delegation had 'unmasked Saudi Arabia as a defeatist and reactionary cell working inside and outside O.P.E.C. against the interest of oil-producing countries and other developing states'.[12] *Rastakhiz*, the newspaper of the Iranian party (of the same name) that had been given a monopoly of political activity by the Shah and was being organized by Amouzegar, commented that 'the Third World and all the anti-colonialist elements in the world express their hatred towards Yamani for selling the interest of his nation to the imperialists'.[13]

Saudi Arabia's dogged stand was greeted with rapturous applause in the U.S. President Ford, who was about to leave the White House, praised the Kingdom and the U.A.E. for exercising 'international responsibility and concern for the adverse impact of an oil price increase in the world's economy'.[14] William Simon, Secretary of the

Treasury, hailed Saudi Arabia as a 'true friend of the West in general and the U.S. in particular'.[15] Cyrus Vance, who had been chosen as the future Secretary of State by President-elect Jimmy Carter, described its action as 'courageous and statesmanlike'.[16] Carter himself responded warmly but with reservation to one remark made by Yamani that was seized upon and magnified by the world's press. Almost in passing, but quite deliberately, he said at his press conference following the agonized meeting of oil ministers that in return for restraint 'we expect a sign of appreciation from the West'.[17]

It was the first time in three years of Saudi resistance to extremist demands in O.P.E.C. that a *quid pro quo* had been asked for. Never before had the great Arab preoccupation so explicitly entered its arena in such a way. Yamani's statement could be interpreted, partly, as an attempt to mollify the anger of the Arab members. Yet it did signify a real desire to use the Kingdom's potential leverage positively. The hints in the summer of 1973 about the use of oil sanctions if there was no progress towards a Middle East peace settlement seemed, in retrospect, like the diffident waving of a stick. Yamani's words conveyed, rather, the flavour of a carrot. Saud al Feisal, who was beginning to project a relatively hawk-like image, was more forthright in saying that the Kingdom's decision to hold down its oil price increase was directly linked to a resolution of the Middle East deadlock. 'We wish to finish once and for all with this conflict,'[18] he asserted. 'I do not believe that the oil price decision should be a factor in the ultimate political decision concerning the Middle East,'[19] commented Carter.

Yamani had stressed that consideration about the world's economic health had again been the predominant motivation – 'three-quarters' responsible – for the Kingdom's willingness to split O.P.E.C. Soon after, Fahd explained the Saudi position wholly in those terms. At the end of January 1977, Khaled described oil as an 'economic and commercial problem'[20] which should be kept out of politics. The Saudi regime sung with several voices, never discordant but with differing emphases. This form of Arabian madrigal had its advantage in creating uncertainty, even doubts, and provoking deep questioning that gave the Kingdom the attention it deserved from Washington but all too infrequently received. At the same time, the approach to the U.S. was convoluted and confused, reflecting both the wish to influence positively and the anxiety not to offend. Almost invariably Fahd reflected the latter concern. As it happens, he was becoming

intensely worried about the threat of Euro-communism. That may
have been the main reason for what many observers regarded as a
noble but dangerous stance from the Kingdom's point of view, and
why some thinking Saudi citizens were uneasy.

Yet, with the coming to power of a new U.S. Administration
headed by an idealistic chief executive, there was optimism in the air.
Carter's choice for the key post of National Security Adviser was
Zbigniew Brzezinski. He had been involved in the preparation of the
Brookings Institutes study[21] on American policy in the region, which
was reckoned to have established guidelines for Carter. It concluded,
as Kissinger had done, that the scope for 'step-by-step' diplomacy was
probably exhausted. The goal should be a comprehensive peace
settlement, it recommended. The study stressed the need for a
'Palestinian homeland' to be established as part of any settlement and
foresaw territorial compromise by Israel on all fronts. Less palatable
for the Kingdom was the concept of peace as a normalization of
relations and the suggestion that the Soviet Union should be associated
with any settlement, if only to impose some moral obligation upon it
not to upset one. However, no sooner had Carter stepped into the
White House than he showed himself thoughtful on the issue,
concerned to evaluate and able to digest the information given to him
by professional advisers. There were encouraging signs, like his
Administration's refusal to supply Israel with cluster bombs or to
permit the Jewish state to sell its American-powered KFIR combat
aircraft to Ecuador.

The question of sovereignty over Jerusalem apart, the House of
Saud showed itself prepared to approve any solution acceptable to the
Arab parties involved, not the least the Palestinians. Despite its
peripheral involvement in the October War and notwithstanding
being branded as a confrontation state by Israel, the Kingdom did not
consider itself directly involved. The leadership in Riyadh appreciated
that the Palestinian movement would have to be more flexible and
show some sign of willingness to accept the existence of Israel. Thus,
it put its weight behind Egyptian and Syrian efforts to influence Arafat
and his colleagues in advance of the critical session of the Palestine
National Council, scheduled to be held in March 1977. In the event,
the assembly did not in any way modify the fundamentalist line of the
Palestine National Covenant of 1964 which called for 'the total
liberation of Palestine'.

Another major factor in the question was the outcome of the

Israeli elections. The Likud bloc led by Menachem Begin, leader of the terrorist group Irgun Zvei Leumi during the days of the Mandate, emerged triumphant from the poll held on 17 May 1977. The accession to power of a political coalition believing the West Bank to be an integral part of Israel, never to be renounced, did not augur well for the future. Yet, surprisingly, hope was still very much alive when Fahd arrived in the U.S. towards the end of May in the wake of other Middle Eastern leaders. Indeed, in an interview given just before his departure, the Crown Prince said: 'I believe that the repercussions of the change of government in Israel are primarily the concern of the U.S.'[22]

Fahd could not have been more amenable to his hosts. In contradiction to the position consistently taken in public by Yamani over the previous year, Saudi Arabia would assist with the American plan to accumulate a strategic oil reserve equivalent to six months' consumption – though only on condition that 'the U.S. should throw its weight behind the achievement of a just settlement of the Middle East problem'.[23] Fahd assured that 'oil will not be used as a weapon' – correcting the assertion given a few days earlier by Ismael Fahmy, the Egyptian Foreign Minister, to the effect that the Kingdom could be carried along with Arab policy and was a 'soft touch'. Fahd was less compliant on one issue disturbing the Saudi Government and straining bilateral relations, which touched very directly on the Kingdom's development plans. That was the legislation passing through Congress aimed at preventing the compliance of American companies with the demands of the Arab Boycott of Israel Office.

As a founder member of the Arab League, the Kingdom had been party to the embargo in force against the Jewish state since its independence. It involved not just a ban on direct trade with Israel. Its secondary aspect extended to the blacklisting of non-Israeli companies adjudged to have contributed (with the emphasis on investment) to the Zionist entity's strength. It had a tertiary dimension in that conditions were placed in contracts with foreign companies whereby they agreed not to employ the goods and services of a blacklisted concern in implementing a project. Israel only began to regard it as a dangerous threat after the 1973 – 4 explosion in oil prices led to phenomenal growth in Arab markets and made the Boycott a potentially potent weapon. Working mainly through the Israeli-American Public Affairs Committee, the Israeli Government had been largely instrumental in building up a great legislative momentum both

on Capitol Hill and in the states aimed at countering the Arab
Boycott. Several federal departments had tightened up administrative
regulations to discourage compliance with Arab demands. But
pressure built up for legislation of a truly draconian nature. Draft
amendments to the Export Administration Act put forward by
Benjamin S. Rosenthal had only missed becoming law at the expiry of
the 94th Congress in 1976 by a whisker, as a result of a filibuster by
Senator John Tower of Texas. They led William E. Leonard, chief of
Ralph M. Parsons, the large engineering company, to say 'the whole
thing is designed to discourage business between the U.S. and the
Arabs'.[24] A senior Mobil executive bluntly charged: 'We believe and
we think you know that the real purport of your bill is to make it a
crime for U.S. companies and their subsidiaries in any way to do
business with Arab nations'.[25]

The whole thrust of the legislative effort seemed to be, and to a
large extent was, directed at Saudi Arabia, not only the richest Arab
market but an oil producer whose friendship with the U.S. gave it
potential leverage over the Kingdom. Carter had promised strong
action during his campaign. For him it was not just a question of the
Jewish vote. It also reflected his concern for human rights, as far as the
racially discriminatory aspects of the Boycott were concerned.

Saudi Arabia had never sought to disguise a traditional policy of
excluding Jews from its sacred confines. But such was its desire for
U.S. technology that it had sought to accommodate Washington by
agreeing to drop the restrictions on the entry into the country of
Jewish personnel. Another concession was made when Congress took
up the question of confronting the Arab Boycott in 1977. The
Administration persuaded the Saudi Government to stop demanding
'negative certificates of origin' (certifying specifically that goods or
services did not come from Israel), which lay at the heart of the
primary boycott, and to request instead only 'positive' ones (giving
the actual origin).

The U.S. Government became alarmed at the possible implications
of the legislation in prospect and the damage it could do to American
political and economic interests in the Middle East. By the time Fahd
arrived in Washington it had nearly taken shape. Passed early in June,
the complexities earned it such epithets as 'a lawyer's relief act' and 'a
legal nightmare'. Months later, businessmen and their attorneys were
still trying to unravel the complications. Much depended on
interpretation of the new law. The considered verdict was 'make it

possible, though very much more difficult, for American companies to do business with the Arabs'.

Meanwhile, Saudi Arabia's act in defying all but one other member of O.P.E.C. may have been courageous and statesman-like but it was by no means successful, at a time of buoyant demand for oil, in terms of bringing down significantly the average increase in price and reducing the output of the eleven maximalists. Over the first six months of 1977 output from the great fields operated by Aramco averaged only 9.1 million barrels a day. It was not easy for customers to switch sources of supply because of contractual commitments and fear of disturbing long-standing relationships. The difference in price after the escalations of previous years was not that great, anyway. Interest in Aramco's cheaper crude was restricted by the limit set by the Saudi Government on the proportion of total production – 65 per cent – that could be made up of its main staple, Arabian Light. There were technical difficulties, in particular the bad weather that delayed loading schedules at the Ras Tanura terminal in the early part of the year.

The trial of strength raised questions about the production that the Kingdom was able to sustain without impairing long-term recovery of reserves. The experience proved it to be at this point very much less than the 11.8 million barrels a day Yamani evidently hoped for. There was talk towards the end of the year of the pressures in the Ghawar field having been harmed. Only Qatar's production was lowered, by about 14 per cent compared with the same period in 1976. Demand for Iranian and Kuwaiti crude was affected. Iraq, and other producers probably, trimmed prices. The Saudi hierarchy may also have been inhibited by the antagonism aroused by the division in O.P.E.C. in fighting out the battle fully in the market place. Politically, its isolation in the organization had not been a happy experience. By midsummer the conditions for a compromise were right. The reconciliation took place at the biannual ordinary conference in neutral Stockholm on 12 – 13 July. Saudi Arabia and the U.A.E. raised their prices by 5 per cent, bringing them into line with other members. The maximalist majority agreed to forego the further 5 per cent increment that they had planned. In return, Saudi Arabia imposed a ceiling of 8.5 million barrels a day on its production. It was also agreed that there should be a freeze on prices until the end of the year.

The Conference on International Economic Co-operation had finally expired in the summer having realized few gains and fallen far short of the aspirations of the Third World. In the event, Saudi Arabia was not disturbed by the lack of progress made. It remained optimistic that Carter's Administration had shown appreciation for its moderation and gratitude had been reflected in the American approach towards the Middle East problem.

Fahd and his peers kept a wary, troubled eye on the Horn of Africa where in Addis Ababa Colonel Mengistu Haile Mariam's Dergue had expelled the U.S. military mission and was being assisted by Cuban advisers. Ethiopia was in conflict with Somaliland whose leader, Siad Barre, had begun a dialogue with Riyadh. But in general the world seemed a more comfortable place for the House of Saud to live than at any time since Ibn Saud's death. In mid-1977 it was at a peak of self-assurance, as well as influence and affluence.

24 The Wages of Wealth

1977–9

If the current Saudi Arabian Government were in fact to underwrite a separate Israeli–Egyptian peace, it would be overthrown by members of the royal family or by some other constellation of forces...It cannot do this. Our 'experts' should know this; the Russians surely do.[1] – JAMES E. AKINS

ON 19 February 1977 Riyadh Radio announced that Abdul Aziz bin Musaid, the cousin of Ibn Saud who had taken part in the capture of the Musmak three-quarters of a century earlier, had died at the age of 100. It was a reminder of a far distant era and the origins of a kingdom that over the past three decades had risen from obscurity to a bewildering prominence in world affairs. At the beginning of the year there had been another death in the family that was to sadden the House of Saud almost as much as that of Feisal. The pious Abdullah bin Abdul Rahman, the founder monarch's brother and counsellor, the man who had played a crucial role in persuading the *ulema* to depose Saud, expired. 'Uncle Abby', as he was rather irreverently known to Westerners resident in the Kingdom, was a less historic link with the past than the warrior Abdul Aziz bin Musaid but one closer to the affections of the family. Only Musaid bin Abdul Rahman remained of Ibn Saud's brothers. Abdullah had thought it a sin not to be in bed by 11 p.m. and considered archaeology sacrilegious. At his funeral, elderly relatives prayed that they would be saved from 'the evils of the present blessings', according to one of his sons, Bandar bin Abdullah. The old man's sleeping habits were certainly not those of most princes, who would sit up late at night telling endless and dubious stories, drinking Pepsi-Cola or more noxious liquids and scratching their toes. Relatively few showed any great dedication to work or sense of public service.

The increasing complexity of government and the decisions involved in the implementation of an enormous development programme made the advice of the technocrats essential. While the King ratified decisions of the Council of Ministers and was sometimes consulted about issues in advance, the easy-going Fahd, as chief executive, showed none of Feisal's over-conscientious inclination to oversee in detail all areas of government, apart from the direction of the Kingdom's relations with the outside world and the formulation of oil policy, which was his special concern. In other spheres of government there was a considerable degree of delegation. Early in 1977, Ghazi Qusaibi, the Minister of Industry and Electricity, could say that there had not been a single decision taken without his initiative since he had taken office in the autumn of 1975.[2]

The administration was able to carry a far greater weight and variety of business through the creation of more departments, the appointment of young, able and ambitious commoners as ministers, and the delegation to them of greater responsibility. Thus, a proportion of the great burden once carried by Feisal, which had become greatly magnified in size, was shifted on to ministers, their deputies and director-generals of departments. They, however, could not delegate, they complained, because to do so would mean no action or decisions. One deputy minister, an old friend, commented wearily to me about this time: 'I have to do everything myself. If I set up any course of action I have to oversee every step or nothing gets done. There's nobody underneath who can do anything – or wants to.' There were lengthy delays in payments. The volume of paper passing across the desks of officials grew inexorably. So, too, did the number of callers cluttering up anterooms and the strain on the tradition of politely receiving visitors increased. The number of official delegations descending on the Kingdom grew by the year, consuming more precious time. The most important ministers would be expected and, indeed, would want to be in attendance for state and official visits to the Kingdom. Pan-Arab organizations and gatherings made greater demands than ever. Ministers had to spend more and more time on their travels. When the recurring annual events of Ramadan, the *Hajj*, and the high summer recess were taken into account, it was amazing that the dislocation was not far worse.

Reliance was placed on a few dozen dedicated men. Their work load and the trust placed in them evidently made some amongst the new generation of ministers believe that power was being shared. In a sense they were right. 'When technocrats agree, their line prevails,' said one of them who, like his colleagues, did not want to be quoted on such an obviously sensitive matter. 'To say our political system will develop new lines of authority is natural,' he went on, mentioning in this context the promised Consultative Council. Another minister seemed to interpret the process more accurately: 'The royal family's command of the structure is not weakened because they have responded to the need for technocrats. They got them into government to keep the system going. The technocrats are grateful for the stability this system provides. It is essential to our development.' Speaking with due deference, he then went on to put the phenomenon into perspective: 'When I accompanied H.M. King Khaled to the south recently, I saw our own democracy in action. We talked to key people there in their own houses and visited the tribes of the area. I saw their way of expressing their allegiance. If it was just a technocrat, a commoner, who went, he would not receive this loyalty.'[3] Fahd's real attitude to the new generation of loyal, well-qualified servants of the Kingdom was probably contained in a remark attributed to him: 'Those Ph.Ds I appointed ministers are good for nothing.'[4]

Princely status did not necessarily mean inviolability, however. Muhammad al Feisal, who was head of the Saline Water Conversion Corporation and had proved himself an able executive, lost his post over a quarrel about appropriations with the Minister of Agriculture and Water, Abdul Rahman bin Abdul Aziz al Sheikh, who was his immediate boss – and, admittedly, came from a privileged family himself. Muhammad al Feisal did not disappear from public view, however. He stayed in the news with his plans to tow icebergs from Antarctica to the Red Sea which, he insisted, could provide a cheaper source of water than desalination, and also in promoting the concept of Islamic banking which would conform to the Koranic strictures against usury by doing away with interest payments.

Nothing more was heard of the Consultative Council. The fortunes of the state still seemed indistinguishable from those of the dynasty and vice versa. In its entirety the clan, including the Jiluwis and other collateral branches, embraced perhaps as many as 20,000 people, with at its heart 2–3,000 princes. In addition, the House of Saud was intertwined with the Al al Sheikhs and Sudairis. The tentacles extended even further through the marriages contracted with the women of the country's other tribes. The tendency amongst Ibn

Saud's descendants had been towards taking fewer brides and marrying close cousins. There was a trend towards exclusivity in this respect. Nevertheless, the size and ubiquity of the family had become greater by the year.

All members of it received royal stipends in accordance with the structured system worked out by Musaid bin Abdul Rahman as part of Feisal's financial reforms carried out in 1963. In setting the scale, there were two basic criteria in the mainstream of the family. First was proximity in direct lineage and by generation to Ibn Saud's great-grandfather Feisal bin Turki. Second was age. Ibn Saud's sons received a lump sum of 200,000 riyals annually and up to 30,000 riyals a month. For junior members of the family, income from the privy purse was one-fifth to one-sixth of that amount. Also receiving regular income from royal funds were members of the Al Sudairi, and the House of Saud's defeated enemy the Al Rashid, as well as descendants of the warriors who had fought with Ibn Saud. Royal privilege entailed responsibilities as well, however. Each senior prince, like European nobility until not so distant times, supported anything from 100 to 1,000 families of retainers, who had to be provided with marriage dowries, medical treatment, education and the like. The extent of the outlays involved was not generally appreciated by foreign observers. More noticeable after Feisal's death was conspicuous spending on new palaces and villas by princes and leading commoners of a kind that the late King would have disapproved of. Manifestations of luxury and greed became reminiscent of Saud's era, but on a more widespread scale. Ib bin Salem, meanwhile, was back in Riyadh.

The Consultative Council may have been quietly forgotten, but there was talk of a reform and rationalization of the haphazard system of provincial government. The plan was to reduce the number of governorates from fifteen to nine and call them states. Governors' powers, which differed, would be defined and also their relationship to central authority. Deputies to them would be appointed, and advisory councils established, which would be partly elected. The traditional aspects of the system remained important, but the need for a more modern approach as well was beginning to be recognized. The calibre of the Governors, all of them royal or Sudairis, varied. Outstanding amongst them and something of a show piece was Khaled al Feisal, third son of the late King, who was Governor of the Asir. He was personally involved in the economic and social development of the region in a way that none of his peers was. At the same time, he

fulfilled the more time-honoured functions of a Governor, holding
three *majlises* daily, at 11 a.m., 1.30 p.m., and 7.30 p.m., after which
he hosted an open dinner for 50-100 visitors every evening, except at
weekends when he would entertain two dozen of his own guests. He
received about 100 requests for financial assistance each year but
declined, modestly perhaps, to say how many of them he granted.
Here, as elsewhere in the Kingdom, the people would ask for medical
treatment, help with educating children, money for dowries.
Explaining his paternal role, Khaled said: 'If you can't put them in
touch with a ministry or other authority, you may have to help them
yourself. By law the Governor is responsible for the welfare of his
province and by religion is the "sheikh of sheikhs" who must see that
all his flock are fed, housed, and prospering under his care. People
expect this service and believe that the Governor is personally
responsible for all – good or bad – that befalls them. He must,
therefore, make sure it is good.'[5]

Khaled was also in charge of the King Feisal Foundation established
in his father's memory with the initial objective of spreading Islamic
education throughout the world. Feisal had left only $1,400 in one
bank account in the U.S., money that had been left over after paying
for his sons' education there. The wealth that he bequeathed had been
in property, mainly his palaces, which were acquired by the
Government. While the estate was being settled, Khaled al Feisal and
his brothers had kept the foundation afloat with their own personal
contributions.

The sense of duty inculcated into his sons by Feisal was exceptional.
Khaled's youngest brother, Turki al Feisal, who was reputedly the
brightest of the old King's seven sons, had been appointed Deputy
Director-General of Intelligence, as understudy to Kamal Adham, in
1973 at the youthful age of 26. Bandar al Feisal was a Lieutenant-
Colonel in the Air Force. This public service seemed to extend even
down to the third generation. One grandson, Muhammad bin
Abdullah al Feisal, had become Assistant Deputy Minister of
Education in 1974 at the age of 27.

Other young princes had risen to prominence in the Armed Forces
and Government administration. Those in the military were
commonly regarded as the 'eyes and ears' of the royal family, placed
there to keep on the lookout for subversive and disloyal elements of
the kind rounded up in 1969. Amongst the half-dozen serving in the
Air Force was Miqrin, Ibn Saud's forty-second son, who had been

born in a tent in 1942. Aviation rather than security surveillance was certainly his first preoccupation, however. For any prince who wanted to serve the state, paternity could be an assurance of a favoured appointment and an aid to advancement, though not a guarantee of it. According to Miqrin, there had been fourteen members of the royal family in the Air Force as officers, but over half of them had not made the grade and had more or less been asked to leave the service. 'Prince Sultan always tells us we'll get nowhere by being a prince,' he said. 'You must work. And we do – harder than others because more is expected of us.'[6] Most prominent of the half-dozen was Bandar bin Sultan, a son of the Minister of Defence. His professional commitment could not be doubted. But he admitted that when it was proposed that he should be appointed a Governor, his father had insisted on his staying in the Air Force at the Dhahran base – where, it will be recalled, a former Commander, Dawood Romahi, had been implicated in the 1969 conspiracy. Altogether, about two dozen princes were in the Armed Forces or the National Guard.

Others like Muhammad bin Abdullah al Feisal performed fairly humdrum tasks in a manner befitting a civil servant. He was frank about his approach to his duty. 'I would like to have done graduate work but Feisal said my generation was needed here,' he said early in 1977. 'Now I don't regret it. We have a great satisfaction from what we are doing now. I have a responsibility for doing something for my country. Without Saudi Arabia, I would not be a prince. Therefore, I have to give something back to Saudi Arabia.' His attitudes to society and Islam were conservative. He expressed the conventional wisdom of the established order when he spoke of 'our job' being to 'prevent money and modernization from undermining Islam.' He was against polygamy and divorce of the traditional, arbitrary kind, but disapproved of the sexes mixing at all before marriage. He took the standard line that all Saudi women cloistered in the bosom of their families were content because they were protected.[7]

Bandar bin Abdullah bin Abdul Rahman (son of Uncle Abby) ran the Ministry of the Interior's department dealing with the provinces. In his opinion it was in the interest of members of the royal family to serve the state. Only if people were satisfied by the way in which the country was run could the clan retain its position – 'which I really like', he was unabashed in saying. In some other respects his views were refreshingly different from those of his peers. He was apprehensive about the effects of segregation in education. He believed a woman's place to be in the home, but added, 'You can't keep women in the house if you educate them.' He questioned the method whereby girls at university were taught by men only through the medium of closed-circuit television. In such circumstances there could be 'no human interaction at all', he pointed out.[8] Just previously it had been decreed that no man whatsoever, not even the dean of the Faculty of Medicine at Riyadh University who had been accustomed to visit the pupils there, could set foot in the compound enclosing the girls' section of the department. Boredom and frustration were such that the daughters of wealthier families would cruise the thoroughfares of the cities getting their chauffeurs to pass notes containing their telephone numbers to male drivers who took their fancies. Assignations could be arranged even in such a tightly segregated society.

Khaled was playing a significant role in keeping the balance between the more outward-looking faction favouring a faster pace of modernization and the conservative camp more deeply rooted in Arabian soil. The importance attached to the King's influence was evident from the anxiety felt by the House of Saud in mid-February 1977 when he was flown to London for major surgery on his left hip. The original complaint had been made worse, it was said, by treatment with a bedouin hot iron. The joint had festered and Khaled was in great pain. Concern was heightened by Khaled's cardiac troubles. The King himself was preoccupied with his health to the point that the subject of heart surgery seemed to interest him almost as much as hawking. A Boeing 747 Jumbo, installed with an operating theatre, had been purchased for him. On arrival at Heathrow Airport he was removed from his aircraft by a fork-lift truck and taken by helicopter to Lord's Cricket Ground before being sped to the Wellington Hospital round the corner in St John's Wood, in view of the mosque and Islamic centre built on the edge of Regent's Park. He and his entourage occupied the whole of the fifth floor of the hospital for six weeks. Sultan, who had travelled with the King, ensconced himself in one of the suites for the full sojourn. After the infection in the bone had been treated the King was operated on twice, on 23 February and 7 March.

King Hussein of Jordan, the Ruler of Kuwait, and the Ruler of Bahrain journeyed to London to see him. A procession of princes, ministers, and other dignitaries flew from the Kingdom to pay homage. At one point half the Council of Ministers was in London.

The Saudi congregation in the British capital demonstrated the affection and esteem in which the King was held. It also reflected the underlying concern felt by the royal family that Khaled should continue at its head.

During his hospitalization there was rumour and speculation that he was about to abdicate, or to retire, leaving Fahd as ruler in everything but name. It was even reported that Sultan was making a bid to have the National Guard incorporated within the structure of the regular Armed Forces and, therefore, placed under his direct authority. The uncertainty created was such that on 16 April Fahd, who was worried that he might be suspected of *lèse majesté,* felt forced to state publicly that Khaled would remain in office and that he enjoyed the confidence of both his family and his people. Nevertheless, the stories persisted through the summer, some of them evidently inspired by an Israeli misinformation campaign aimed at sowing doubts about the stability of the regime. There was, however, some substance to the speculation, according to reliable informants, in that discussions did take place within the family about who should be the next in line after Fahd. He did not like Abdullah and favoured Sultan as third man in the hierarchy. In some respects the Minister of Defence seemed a far more suitable candidate than Abdullah. But the all-important consensus in the royal family was not prepared to contemplate the prospect of Fahd and Sultan holding the top two positions. The question was a difficult, as well as contentious, one because Abdullah at this point apparently had some reservations about being designated the next Crown Prince after Fahd. Essentially a man of the desert, at home with tribesmen and hawks, Abdullah may have looked askance at the tasks and responsibilities involved in being heir apparent or sovereign, as the less forceful Khaled had before him. Another problem was who would replace him as Second Deputy Premier and, more important, as head of the National Guard, or indeed whether he would be prepared to relinquish his command of it.

In the summer of 1977 an incident occurred that threw into sharp relief the differing attitudes of the 'progressive' modernizers led by Fahd, and the conservatives, whose focal point was Muhammad. This was the execution in Jeddah at the end of July of Mishaal bint Fahd bin Muhammad, the grand-daughter of the founder King's eldest surviving son, and her lover Khaled Muhallal, a nephew of General Ali Shaeir, Saudi Ambassador to the Lebanon. The wider world did not know of the affair until more than five months had elapsed when a version of the story was published in the *Observer* newspaper of London on 22 January 1978. In some details the tale reported there proved to be wrong. It said that Mishaal had met Muhallal while studying in Lebanon, that she had been summoned home when her family learned of her romance with him and had been told to marry a man chosen for her, and that she had been caught with her lover at Jeddah airport as they attempted to escape the country. That last detail proved to be correct. But as Rosemarie Buschow, a German-born nanny who had worked in Muhammad's household revealed in an interview with the *Daily Express* on 28 January, Mishaal was already married – to her first cousin, who had left her.

Some expatriate residents in the Kingdom had been aware of the executions. The killings had, with good reason, been the subject of comment and concern. They had about them several unusual features that were bound to raise eyebrows locally. They took place on a Saturday rather than the customary time for judicial punishment which was after the noon prayers on a Friday, and they were carried out in a parking lot off King Abdul Aziz Street instead of in the Bab al Sherif Square where official acts of justice were normally carried out. There was no prior announcement of the executions as was usual when a verdict had been delivered by a court. The princess was shot in the head six times, a method of punishment neither decreed in the *sharia* law nor laid down in the temporal regulations governing its application. Police were present, as they always were when justice was meted out in the Bab al Sherif Square, but the swordsman who beheaded Khaled Muhallal was not a public executioner. Moreover, the five strokes taken to perform the task failed to sever the head and indicated a certain lack of professionalism.

It was obvious to Saudi onlookers and citizens of Jeddah that Mishaal and her lover had not been tried by due process of law. The truth was that Muhammad had exercised his 'right' as the head of the family to decree the death of a member who had brought shame upon it. Such a principle was deeply rooted in tribal society. This rough form of justice was understood by Saudis cast in a traditional mould and in public, at least, they defended the act. Privately, many educated and more liberal Saudis, especially Hejazis, regarded the executions as a brutal assertion of royal prerogative. Mishaal's most serious offence in her grandfather's eyes was to have been caught escaping with her lover.

The King refused to sanction a public execution, but Muhammad was not prepared to have the act of retribution carried out privately behind palace walls, in spite of the fact that in going against Khaled's wishes he only attracted more attention to the dishonour brought upon his family by Mishaal. Muhammad's right to dispose of his grand-daughter was recognized, however reluctantly, by Khaled and other members of the royal family. But it is difficult to see how it could be extended to her lover. Under *sharia* law, proof of adultery would have required a confession, made three times, or the evidence of either four male or eight female witnesses to the sexual act given before a *qadi*, or judge. If convicted, she, as a married woman, should have been stoned to death, and he, as a bachelor, sentenced to 100 lashes. One can only speculate what, if any, Muhammad's calculations were on this score and whether he was in one of his fabled rages when he condemned his grand-daughter to death. Such was his power and the respect for his seniority that the King felt impotent to prevent an act that could only arouse the criticism of liberals at home and damage the image of the royal family abroad. Fahd was deeply angered, according to reliable informants.

The affair provided pro-Israeli lobbyists everywhere, but especially in the U.S., with a convenient stick with which to beat the Kingdom. Diplomatic repercussions were soon felt. In London, a Foreign Office spokesman described the death of Mishaal as a 'tragedy', technically a not unreasonable description of the sad, as well as horrifying, episode. This prompted a Saudi Government statement asserting that the executions had taken place in accordance with Islamic law and that the two lovers had paid the penalty for adultery. David Owen, the Foreign Secretary, then sent an apology to the Saudi Government for the comment made on behalf of his department, which was delivered by an intensely embarrassed Sir John Wilton, the U.K. Ambassador. Owen then found himself in deeper water. Questioned on the subject at a meeting of the Zionist Federation of Great Britain, he explained that though governments should not keep silent on cases abroad concerned with human rights, they should be careful not to do 'unnecessary damage' to diplomatic relations. A group of Labour M.P.s tabled a motion saying that the House of Commons was ashamed of the apology, which it described as incompatible with Owen's position on the question of human rights.

*   *   *

At the beginning of 1977 a new external threat to the Saudi regime had become fully apparent. Since the overthrow of the Emperor Haile Selassie in September 1974, turbulence had been growing across the Red Sea in Ethiopia. In Addis Ababa the political drift of the military junta, known as the Dergue, towards the left and Moscow had been inexorable. Reversing the country's long-standing and close relationship with the U.S., the Ethiopian regime had sought assistance from the Soviet Union in December 1976 and been pledged $386 million in military aid. Early in 1977 the power struggle within the Dergue was resolved in favour of the hard-line Marxists when Colonel Mengistu Haile Mariam emerged as undisputed leader after brutally executing seven of his former colleagues. In April, Mengistu summarily expelled the American military mission and withdrew the facilities extended to the U.S. Air Force at Asmara.

Somalia at the tip of the Horn of Africa and commanding the southern approaches to the Red Sea had long been dedicated to 'scientific socialism'. Since Siad Barre's regime had come to power in 1969 it had collaborated closely with the Soviet Union, which enjoyed extensive military facilities in return for its aid, including the use of the naval base at Berbera developed by it and the air base at Uanle Uen. On the northern side of the entrance to the Red Sea the P.D.R.Y., another of Moscow's client states, provided the Soviets with refuelling facilities for both vessels and aircraft. It seemed as if the Soviet Union would soon be able to throttle one of the world's most important waterways and the Red Sea's historic name would prove to be ideologically apt.

Fahd judged that the Kingdom, with its vast oil resources, was a target of this communist thrust. Special concern focused on Djibouti, a French dependency since 1888, officially entitled the French Territory of the Affars and Issas. It bordered the southern side of the Bab el Mandas Straits. France had promised it independence by mid-1977. Its future had been a topic of particular concern when Giscard d'Estaing had visited Riyadh in January 1977. Like most, if not all, colonial territories in Africa, Djibouti had boundaries that had been drawn with scant regard for ethnic realities. The Affars had tribal connections with Ethiopia, and the Issas were of Somali stock and had been generally favoured by Paris. Ethiopia's only railway outlet to the sea ran through Djibouti, providing 70 per cent of employment for the colony, whose commerce had languished since the closure of the Suez Canal. Ethiopia's natural interest in being the predominant influence over the state could only be encouraged by Moscow's great design. Similarly, Siad Barre's regime regarded Djibouti as part of a 'Greater Somaliland'.

Somali-Ethiopian rivalry was one aspect of a basic tension at the heart of a complex situation that Saudi leaders, like Sadat of Egypt and Jafaar Nimairi of Sudan, saw could be exploited in countering Soviet expansionism. The differences over what the two regimes in power in Mogadishu and Addis Ababa perceived as their national interests were greater than their ideological similarities. The West Somali Liberation Front, which had been created as early as 1960, was witness to Mogadishu's claim to the Ogaden, a desert region of Ethiopia inhabited mainly by nomadic Somali tribesmen. The first serious clashes there occurred in February 1977. Barre felt encouraged to embark upon an aggressive policy by developments elsewhere in the Dergue's territory. The Arabic-speaking people of Eritrea, the Ethiopian province bordering the Red Sea, had been in conflict with the central authority since 1964. The revolt had taken on much larger proportions since the death of Haile Selassie. Syria had from the beginning supported the secessionist movement and its aid still went to the mainstream Eritrean Liberation Front (E.L.F.), to which the P.L.O. and Tunisia also gave assistance. Iraq backed the Eritrean People's Liberation Front, a Marxist faction. Saudi Arabia looked askance at the radical elements of the movement and provided aid to the more moderate E.L.F. In the west of Ethiopia two other, non-secessionist, rebellions were stretching the Dergue still further. It seemed as if the country might disintegrate. Moscow was quick to see the dangers to its strategy in the region, and Riyadh equally alert to the possibilities created by what appeared to be an inevitable Somali-Ethiopian conflict.

At the Kremlin's request Fidel Castro toured Africa in March 1977, where his troops were fighting in Angola to bolster the communist regime against the Western-backed U.N.I.T.A. guerrillas. His first stop was Libya. Qaddafi chose his visit to proclaim that his country would henceforth be known as the Libyan *Jamahiriya* (literally translated as the 'state of the masses') and government would derive from 'direct popular power'. The Koran would still be the basis of the system, however. With his half-baked 'Third International Theory' Qaddafi believed that he had synthesized the best of Marxism in a new Islamic political philosophy and therefore could co-exist ideologically with the Kremlin, which was actively assisting him.

On 12 March the Cuban leader arrived in Aden for talks with the leaders of the ruling National Liberation Front. The joint communiqué was hardly such as to reassure the Saudi leadership, which was gingerly, but with little idea of how to go about it beyond flourishing petrodollars, trying to lead the P.D.R.Y. into the path of moderation. In the statement Cuba declared its support for and solidarity with 'the struggle of the heroic Omani people under the leadership of the Popular Front for the Liberation of Oman for total national liberation and the liquidation of the foreign and Anglo-Iranian bases and all forms of alien influence in their territory'.[9]

Castro's main concern in this part of Africa, however, was to maintain socialist solidarity between Somalia and Ethiopia to consolidate the Soviet hold on the Red Sea and strengthen its presence in the Indian Ocean. The bearded cat's paw of the Kremlin proceeded from Aden for talks with Siad Barre in Mogadishu and Mengistu's junta in Addis Ababa. Castro succeeded in arranging a secret meeting in Aden between Barre and Mengistu, but to no avail. 'Unfortunately we could not agree,' the Somali leader said two months later when he confirmed the assignation.[10] He had rejected a proposal by Castro for a federation embracing Somalia, Ethiopia, and the P.D.R.Y.

The P.D.R.Y. and Libya apart, Arab sympathy was wholly with Somalia, just as it had been with the Eritrean rebels, and more recently, with the independence of Djibouti looming, the Issas. Somalia was a Muslim country with much in common culturally with the Arab world, though a Latin alphabet and script were chosen in 1972 for the first official written version of the language. The feeling of identity was such that Somalia had applied for membership of the Arab League and been granted it in 1974. Above all, Arab control over the outlet from the Red Sea to the Indian Ocean was considered a strategic essential in the event of another Middle East conflict and a means of putting pressure on Israel. This objective was one to which Saudi Arabia openly subscribed, even though espousal of such a doctrine made the vulnerable Kingdom even more liable to attack by Israel if hostilities broke out.

The moderate littoral states rallied to face Soviet expansionism. After two days of talks in Khartoum, Sadat of Egypt, Assad of Syria, and Nimairi of Sudan had declared on 28 February that 'unity of the Arab nation' was the 'shield' which would 'foil the hostile plots and schemes which surround it'.[11] Saudi Arabia left it to Nimairi to take an open diplomatic initiative, and is believed to have prompted him to do so.

The Kingdom's Exchequer was bank-rolling the Sudanese leader, who had survived a well-organized coup attempt backed by Libya and Ethiopia in the previous year. In the wake of Castro, Nimairi went to Oman and the P.D.R.Y. After his mission the bluff ex-soldier felt confident enough to say that they had mended their differences over Aden's attempts to wrest control of Dhofar through the Popular Front for the Liberation of Oman. He called a meeting of heads of Arab states bordering the Red Sea and its approaches. The gathering was something of an oddity. It took place in Taiz, the second city of the Y.A.R., on 22 – 23 March and was attended by Ibrahim Hamdi of the host country, Salim Rubai Ali of the P.D.R.Y. and – somewhat unexpectedly to observers at the time – Siad Barre of Somalia. But no representative of Saudi Arabia went. The participants declared the Red Sea to be a 'zone of peace and harmony'. The communiqué issued after the conference took care not to offend Addis Ababa, largely out of deference to the P.D.R.Y. whose loyalty to Ethiopia, its fellow Marxist state, and the Soviet Union could not easily be shaken by the call of Arabism.

Saudi Arabia approved the proceedings from a lordly distance. Yet the quiet, mysterious ways of its diplomacy belied a great sense of urgency. Saud al Feisal, better attuned than his elders to dealing with regimes adhering to left-wing ideologies, had held talks with Barre during the Afro-Arab summit held in Cairo on 7 March at which he had also pledged $1,000 million on behalf of the Kingdom in aid for African countries. He followed up the meeting with a visit to Mogadishu early in April. These initial contacts were tentative and inconclusive, but promising. Barre remained cautious. Fahd, however, was confident that Somalia could be lured into the moderate Arab fold.

From Mogadishu the youthful Saudi Foreign Minister went to Sana and Aden. The Y.A.R. continued to be a blind spot in the context of Saudi Arabia's increasingly sophisticated diplomacy. Its subventions to the big tribal confederations in the north and money-lubricated influence in the capital tended to undermine rather than cement the stability that Hamdi, an able and generally popular president, was trying to create. The bogey of a more populous, vigorous Arab neighbour and, even worse, the possibility of its merger with Britain's radical bequest in Aden, continued to haunt the consciousness of the House of Saud. In Aden, its cheque-book diplomacy was still failing to undermine the rooted dogma of the N.L.F. or to weaken the position of its Secretary-General, Abdul Fattah Ismail. The extent of Saudi inhibition and hesitation could be seen from the fact that only in the first week of April 1977 did the Kingdom's first Ambassador to the P.D.R.Y. present his credentials, over a year after the agreement to establish diplomatic relations. Fahd and Sultan were angered that the P.D.R.Y. allowed Aden to be used as a staging post for the supply of armaments to Ethiopia. The 'Voice of Oman Revolution' continued to broadcast periodic communiqués about the Popular Front for the Liberation of Oman, which had not in fact resumed its activities since the end of 1975, though the defections across the border showed that Aden was keeping the movement very much alive. In May, Mengistu publicly cited the P.D.R.Y. as being Ethiopia's only friend in the region. An abortive attempt was made in June by the P.D.R.Y., on behalf of Moscow, to mediate between Addis Ababa and Mogadishu. There remained some ambivalence in Aden's position, however, reflecting the fact that the more moderate faction of the regime led by Rubai Ali was not unamenable to the inducements offered by Saudi Arabia.

It was bizarre that, apart from Libya, Ethiopia's only other ally in the Middle East was Israel. Determined to prevent total Arab control of the Red Sea it extended support to Addis Ababa, thereby indirectly becoming an ally of the communist world in its attempt to obtain control of the Horn of Africa. Regular Israeli violations of Saudi air space, not the least over the Tabuk base, emphasized its interest in the struggle in a manner humiliating to Saudi Arabia.

From the summer of 1977 onwards, Somalia was more or less in a state of open, undeclared war with Ethiopia as the West Somali Liberation Front, given the best equipment possessed by the Army and strengthened by 'volunteers' from the regular forces, thrust deep into the Ogaden. In this situation, Mogadishu's relations with Moscow grew more strained, and Saudi Arabia's courtship with Somalia progressed more rapidly. Barre was responsive. Somalia had no designs on Djibouti nor did the naval facilities accorded to the Soviet Union amount to a base, he assured visiting Western newsmen. Fahd proposed that the U.S. should supply weapons to Somalia, to be paid for by Saudi Arabia, during his talks with Carter in Washington at the end of May, but the President demurred. Barre was still reluctant to burn his creaking bridges with Moscow, however. At the end of June he was at pains to tell the West that there was no quarrel between Somalia and the Soviet Union. But he warned that Somalia would take an 'important historic decision...if it should transpire that the arms sent by the Soviet Union to Ethiopia constituted a threat to Somalia'.[12]

Djibouti duly celebrated its independence on 27 June under the presidency of Hassan Gouled, whose inclinations were towards Somalia rather than Ethiopia. Apparently irritated by the manner in which he lionized the Somalis present at the celebrations, the Ethiopian delegation walked out in pique. Nevertheless, there was still a jittery uncertainty about the fate of the territory. Barre went to Saudi Arabia on 12 July for talks with Khaled and Fahd. No communiqué was issued. Although Barre spoke on his return to Mogadishu of a complete identity of views with the Saudi leadership, the visit was presented in terms of an expansion of Somalia's contacts in the region. Fahd had dangled before him the prospect of substantial aid if he could extricate Somalia from the socialist camp. On 26 July the U.S. announced that it was willing to supply it with arms. On the following day Britain said that it, too, would provide 'modest quantities of weapons for defensive purposes'. As far as Somalia was concerned, Saudi policy appeared to be slowly achieving its aims.

The P.D.R.Y. presented a more difficult challenge. At the end of July, Rubai Ali, accompanied by Muhammad Saleh Mutia, the Foreign Minister, arrived in Taif for talks with Fahd, Sultan, Saud al Feisal, and Suleiman Suleim, the Minister of Commerce. The encounter was a wary one, but for the moment at least the warmth of Arab brotherhood seemed to triumph over fundamental ideological differences and engrained suspicions. The communiqué spoke of the two Governments' 'high level of satisfaction with their relationship'. It 'affirmed the importance of making the Red Sea a zone of peace and stability and keeping it out of any conflict and clear of international ambitions'.[13] The Saudi Development Fund had just granted the P.D.R.Y. a second loan. A bigger financial commitment depended on some move on the part of the P.D.R.Y. to reduce or remove altogether the large Soviet and Cuban presence on its soil. But the Saudi success indicated by the communiqué proved illusory. Only a few days before, Muhammad Nasser Ali, the P.D.R.Y. Prime Minister, had been to Moscow for discussions with the Kremlin. The P.D.R.Y.'s commitment to Ethiopia, as it struggled to preserve its territorial integrity, became stronger than ever.

By the late summer it looked as if Ethiopia might be dismembered. On 14 September, the Somali insurgents had captured the important tank base of Jijiga and were besieging Harar, Ethiopia's third largest city, situated only 150 miles from Addis Ababa. The bulk of the Ethiopian Army was engaged in desperate fighting in Eritrea, where it was barely maintaining a foothold in the provincial capital Asmara, the ports of Assab and Massawa, and Barentu on the road from Asmara to the Sudanese border. It seemed as if Barre was on the verge of triumphing. The fall of Harar would almost certainly lead to the fall of Mengistu. Somalia, however, had received no arms supplies from the Soviet Union since May. Barre returned empty-handed from a trip to Moscow where Alexei Kosygin, the Premier, demanded nothing less than total Somali withdrawal from the Ogaden. He turned in vain to the Arab League, which had little option but to support the sacred tenet of the Organization of African Unity of respecting the borders inherited from the colonial era. After sober reflection, the U.S. and Britain decided that they could not assist what was, in fact, direct Somali aggression. The Saudi leadership was exasperated, especially Sultan, who could see no reason why the U.S. should not invade Cuba and overthrow Castro.

In its own turbulent backyard, Saudi apprehension about seeing a strong Y.A.R. on its western flank was as acute as ever and policy towards it was still one of divide and rule, while subjugating the central government in Sana through financial assistance. Payment of some 30 per cent of the Y.A.R.'s budget gave the Kingdom some leverage over Hamdi, a form of control resented by him. The first consignment of U.S. armaments financed by Saudi Arabia had arrived early in 1977. It included tanks, trucks, anti-aircraft guns, and howitzers. But shipments were hardly on a scale sufficient to allow Hamdi to reduce his armed forces' dependence on the Soviet Union. Saudi Arabia took a grim view of the continued presence of Soviet military advisers and others working on development projects. In the early part of 1977 their number probably increased. The ruling hierarchy in Riyadh evidently had misgivings about the tough, modernizing Hamdi, not the least because of the wide measure of popular support enjoyed by him – which did not, however, extend to the Hashed and Bakhil tribal confederations whose independence he wished to restrict. The Kingdom was also nervous that Sana's growing rapprochement with Aden – which, paradoxically, had been encouraged by Saudi attempts to moderate the P.D.R.Y.'s regime – would lead to a serious attempt at unification. Hamdi had no strong tribal links and quite genuinely believed in Yemeni unity if it could be brought about on terms acceptable to him.

Saudi subventions made it possible for the Hashed and Bakhil to enjoy a large measure of independence. Hamdi's view of his powerful neighbour was not improved by what amounted to a Hashed uprising in the summer of 1977 under the leadership of Sheikh Abdullah bin Hussein al Ahmar, who had close links with Sultan. Towards the end of July, Ahmar rallied an estimated 40,000 armed men and occupied the town of Saada, forty miles from the Saudi border, as well as the surrounding region and Khamir, some sixty miles north of Sana. With ample money at his disposal, Ahmar was able to offer troops of the regular armed forces 2,000 Saudi riyals a month if they would desert to him. Hamdi was quick to take action. He first ordered the Y.A.R. Air Force's MiG 17 fighter-bombers to strafe and bomb the rebels, then sent in an armoured column to force their retreat. In the negotiations that ensued he was strong enough to make Ahmar renounce his plan for an autonomous fiefdom. Ahmar was said to have made a demand, which may have been put forward at Saudi Arabia's behest, that Hamdi should purchase no more Soviet weapons. The President of the Y.A.R. was reported to have replied: 'We are already diversifying our sources, but it will take a few years and, besides, we had asked for U.S. F-5s, but the Saudis denied them to us.'[14]

The tension in relations between Sana and Riyadh was indicated in an interview given by Hamdi about this time. He was asked what he felt 'about a powerful neighbour who is rigorously opposed to the influence of socialist countries in the Arabian peninsula, like Saudi Arabia'. He replied: 'We know how to accept advice from others and we also know how to give it. We do not accept prevarication. We maintain our autonomy, and our relations with others are our own – and as such are constant, because they are based on the principle of friendship of people and reciprocal respect.'[15] Yet the price demanded for a reconciliation with the tribal confederations was believed to be a high one – the appointment to the ruling Military Command Council of Ahmar, with responsibility for tribal affairs, and of Sheikh Sinan Abu Luhum, a Bakhil chieftain, who was to be in charge of economic affairs. Like Ahmar, Abu Luhum, who enjoyed almost feudal powers as Governor of Hodeida, was a long-standing opponent of the central authority and a supporter of Saudi Arabia.

On 13 October Hamdi was scheduled to fly to Aden for talks with the ruling N.L.F. there. At midnight on 11 October it was announced on Sana Radio that he and his brother, Lieutenant-Colonel Abdullah Hamdi, the Commander of the Army's elite Parachute Regiment, had been murdered at the home of the latter. Also killed was Lieutenant-Colonel Ali Kannas, Commander of the Armoured Brigade. On the following day the remaining members of the Military Command Council met and announced that Hamdi's right-hand man, Lieutenant-Colonel Ahmed Hussein Ghashmi, previously Chief of Staff, would succeed him as Chairman of the Military Command Council, which was now reduced to three in number, compared with twelve when Hamdi had seized power in 1974. Ghashmi felt it prudent to consult with Riyadh before appointing new ministers, and in return received a pledge of $570 million for the five-year development plan being drawn up by the Y.A.R.

Following Hamdi's burial it was announced that a commission of inquiry would be established to investigate the killings. An official statement given on 16 October said that the Hamdis had been gunned down by miscreants with a record of murder and other crimes. They were not identified. Perhaps sensitivities were such that they could not be named even if the authorities knew who they were. There was a bizarre dimension to the affair. The mutilated bodies of two highly-paid French prostitutes, Veronique Troy and Francoise Scrivano, were found in Abdullah Hamdi's house, apparently dumped there. They had flown out on a ten-day 'modelling assignment' on 2 October. Intensive police and journalistic investigations in Paris never established who had arranged their ill-fated journey. The intention, obviously, had been to use their disfigured corpses as a means of discrediting the Hamdi brothers as debauchees. It seemed unlikely in the extreme that the deceased President was responsible for the two girls coming to Sana. He was known to be a homosexual. The general consensus amongst impartial observers was that disaffected tribal elements, specifically the Hashed and Bakhil, had in some way been behind the assassinations, and that it was a brutal protest in their long rearguard action against the encroachment of the central Government. The people of Sana were convinced that the killings had been stage-managed by Saudi Arabia. Western intelligence men and diplomats tended to discount such a theory. Neither Sultan's apparatus nor the Saudi intelligence services had the competence to do the job themselves. The possibility that the impetuous Sultan might, in a fit of rage, have said to one of his Yemeni contacts, 'Who will rid me of this cursed Hamdi?', or words to that effect, could not be ruled out, however. The commission of inquiry never reported at all, but Ghashmi's Government did announce on 31 October that the families of the two French girls would be paid compensation.

The Saudi Government sent Abdul Aziz Thunayan, a senior Foreign Ministry official, on a tour of the Gulf states. It was believed that one of the prime purposes of his mission was to give reassurances about the non-involvement of the Kingdom in the killings. Turki, the Deputy Minister of Defence, was sent to Sana for talks with Ghashmi. The National Democratic Front, an Aden-backed opposition group in the Y.A.R. whose strength was in the south of the country and whose membership belonged almost wholly to the Shafei sect, accused Saudi Arabia and the U.S. of organizing the killings, prompting a strong denial from Riyadh. The P.D.R.Y. made no allegations as to who might have perpetrated the foul deed but deplored the assassination of Hamdi as a 'criminal and despicable act'. The closer relationship with Aden that Saudi Arabia had so assiduously tried to nurture in the course of 1977 disintegrated almost overnight. The P.D.R.Y. came out in more open support of the Ethiopian military junta than before and in the U.N. General Assembly it delivered a stinging onslaught against Oman, to which Riyadh was now extending massive aid in an attempt to shore it up against communist subversion and attack. On 14 November, the P.D.R.Y. and Saudi Arabia withdrew their respective Ambassadors.

In Somalia, however, the Kingdom at last had its reward for its persistent diplomacy backed by the persuasive power of its mounting financial surplus. Barre had become exasperated by the Soviet Union's failure to supply the armaments and spares it required. The final straw came in the form of a demand from Moscow that Somalia should cease 'all interference in the internal affairs of Ethiopia'. On 13 November, Barre gave the 1,500 Soviet military advisers a week to quit, withdrew permission for the use of naval and air facilities, and abrogated the treaty of friendship and co-operation signed little less than three years before. For good measure he demanded a reduction in the Soviet Embassy staff in Mogadishu. In recognition of Castro's services to the Dergue, Barre severed relations with Havana, allowing the Cuban diplomats only forty-eight hours to pack their bags and leave.

The Saudi hierarchy was delighted, but characteristically concerned at the same time about the danger of Somalia asserting its other territorial claims, in particular to Djibouti and possibly also to Eritrea. Riyadh is believed to have made an immediate commitment of $400 million in grant aid. East European weapons, purchased through middle-men in Switzerland and elsewhere, were air-freighted via Jeddah on their way to Mogadishu. By the turn of the year a stream of vessels carrying armaments paid for by Saudi Arabia were on their way to Somalia. They were urgently needed. The Somali offensive in the Ogaden had become bogged down, literally, as a result of seasonal rains and also because of better-organized defence by the Ethiopian Army which was now backed by at least 800 Cuban and 500 Soviet military advisers, according to U.S. estimates. The tide was turning and Barre's appeals to the West were taking on an urgent tone.

The Arab-Israeli conflict, meanwhile, had reached a critical turning point. In the second half of 1977 Carter had made a comprehensive peace settlement in the Middle East a primary foreign policy objective. The U.S. Administration's initial approach to the problem gave the Arabs some reason for guarded optimism and seemed to justify the confidence placed by the Kingdom in it. Particularly encouraging was a statement giving the U.S. Government's official position, which was read out at a State Department press conference on 27 June. It expressed the view that 'progress toward a negotiated peace in the Middle East is essential this year if future disaster is to be avoided'. Moreover, the Administration's interpretation of U.N. Security Council Resolution 242 was that it meant 'withdrawal on all fronts of the Middle East – that is, Sinai, Golan, West Bank and Gaza – with the exact border and security arrangements being agreed in the negotiations'.[16] Carter's belief that the Palestinian issue was an essential element of a peace settlement was reiterated. On the following day, for good measure, Alfred Atherton, Assistant Secretary of State for Near East and South Asian Affairs, told a conference of editors and broadcasters that 'the ultimate question of Jerusalem, how it will be administered, the sovereignty issue, and everything else, must be part of the negotiating process'.[17] From the Arab point of view it was encouraging that the thrust of the U.S. initiative was directed at the reconvening of the Geneva Conference that had met briefly at the end of 1973.

On 1 October, a joint U.S.-Soviet statement was issued. It was generally welcomed by the Arab camp because it acknowledged the principle of Palestinian self-determination, but for the same reason it provoked an angry Israeli reaction. So fast did the Israeli lobby react that Carter told a group of Congressmen a few days later on 6 October that he would 'rather commit political suicide than hurt Israel'.[18] One hundred and fifty elected representatives signed a letter to the White House attacking the statement. It was not sent, however, because on 6 October Moshe Dayan, the Israeli Foreign Minister, reached agreement with Carter on a draft 'working paper' relating to the Middle East peace negotiations. Acceptance of the U.S.-Soviet statement was not a prerequisite for the conducting and convening of the Geneva Conference, it said. Even before the intervention by the Congressmen and Dayan, however, momentum towards the reconvening of the Geneva Conference had slowed down and almost drawn to a halt as a result of the impasse over P.L.O. representation. Arafat and his colleagues had once again been incapable of extending an olive branch that was not tied to preconditions about the recovery of all Palestine. The working paper also adopted the principle of bilateral dealings between Israel and any one of its three adversaries, and made it possible for any of the parties to the conflict to exclude the P.L.O. from the negotiations.

Zuhair Mohsen, the leader of Saiqa, told a rally in Beirut that 'any acceptance of the working paper would be pusillanimous and unpatriotic'.[19] He was clearly referring to Egypt and speaking on behalf of Assad's regime, which completely controlled the guerrilla group led by him. Relations between Cairo and Damascus had become strained. Assad was wary and suspicious of Sadat, who gave the impression of being compliant over the principles of Palestinian rights and P.L.O. representation. Pan-Arab diplomatic activity became tense in the course of October and the criss-crossing movements of envoys bewildering, as if the Arab community of states knew something momentous was about to happen but was confused because it did not know what. Saudi Arabia tended to favour Egypt's more accommodating attitude to the U.S.-Israeli working paper. But it wanted, as always, consensus. An added sense of urgency was given by the suggestions emanating from Israel that the Kingdom would not be spared in any future Middle East conflict and its oilfields would be a legitimate target. Saudi Arabia had become alarmed at the possibility of being caught up in a war. In February 1977, the air base at Tabuk in the north-east, only 250 miles from Tel Aviv, had been completed and a squadron of Lightnings stationed there. Two months later five radar blips indicated the approach of aircraft from Israel. The pilots of two Lightnings who were in the air at the time panicked and both landed their aircraft, one forgetting to lower his undercarriage and belly-flopping. Five Kfir jets roared over the runway in a simulated bombing run. It was a sobering experience. So, too, was a statement made soon after by General Mordechai Gur, Israeli Chief of Staff, that he considered the Kingdom to be a front-line Arab state.

Desperately anxious to preserve a façade of unity, Saud al Feisal went to Washington, where he was uncompromising in his insistence on the principle of P.L.O. representation in any peace negotiations. He warned that the time for diplomacy was running out and that in the event of war Saudi Arabia 'would not only sacrifice its oil and money but the blood of its sons as well'.[20] Kamal Adham made one of his unobtrusive visits to Cairo on 23 October carrying a message from the King to President Sadat that was believed to be about the American-Israeli working paper and also the Egyptian leader's forthcoming visit to Riyadh, scheduled to take place early the following month. Before going to the Saudi capital, Sadat held talks from 29 to 31 October with President Nicolae Ceausescu of Romania in Bucharest, where Menachem Begin, the Israeli Prime Minister, had been little more than three months before. Sadat's talks with Khaled and Fahd on 2 November concentrated on co-ordination of policy in advance of Carter's visit to the region scheduled for late November, though the U.S. President had warned that his trip might have to be postponed if Congress thwarted his energy legislation. King Hussein of Jordan arrived in Riyadh hot on Sadat's heels fresh from consultations with Assad. On 7 November, Saud al Feisal flew to Damascus with a message from the King to the Syrian leader.

Two days later Sadat told the People's Assembly that in the pursuit of peace he was 'ready to go to the ends of the earth'. Then came what sounded like, and proved to be, a bombshell. The Egyptian leader proclaimed: 'Israel will be astonished when it hears me saying now, before you, that I am ready to go to their own house, to the Knesset itself, to talk to them.'[21] It seemed to many observers a flamboyant, though calculated, piece of rhetoric primarily aimed at softening Israel's position. Begin was quick to respond, however, saying that Sadat would be welcomed in Jerusalem. That was the response Sadat wanted, it became clear subsequently. His plan for the bold initiative had crystallized as a result of his talks with Ceausescu, Muhammad Heikal told me. The Romanian leader had warned him that Begin was a wild and dangerous man, quite capable of launching a 'pre-emptive war' if the Arabs veered towards confrontation rather than peace. But if Sadat was prepared to hold direct talks, then Begin would have much to offer.

Arab foreign ministers meeting in Tunis from 12 to 14 November were not convinced about Sadat's resolve, though some were alarmed. But there could be no doubt about his serious intent when on 16 November he went to Damascus in an attempt to win Assad's support for his initiative. Sadat failed to obtain it. On 17 November he announced his decision to go to Jerusalem, and four days later the Egyptian leader, who had flirted with the Muslim Brotherhood and Nazi Germany, was performing prayers at the Al Aqsa Mosque. There the mullah warned that the surrender of Al Quds (Jerusalem) would be like 'abandoning Mecca'. In his address to the Knesset Sadat said that there could be no bilateral peace, no argument about withdrawal from occupied territories, including 'Arab Jerusalem', and no settlement without a resolution of the Palestinian problem. He made no reference, though, to the P.L.O.

Sadat's break of what he called the 'psychological barrier' threw Arab ranks into consternation and confusion. Only Morocco, Oman, and Sudan were prepared to endorse his move. At the other extreme, Qaddafi judged it an 'unpardonable crime'. Iraq, Algeria, the P.D.R.Y., and the extremist left-wing Palestinian groups outside the fold of the P.L.O. adopted a similar stance of outright rejection. A meeting between the P.L.O. and a delegation of the Syrian ruling Baath Party condemned Sadat's journey as 'traitorous'. Assad rejected the invitation issued by Egypt to meet in Cairo with representatives of Israel, the U.S., the Soviet Union, and the U.N. Nevertheless, he refused to characterize the rift with Sadat as a divorce. In reality, both Syria and the P.L.O., though they were very dubious, waited to see if Sadat's initiative offered anything for them. Jordan was non-committal.

As cautious and circumspect as ever, the Saudi leadership was disturbed at the prospect that the modicum of unity at the heart of the Arab world, very much the work of the Kingdom's patient diplomacy in 1976, should now be about to fragment. It was deeply apprehensive about Sadat's isolation and the implications for his regime. Riyadh's initial reaction was restrained to the point of being almost Delphic. An official statement issued on 18 November expressed surprise and said little more than that any Middle East peace initiative 'must emanate from a unified Arab stand'. The reserve disguised the hierarchy's belief that Sadat had committed a serious error through his unilateral initiative. Khaled and the Saudi leadership were also sullen, feeling that they, as much as the Arab cause, had been betrayed. Sadat had revealed nothing to them about his initiative when he had come to Riyadh on 2 November. He did tell the Saudi leadership, however, that he had information of a planned Israeli strike against Arab oilfields in the Gulf. There had also been reports, apparently inspired by the

C.I.A., that Israel was planning a pre-emptive strike to shatter the armed potential of Egypt and Syria for another ten years.

Sadat was later to tell Saudi friends that Fahd must have understood his reasons for going to Israel even if he had not been informed. Yet the well-connected Lebanese journalist and publisher Salem Lozi, who was present in Riyadh in the immediate aftermath, described the embarrassed confusion that he found in the Saudi capital where, he noted, the principle adopted at times of crisis was 'that the politician remains master of the statement he has not made and slave to the one he has made'. Lozi wrote: 'When I met Crown Prince Fahd and talked briefly about the Sadat visit, he stuck to generalities, as he had with the Palestinian delegation which he met on the same day. He talked with the Palestinians in exactly the same way that he talked to journalists – he spoke at length without revealing anything, and without committing himself. It was the same with Prince Abdullah, number three in the Saudi regime, with Prince Sultan, the knowledgeable diplomat, and Prince Salman, who is generally known as the Royal Family technocrat. But while reserved in their public meetings, Saudi officials talk openly in private. The first point they raise is that they were taken by surprise, and that they were just as offended as the Syrians, if not more. Egypt and Syria had been at loggerheads several times over the past five years. But never has Egypt fallen out with Saudi Arabia, even when Egyptian policies differed from those of Saudi Arabia.'[22]

Equally galling to Saudi Arabia was the fact that Sadat, after he launched his initiative with his speech to the People's Assembly, had sought Syria's support, not the Kingdom's. Especially offensive to the Guardian of the Holy Places and his kin was the manner in which Sadat had prayed at the Al Aqsa Mosque, the third holiest shrine of Islam, while the Old City of Jerusalem was still under Israeli occupation. It was humiliating to be treated in such a way after the Kingdom had poured thousands of millions of dollars into Egypt over the previous six years with the aim of consolidating Sadat in power and retaining Egypt's friendship. The limits of petrodollar diplomacy had been exposed. The hierarchy must have felt that it had been taken for granted by the Egyptian leader who, it could only be concluded, had assumed that the Kingdom would be carried along by his initiative. Irritation was caused by a report in the Cairo newspaper *Al Ahram* that the Saudi King had sent Sadat a message extolling the sacrifices made by the Egyptian people – which seemed to imply Riyadh's endorsement of his action. The Saudi Press Agency responded by saying that the Kingdom's position had been outlined in the official statement of 18 November.

Saudi Arabia's response to the crisis caused by Sadat was to join forces once again with Kuwait in a joint diplomatic mission dedicated to restoring some measure of unity in the Arab world, or at least minimizing the damage caused by the Egyptian leader. Adham was held responsible for the lack of prior knowledge of Sadat's initiative and was consequently under a cloud of opprobrium. Henceforth, control over the General Directorate of Intelligence was placed in the hands of his deputy, Turki al Feisal.

Saudi Arabia and the other conservative Arab oil-producing states of the Gulf and Jordan sat cautiously on the fence. They stayed away from the summit called by Qaddafi which took place in Tripoli from 2 to 4 December and was attended by Syria, Iraq, Algeria, the P.D.R.Y., the P.L.O., and the radical Palestinian groups. The Tripoli Declaration resulting from the meeting castigated Sadat's trip to Jerusalem as a 'flagrant violation of the principles and aims of the pan-Arab struggle against the pan-Arab enemy'. The summit decided to freeze relations with Egypt, 'suspending dealings with it on the Arab and international levels', rather than rupturing them totally. Negotiations were not referred to and, therefore, not ruled out. The declaration went above Sadat's head in paying tribute to 'our Arab people in sisterly Egypt, especially its nationalist and progressive forces which reject the policy of surrender executed by the Egyptian regime'.[23] Egypt responded by breaking off relations with the five states involved. Saudi Arabia agreed to look after Egypt's interests in Iraq, Syria, and the P.D.R.Y. All the signatories to the Tripoli Declaration, with the exception of Iraq, formed the 'Steadfastness Front' whose purpose was to resist Sadat's initiative.

The Kingdom's position was now pivotal. As the main contributor of funds to Egypt and the power from which the other oil-rich donors in the Gulf took their cue, Saudi Arabia had the greatest potential influence and leverage over Sadat. It would inevitably be subject to pleas and pressures from Syria and the P.L.O. on the one hand, with Iraq as always lurking ominously in the background, and the U.S. and Egypt on the other. The Carter Administration's initiative aimed at the reconvening of the Geneva Conference had, meanwhile, been undermined by Sadat's – which Washington now had little choice but to sponsor as an avenue towards a comprehensive settlement.

The possibility of it leading to a full forum presided over by both super-powers looked very remote, as representatives of only Egypt, Israel, and the U.S. sat down at the Mina House Hotel near Cairo for the gathering which Sadat had called as a preparation for the elusive Geneva Conference. It was clear from a brief, inconclusive meeting that Carter would henceforth give priority to the conclusion of a bilateral Egyptian-Israeli peace agreement.

In September, Yamani had warned again that Saudi willingness to raise oil production to satisfy expansion of long-term demand in future would depend on the achievement of a satisfactory peace settlement in the Middle East. Amongst the Arab producers a ripple of anger was caused when James Schlesinger, who was now U.S. Secretary of Energy, implicitly threatened military intervention when he said that the U.S. might yet be forced to ensure the security of oil supplies in the Middle East. Then suddenly the Shah, almost overnight, became a convert to moderation as a result of a visit to Washington in November, declaring himself in favour of a prolongation of the price standstill decided in Stockholm. Expressing concern about the economic health of the West, he spoke the language of sweet moderation that the world was accustomed to hearing from Saudi Arabia. It became clear subsequently that his change of tune – and policy – was the *quid pro quo* exacted by Carter in exchange for U.S. willingness to supply Iran with sophisticated weaponry and for not making an issue in public of the Shah's violation of human rights, though on the latter score he made, confidentially, certain commitments. Moreover, the oil market had softened and O.P.E.C.'s commanding position had been weakened by the coming on-stream of supplies of Alaskan oil and the build-up of North Sea production. Thus, the outcome of the conference held in Caracas in December 1977 was a foregone conclusion. The price freeze was extended. There was some substance to the charge made by Izzedin Mabrouk, the Libyan Minister of Oil, that the decision of the conference had been a politically motivated one – on the part of Saudi Arabia and Iran. It was Iran, however, who received the credit in Washington for the alliance enforcing price restraint rather than Saudi Arabia, which had risked much in the cause over a period of three and a half years.

The importance of Iran to Washington was further highlighted by a report drawn up on behalf of Senator Henry Jackson, the passionate Congressional champion of Israel and Chairman of the Senate Committee on Energy and Natural Resources, which was published just prior to the O.P.E.C. meeting. It read: 'If Iran is called upon to intervene in the internal affairs of any Gulf state, it must be recognized in advance by the U.S. that this is the role for which Iran is being primed and blame cannot be assigned for Iran's carrying out an implied assignment.'[24] To many ears in Riyadh such words seemed like a semi-official recognition by Washington of the guardian role assumed by the self-appointed gendarme of the region, one which hurt the sensitivities of all the Arab states of the Gulf littoral, and was a slight to the Kingdom in view of the determined pro-Western stance taken by it.

At the end of the year Carter set out on the tour abroad that had been postponed in November. He expressed confidence to the reporters travelling on U.S. Air Force One about the chance of Jordan and Saudi Arabia endorsing Sadat's initiative. He met Hussein in Tehran, where the King was on a private visit. The Hashemite rejected out of hand Begin's proposals for a limited autonomy for the Palestinian inhabitants of the West Bank and Gaza Strip which would have given them a choice of Jordanian or Israeli citizenship. Carter also received a cool, albeit polite, response in Riyadh during his brief stop-over on 3–4 January. His exchanges with Fahd, Sultan, and Saud al Feisal were sticky. The Saudis were not prepared to give any blessing to Egypt's new approach to a settlement of the Arab-Israeli conflict. After the President's departure, the Saudi Foreign Minister pointedly issued a statement to the effect that any Middle East settlement must be based on complete Israeli withdrawal from occupied territories, including Arab Jerusalem, and fulfilment of the legitimate rights of the Palestinian people. The Saudi leadership, Fahd and Sultan in particular, was also perplexed over what it regarded as Carter's pusillanimity in the face of Soviet expansion. Washington took for granted Riyadh's anti-communist stance but did nothing to counter the Marxist threat to the Kingdom. It assumed the House of Saud's reliance on the U.S. as protector of last resort without being prepared to take any action to guard it against malevolent forces.

Another issue, as it happened, was beginning to strain Saudi-U.S. relations. When Fahd had been in Washington in May of the previous year he had raised the question of Saudi Arabia purchasing the McDonnell-Douglas F-15 combat aircraft, a request not unrelated to Israel's violations of Saudi air space but also very much connected with the communist encroachment so much preoccupying the regime. The Kingdom wanted six F-15s, of which Israel had ordered twenty-five.

Harold Brown, U.S. Defense Secretary, and the State Department, were in favour of the deal, estimated to be worth some $1,500 million. On 14 February Carter gave his approval to the sale as part of a package that included also F-5Es to Egypt, as well as fifteen more F-15s and seventy-five of the cheaper General Dynamics F-16s to Israel.

The Israeli Government mobilized its lobbyists to block the transaction. The Saudis invested $365,000 in a campaign to counter the formidably powerful Zionist machine. Bandar bin Sultan, the articulate son of the Minister of Defence and a Captain in the Royal Saudi Air Force, had been drafted to Washington to fight the cause. He and Ali Alireza, the Saudi Ambassador, lobbied dozens of Congressmen. A public relations company recommended by John C. West, U.S. Ambassador in Jeddah, showered opinion-makers with hundreds of booklets explaining the Kingdom's case. West himself warned the Senate Foreign Relations Committee that Saudi Arabia would drop its resistance to extremist demands within O.P.E.C. if agreement on the F-15s was not forthcoming. That was an impression which Riyadh did not refute. It was also happy to foster speculation that the Kingdom would turn to France for an alternative if denied the American aircraft. Khaled's forthcoming visit to Paris, set for the end of May, hung over the whole issue. The Saudi Government postponed without explanation a scheduled visit by Alfred Atherton, Assistant Secretary of State with responsibility for the Middle East.

The F-15 question preoccupied Riyadh as much as the halting U.S. diplomatic moves to build on Sadat's initiative and the increasingly aggressive Israeli air strikes against the Palestinian presence concentrated in southern Lebanon. The Egyptian leader's unequivocal commitment to peace and Cairo's détente with Jerusalem had only intensified the guerrillas' desperate activity against civilian targets in the Jewish state and the weight of Israeli retaliation. Tension in the region rose when Israeli forces invaded the south of Lebanon on the night of 14–15 March after a Palestinian terrorist attack on a bus had left twenty-eight Israelis dead and many more wounded. The King urged Carter to bring Israel's 'flagrant aggression' to a halt. Riyadh also complained privately to Washington about regular over-flights by Israeli aircraft which were practising bombing runs on the Tabuk air base. Possession of sophisticated aircraft might discourage such intrusions. But the Kingdom did not in any way contemplate becoming involved in a conflict with Israel that could be calamitous for the regime. It saw the F-15s as necessary for the protection of its south-western flank and its priceless oilfields. Hence the exasperation and feeling of desperation. Saudi Arabia's approach was genuinely defensive. The Government was prepared to forego the auxiliary fuel tanks and multiple ejection bomb racks required if the F-15s were to have a truly offensive capability. It undertook not to base the aircraft at Tabuk or make them available to any other Arab state. These compromises were explained by Brown in a letter to the Senate Foreign Relations Committee. A motion on 11 May to stop the deal was tied but defeated four days later by a full session of the Senate. The date for the delivery of the first F-15s to Saudi Arabia was set for 1982.

U.S. approval of the aircraft sale was timely as a gesture of friendship and of Washington's commitment to the Kingdom's security. Only a fortnight before the crucial Senate vote, President Muhammad Daoud of Afghanistan, who had been Khaled's guest in Riyadh only a few weeks before, was overthrown in a military coup and his regime replaced by one of a distinctly Marxist hue under Nur Muhammad Taraki, which soon expressed its desire for closer relations with Moscow. The event caused little concern in the West. But Saudi Arabia's leaders were immediately and apprehensively aware of its implications. They felt that the Kingdom was the target of a communist pincer movement.

There had been a dramatic reversal of fortunes in the Ogaden where the Somalis, having failed to capture Harar in January, were put to flight by the Ethiopians who, under the direction of senior Soviet officers, had been provided with new equipment and reinforced by Cuban units. Barre virtually admitted defeat on 9 March when he announced the withdrawal of all regular forces from the disputed region. In June, the Dergue unleashed its revitalized armed forces against the Eritrean guerrillas, who were quickly forced from their hard-won positions threatening the main cities.

On 24 June the Kingdom felt the shock waves from a tremor much closer to its borders. Nine months after Ghashmi had assumed power in the Y.A.R. he was killed when a briefcase carried by an envoy from Aden exploded within two minutes of the Yemeni entering the President's office. Viewed from Riyadh the outrage seemed directed as much against Saudi Arabia as the regime of the unruly republic. Ghashmi had proved much more amenable than Hamdi to the Kingdom's efforts to assert its hegemony over the Y.A.R. The complicity in the assassination of the extremist faction in the N.L.F.

regime was obvious. Two days later Rubai Ali, on whom Saudi Arabia had largely placed its hopes of a rapprochement with the P.D.R.Y. in 1976–7, was executed in Aden by firing squad, accused of 'terrible crimes' against the people and constitution. Ali Nasser Muhammad succeeded Rubai as head of state but the real power in the land was the hard-line servant of Moscow, Abdul Fattah Ismail. It was reported that 500 Cuban troops had flown from Ethiopia two days after Rubai's execution to suppress a rebellion by troops loyal to the dead President. The Soviet Union was more heavily ensconced in the P.D.R.Y. than ever, while the modicum of stability achieved in the Y.A.R. had been severely shaken.

The Kingdom was the prime mover in the convening of an emergency meeting of the Arab League on 2 July that suspended political and economic relations with the P.D.R.Y. Tension rose as Aden accused Saudi Arabia, without any justification, of massing troops on its frontier. The Soviet Union warned against any intervention in the dispute between the two Yemens. It was of little immediate compensation to the Saudi leadership that Ali Abdullah Saleh, who emerged as the Y.A.R.'s new leader on 17 July, was very much its man. Further afield the Kingdom was doing what it could to combat international communism. It financed the airlift of 1,500 Moroccan troops to Zaire in June and their operation against the left-wing insurgents backed by the Marxist regime in Angola which had invaded the mineral-rich province of Shaba. But the power of the Kingdom's purse strings seemed desperately inadequate in the face of the red peril.

Saudi Arabia had spent lavishly in assisting its neighbours to counter Soviet designs on the region, in fortifying the Arab states confronting Israel, and in aiding other Arab countries with the aim of strengthening those traditional values that the Kingdom, whatever the conduct of some of its rulers, saw as the ultimate defence of a God-given order against the forces of evil arraigned against it. Many of its disbursements in the form of grant aid were not recorded and the total defied calculation. But by 1978 the volume was reckoned to be running at an annual rate of $5,000 million. The return on the outlay looked somewhat meagre. Saudi Arabia's environment seemed very much less secure and stable in mid-1978 than it had a year previously.

There had been some reassuring developments, however. A visit by Sultan to Iraq in April had created a new basis of confidence with Baghdad. Riyadh was in no way responsible but looked on approvingly in the following month when the ruling Baath Party cracked down on the Iraqi Communist Party, theoretically its junior partner in the Progressive National Front, and executed twenty of their leading members. Even more pleasing was Vice-President Siddam Hussein's threat to break diplomatic relations with Moscow because of its support for Ethiopia in its war against the Eritrean secessionists. Saudi Arabia expressed its willingness to co-operate with Iraq and Iran in ensuring the defence of the Gulf. More unobtrusively, Naif was arranging a system of collaboration linking the Kingdom's security apparatus with those of the other traditional regimes in the area – Kuwait, Bahrain, Qatar, the United Arab Emirates, and Oman.

Strains in O.P.E.C., meanwhile, were minimal. There was a glut of oil on the market, where demand was still weak. Saudi production had fallen well below the ceiling of 8.5 million barrels a day imposed in the middle of 1977 to a little over 7 million barrels a day a year later. Yamani was of the opinion that demand would not recover until towards the end of 1979. In this situation an extraordinary conference was called to discuss long-term strategy and met in Taif on 6–7 May. A ministerial committee was appointed under Yamani's chairmanship to draw up a report. When ministers met in Geneva from 17 to 19 June for the regular ordinary conference, members were resigned to the fact that the market could not sustain a real increase. Some pressed for a switch to the pricing of oil on the basis of a unit of account other than the dollar, the value of which had steadily eroded over the previous year. The proposal was resisted by Saudi Arabia, amongst others, on the grounds that such a device would only weaken the dollar further by raising the price of oil in real terms while also further depressing the world's economy. The meeting broke up amicably enough.

Saudi Arabia shared the hopes of other Arab states that Egypt might soon be brought back into the common fold. Its worst fears of Sadat's initiative leading to grave instability, and even his overthrow, had not materialized – not the least because of the aid given by the Kingdom and the other oil producers. By the summer it looked as if his bid for peace had foundered, largely over Begin's refusal to contemplate anything more than a very limited form of autonomy for the West Bank and the Gaza Strip, and his clear determination that Israel should keep actual control of the territories. The deadlock appeared to be total by mid-summer. At the end of July, Fahd felt sufficiently confident to embark on a mission to Cairo and Damascus in the hope of reconciling them. It was a tentative and premature move. On 8 August the White House announced that Sadat and Begin had agreed to attend a summit at Camp David, the presidential retreat near Washington.

Carter staked his personal prestige on a successful outcome to the talks in a bid to brighten up the tarnished image of his presidency, but he neglected to consult Saudi Arabia, the only major Arab ally of the U.S., apart from Egypt. There were two parallel and related sets of negotiations, one concentrating on the principles of an Egyptian-Israeli peace agreement and the other on a general Middle East settlement focused on giving the Palestinians autonomy in a homeland of their own on the West Bank and in the Gaza Strip. Sadat obtained most of what he wanted in terms of the bilateral accord, in particular a commitment by Israel to military withdrawal from all of Sinai, with the date of the final pull-out to be agreed later. Otherwise, he was sympathetic to Carter's predicament and made the most concessions, allowing the goblin-like Begin to triumph in what became entitled 'The Framework for Peace in the Middle East'. Sadat was prepared to leave loose ends in his naive confidence that in the last resort the U.S. would persuade or pressurize Israel into submitting to a solution of the Palestinian problem acceptable to a pan-Arab consensus. Final agreement on the basic principles for a framework hinged on whether or not Begin would agree to a freeze on the establishment of new settlements on the West Bank and in the Gaza Strip. Carter managed to exact an undertaking on this score from Begin who promised to confirm his commitment in writing on the following day. Having been assured that the Israeli Premier had made such a pledge, Sadat signed the Camp David accords on 17 September.

Carter was deeply disturbed and Sadat angered when they saw the text of the letter. Begin promised a moratorium of no more than three months, the proposed period set for concluding a bilateral Egyptian-Israeli peace treaty. Hermann Eilts, who was present at the Camp David talks as U.S. Ambassador to Cairo and had previously served in the same capacity in Jeddah, subsequently wrote: 'The costs were high. In the eyes of Europeans and friendly Arabs, such an ignominious collapse of the U.S. position on that issue was inexplicable. It destroyed the last chance of obtaining moderate Arab support.'[25]

Apart from the question of the freeze on Jewish settlements on the West Bank and Gaza Strip, the shortcomings of the framework were immediately apparent to the wider Arab world, including the

Kingdom. The accord envisaged nothing more than a self-governing Palestinian body for the territories over a five-year transitional period, with the eventual status of the Palestinians left unclear. The extent of Palestinian autonomy contemplated was severely limited to administrative affairs alone. Moreover, the proposal was that the status of the Palestinian authority would derive from agreement involving not only Egypt and Israel but also Jordan, which had not taken part in the negotiations, in fact had not been consulted at all. Nor for that matter had the Palestinians whose future was at stake. The delicate issue of Jerusalem was avoided altogether. No reference was made at all to Syria, part of whose Golan Heights was still under Israeli occupation. There was no direct, only an implied, link between implementation of the two agreements, so that a bilateral Egyptian-Israeli peace treaty was not made dependent on a solution to the Palestinian problem. The text of the accord concerning the latter merely seemed to confirm Israel's determination to keep control over the West Bank and Gaza Strip at all costs.

To assume the endorsement of Jordan for the restrictive framework and its participation in any future negotiations was both a miscalculation and a presumption. So, too, was Carter's private assurance to Sadat that he would ensure Saudi Arabia's approval. He had no justification for giving such a pledge. Although the Camp David accords came as a surprise because the negotiations had been widely expected to fail, the Saudi Government was uncharacteristically quick to respond. After a three-hour meeting of the Council of Ministers on 19 September chaired by the King, a statement was issued in his name. The agreement relating to the West Bank and Gaza Strip was described as 'an unacceptable formula for a definitive peace'. It had not made 'absolutely clear Israel's intention to withdraw from all Arab territories it occupied including Jerusalem', and had failed to put on record 'the right of the Palestinian people to self-determination and to set up their own state in their own homeland and on their own soil'. Moreover, it had failed to recognize the P.L.O. as their sole representative. The Kingdom had no right to intervene in an attempt by any Arab state to restore its sovereignty over territory occupied by an enemy, it went on to say out of deference to Egypt. The statement politely expressed appreciation of American efforts to arrive at a peaceful solution but added that Saudi Arabia could not accept formulas that 'contradict higher Arab interests'.[*] The Jordanian Government pointed out on the same day that it could not be a party to an agreement which it had not helped to negotiate. Its convoluted statement said, in effect, that the accords were incomplete and inadequate. The Hashemite King was more forthright in a subsequent interview in which he dismissed the accords as 'a fig leaf' and 'pure sugar coating'.[27]

The reaction of the monarchies would probably have been no different if the accords had not been immediately and roundly condemned by the P.L.O. on 18 September as the most disastrous Arab sell-out of the Palestinian cause ever. This attitude was reflected in the waves of protest in the occupied territories. An equally virulent denunciation came from the 'Steadfastness Front' – Algeria, Libya, Syria, the P.D.R.Y., and the P.L.O. – which had been formed a year earlier in the aftermath of Sadat's mission to Jerusalem.

Carter immediately despatched Cyrus Vance, Secretary of State, on a thankless mission to convert Saudi Arabia and Jordan. Straight after the Camp David talks Carter had telephoned Hussein urging his support, but received a non-committal reply. In Amman, Vance adopted a heavy-handed approach, warning the Hashemite King of the possible consequences for U.S.-Jordanian relations if he did not endorse the accords. The assumption was that the U.S. could exercise leverage on him because it was a major donor of aid and arms to Jordan. Hussein politely held his ground, asking for answers to a number of queries about the accords. Vance could hardly treat the Kingdom in the same manner, though the Administration believed that it could expect some co-operation in return for having agreed to supply it with F-15s. When Vance arrived at Riyadh airport the Saudi authorities denied him the opportunity to deliver a prepared statement by not providing a microphone. In conversation with the Secretary of State, Fahd kept Saudi Arabia's options open and left the impression that the Kingdom was not fundamentally opposed to the Egyptian-Israeli negotiations. Assad, to the delight of the U.S. Administration, had agreed to receive Vance, who described his exchanges with the Syrian President as 'useful'. There still seemed a remote and very slender chance that the Camp David accords might make the basis of a comprehensive settlement. Saudi Arabia harboured hopes that they would.

Apart from Jordan and Syria, the Kingdom was the most critical factor in the complicated, shifting Arab equation. In seeking Saudi Arabia's endorsement of the accords, Washington was banking on its reliance on the U.S. as an ultimate protector and its continuing preoccupation with the maintenance in Egypt of a moderate, pro-Western regime. But the American Administration's assumptions failed to take account of the declining credibility of the U.S. as a guardian of its friends. The Kingdom felt caught in the midst of an Arab world that was in danger of being polarized by the Camp David accords.

Iraq had called an Arab summit for the beginning of November to discuss them, a meeting to which it did not invite Sadat. Baghdad also proposed establishing a $9,000 million fund, provided by the oil-producing states, to help 'meet Egypt's heavy financial burden, in the event that the Egyptian Government is prepared to renounce the Camp David agreements'[28] and to bolster Arab resistance generally. The plan was that the lion's share, $5,000 million, should go to Egypt, and the balance to Jordan, Syria, Lebanon, and the P.L.O. The Saudi desire to lure Egypt back to the path of Arab righteousness was mixed with misgivings about undertaking yet more financial commitments to a country that was proving to be a perpetual drain on Arab producers' surpluses. To everyone's surprise, not the least the Saudi Ministry of Finance's, Government expenditure was actually catching up with revenue as a result of the fall in the Kingdom's oil output. Nevertheless, in advance of the summit Saudi diplomacy concentrated on trying to prevent the premature ostracization of Egypt.

Early in October, detailed Egyptian-Israeli negotiations on a bilateral peace treaty and also the general framework started smoothly enough at Blair House, Washington, but stalled towards the end of the month. There seemed a possibility that the talks might break down completely, opening the way for a general Arab reconciliation with Sadat. Riyadh kept open its lines of communication with Cairo. On 11 October, Kamal Adham flew to the Egyptian capital for talks with Sadat on the forthcoming summit and the future of Saudi aid to Egypt if and when a peace treaty was concluded. Saud al Feisal toured the Gulf oil-producing states from 21 to 24 October, seeking to rally them behind the Kingdom in a conciliatory and flexible approach towards Egypt. He insisted that the purpose of the summit was not to isolate Sadat but rather to restore trust amongst the Arab family of states. Increasing turmoil in Iran, where the Shah's will to survive was declining by the day, made it even more imperative to maintain maximum Arab solidarity.

To Syria and Jordan, the Kingdom's policy seemed like one of vacillation. They were also irritated by the manner in which Saudi Arabia turned the financial tap on and off. The entente with Iraq was holding well, but a big question mark over its future was raised when towards the end of October Assad of Syria visited Baghdad for the first time in five years and there signed a 'National Charter for Joint Action' after two days of intensive discussion. As a result of Camp David the two deadly Baathist rivals agreed upon a programme of fulsome collaboration, including military union. Since 1973, Iraq had maintained a stance of sultry, rejectionist isolation as far as the Arab-Israeli conflict was concerned. It now appeared to be aligning itself with the Steadfastness Front. Saudi Arabia might be left floating in limbo.

At the customary foreign ministers meeting prior to the summit, Saudi moderation was soon confronted by Palestinian militancy. Rejecting Saud al Feisal's arguments against immediate condemnation of the U.S. and punishment of Egypt, Farouk Kaddoumi, head of the P.L.O.'s Political Department, asserted: 'Even now Mecca is threatened by Israel's expansionist plans.' The Saudi Foreign Minister replied: 'God is there to protect the holiest places of Islam.' Kaddoumi snapped back: 'God was there when Jerusalem fell into Israeli hands and God failed to do anything.'[29]

At both the preliminary meeting and the four-day summit, which began on 2 November, Iraq played a mediating role, insisting that the aim should be to achieve the 'minimum acceptable' (to all the participants) or, put another way, the highest common denominator possible. The militants of the Steadfastness Front pressed for immediate sanctions against Egypt, its expulsion from the Arab League, and the transfer of the organization's headquarters from Cairo. Saudi Arabia campaigned for a stay of execution to give Egypt a chance to return to the fold and opposed the condemnation of the U.S. which, Fahd believed, might yet – under a polite or implied threat of Arab economic pressures – be persuaded to make Israel submit to Arab demands. 'Our war is with Israel and not with Egypt,' he told the meeting. 'Saudi Arabia will back up any measures aimed at Israel – military, economic and political – that can eventually restore Arab land and the rights of the Arabs to a homeland. There is no way we could let the Egyptian people suffer as a result of a boycott. They have gone through enough suffering.'[30]

Saudi and Iraqi moderation won the day. The declaration issued after the meeting was restrained. It did not condemn Egypt's signing of the Camp David accords, but recorded instead the summit's decision 'not to endorse these two agreements, not to co-operate with the results arising from them, and to reject the political, economic, legal and other consequences to which they have given rise'.[31] An appeal was made to Sadat to renounce his signature. No overt threat was made, but secret resolutions were adopted that provided for the expulsion of Egypt from the Arab League and for sanctions to be imposed should a peace treaty be concluded. The $9,000 million support fund proposed by Iraq was reduced to $3,500 million, of which the Kingdom's share was set at $1,000 million, to be allocated over a ten-year period. Egypt was still standing aloof and therefore would not be a recipient of the $5,000 million originally intended for it.

Fahd telephoned Sadat twice during the summit and was left with the impression that a mission bearing a message in the name of President Hassan Bakr of Iraq, host of the conference, would be received by him. A four-man delegation duly arrived in Cairo, but the Egyptian leader refused to receive it. Sadat lashed out in contemptuous anger. 'In Egypt we don't respect those who bribe with their money,' he told the Kuwaiti newspaper *Al Siyassa* in an interview published on 8 November. The 'rich oilmen' kept their money in foreign banks controlled by Jews from whom other Arab countries like Egypt had to borrow at rates of 12–15 per cent, he asserted.[32] For good measure, Ali Mansour, the editor of the weekly magazine *October*, who was close to Sadat, wrote on 11 November that Saudi Arabia had 'opted for the Baath Party, the Warsaw Pact and the Soviet Union'.[33] It was the first criticism, very gauche and hurtful in nature, directed at Saudi Arabia from the banks of the River Nile since Sadat had come to power eleven years earlier. But the Kingdom continued to give aid to pay the U.S. for the F-5E fighter-bombers being supplied to Egypt. The only concession made to those who wished to cast Sadat into outer darkness was the unusual public announcement over two months later, on 19 January 1979, to the effect that Adham, who had been chiefly responsible for the entente formed with Egypt in 1970–1 and the main link with Sadat, had been dismissed from his post as Royal Adviser. Normal practice with such an important personage would have been to say that he had retired or resigned. His 'sacking' was a sacrificial offering to Arab militancy. Adham remained an influential figure, however. Preoccupied with his proliferating business interests, he was happy.

Khaled had flown to Cleveland, Ohio, at the end of September to undergo a second heart operation. Accompanied by a retinue of about 200, he made a bigger impact locally than he had in London a year and a half before when his hip joint was treated. His hospital and hotel bills were estimated by local newspapers to have amounted to over $2 million. Sultan was once again in watchful attendance, while Fahd held the fort at home. The operation was carried out successfully on 3 October. Just over a week later Khaled was in a fit enough condition to receive Cyrus Vance and the Secretary of State for Defense, Harold Brown. In the wake of Camp David their call was not just the kind of courtesy normally accorded a visiting head of state. The important role of the King, who had been regarded as just a figurehead in the early days of his reign, was also highlighted by the arrival in Cleveland of General Hassan Ali Kamel, Egyptian Minister of Defence, and Boutros Galli, Acting Foreign Minister. No one in the hierarchy feared the King's receptiveness to Egyptian blandishments, however. Khaled had developed a dislike for Sadat well before Camp David.

The King's sense of duty, it was clear, was strong enough to ensure that he would persevere as head of state as long as he was physically able to. His condition before the operation had been such, however, as to trigger off another debate within the royal family about the succession. It was not just a question of the possibility of the King's death or incapacity that made a clearer understanding necessary. The communist penetration and the growing instability of the region made a settlement of the issue even more vital. Thus, shortly before Khaled's departure for Geneva en route for the U.S., another in a series of conclaves involving senior members of the ruling hierarchy was held, according to well-informed Government officials. A definitive decision was said to have been made at this point that Abdullah should be the third in line and the next heir apparent after Fahd. Yet, once again, the question of who should command the National Guard, the royal family's main insurance against a military coup, and who would replace him as Second Deputy Premier in the event of his elevation was not resolved.

For some of the traditionalists the prospect of Fahd holding the first position in the hierarchy and Sultan the third was, evidently, too much. As a group, the seven sons of Hassa bint Ahmed al Sudairi were more formidable than ever as a result of the greater influence asserted by Naif and Salman since the death of Feisal. The Al Fahd branch of the royal family could not be described as the 'progressive' faction.

Naif and Ahmed, who ran the Ministry of the Interior, showed a conservative cast of mind. But collectively the seven full brothers were a strong force precisely because the differing emphases of views amongst them did not in any way undermine the loyalty that they felt towards each other. They were weakened, however, in the autumn of 1978 by the fall from grace of Turki, the Deputy Minister of Defence. It was announced that he had resigned for reasons of ill health, though his only conspicuous malady appeared to be a partiality to alcohol, a failing that did not disqualify the crapulous Fawwaz and Abdul Mohsen from being Governors of the two holy cities of Mecca and Medina. Turki's real sin was matrimonial. He had been married to a sister of Khaled bin Abdul Rahman, but had cast her aside in taking as his wife the daughter of Muhammad Fassi, a Jeddah physician of Moroccan origin. They had cohabited for some time before marrying, which was perfectly acceptable to the elders of the family. A marital relationship between them was not, however.

The Fassis were not highly regarded at the best of times. Worse, the name had become an embarrassment because of the flamboyant behaviour of one son of the family whose activities in California had caught the attention of the American press earlier in the year. Muhammad Fassi and his wife had scandalized neighbours on Sunset Boulevard by painting their $7-million home lime green, placing plastic flowers in Romanesque urns, and painting white statues with skin tones, picking out the erogenous zones in bright colours. A lavish party held to reconcile wealthy residents of the area had duly been reported in the *New York Times*. There were limits to the tolerance of the House of Saud at a time when it was felt that members should close ranks in a display of firm respectability in the face of the Godless disorder closing in upon the Kingdom.

Not since the outbreak of the civil war in the Yemen over seventeen years before had there been such a premium on maintaining maximum solidarity in the senior echelons of the royal family. The Kingdom had lavishly spread its financial resources and strained its diplomatic skills to create a secure and stable environment. But as 1979 dawned the barricades seemed to be collapsing around Saudi Arabia. The protection afforded by a firm, close friendship with Egypt, its most desired partner in the Arab world, had been weakened by the Camp David accords, depriving Riyadh of an assurance that it had invested billions of dollars to obtain. In Ethiopia, the Dergue, with massive Cuban and Soviet military assistance, had asserted control over virtually the whole of the country and had ruthlessly suppressed opposition. To the north-east, Afghanistan was in the hands of Moscow's obsequious subordinates. On the periphery of the Kingdom's immediate horizon, Turkey and Pakistan, which Saudi Arabia was also trying to buttress with aid, were both unstable and looking a possible prey to Soviet designs. Closer to home, Saudi efforts to tame Aden had failed. The Marxist regime there, still bent on subverting the Sultanate of Oman and virulently antagonistic to the Kingdom, was a stronger launching pad for Soviet expansion in the region than ever.

Worst of all, the authority of the Shah had been rapidly crumbling from the middle of 1978 in the face of a rising tide of violent and open protest that was undeterred by bloody repression, and combined forces ranging from liberal democrats to Marxists. But the upheaval derived its real strength from the inflammatory zeal of Shiite religious leaders opposed to the secular pretensions and Westernizing ambitions of the 'King of Kings'. Alternating measures of liberalizing reform and harsh retaliation only encouraged the groundswell of insurrection. Martial law had been imposed in four cities in August 1978 and was extended to twelve other urban centres after 2,000 demonstrators had been killed on 9 September, which became known as 'Black Friday'. According to a senior Iranian diplomat of the old regime, it was at the request of the dreaded secret police S.A.V.A.K. that the Iraqi authorities in October released Ayatollah Ruhollah Khomeini, the Shah's most implacable foe, from his captivity in Najaf where he had been exiled since 1964. But despite being under virtual house arrest there, he had been able in 1979 to incite the mobs in Iran. Erroneously, both the Shah and the Baathist regime in Baghdad, which was fearful of the discontent amongst its own Shiite population, believed the charismatic sage would be less harmful at a greater distance and allowed him to go to France, which gave him asylum. But in fact Khomeini was better able to orchestrate insurrection from there. Soon after his arrival in Paris workers in the Iranian oilfields took industrial action, drastically reducing the flow of oil exports. In desperation the Shah appointed General Gholam Reza Azhari as Prime Minister at the head of a ten-man military government on 5 November.

The tremors from across the Gulf were sending shivers down Saudi spines. Bankers noted an enormous outflow of private funds, estimated at several billion dollars. The hierarchy in Riyadh was alarmed and aghast at what was happening. The House of Saud had

never had any love for the Shah, even if a closer understanding had
been reached with him in recent years. But it regarded the fanatic
Shiite mullahs with revulsion. There were apprehensions that the
contagion might spread to the adherents of the despised heretical sect
in Hasa. The crisis heightened the awareness of the elders of the royal
family about the discontent that could arise from too rapid economic
and social change. The Saudi Government expressed its support for the
Shah's continued leadership but could do nothing to help him.

The Kingdom did what it could for oil consumers, however. The
Supreme Petroleum Council allowed output to rise above the 8.5
million barrels a day maximum set in the middle of 1977 to
compensate for the shortfall in Iranian shipments. Production surged
to about 10 million barrels a day in November and December. It
looked as if Saudi Arabia's ability to restrain the pressure for a price
increase of 15 per cent sought by the majority of O.P.E.C. members
had been badly impaired by the tighter market created by the Iranian
crisis. In the event, when ministers met in Abu Dhabi on 16–17
December, Yamani was able to obtain a very reasonable compromise.
It was agreed that there should be four quarterly increments in the
course of 1979, starting with one of 5 per cent, which would raise
prices to 14.5 per cent above the 1978 level by the end of the year.

By the end of December the Shah seemed ready to admit defeat. As
riots engulfed Iran and the country slid into greater chaos, he
dismissed General Azhari and appointed in his stead a civilian Prime
Minister, Shahpour Baktiar. Riyadh still hoped for a miracle that
would redeem the Shah. Fahd told the local newspaper *Al Jazirah* that
the Kingdom supported any legitimate regime and that the Shah
'enjoyed legality'. The situation in Iran, he said, was 'not in the
interest of Islam, Muslims, or Middle East stability'.[34] His voice was
hardly heard, if at all, in the tumult, and can have been no
encouragement to the Shah, who departed Iran on 23 January 1979
leaving Baktiar, a Regency Council, and the Armed Forces, whose
loyalty was now very suspect and unity fragile, to salvage what they
could from the debacle. Khomeini returned to Iran in triumph on 1
February.

The U.S. Administration had not only failed to save the Shah, but
had probably hastened his downfall by encouraging him to liberalize. Its
credibility was further eroded in the minds of the ruling hierarchy in
Riyadh. Even the termination by the U.S., in the interests of its new
opening to China, of its relations with Taiwan, with which the

Kingdom had developed close ties, was regarded as a betrayal by Fahd and his colleagues. Washington was conscious of the disillusionment. Its response was to offer to send the Kingdom a squadron of F-15s, together with some 300 support personnel, as a gesture of its concern for the safety of the country. There was a measure of theatrical absurdity about the operation. The aircraft would not carry missiles or bombs, it was stressed, though armaments could be flown out rapidly if required. It was also a measure of the fright felt in Saudi Arabia and also of the regime's confusion that it should have agreed to the despatch of the squadron, which arrived in mid-January.

In Riyadh at the beginning of 1979 there was a discernible lack of authority and direction. One reason was that Fahd was listless, tired, and unwell. He was overweight and suffering from diabetes, both of which conditions were aggravated by his consumption of up to sixty cups of sweet tea a day. In addition, he was afflicted with a slipped disc. The vacuum in the leadership became even more marked when the Crown Prince elected in the middle of January to take a young bride, the daughter of the Emir of Baha in the Asir, and to absent himself from the capital as he honeymooned in the countryside for two weeks. He planned also to go to Marbella in March for treatment and a rest, and was less than pleased when the U.S. Government announced that it was inviting him to Washington some time that month. The solicitation was obviously related to Carter's relentless pursuit of an Egyptian-Israeli peace agreement, for which Saudi Arabia's blessing was still wanted.

Fahd had to return to Riyadh to receive Juanita Kreps, U.S. Secretary of Commerce, who arrived as the F-15s, having shown the flag, departed on 25 January. Far from giving an additional boost to Saudi confidence she irritated her hosts with a querulous plea about easing procedures for granting visas to American businessmen and exhortations about the maintenance of a high Saudi oil output. Kreps gave the Saudi Government particular cause for anger when she told the Jeddah newspaper *Arab News* that Saudi officials had assured her that production would be maintained at a rate of some 10 to 10.5 million barrels a day 'for some time' when actual policy was to limit it to an average of no more than 9.5 million for each month of the first quarter of 1979.

Confidence in the U.S. was further shaken when Frank Church, Chairman of the Senate Foreign Relations Committee and an ardent supporter of Israel, called for a review of the decision to sell F-15s to

Saudi Arabia and for a fundamental reassessment of policy towards it because of its lack of co-operation over the Camp David process. The White House disassociated itself from his statement. But the strain in U.S.-Saudi relations was such that Igor Belyayev, an authoritative Moscow commentator on Middle East affairs, felt able to claim in the *Literaturnaya Gazeta* that Saudi Arabia was 'not as anti-Soviet as portrayed in the Western press' and to suggest that it might be 'ready for a relationship with Moscow'.[35] Disenchantment with the U.S. was such that the Saudi leadership did not discourage speculation that it might respond to Moscow's overtures.

The U.S. Administration was now aware of the need to satisfy its friends in the region. Defense Secretary Brown set off for the Middle East in the second week of February and made Riyadh his first destination on a tour that included also Jordan, Israel, and Egypt. The Saudis turned down an embarrassing proposal that the U.S. should establish a base in the area. In his talks with Sultan, Brown agreed to the supply of $300 million worth of arms to the Y.A.R., now firmly established as a Saudi client state, including F-5E fighter-bombers, M-60 tanks, and 100 armoured personnel carriers, with the Kingdom paying the bill.

Aden had been actively fomenting trouble for Sana since the turn of the year, backing the rebellion of the National Democratic Front in the south of the Y.A.R. On 19 February, P.D.R.Y. armoured units of Soviet-supplied T-55 tanks punched through its neighbour's ill-fortified and ill-equipped defences under the cover of air strikes by MiG 21s and artillery directed by Cuban advisers. The Y.A.R. was slow to deploy its inadequate forces and had to resort to the expensive expedient of mobilizing Hashed tribesmen to plug some of the gaps. The Marxist enemy quickly gained positions nearly twenty miles inside the Y.A.R. Aden represented the incursion as an uprising by an indigenous liberation movement, protesting that it was not interfering in the Y.A.R.'s internal affairs. The Saudis correctly judged the aggression to be aimed indirectly at them, though initially its purpose was to overthrow Saleh's regime. The Kingdom's slow response was to announce on 28 February that the Saudi Armed Forces were on alert, all leave being cancelled, and to order the withdrawal of the 1,500-man contingent serving with the Arab Deterrent Force in Lebanon.

Washington now showed a greater sense of urgency. Steps were taken to expedite the supply of weapons promised to the Y.A.R. via the Kingdom, and the U.S. took a higher profile in the region as a naval flotilla headed by the aircraft carrier U.S.S. *Constellation* was sent from the Pacific to the Arabian Sea. Saudi Arabia readily agreed to the U.S. Air Force operating two Airborne Warning and Control Systems from its territory so that the Yemeni conflict could be monitored. Washington again offered to put a squadron of F-15s at the Kingdom's disposal. This time the Saudi Government declined. Because of the Camp David accords it was acutely embarrassing for the Kingdom to ask for or receive U.S. backing now that it most needed it. A flurry of diplomacy led to the Arab League mediating between the two Yemens.

The Egyptian-Israeli talks had progressed as far as approval by the participants of the basic text of the peace treaty, which Carter regarded as '95 per cent' complete. But negotiations had become bogged down on annexes and accompanying letters relating, in particular, to two major issues. One was Sadat's insistence that the implementation of the treaty should be linked to a timetable for concluding agreement on Palestinian autonomy within the context of the framework agreed upon at Camp David the previous September. The second was Begin's adamant demand that the treaty should take precedence over any of Egypt's commitments to other Arab states. Washington called over Mustapha Khalil, the Egyptian Premier, who had recently been given the Foreign Affairs portfolio as well by Sadat, and Moshe Dayan, Israeli Foreign Minister, in an attempt to break the deadlock. Begin was then invited to Washington, where he arrived on 1 March. He proved to be as intransigent as ever. Carter, whose ratings had slumped in U.S. opinion polls, badly needed a foreign policy success. His determination to pull off a diplomatic coup was such that on 7 March he travelled to Israel and Egypt with new proposals for talks with Sadat and Begin. It was a breathtaking gamble as far as his political future was concerned.

Fahd declined the opportunity to influence events. The Saudi Government informed the State Department that the Crown Prince would not be taking up the invitation, which had been provisionally accepted, to pay a visit to Washington beginning on 11 March. The Administration was taken aback. Its main hopes of winning Saudi Arabia round to the Camp David accords rested upon Fahd who, like his father, believed that friendship with the U.S. was essential for the *umma al arabia,* or wider Arab community. There was some confusion as to the reason for the visit being called off. A State Department spokesman said that it was 'because of health concerns that may require hospital tests'. But Issa Nuwaiser, the Saudi Chargé d'Affaires, pointedly stated in answer to a press inquiry: 'One thing I can assure you of is that the health of Prince Fahd is perfect.'[36] The State Department's explanation was not wishful thinking, but neither was Nuwaiser's rejoinder completely true. Fahd was unwell, and for that reason alone probably did not want to make the trip to Washington. At the same time, others in the Saudi hierarchy would have been relieved not to see him subjected to Washington's pressure at this critical conjuncture. Fahd's policy of conciliation with Egypt over its initiative had become virtually untenable in the pan-Arab context. The prospect of the peace treaty being signed increased his fear of a complete rupture with Sadat and further estrangement from the U.S. He believed that a compromise might still be reached if the Egyptian leader agreed to submit the treaty and associated accords to foreign ministers of a representative sample of Islamic states to judge whether they might provide the basis of an acceptable settlement. Communications with Sadat indicated that he was not unenthusiastic. But Carter was far too impatient to risk losing the treaty by letting it be submitted to such a vetting procedure, which would certainly have been unacceptable to Begin.

The Saudi Government made one belated move to forestall the peace treaty. On 8 March Saud al Feisal summoned West, the American Ambassador, for an urgent meeting. He told the envoy how gravely Riyadh regarded Carter's initiative. The Saudi Foreign Minister expressed his Government's anger that it had not been informed of Carter's latest proposals. For that matter, West himself had not been informed. He was asked by Saud al Feisal to deliver personally an invitation to the U.S. President, who was still in Egypt, to come to the Kingdom before proceeding, as he planned, straight to Jerusalem. Amongst the suggestions that the Saudi leadership wanted to put to Carter was Fahd's proposal for an Islamic review of the terms of the pact. West was brushed aside.

On Carter's return to Washington he despatched his National Security Adviser, Zbigniew Brzezinski, accompanied by the President's son 'Chips' and General David Jones, the Chairman of the U.S. Joint Chiefs of Staff, to Saudi Arabia and Jordan. It was an eleventh-hour bid to win their approval for the prospective peace treaty. Brzezinski was received by Khaled and Saud al Feisal on 16 March. Fahd was conspicuous by his absence. The U.S. sought to play on the apprehensions of both Arab monarchies, offering them a place under an American umbrella of security in return for compliance. The National Security Adviser argued that the Camp David accords still provided the basis for a comprehensive settlement. The Saudi leadership's answer was summed up in a statement issued by the Foreign Minister after Brzezinski's departure: 'Peace and stability can come only through a comprehensive, just settlement which the Arab and Islamic nations and the Palestinian people desire.'[37] King Hussein pre-empted Brzezinski the day before he set foot in Amman by issuing a joint communiqué with Arafat which not only supported the resolutions of the Baghdad summit held in November the previous year but also committed them to helping the inhabitants of the occupied territories to resist the plan for self-rule envisaged by the Camp David accords.

Fahd seemed to have opted out of active participation. He did, however, give an interview to *Newsweek* magazine, his favoured medium for communicating with public opinion in the U.S. and the West. In it he deplored 'separate initiatives', without referring to Sadat, but expressed 'our deep thanks and appreciation for the incessant and enormous efforts by President Carter to secure Israeli withdrawal from all occupied territories'. He was asked why the P.L.O. did not try to find out if Israel was serious about a Palestinian solution by becoming involved in the negotiating process. Fahd replied: 'The P.L.O. makes its own decisions, but it seems to us that there is still an opportunity to reinforce the link [in the draft peace treaty] to a just solution for the Palestinian people.' Then he suggested that 'Islam could, indeed, provide a bridge to the solution'. But by then it was far too late for him to pursue that idea.

The peace treaty was signed on the White House lawn the day the Crown Prince's interview was published, 26 March. Egypt was to recover by 1982 all of the territory lost nearly eleven years before. Sadat had finally given ground on the question of linking the bilateral pact closely to the plan for Palestinian self-rule by accepting a vaguely worded formula contained in an exchange of letters attached to the treaty. They merely said that negotiations should start – with or without the participation of Jordan and 'Palestinians from the West Bank and Gaza, or other Palestinians as mutually agreed' – within a month and proceed in the 'hope' of completion within one year of commencement. Egypt, now a significant oil producer, also undertook to supply Israel's requirements. The question of prior treaty obligations was semantically fudged. Effectively, Sadat had entered into a deal which in no way bound Israel to the emergence of a Palestinian homeland. Egypt, as well as Israel, received the promise of massive American aid that would lessen Sadat's financial dependence on the oil producers of the Gulf.

Saud al Feisal represented the Kingdom at the Arab summit conference which convened in Baghdad on 27 March to consider sanctions against Egypt. It was from Khaled, Sultan, and Abdullah, rather than Fahd, that he took his instructions. The Crown Prince had departed for Marbella on 22 March to undergo a form of treatment for reducing weight whereby the patient is injected with cells scraped from a lamb's foetus. Prior to the conference Hussein had obtained the agreement of the Saudi King to the suspension of economic assistance to Egypt. Sudan and Oman, which backed Sadat, were the only absentees from the conference, apart from Egypt. Arafat took the initiative in calling for an economic embargo against the U.S., as well as Egypt. The proposal had little or no support on the grounds that it would only damage the interest of the Arab states themselves. One sanction agreed upon which had not been included in the original secret package was that Egypt should be suspended from membership not only of the Arab League but of the various economic and other organizations sponsored by it as well.

The critical point for Saudi Arabia came on the second day when the complete severance of diplomatic relations with Egypt was discussed. Saud al Feisal expressed the Kingdom's opposition to a total rupture in line with the Kingdom's traditional policy of keeping all options – and lines of communication – open. At this point the conference became heated as Arafat accused him of being 'a disgrace to his father'. The Chairman of the P.L.O. was also reported to have said: 'Do not force us to become a band of assassins.'[38] The meeting was adjourned after Arafat stormed out, followed by the Libyan and Syrian delegations. The deadlock continued when the meeting was resumed on the third day, and a twenty-four-hour adjournment was agreed. Saud al Feisal left Baghdad with his colleagues from Jordan, Kuwait, Bahrain, Qatar, and the United Arab Emirates for consultations amongst themselves and, it was assumed, with their Governments. The compromise reached the following day, for which Kuwait was largely responsible, was a victory for the radicals. The summit agreed on a recommendation that participating states should break off relations with Egypt within one month. The period of grace was a cosmetic concession to Saudi Arabia, but in practice the course of action was reckoned to be obligatory. The economic sanctions to be imposed were far reaching, though it was agreed that the Suez Canal would not be boycotted and Egyptian workers in other Arab countries would continue to be allowed to send their earnings home, both of which were vital sources of foreign exchange for Cairo.

The House of Saud had been placed in a cruel dilemma by the extremist pressures for the total ostracization of Egypt. It was clear that there had been an agonized debate within the ruling hierarchy. The consensus, strongly influenced by the King, was that the interests and security of the Kingdom would be best served by identifying with the Arab mainstream, which it could still be in a position to modify. The rupture of relations with Cairo was painful, however, and a step that Riyadh was extremely reluctant to take. In the event, it applied the letter of the Baghdad resolutions, even cancelling payment for the F-5Es that the U.S. was supplying Egypt. Its adoption of economic sanctions in the face of U.S. displeasure seemed like an abrupt change of policy, but was, in reality, the culmination of a steady shift. Fahd, the pro-Western 'dove', had lost out and appeared to be licking his wounds in Spain. He re-emerged invigorated on 12 May, converted to taking a hard line against the peace treaty, and resumed his role as the strong man of the regime.

Despite some signs of irritation, the U.S. Administration's main concern was that there should not be a backlash in Congress of the kind threatened by Church's statement in February. Yet American press reports concerning divisions and rifts in the family soon became the focus of Saudi anger. Khaled denied categorically on 1 April a report attributed to U.S. intelligence sources that he was planning to abdicate in six months. Great annoyance was caused by an article in the *Washington Post* on 15 April headed 'U.S. Said to Fear an Apparent Shift in Saudi Power'. It speculated about the Crown Prince's absence at such a crucial time and suggested that 'the Fahd problem may be part of a potential crisis in Saudi leadership that could shake some of the basic assumptions of U.S. foreign and energy policies'. It cited U.S. intelligence reports as being partly responsible for a 'spreading impression in official Washington of a new fragility and uneasiness within the 2,000 member royal family that rules the world's largest petroleum reserves'. The Saudi Government reacted angrily. It was assumed that the informant had been George Cave, the C.I.A.'s head of station in Jeddah, who had – it seems – been somewhat indiscreet in his inquiries about the views of prominent princes. Turki al Feisal summoned the hapless agent for an interview and confronted him with unaccustomed fury. Cave was given twenty-four hours to leave the country, but the Saudis politely did not declare him *persona non grata*. The actual source of the *Washington Post* article was James Schlesinger, according to reliable sources.

Early in 1979 Yamani had warned that the Iranian crisis had seriously diminished Saudi Arabia's ability to moderate the price of oil. His cautionary words were justified. Iranian exports had slumped from an average of over 5 million barrels a day during September 1978 to about 1.5 million in December. Then the flow stopped completely. Stocks and oil in transit were sufficient to satisfy consumption in the non-communist world for a couple of months. But panic about the availability of supplies set in. Rates paid on the 'spot market' for marginal volumes of crude soared to $23 per barrel. Exploiting the psychosis, all members of O.P.E.C., with the exception of Saudi Arabia, slapped on what were euphemistically called surcharges and premiums but were, in effect, permanent increases. The schedule for graduated, quarterly increments agreed at the O.P.E.C. meeting in Abu Dhabi in December 1978 had completely fallen apart by the beginning of March when Iran resumed shipments. Prices being charged by members already exceeded the levels that had been set for the last quarter of 1979. The 'leap-frogging' by members continued, with militant, revolutionary Iran setting the pace.

An extraordinary conference of O.P.E.C. was called to review the situation. The outcome of the meeting held on 26–27 March in Geneva, which fatefully coincided with the signing of the Egyptian-Israeli peace treaty and the beginning of the Baghdad summit, was an untidy arrangement, another 'agreement to disagree', a multi-tier system hardly amounting to a compromise. Saudi Arabia set a base price of $14.54 per barrel. Other members were left free to obtain what they could on the market, the North African producers charging over $18 per barrel for their premium crudes. Saudi Arabia's will, as well as its ability, to resist had been weakened by its dissatisfaction with the Egyptian-Israeli peace treaty and its disillusion with the U.S. Despite the turmoil in the market and continued concern about supplies from Iran, the Kingdom decided to reimpose the ceiling of 8.5 million barrels a day on the output from Aramco's production for the second quarter of the year. The escalation in prices went on. When O.P.E.C. met again at the end of June in Geneva, Saudi Arabia raised its price to $18 per barrel and other members agreed to set a ceiling of $23.50, with a maximum surcharge of $2. Early in July the Kingdom lifted the maximum limit on its production back to 9.5 million barrels a day. By the autumn the Geneva agreement had disintegrated. Saudi Arabia lost control of prices completely in 1979.

The Kingdom was still trying to exercise some restraint, however. Its own price levels consistently lagged behind those of other O.P.E.C. members. Meanwhile, the extra revenue was not unwelcome. There had been a temporary, but still surprising, turnabout in Saudi Arabia's financial fortunes as a result of the erosion in the value of oil revenues in the 1975–8 period and a fall in petroleum exports in the last of those years. In the financial year 1978–9, the Government's expenditure actually exceeded the budgeted 141,000 million riyals (nearly $42,000 million) by some 14,600 million riyals ($4,345 million). Defence and security allocations had amounted to no less than 33,300 million riyals (nearly $10,000 million), only marginally less than what the Shah had planned to spend in the equivalent Iranian financial year. Saudi Arabia now ranked seventh in the world in terms of military expenditure after the Soviet Union, the U.S., China, West Germany, France, and Britain.

Having grown accustomed to large surpluses, the Government reacted almost with alarm at its financial deficit, small though it was in relation to the country's accumulated reserves. In the autumn of 1978 a blanket order was issued to the effect that no department could spend more than 70 per cent of allocations without referring back to the Ministry of Finance. It was interpreted in varying ways by different ministries and state agencies with resulting confusion. Those who suffered most were contractors, some of whose payments were months in arrears in the spring of 1979. Another Royal Decree said that all contracts valued at more than 100 million riyals should be approved by the Ministry of Finance. No sooner had it been issued than Sultan and Abdullah placed orders for the Ministry of Defence and the National Guard far in excess of the limit without authority.

Saudi Arabia was a fraught place in 1979. The upheaval in Iran raised the question whether or not the traditional Saudi system could withstand the pressures of modernization and the foreign influences associated with it. One symptom of the prevailing unease was the growing obsession about the size of the expatriate community in the country. A clampdown on illegal immigrants was imposed in the summer of 1978. The authorities ruthlessly searched them out following the *Hajj*, when 800,000 Muslims from abroad arrived to perform their religious duty. Early in 1979 fifty-two African Muslims, who had stayed on after the *Hajj* in search of employment, were reported to have been shot dead in a clash with the police.[39] Saudi Arabia seemed to be becoming xenophobic as a large part of the blame for anything that went wrong, from the carnage on the streets as a result of traffic accidents to the high cost of projects, was placed on foreigners.

It seemed symptomatic of the irritation and resentment against expatriates when Hyundai, the South Korean construction company which had grossed more in contracts than any other, fell foul of the Government as a result of an attempt at bribery of a kind not uncommon in the Kingdom. To further a bid for an important sub-contract, an executive of the company deposited $2.5 million in cash with a senior official at the Military Works Directorate of the Ministry of Defence and Aviation. It appeared that the sum was not enough. He returned accompanied by six employees carrying another $5.5 million in $50 and $100 bills. The military police burst in and in the ensuing fracas one of the Koreans was shot in the leg. The executive was imprisoned and Hyundai was banned from bidding for any more contracts for a while.

The scale of corruption in the Kingdom was as great as ever. It even reached into a sphere that had previously been untouched by it – the sale of crude oil. In November it was revealed in the Italian parliament that an agreement whereby Petromin undertook to sell 90 million barrels of oil to the Italian oil company A.G.I.P. involved a premium over and above the official price of $1.27 for each barrel, or $120 million in total, to be paid into a Panamanian bank account. The volume of direct oil sales by Petromin had risen in the course of 1979 to 1.5 million barrels a day. It emerged that 'disguised commissions' of up to $4–5 per barrel were being paid on some of the contracts signed by it. In one case investigated by the *Financial Times*,[40] an influential prince of the second generation, who was not named, was to receive $2.40 out of the total commission of $4.17 per barrel on a total of 100,000 barrels a day for a period of a year. The balance was shared by his Saudi business associates and a European intermediary. In the international oil industry it was accepted that members of the royal family were the chief beneficiaries of these Petromin deals.

As far as corruption was concerned, the House of Saud seemed to have drawn few conclusions from the fate of the Shah's regime. Yet the taunts of Ayatollah Khomeini about the deviation of the Kingdom from the path of true religion, as well as the apprehensions created by Western influences within the Kingdom, led to increasing religiosity on the part of the authorities in the course of 1979. It took petty forms, such as the intensification of a campaign, which dated back to 1977, to prevent badly needed female secretaries from working in the same offices as males. Naif and Ahmed, puritans themselves, responded readily to the promptings of the *ulema* in enforcing periodic purges, including the ban on mixed bathing in hotel swimming pools. The learned men of religion even went so far as to deliver a *fatwah* forbidding the game of table football on the grounds that it was not conducive to morality.

25 The Return of the Ikhwan

1979

Your old men will dream dreams, your young men will see visions.
JOEL II 28

EVEN in a country not known for celebration, the morning of 20 November 1979 was something of an occasion. For the world's Muslims, it was New Year's Day; more than that, it was the first day of the year 1400 and, by popular reckoning, the commencement of the fifteenth century since the Prophet quit the Godless of Mecca for the more hospitable ground of Medina. This was a day to pray at the great shrine in Mecca, the holiest place on God's earth.

The nearest non-Muslims may expect to come to this focus of 600 million souls is a tiny roadblock on the Mecca Road, shaded by dusty tamarisks and enlivened by billboards announcing in seven languages (including Korean) that they may not proceed. A policeman will bellow and unbutton his (empty) holster if they dally a second before taking the long by-pass to Taif.

All was not completely well, however. The North Yemenis were once again turning to Moscow for arms. The news from Afghanistan was bad; *ulema* were being persecuted and the Americans seemed unconcerned about the scale of the Soviet commitment to the Kabul regime. All summer, too, the noises from Iran had been threatening. If the shakiness of the country's institutions was not frightening enough, the Iranian *ulema* were now talking about 'exporting Islamic revolution'. They had already singled out Bahrain as a place where Shia brethren were suffering oppression. Hasa, too, had a Shia minority which the Al Saud had still not quite learned to trust or like.

The King acted in characteristic fashion. There must be consultation. That June he had called a conference of traditional Gulf rulers, not just those families like the Al Sabah of Kuwait or the Al Khalifa of Bahrain that the House of Saud respected, but even the Sheikhlings of the lower Gulf that they did not. They gathered among the junipers of Khamis Mushayt and watched a fly-past by Royal Saudi Air Force F-5Es and a march-past by 40,000 Asir and Qahtan tribesmen. In terms of reminding one another what they had in common, the durbar was a resounding success. In September, the foreign ministers met in Taif to discuss what they could actually do about protecting themselves and the single, vulnerable outlet for their oil in the Strait of Hormuz. Oman received the least perceptible of nods to its plan to allow American military use of its bases and the old project for a new pipeline south of the Strait of Hormuz was aired again.

The *Hajj* was now upon the country and with it the largest delegation of pilgrims that Iran had ever sent, arrogant with revolution. Reports reached Turki al Feisal that there were plans to agitate in the Shia villages of Qatif, Safwa, and Seihat. There would surely be demonstrations at the pilgrim encampment of Mina, outside the town, and all the misguided, offensive and noisy paraphernalia of their revolution. With nearly 2 million men, women and children expected for the rites, Turki could well feel anxious.

As it turned out, the demonstration, marches and handing out of Khomeini's sermons passed without serious incident. Another *Hajj*, free of any ailment worse than flu, was over. It was therefore with some relief, perhaps, that at 5.20 a.m. on 20 November, the Imam of the Grand Mosque, Sheikh Muhammad bin Subayyal, approached the microphone to lead off this most special of dawn prayers. But no sooner had the pious and prudent Imam completed the rite than a rough hand pushed him out of the way, a shot was fired, then two more, and an acolyte fell dead. It was Juhaiman bin Muhammad Utaibi who had fired the shots and Muhammad bin Abdullah al Qahtani now shouting into the microphone that he was the Expected Mahdi as was written: 'The Madhi and his men will seek shelter and protection in the Holy Mosque because they are persecuted everywhere until they have no recourse but the Holy Mosque.'

But pandemonium had broken out. A few of the worshippers stood transfixed by these wild-eyed baptists; most of them ran for the gates in terror at the shooting and the desecration of the Mosque. Eye-witnesses were later to tell of armed men moving to bar the gates and dispatching the handful of security men. In the chaos, with shots rattling off the marble pavement, many of the pilgrims were hit. .Sheikh Muhammad, according to his own account,[1] threw off the robe by which he was known and slipped away to call for help from a public telephone.

There began for the House of Saud the two most frightening weeks since the revolt of Feisal al Duwish and the other rebellious Ikhwan in 1928/9. In a typical response, Naif ordered the Canadian contractors at the telephone company to cut all links with the outside world; even so, vague and confused reports soon surfaced in Tunis, where Fahd and Turki al Feisal were attending an Arab summit meeting, and in Washington, which enjoyed its own direct line to Jeddah.

The Muslim world was shocked. 'No, not in Islam' was the banner in the Dammam newspaper *Al Yom* when it was finally permitted to publish what little it knew. Yet it was not the fictive peace of Islam but the prestige of the House of Saud that had been shattered. As the days passed, attack after attack was repulsed, while far away in Hasa, the Shia of Safwa and Qatif rioted when nervous security officers intervened in the ceremonies traditionally held on the Ashura, the tenth day of the new month, Muharram, and the climax of the sect's period of mourning for Hussein, Grandson of the Prophet, who was slain by Sunnis. Government announcements were sparing, contradictory, and occasionally hysterical. In the cities, the educated chafed at their exclusion from any role in the crisis.

*   *   *

Even the day he died, Juhaiman bin Muhammad Utaibi never seemed entirely human. The word on the lips of every Saudi was *wahash,* like a wild beast. The bedouin of his great tribe − Doughty called it a nation − still speak of him with awe and will tell you he spoke not a word under interrogation.

Myths are made quickly in Islam and it was but a matter of days before Juhaiman had been cast into the lowest circles of the Muslim hell. Every day there were new stories of the abominable acts of his Ikhwan: that harried by security forces into the warren of storehouses and retreats beneath the Mosque, they had burned pages of the Koran to disfigure their dead comrades' faces and that, in the terror of that week-long night, they had solaced themselves with sex. These were defilements of God's house as horrid as the shedding of blood.

The spokesmen of the House of Saud were later to claim that Juhaiman's revolt was a bizarre exercise in radical theology and of no political or social concern. Many observers of the scene, including Saudis, thought the revolt as ominous for the Al Saud as the great demonstrations in Tehran had been for the Pahlavi family. It was never that. Only a handful of men outside the Grand Mosque sympathized with Juhaiman's violence or his strange and exact programme for the redemption of the earth by his Expected Mahdi, Muhammad bin Abdullah al Qahtani. These were anyway scattered about the country in the tiny cells Juhaiman had formed in the towns and bedouin settlements.

A better assessment was that of the Lebanese journalist, Salem Lozi (who was murdered in Lebanon by Syrian agents soon after the siege, but for other reasons). He wondered if the Ikhwan were not the poison that must be secreted in a body so racked as modern Saudi Arabia. Naif approached the heart of the matter. At an interminable press conference 'to explain the events of Mecca to the world', the Prince was asked, not without seriousness, whether he would crack down on men with long beards. These had been a mark not only of Juhaiman's band and the Ikhwan of Ibn Saud, but also of the state *ulema* and their 'lay' counterparts, the *mutawain* of the Public Morality Committees. 'Certainly not,' replied Naif. 'Half the Kingdom's population is bearded. We will ever respect the marks of piety on a man's face.'

Juhaiman's band drew its strength from the same source as the Saudi state, from a militant piety almost as old as Islam itself. Here small communities react to the corruption of the present by puritanical behaviour and violent disdain. This was the pattern of life in the Nejd of Abdul Wahhab and among the Ikhwan settlements of the 1920s. The battle of 1929 – 30 showed just how incompatible was this violent zeal with even Ibn Saud's vision of a modern state.

When Juhaiman was born, around the year 1940, in the small Utaibah Ikhwan settlement of Sagir (or Sajir) in Qasim, the memory of Ibn Humaid's revolt and defeat was still green. As a young man, Juhaiman needed only to travel a few score of miles to Ghatghat to see the shattered mud ruins of Ibn Humaid's base. Even thirty years later, the handful of old warriors still living around the settlement would draw out a bullet-ridden Ikhwan banner for the edification of visitors.

It is not known if the boy named by his father the 'little scowler' was linked by blood to Sultan bin Bijad bin Humaid's rebellious

Barka Utaibah. Certainly, Juhaiman had strong sentimental links with the early Ikhwan, rebellious or loyal. He borrowed their name and imitated their beards and the cut-off, shin-length robe which they affected as a sign of their simplicity. Like the Ikhwan leaders, Juhaiman was to reject the innovations the House of Saud had sponsored. His castigation of television and photographs evoked their fierce opposition to wireless telegraphy. He attacked relations with infidel powers and the presence of non-Muslims in the Kingdom no less bitterly than they. The Kingdom was in disorder. But this had not begun with the Second Five-Year Plan but 'when the people accepted Abdul Aziz bin Saud'[2] at the oath of allegiance in 1932.

At the congresses called throughout the late 1920s to reconcile Ibn Saud and the rebellious Ikhwan, the *ulema* sided with their Imam. On the crucial issue of *jihad*, so heady with Paradise and plunder, the *ulema* had ruled in 1927 that the question was to be left to the Imam. Feisal al Duwish and Sultan bin Bijad, who never even bothered to attend the congresses, thought otherwise. Hafiz Wahba wrote that they 'proclaimed throughout the *hijrahs* that they, and they alone, were the defenders of the True Faith and the supporters of the Law which Abdul Aziz was attempting to destroy'.[3] For Juhaiman too, *ulema* and state had combined in a truly unholy alliance. The Nejd *ulema* had been bought, he wrote in a pamphlet that appeared in 1978.[4] Where is it that the *ulema* and sheikhs find their money, except through corruption?

His scorn falls most heavily on Sheikh Abdul Aziz bin Baz, the blind theologian who argued for a geocentric universe. Juhaiman writes that Baz had at first encouraged the group but was now in the pay of the Al Saud, little better than a tool for the family's manipulation of the people. Other *ulema* had warned of the corruption of the royal family, but Baz had silenced them. 'Ibn Baz may know his Sunna well enough, but he uses it to bolster corrupt rulers,'[5] Juhaiman wrote. The Sunna is the body of Traditions relating to the Prophet's conduct and ranks in importance second only to the Koran.

The Nejd *ulema* were not the only religious force in Saudi Arabia nor the only teachers to feel Juhaiman's scorn. For a century, the Muslim world had been in the grip of a debate about religious fundamentals. Extreme ideas that had sprouted in the more fertile ground of Egypt, Sudan and the sub-continent had been wafted by the winds of trade, exile and pilgrimage to seed themselves in the Holy

Land. Juhaiman mentions the Muslim Brotherhood, founded in Egypt, the Jamiat al Tabligh, an anti-imperial movement from India, the Ansar al Sunna of Sudan, and the Jamiat al Islah of Kuwait. He remarks that he corrected their errors and recruited their adherents – almost all of them non-Saudi – to swell his band. Other sects which he does not mention but clearly provided willing listeners were the Black Muslims of the United States and the Takfir Wa Hijra group from Egypt, which had been responsible for the death of one of President Sadat's ministers in 1978. Men of these militant, and often violent, persuasions were to be found in almost every large town in the Kingdom; after all, by 1978 there were some quarter-million Egyptians and as many Pakistanis in the country. But their focus was the Haramein – the shrines of the *kaaba* in Mecca, and the Prophet's Mosque and Tomb in Medina.

In earlier days, the shrine at Mecca had been known as 'the university of all Islam' from the vast numbers of believers from all parts of the world who came to study and contemplate in the shadow of the sacred meteorite. Thanks to the munificence of King Saud – who spent over a billion riyals – and of his successors, the Grand Mosque could accommodate over a quarter-million worshippers. Beneath the Mosque courtyard, in the warren of nearly 300 *khalawi*, or hermitages, burrowed out of the grey lava when the shrine was the home of Mecca's rain and fertility deities, the devout could retreat to pray, living like the Prophet on milk and a handful of dates.

Of the two shrines, Medina was the more radical. The Islamic University there had been founded in the early 1960s by Muslim Brothers, who had been driven into exile by Nasser's secular policies. They had persuaded the Al al Sheikh and the King, who needed little convincing, that Nasser was tampering with the sacred curriculum of Cairo's Azhar University, the fount of Muslim learning since the tenth century.

The Muslim Brothers had first become active in the 1920s. At the heart of their message, refined over the years, was a return to the faith of the 'Salaf', the early Muslims. They balanced a deep distrust of the science of theology *(kalam)*, which had grown up under the dialectical influence of the Greek philosophers, with a devotion to the study of *hadith*. These are 'Traditions' of the Prophet's conduct and sayings, usually in the form of: 'A heard from B, who heard from C . . . who heard from the Prophet's daughter that . . .' Such is the nature of Islam that these *hadith* were used to legitimize political movements and ideals right through the factional squabbles of the Middle Ages. A vast number are palpable forgeries. Most Sunni Muslims, however, accept a corpus of 'Sound Traditions', notably those of the ninth-century scholars Muslim and Bukhari.

The Muslim Brothers were also known for the violent defence of their beliefs, not only under British rule but also in the Nasser era. They were not alone at Medina University; Feisal's strong missionary activity throughout the 1960s naturally attracted large numbers of foreign students, many of them, for obvious reasons, opposed to the regimes in power in their own countries. In 1969, Baz became rector of the university and further impressed his primitive vision of Islam, his meticulous emphasis on the letter of the Koran and Sunna, and his rejection of innovation, onto the teaching. A royal adviser was later to remark sadly that if you teach a 1,000-year-old curriculum, you must expect the behaviour of 1,000 years ago.

It is hard to give a coherent account of Juhaiman's life, because the evidence is so conflicting. Little is known of his career except that he had joined the National Guard in his late teens and risen to the rank of corporal in a platoon of Utaibah *mujahiddin,* or levies. It would be foolish to speculate that like the Prophet or Abdul Wahhab he had passed through an intense mental crisis. Perhaps it was simply that his fierce piety sat ill with the military demands, however light, of the National Guard; he was later to reject any form of state service as innovation. Whatever occurred, he was released from service at the National Guard base in Qasim. It is said he attended the Faculty of Law at Medina but had been Baz's pupil for no more than two years when he broke with the venerable divine, in 1974, and set off with ten students back to Qasim. Juhaiman was probably illiterate, as the Saudi authorities insist, but he could still have attended Baz's lectures because they were held in the mosque to which all Muslims had access. It is hard to know what happened in such a strange and distant intellectual world. Sheikh Hamoud bin Saleh al Uqail, the prayer leader of a Riyadh mosque, gave a rather defensive account: 'An atmosphere favourable to Islamic heresy sprang up at Medina because of the presence of large numbers of foreign students.'[6] He does not mention Baz by name, but speaks of a group which called for the rejection of 'interpreted' doctrines in favour of a return to the basic Koran and Sunna. 'The authorities expelled the foreigners and rehabilitated the Saudis, but after the death of King Feisal the heresies started up again.'

It is probable that Juhaiman absorbed Baz's fierce advocacy of pure Islam and extended the arguments to embrace the Al Saud; he was later to claim that Baz never objected to the Ikhwan's doctrine but only to the fact that they singled out Saudi Arabia as a corrupt state. It is easy to imagine Baz, loyal to the Wahhabite alliance and fearful for his position, ejecting the students. It was a crucial step. Juhaiman wrote later that 'the Ikhwan could not find religious education in the schools and thus had to educate themselves'.[7] This self-education was to be in the dangerous chemistry of the Traditions.

Among the palm plantations and zealots of Qasi, Juhaiman again passes out of view. We can be sure that he was preaching among his own kind, for in 1976 his band, considerably swelled in numbers, moved to Riyadh, a town being transformed more rapidly than any in history, a modern Klondike. Like every other property speculator in the town, Juhaiman bought a house with money provided by a rich member of the group, one Yusuf Bajunaid, a pious young man from a Jeddah family of wholesale merchants. Men were later to suggest that it was Yusuf who actually wrote the pamphlets signed by Juhaiman. It is possible. He certainly financed the printing, at the Talia Press in Kuwait, of the group's first major essay. It was called 'Rules of Allegiance and Obedience: The Misconduct of Rulers', and is a naked attack on the *ulema* and on the legitimacy of the Al Saud. The argument, put simply, is that Muslim rulers who do not follow the Koran and the Sunna deserve no obedience but must be opposed. The Saudi royal family are rulers of this class; they are corrupt and avaricious, turn the law to support their own ends and consort with unbelievers and atheists.

The pamphlet carried no date, but it was surely these treasonable sentiments that forced Naif's intelligence service, the *mubahath,* to act in the summer of 1978. Juhaiman and ninety-eight other persons were arrested in Riyadh. They were never made subject to the political embarrassment of a trial. Instead, Baz was summoned from Medina and questioned them but he could not, or would not, find treason in their preaching. After six weeks in a sweltering Riyadh gaol, the band was released against a promise of no further subversive preaching or public activity.

In the agony of self-questioning after the attack at Mecca, many Saudis were to question Baz's motives. He was not alone in his interest in the band. One of the most forceful and ambitious of the *ulema,* Sheikh Saleh bin Lehedan, also encouraged the Ikhwan.

Juhaiman characteristically writes him off as a spy. It is as well, perhaps, to remember that this penultimate phase of Juhaiman's career coincided with a flood of legislation from Naif's office. Women were banned from travelling alone and from working among men; the Minister of the Interior promised a dress code would be provided. An increasing anti-foreign feeling culminated in Ramadan of 1979, when notices were plastered over the *souk* warning foreigners of the consequences of immodest dress, the wearing of crosses, or couples holding hands in public. It is not fanciful to suppose that many of the *ulema* and grass-root zealots were content, to say the least, to use Juhaiman as a stick with which to beat the Al Saud and that Naif himself was disturbed by the groundswell of opinion.

It is in this first pamphlet, too, that Juhaiman alludes tentatively to the Mahdi: 'Even a false Mahdi (or Anti-Mahdi) is preferable to a false Imam.' He developed his ideas more fully in the 'Call of the Ikhwan', the pamphlet numbered three that appeared in Ramadan of the next year.

The doctrine of the Expected Mahdi – the one who is Rightly Guided – is essentially a borrowing from Jewish and Christian concepts of the Messiah and is somewhat alien to strict Sunni Islam. Most non-Muslims know little more of the doctrine than the tale of Muhammad Ahmed bin Abdullah of Sudan, the siege of Khartoum, and the death of General Gordon in 1881. The Sudanese movement owed something to the Wahhabites – the Mahdi banned smoking and the worship of saints – but drew its inspiration primarily from Sufi mysticism, itself much influenced by Christianity, and from the Shia. In Shia Islam, the Mahdi is the Twelfth Imam or direct descendant of the Prophet's son-in-law, Ali bin Abi Taleb. He has gone into hiding until the time is ripe for his triumphant return. For the Sunnis, the best 'Sound Traditions' scorn the doctrine, but the popular mind was ever drawn to the hope it extended in troubled times.

The great Dutch scholar, C. Snouck Hurgronje, was living at Mecca during the Sudanese events and watched with disdain how 'the greatest lights of science' in the town discussed the 'rule of the Mahdi, its overthrow by the Antichrist, the second coming of Christ, the accumulating abnormal phenomena in nature and human society which will announce the approach of the Resurrection . . .' For the fastidious Snouck, this was merely 'strange stuff that the Moslim fancy has piled up out of Christian, Jewish and Persian traditions'.[8]

Nevertheless, this strange stuff had been enshrined in Traditions of more or less dubious authenticity and the Ikhwan had studied them assiduously. A good example, and an early one, occurs in Ibn Khaldun's great social history, the *Muqaddimmah:* 'There will arise a difference at the death of a Khalifa and a man of the people of al-Madina will go forth, fleeing to Mecca. Then some of the people of Mecca will come to him and make him go out (apparently rise in insurrection) against his will and they will swear allegiance to him between the *Rukn* (a courtyard in the Grand Mosque) and the *Makam* (a small building beside the *kaaba* holding a stone with the purported imprint of Abraham's foot). And an army will be sent against (or "to", *ila*) him from Syria but will be swallowed up in the earth in the desert (al-baida) between Mecca and al-Madina'.[9]

The *fitna,* the topography of the Mosque, and the divine destruction of the army from Syria, or in other accounts from the north, are features of almost all the Sunni Mahdist Traditions. Others record that the Mahdi will bear the name and patronymic of the Prophet (Muhammad bin Abdullah), will be of his tribe (the Quraish) and will appear some six to ten years before the Anti-Mahdi or Antichrist, who will be destroyed by Jesus Christ when he descends to restore the peace of Muhammad. These Traditions occupy the bulk of the thirty-six printed pages of the 'Call of the Ikhwan'.

Among Juhaiman's band, with its assortment of Egyptians, Kuwaitis, Yemenis, Pakistanis, and Americans, there must have been men to whom this bizarre world of ancient (and partly spurious) prophecy meant little. There can be no doubt that for Juhaiman, and most of his followers, the centuries had rolled away, the Riyadh of automobiles and garish villas and office blocks crumbled, and what was written was about to come to pass.

Had the Mahdi already been revealed? Did wish precede fulfilment? We know only that one night a woman of the group, either Juhaiman's wife or sister, dreamed that the Mahdi was there among them in the form of a 27-year-old student from Riyadh's Islamic University. His name was Muhammad bin Abdullah al Qahtani. Was his name not as was written, and the Qahtan, were they not descended from the Quraish? And look around you, is this not a time of disorder? By God, it started when the people accepted Abdul Aziz bin Saud.

On his release from gaol, Juhaiman disappeared. It is said that he visited his old haunts in Qasim and Medina, consolidating his support, then set out preaching the length and breadth of the country like a modern Abdul Wahhab. The security forces lost touch. Naif was later to say that the pamphlets 'contained matter that showed the abominable clique had deviated from Islam', but the security forces had no idea they would mount an insurrection.

A careful reading of 'Call of the Ikhwan' would have shown him that the band was looking to Mecca and expected violent opposition; but Naif said it was not until the *Hajj* month, the last before the New Year, that Juhaiman had fixed the date for the Mahdi's epiphany. Interrogation of the survivors revealed that Juhaiman had given the Ikhwan just two weeks to gather their weapons and come to Mecca in the last five days of the month. Yusuf Bajunaid sold a property in Jeddah to cover the cost of the weapons.

As put on show by Naif, the armoury was extraordinary. It consisted largely of semi-automatics of East European and Soviet make, but ranged from nickel-plated Ak-47 rifles to Spanish shotguns and Yemeni cutlasses. There were even giant chains and padlocks to bar the gates. Many observers jumped to the conclusion that the weapons had been supplied from Syria or the P.D.R.Y. in a deliberate, Soviet-sponsored effort to undermine the House of Saud. This appeared to be confirmed, at least in Washington, when Bandar bin Sultan, the R.S.A.F. major who had played such an important part in the 1978 F-15 sale, told U.S. officials that the Ikhwan had learned their devastating accuracy of fire in a P.D.R.Y. training camp. Such an assertion by a sophisticated young prince indicated the desperate desire of the House of Saud to secure U.S. support. After the siege Naif denied the P.D.R.Y. connection flatly and remarked, somewhat hopelessly, that the weapons were easily to be found among the bedouin and they were good shots anyway.

There is no reason to disbelieve him. Indeed, with three-quarters of the equipment of the Yemen Arab Republic Army, and almost all that of the P.D.R.Y., supplied from the communist bloc, these rifles glutted great arms *souk* at Saada, just over the border from Asir. The bedouin of Saudi Arabia's south-west still regard the Ak-47 as a symbol of pride par excellence. With a lull, too, in the murderous Lebanese civil war, large volumes of arms and ammunition were being dumped on an avid, and increasingly liquid, bedouin market, where profits of up to 1000 per cent were to be had.

Juhaiman was accompanied at the *Hajj* by Yusuf and Muhammad bin Abdullah, who had quit his family home in Riyadh, taking his sister and mother with him. Messengers slipped out to Qasim, to the north, and to the south-west. The security forces, obsessed with the noisy Iranians and the Shia villages, had no eye for them. As the month, and the century, drew to a close, Ikhwan trickled into Mecca. Some were arrested in routine road checks in the south, on the Medina Road, and at the entry to the Grand Mosque itself; but still the authorities did not act.

A Toyota station-wagon laden with arms and ammunition was brought in and secreted in the cellars; with it, a station-wagon full of rice and dates. Coffins were made to conceal not bodies for the ritual washing in this sacred place but automatic rifles for the first assault. On the last night of the century, the band, now numbering over 200, settled down in the dark and crowded Mosque. Perhaps, like Ibn Saud outside the Musmak, they spent it in quiet prayer.

*   *   *

After the first gunfire there was silence awhile. But by the noon prayer, when Juhaiman again came forward to introduce the Mahdi to the bewildered pilgrims, the *souks* of Jeddah and Riyadh were alive with rumours. The garbled accounts of the pilgrims who had escaped spread and multiplied: the whole of Mecca was in the hands of assailants; Medina, too, had fallen; a bomb had exploded at the King's palace in Riyadh.

As better intelligence filtered in and it became clear that the attackers, whoever they were, were confined to the Mosque, the authorities may have regretted their hasty action with the telephone and telex lines. While the King fretted, Sultan took charge and resolved that the safest course was massive action. Units of troops, police and National Guard were despatched from Riyadh. The skies were loud with the lumbering of C-130 Hercules transports, despatched to Tabuk and Khamis Mushayt to ferry men. The 600 men of the élite Special Security Forces were in Mecca by mid-afternoon.

The House of Saud must carry the *ulema* with them. That same afternoon the King called a meeting of the senior divines, including Baz, and asked for a ruling on the legality of clearing the Mosque by force. Although the *fetwah* was not to be published for several days, the religious men assented immediately, finding their authority in the verse: 'Do not fight them near the Holy Mosque until they fight you inside it, and if they fight you, you must kill them for that is the punishment of the unbelievers.'

Muhammad Abdo Yamani, the Minister of Information, did not know whether he should tell the world or wait and could find nobody to listen to him. Naif and Sultan had already left for Mecca, where they were joined by the head of external intelligence, Turki al Feisal; even at this stage, many found it hard to believe that there was not some foreign hand, Zionist or communist, behind the attack. As Director of the Foreign Liaison Bureau, Turki was also expected to work with the various friendly intelligence services. Fahd, however, elected not to allow his fellow Arab leaders opportunity to crow and remained in Tunis. Abdullah, too, alarmed perhaps at this insurrection amongst his bedouin, felt it more prudent to continue his holiday in Morocco.

By the evening, all power and electricity had been cut off to the Mosque and a cordon tightened round it. Without its usual string of lights, the Grand Mosque looked heavy and forbidding. Juhaiman's snipers were picking off the police and soldiers milling around the main gates; Fawwaz, the crapulous Governor of Mecca and sometime 'liberal' prince, rushed in to offer assistance only to see his driver shot in the head.

The troops and *mujahiddin,* fired by the horror of the crime and avid, if not for Paradise, at least for the Government's bounty, were proving hard to restrain. Sultan decided to launch an attack on the Marwah spur and he ordered an artillery barrage round the Marwah, Salam and Osman gates. But as troops advanced across the lighted ground towards the gates, they were cut up badly by heavy fire from the upper stories and minarets; although they occupied the surrounding positions, ensuring there would be no sallies or reinforcements, they could not proceed into the Mosque proper. The Government forces had not expected such fierce resistance or, for that matter, such good shooting. A plan was needed that would keep the casualties down and also keep damage to the Mosque and 'innocent' deaths to a minimum; the King, in particular, had insisted on this.

It is a curiosity of the siege that, as the casualties mounted and attack after attack was repelled, the youngest of the princes present, Turki al Feisal, almost the youngest of the old King's sons, was increasingly deferred to by his uncles. Turki had selected the General heading the Special Forces to command the attack; as he remarked later, 'someone had to take charge'. The problem was that with so many princes, generals and services present, it was becoming impossible to work out a coherent strategy or see it followed through.

There was no questioning the enthusiasm of the troops; indeed, it was rather the opposite. But the traditional Al Saud mistrust had ensured that the services had never worked together. The National Guard, too, might they not be suspect? Abdullah was later to claim that the Utaibah and Qahtan *mujahiddin* had particularly distinguished themselves. But, in a typically suspicious move, officers were brought in from the mechanized battalions to provide a higher command and this upset the bedouin. The foreign military missions, from the Americans to the Pakistanis, were prodigal with advice, much of it conflicting.

Partly because of the difficulty of co-ordinating attacks on the gates, Sultan approved a helicopter assault into the courtyard itself. This might have made sense at night but the Saudis lacked the experience for such an operation. Troops were winched down into the courtyard in broad daylight and the losses were appalling.

Every day that passed caused the prestige of the House of Saud to suffer, as Sultan was well aware. Despite ludicrously optimistic bulletins from Naif and Muhammad Abdo Yamani, the Muslim world was becoming restive at the delay. By Friday, the tide of battle had turned. The Government forces had gained a foothold inside; they established themselves in the upper stories and dislodged the rebels from two minarets. To allow them to advance along the upper stories and across the courtyard, Turki chose to use gas provided through the good offices of the U.S. Embassy. Troops backed up by snipers raked the exposed positions of Juhaiman's men, and though the Ikhwan put up barricades of mattresses and prayer mats, they were slowly beaten back and down towards the entrance to the basement. By the Monday, the whole Mosque above ground was in Government hands.

Deep below the courtyard, the Ikhwan barricaded themselves round their supplies. Those who had surrendered above had revealed their modest numbers to Naif's interrogators; but the Government was under no illusion that they would surrender in the utter hopelessness of their position. Troops with bullet-proof vests lobbed in gas shells, but the main passages were barred against them; in the eerie, steaming darkness the rebels seemed to be everywhere and might not some innocent pilgrims be amongst them?

On the Wednesday, cleaners had managed to remove some of the blood and debris from the courtyard and a hurried prayer ceremony was shown on television. But that evening, confused reports arrived from Hasa that what had always been feared had happened: the Shiite villages were aflame.

That evening was the beginning of the Ashura, the tenth day of the month Muharram. Like the Shah, the Jiluwi had hitherto always banned the extravagant mourning ceremony where, in the past, men would whip themselves or pierce their flesh with knives in a paroxysm of sectarian grief. This year the Shia leaders, emboldened by Khomeini's revolution and unrest among the Shia in Iraq, had let it be known that marches at least would be staged. In the small village of Safwa, a crowd rioted when a policeman struck a demonstrator and, throughout the next day, a mob rampaged through Qatif. By the evening of the Thursday, cars had been burned, shops looted, a bank attacked, and seventeen people killed by security forces. It was as if the pent-up frustration of fifty years had at last found expression.

With Naif in Mecca, his deputy Ahmed was dispatched to Qatif with instructions to contain the situation at all costs. Now in his early forties, Ahmed had long been obliged to defer to his elder brothers, Fahd, Sultan, and Naif. Foreigners found him as strict as Naif in his personal life but more accessible to ideas and more articulate in expressing them. He toured the scene of the riot in Qatif and visited a school. In conversations with townsmen, he admitted that the Shia had suffered – the first member of the Al Saud ever to do so. In the past, he said, they had not prospered as much as others in Hasa; they had not done so well in state service; even their womenfolk had not, on occasion, received the respect Muslim women should expect. No doubt these schools and hospitals were inadequate. But, Ahmed continued, their prosperity was beginning, the Eastern Province was booming and Qatif too. The past was the past, and if only they would co-operate in this crisis for the Kingdom, all would be well.

His mission was a success, but the rioting had showed the commanders in Mecca that the siege must be brought to an end soon. Turki must have been near to despair. The Binladen Company, which had carried out King Saud's renovation, was called in to produce their original plans; gas, water, burning tyres had been tried in vain. In the heart of the basement, beside their stocks of ammunition and dates, Juhaiman and his followers, including women and children, remained. Perhaps they still believed that God would cause the ground to swallow up their enemies?

Was Muhammad bin Abdullah still alive? Saudi newspapers later published a picture of a young man with his jaw blown away and said that this was the Expected Mahdi, killed in the early fighting in the basement. Some writers claimed that Juhaiman had killed his friend in the depths of despair.

Perhaps the ammunition had run out. We may never know what the Ikhwan were feeling in the last terrible days, weak with exhaustion and thirst, many of them, including Juhaiman, wounded. But at 1.30 in the morning of Tuesday, 5 December, exactly two weeks after the attack had been launched, Juhaiman led out the last 170 of his band into the clean air.

It is said that as they emerged, many weeping and too tired to stand, muttering constantly, spat on and reviled, one of the band turned to a National Guardsman and asked: 'What of the army of the north?'

26 | The Ruins of Diriya

THROUGHOUT the Kingdom and Islam there was general applause when sixty-three heads rolled in the dust on 9 January 1980 in eight different towns of Saudi Arabia, the violators of the Grand Mosque having been condemned by due process of law. The final act of justice was spread around the Kingdom so that the message of retribution would be more widely and effectively spread. The outrage to the holiest shrine of Islam was abhorrent to all but a small minority of citizens and Muslims in the greater *umma*. Those who dissented from the common view kept quiet. Even so, the confidence of the custodians of Mecca and Medina was badly shaken. Juhaiman and his fellow zealots had sought to redeem not only the revered place of worship but also the world from the corruption and laxity that they associated with the House of Saud. The seizure of the holy shrine was an abrupt reminder that the legitimacy of the dynasty was based upon its guardianship of the Holy Places and upholding the true faith. The affair was hardly likely to evoke sympathy amongst the devoted for the rulers entrusted with the safe-keeping of the *kaaba*. In recent times only Ayatollah Khomeini in Shiite Iran and Colonel Muammer Qaddafi of Libya had challenged the House of Saud's moral title to be the protector of the Holy Places. Now it had been defied, violently, in a manner that was bound to disturb the assumptions of many within and without the Kingdom. There had been no open, indigenous protest motivated by religious fervour since Khaled bin Musaid and his followers had attempted to attack the new television transmitter in 1965.

Amongst those decapitated were forty-one Saudi citizens, ten Egyptians, six South Yemenis, three Kuwaitis, one North Yemeni, one Sudanese, and one Iraqi. The nationality and identity of many of the seventy-five rebels found dead in the Mosque, in addition to twenty-five pilgrims, were never established. Twenty-three women and thirteen boys were amongst those who had trooped out when the dishevelled insurgents surrendered. They had been part of the cover for the rebels and were detained for corrective education. The interrogation of the insurrectionists prior to their execution was intensive. The authorities established to their satisfaction that there had been no conspiracy masterminded externally. The whole affair would, perhaps, have been less disturbing if one had been discovered.

In November 1979 the frenzy of the Ashura had developed into an open show of defiance against the authorities by the Shiites in Qatif. Even more ominous were the less spontaneous riots there in February 1980. The despised sect, numbering about 275,000 according to reliable estimates, constitutes a substantial minority in the whole country but predominates in the vital oil-bearing region of Hasa. The spectre of disaffected Shiites throttling the lifeblood of the Kingdom by attacking vulnerable oil installations, perhaps aided and abetted by their fellow Shiites in other countries, was alarmingly conjured up by the demonstrations in Qatif.

Yet the siege of the Grand Mosque raised more fundamental questions relating to the legitimacy and credibility of the dynasty. Juhaiman and his hirsute followers may have belonged to the lunatic fringe of Islam but they were representative of a current in the country towards a reversion to fundamentalist values and against the West and its technological innovations. Moreover, their revolt, however much it may have been stimulated by the example of Khomeini, arose from the Sunni mainstream of Islam, the tendency of which is to respect the established order – until it is shown to be demonstrably unfit to govern and collapses or is overthrown.

The whole affair, inevitably, focused upon the historic but still very valid and relevant foundations of Saudi authority. The right to rule still, to a large degree, rests on traditional and tribal discipline, instinctively understood by the people, governors and governed alike. Saudi Arabian society is a loose, cellular structure with the royal family at its apex. To a surprising degree the network of human groups and relationships binding the fabric of Saudi society has remained unchanged by urbanization, development and modernization. It is a system where personal contact counts for more than bureaucratic institutions and procedures. Ease of access to the governors by the governed is an essential part of what is really a super tribal system. Hence the continuing importance of the *majlis,* whether it be the

King's or a Governor's. In such a scheme of things the concepts of *shurra* (consultation) and *ijma* (consensus) are, or should be, part of the wider governmental process, the means by which the House of Saud obtains the people's consent to its rule.

The system is buttressed by the tenets of Islam, which have the force of law and are to all intents and purposes inseparable from the business of every-day living. The effect has been to lead to an unthinking acceptance of the status quo. At the same time, the House of Saud is responsible for the preservation of the religious and social values of the country. An attack on one is, necessarily, a challenge to the other – as long as they are in tune.

The unification of the peninsula by Ibn Saud, the consolidation of his power, and the imposition of stability also confer a legitimacy on the ruling family. The Saudi Arabian mentality, with its theocratic view of government, does not allow for political theorizing, even of an idle kind, by Western-educated intellectuals. The majority of Saudis want a physically secure environment, whatever their social existence. Before Ibn Saud established order, life in the Kingdom could, indeed, be 'nasty, brutish and short'.

The seizure of the Grand Mosque and the Shiite disturbances brought these foundations of Saudi authority into question. Their immediate threat to the established order and implications for stability in the short term should not be exaggerated, but they did raise disturbing questions, which should have prompted a great deal of heart-searching in the palaces of Riyadh. Juhaiman's insurrection indicated that the upper branches of the great tree constituted by the House of Saud had lost touch with the grass roots from which it had grown. Some in the higher reaches of the foliage were evidently unaware of the strains of the traditional social fabric caused by rapid development. Indeed, the planners were only beginning to acknowledge privately the problems caused by the disintegration of the bedouin's pastoral existence, the movement of people from the land to the main cities and towns, the growing disparity between rural and urban standards of living, and the failure to distribute more evenly the state's wealth. Foreign analysts and sociologists trying to interpret the phenomenon of the deracinated Bedou Juhaiman and his followers saw these factors as an explanation of their behaviour. More clearly, their bid to restore Islam to pristine purity reflected a protest against modernization and Western influences.

Khaled, Fahd and their colleagues could not be accused of not heeding the *ulema* and ignoring religious sensitivities, however. The enforcement of orthodox practice had become marked from 1978 onwards. The segregation of the sexes had become more rigid. Mishaal bint Fahd bin Muhammad's first meeting with her lover had taken place in a boutique, according to the controversial television film *Death of a Princess*. The best part of a year before the furore over the screening of it in Britain in April 1980, Naif had ordered his plain-clothes, thobe-clad secret policemen to keep a watchful eye on dress shops to prevent such assignations taking place. Following Mishaal's execution, Saudi girls had been forbidden to study abroad.

The leadership seemed responsive enough to the moral preoccupations of the religious elders. But the *ulema*, who had become very much part of the temporal order and an arm of the state, were clearly not being as insistent in their zeal as many citizens, apart from Juhaiman and his friends, would have liked. Criticism of the personal conduct of members of the royal family and the enormous fortunes being made by them out of state contracts reached far beyond wild visionaries like Juhaiman. Its extent was indicated by a literate, well-presented document recounting in some plausible detail the lax and corrupt practices of the prominent senior and junior princes, which was posted to Fleet Street newspapers in the early summer of 1980. Production and distribution of the indictment, signed simply 'Gratefully yours, The Young Revolutionaries', was said by Western intelligence men to be the work of Saudi students in the U.S.

The most disturbing aspect of the Grand Mosque seizure was the spectacular nature of the breakdown in law and order in the most sensitive of all places in the Kingdom. The labour unrest at Dhahran in Hasa had been peacefully resolved in 1953. The oil workers' protests, in 1956, inspired by Nasser's revolutionary rhetoric, were ruthlessly suppressed by the tyrannical Saud bin Jiluwi. The spontaneous demonstrations triggered off by Nasser's 'big lie' in 1967 had posed no serious threat and been handled efficiently without bloodshed or retaliation. But the violation of the holiest shrine of Islam was a security lapse that must have caused major misgivings in the hierarchy. The regime employs a high proportion of its limited manpower in discreet, unsinister, but pervasive surveillance. Juhaiman was not, after all, unknown to the authorities. It seemed incredible that the intended putsch inspired by Allah had not come to the attention of the authorities. The damage to the credibility of the super-tribe in terms of its competence to provide stability was difficult to calculate, but it was not negligible.

Disturbing also as an open act of defiance were the riots in Qatif. They had nothing to do with labour grievances or wounds inflicted on Arab nationalist pride. Underlying those disturbances was Shiite resentment, stimulated by the emotive rhetoric of Khomeini and the Iranian revolution. Historically, the eruptions should be seen as a reaction to neglect and suppression born of Sunni prejudice existing throughout Islam but present in the Wahhabite Kingdom in particularly primeval form. The leaders of the House of Saud should have been alerted to the dangers arising from the existence of a disaffected community long before. Feisal, who in the later years of his reign never visited Hasa at all, and his successors did nothing to overcome discrimination against the Shiites and made no effort to integrate them into society. This failure is only understandable in the context of religious tunnel vision, deeply rooted in history. The blinkered outlook was even more lamentable, in terms of self-interest if nothing else, in that the Shiite community has proved relatively fertile ground for the implantation of radical ideology. It has also led to a waste of badly needed human resources. Aramco's experience has been that members of the sect show a greater willingness to undertake manual tasks, skilled or otherwise, and a stronger desire for self-improvement through education, than Sunnis of bedouin origin. But by 1981 only one Shiite had ever held ministerial rank.

The Shiite community is and will continue to be the most divisive element in the Kingdom. Otherwise, its society is more homogeneous and cohesive, ethnically and religiously, than that of any other Arab state of significance except Egypt. The pattern is by no means one of uniformity, however. Cosmopolitan Jeddah is still a refreshing contrast to Riyadh, a city overladen by a heavy pall of Wahhabite gloom, even if it is the form rather than the substance of Abdul Wahhab's puritanism that is now observed. Hejazi resentment over Nejdi domination is by no means dead. It seemed significant that the instinctive reaction of one Jeddah businessman on hearing the first confused news of the takeover of the Grand Mosque was to mutter one word – 'Taweel', the central figure in the conspiracy exposed in 1969. In reality, Hejazi separatist feeling ceased to be a threat when oil revenues gave the people of the province a vested interest in the stability of the territory carved out by Ibn Saud. The people of the Asir, cultivators rather than pastoralists, are mainstream Sunni in outlook and their adherence to Wahhabite forms is as thin as a diaphanous veil. But the former province of the Imamate of Yemen has been integrated well enough into the realm even if Riyadh was slow to improve the lot of its people.

Because the Saudi Arabian mentality is so heavily imbued with a fundamentalist Koranic view of the world, it has not been responsive to radical ideology. The appeal of Nasser to the first generation of educated, politically articulate young men derived from their identification with pan-Arab nationalism and revulsion against the excesses of Saud's incompetent regime. There was evidence that the Baathist movement had established cells in Hasa, but the wholesale round-up of suspects at the end of 1969 was an over-reaction on the part of the authorities. The main vehicle of active opposition in Hasa is the Islamic Revolutionary Organization, which claimed responsibility for the Qatif disturbances. A shadowy movement, it is almost certainly directly affiliated to Tehran. In the recent history of the Arab world the Kingdom has been remarkable for not producing any indigenous left-wing dissident movements that in themselves constituted a threat to the established order. The mysterious organizations which have attacked the regime and made claims of active subversion over the past few decades have been the creatures of hostile Arab powers and Moscow, without having any base in the Kingdom. The only Saudi dissident movement that has survived in any form has been the Arabian Peninsula People's Union, the exiled opposition group founded by Nasser Said after the labour troubles at Dhahran in 1953. Several Beirut newspapers quoted him taking some credit for the capture of the Grand Mosque, a claim that could not be taken seriously and seemed like the bleat of a forgotten opposition leader. He was reported to have been abducted on 17 December 1979, probably to Riyadh. A handful of anonymous exiles constitute an embryonic Saudi Arabian Communist Party, which apparently has no cells functioning in the Kingdom. The Vanguard of the Arabian Revolution circulates its Arabic language *Sawt al Talia* mainly to students and military officers abroad, but has also managed some covert distribution within the Kingdom. Its base is outside the country, but where exactly is apparently not known to Western Intelligence and security services. A few young men, like the unbalanced Feisal bin Musaid, may come back from seats of higher education abroad with half-baked revolutionary notions. Nearly all, however, return home and settle down (unlike their Iranian counterparts in the days of the Shah), conforming to those 'traditional values' to which an almost parrot-like lip-service is now *de rigueur*.

A certain duality is apparent from the manner in which a Saudi's mental outlook alters when he sets foot on his native soil and changes his Western suit for the thobe, gutra and ageel.

In the social setting of Saudi Arabia, the regime's only possible response to the seizure of the Grand Mosque must be some measure of reversion to original values, a return to fundamental practices, and a concession to xenophobia and hostility to innovation. Juhaiman cast a great shadow of doubt on the official Saudi shibboleth – repeated almost as an article of faith by senior princes and ministers – about the total compatibility of rapid economic development and the maintenance of traditional and religious values. But too big a retreat by the House of Saud could lead to opposition from the educated, articulate 'progressive' elements committed – with due regard to society's capacity to undergo change – to maximum possible development, economic diversification and modernization, including a measure of liberalization and reform. They would include Western-educated technocrats, some officers and N.C.O.s in the Armed Forces, merchants and businessmen, and, not the least, many of the princes – apart from those in the inner circle – with a vested interest in seeing state expenditure maximized. Together they could coalesce into a broad front of opposition.

Saudi Arabia's 35,000-strong military establishment is an unknown quantity politically. The Armed Forces seem unlikely to move against the established order unless in reaction to a pan-Arab catastrophe like the military humiliation suffered at the hands of Israel in 1967. But the House of Saud is acutely conscious of the fact that regular armed forces have been the instrument of most change and revolution in Arab countries since the end of the colonial era. The regime distrusted its own well before the discovery of the conspiracy in 1969. In the Kingdom, the military is a potential source of instability because those entering the ranks are largely detribalized. With money no object but recruitment a problem and loyalty always a matter of concern, Saudi Arabian servicemen, especially the officer corps, have been pampered and given a number of fringe benefits, such as grants of land, apart from high salaries. In the spring of 1981 pay rates were doubled, giving a lieutenant up to 8,209 riyals ($2,445) monthly, and a general up to 21,561 riyals ($6,420). The National Guard, now including highly trained units with modern weapons, is still regarded as an essential counterpoise and the defender of the regime against a coup attempt.

It would be an over-simplification to portray Saudi Arabia as having become polarized between 'reaction' and 'progress'. There is, certainly, a marked dichotomy between the wider perspective of the well-qualified, Western-educated Ph.Ds and the obscurantism of the restricted learning of Baz. But the intelligentsia had become increasingly preoccupied, and confused, over the question of how fast the pace of development should be well before the 1979 eruption. In the debate, the conservative, traditionalist side of the national character seems to be at least holding its own and probably prevailing over the liberal, progressive one, even amongst the 'Californian Mafia' of Ph.Ds holding significant offices. Ministers, senior officials, and academics favour, on balance, a slower economic growth and more moderate social change. A concomitant of such thinking is a hardening consensus amongst the technocrats that petroleum resources should be conserved and not depleted at such a fast rate.

The energy crisis caused by the Iranian revolution led to an escalation in oil prices, despite continued Saudi efforts to restrain them, bringing the official selling rate for a barrel of Arabian Light to $32. As a result, Saudi Arabia is accumulating surplus revenue at the rate of $1,000 million a week with the prospect of foreign assets mounting to $180,000 million by the end of 1981. Concern about S.A.M.A.'s massive investments becoming a hostage to fortune was heightened by the U.S. freeze of Iranian funds after its diplomats were taken hostage in Tehran in November 1979. In addition complaints can be heard about the low rate of return from them and their depreciation in real terms, by about 5 per cent a year, because of inflation. Doubts are even expressed about the wisdom of the Kingdom's heavy industrialization programme in the two relatively isolated and self-contained urban centres being developed at Jubail and Yanbo. The projects can only increase the country's dependence on foreign manpower for the indefinite future. More than ever, resentment and apprehension focuses on the huge expatriate community that by the middle of 1980 was probably in excess of 2 million rather than the official, and very precise, estimate of 1,059,800.

Preoccupation about the size of the foreign presence is one reason for the audible criticism of the scale of the Kingdom's military spending and weapons procurement programme. It is not just a matter of the expense. Reliance on expatriate, and particularly American, skills not only for maintenance but also training and the operation of sophisticated systems is as great in the sensitive field of security as elsewhere. An undisclosed but significant proportion of the 40,000 or so Americans in the country works on defence contracts.

For the most part, doubts and criticism are expressed in private. When aired in public, they do no more than touch upon the rate of development, the level of oil production, and the financial surplus. This is healthy debate, however, by the standards of most Arab countries. Discontent about the proportion of the Kingdom's wealth enjoyed by the ruling clan is easily discernible, however. It is not a matter of the stipends paid to members of the royal family. They are regarded as the legitimate entitlement of those in authority. The jealousy is over the proportion of the national income enjoyed by the clan. The issue is not 'corruption', however. The concept, as understood in the West, is hardly relevant to Saudi Arabia. Most of its inhabitants, from the humblest bedouin to the most urbane and worldly-wise merchant, consider anyone who does not profit from his status or ability to peddle influence foolish or incompetent. To an extent, the practices whereby the state's revenue is distributed to commoners, in office or in business, through channels not recognized by official budgetary procedures, strengthens the system by widening the vested interest in it. As the U.S. State Department's foremost expert on Saudi Arabia put it: 'The greatest destabilizing aspect of corruption in Saudi Arabia is probably psychological. The practices are not unequivocally condemned, indeed they comprise the national pastime. On the other hand, if the extent of royal involvement in "corrupt" practices were perceived, this could serve to undermine the reputation for strict Wahhabite integrity which has over the years been a hallmark of the regime. It would be far easier for politically disaffected Saudis to accuse a regime sullied with the reputation of corruption of being the cause of their frustration than to look to the real though complicated root causes – the impact of technology and modernization on traditional Islamic society.' The commissions on defence contracts might have been eliminated, but the 5 per cent ceiling set on agents' fees is certainly being substantially exceeded. For instance, the rate charged by one of the younger princes, known for his exponential greed, on a multi-billion dollar engineering contract awarded to an American company in 1978 was known to be 17 per cent. It is hardly reassuring, and a reflection of investment opportunities in the Kingdom, that a significant proportion of the accumulated gains of many princes and merchants is being put to profitable use abroad.

The House of Saud is not in any way – unlike the late Shah of Iran – blinded by delusions of grandeur or convinced of its own infallibility, even if it does sublimely assume its right to rule. Yet the royal leadership's haphazard and archaic ways of consulting its variegated subjects look desperately inadequate. Even able technocratic ministers and others in charge of vital sectors of the establishment cannot assume that they will have any decisive effect on the formulation of the policies they execute. In the final analysis, the most senior and influential commoner is subordinate to any princeling. Neither proven ability nor status achieved through performance outweighs royal prerogative. Confidence in the system has sunk to such an abysmally low level that many of the bright technocrats are making provision for an uncertain future in anonymous Swiss and Caribbean bank accounts. The House of Saud will have to adapt if it is to survive. The challenge facing it is to revive new methods of consultation and participation in the running of the state by the emergent educated elite, to which relatively few members of the royal family belong, while still respecting and preserving tradition. The dilemma may prove insoluble, not the least because of the decadent leadership's inability to define or confront the problem.

The regime's response to defiance from the discontented atavistic elements in society represented by Juhaiman has not been encouraging, indicating indecision and dilatoriness. In the aftermath of the occupation of the Grand Mosque, Fahd assured in January 1980 that the long-promised Consultative Council, or *Majlis al Shura*, would be established within two months in accordance with 200 basic provisions derived from the *sharia*. They had been under preparation for a long time, the Crown Prince said, in line with the regime's determination that they should be applied 'only after careful and complete study'. Two months later a committee under Naif was appointed to consider the necessary legislation for a new system of government. Fifteen months later, in mid-1981, its findings still had not been announced. The Government did, however, launch new development schemes in the Qatif area in the course of 1980 and Khaled visited the town half way through Muharram in November of that year.

The nervousness of the regime was highlighted rather than disguised by the anger with which it reacted to the showing on British commercial television of the film *Death of a Princess*. The House of Saud was deeply offended by what it saw as an intrusion into its family affairs. Yet its action in demanding the suspension of James Craig, the British Ambassador, and its implied threat to cut the U.K. off from Saudi Arabia, one of its richest export markets, was aimed largely at impressing the domestic constituency. Twisting the tail of a toothless old lion was an easy way of demonstrating the muscle with which the hierarchy could deal with impertinence, at home as well as abroad. It only succeeded in drawing more attention to the film, ensuring a boom in sales of video cassettes of it. Craig was allowed to resume his duties at the end of the summer after Lord Carrington, the Foreign Secretary, had made due obeisance. Somewhat estranged from the U.S. the Kingdom could hardly afford a serious rupture with Britain, one of the few European powers able to provide an alternative source of arms and of some importance in diplomatic efforts being made to bring about a solution to the Palestinian problem.

In 1980 Saudi Arabia felt it necessary to distance itself from Washington as much as possible in public. Following the Soviet invasion of Afghanistan, it adopted a formal position, in common with other non-aligned Islamic states, of rejecting any interference in the region by super-powers. It sought to strengthen relations with the other conservative states of the Gulf and also Iraq. At the outset of the Gulf war in September 1980 it exulted in Iraq's initial successes and offered 'safe haven' to its forces. Enthusiasm was soon tempered, however, by Ayatollah Khomeini's threat of retaliation and the possibility of war reaching the southern littoral of the Gulf.

At the end of January 1981 Saudi Arabia played host to the Third Islamic Summit Conference at Taif, the sprawling and unlovely summer capital and resort. Delegates gasped at the opulent Government guest palace and conference centre, built at frenetic speed and at the cost of 450 million riyals ($135 million) in the previous year, with most of the construction materials having been flown in from abroad. The Kingdom was almost obliged to hold the summit, having called a *jihad* against Israel the previous summer in response to a law declaring undivided Jerusalem the 'eternal' capital of the Jewish state. In the absence of Khomeini or any representative of him, the gathering was suffused with the warmth of Islamic brotherhood. It remained to be seen what would become of the summit resolutions calling for the liberation of Jerusalem, Soviet withdrawal from Afghanistan, and an end to the Iraqi-Iranian conflict. Saudi Arabia, however, emerged with its prestige enhanced. The heirs of Ibn Saud appeared to have recovered a great deal of confidence and composure.

Having witnessed the spectacle and been impressed by it, I went to Riyadh, pondering upon the future of the dynasty to which Islam, the West, and the rest of the civilized world owed so much. The maintenance of maximum unity within the all-important inner circle of the royal family was essential to its survival, I knew. The question of the succession had been raised again in February 1980 when Khaled had suffered a minor heart attack. There would be a major problem when the time came for the leadership to pass to a son of a younger generation. Apart from that, could the cumbersome, arcane decision-making process strike the necessary, finely-tuned balance between tradition and modernization? Would it be able to maintain a close friendship with the West, particularly the U.S., while fulfilling its important Arab and Islamic roles? Could a consensus with sufficient flexibility and foresight emerge to broaden the system to embrace some of those subjects of far greater ability than the billionaire princelings? I doubted the ability of the lugubrious House of Saud not to squander the political legitimacy and moral authority bequeathed by its forefathers, the wide measure of popularity that it still enjoyed, and the negative but very valuable surety deriving from the widespread feeling in the realm that any other regime would be very much worse. The dynasty, undoubtedly, had become progressively more exposed to malevolent external forces beyond the control even of its ample finances. The U.S., or more precisely the democratically elected Congress to which it is beholden, could prove the fatal undoing of the regime in its blundering policy in the region. So could the 'homicidal dwarf' Begin[1] through his confrontational strategy with the Arab power possessing the greatest potential leverage over Washington. I reckoned that within five years Saudi sovereigns could have had their last page in history.

As I surveyed Riyadh's landscape of soaring structures, glossy and elegant, from my hotel room, I was reminded of the reflections of D. van der Meulen after he had visited the remains of Diriya, ancestral home of the House of Saud, in 1952. His words rang true, to me at least, twenty-nine years later: 'Wahhabism was in ruin. The capital, bigger, wealthier and richer in palaces than any town in Central Arabia had ever been before, was witness of a ruin that was greater, immeasurably greater, than that first ruin because this time the ruin was spiritual.'[2]

But the Kingdom is still in being.

Bibliography

ALICE, Princess, *For My Grandchildren*, Evans, London 1966/World Publishing Company, Cleveland 1967.

ALMANA, Mohammed, *Arabia Unified: A Portrait of Ibn Saud*, Hutchinson, London 1980.

ANTONIUS, George, *The Arab Awakening: The story of the Arab national movement*, Hamish Hamilton, London 1938/Putnam, New York 1938.

Arab Report and Record, London.

ARMSTRONG, Harold C., *Lord of Arabia: Ibn Saud, an intimate study of a king*, Penguin, London 1938.

ARNOLD, José, *Golden Pots and Swords and Pans*, Gollancz, London 1964/Harcourt, Brace and World, New York 1963.

BADEAU, John S., *The American Approach to the Arab World*, Harper and Row, New York 1968.

BAROODY, George (ed.), *Saudi Arabia and Arabian-American Oil Company, First Memorial*, privately published by the Arabian-American Oil Company, 1958.

BELING, Willard A. (ed.), *King Faisal and the Modernization of Saudi Arabia*, Croom Helm, London 1980/Westview Press, Colorado 1979.

BELL, Lady (ed.), *Gertrude Bell, Letters*, 2 volumes, Ernest Benn, Lonon 1927/Boni and Liveright, New York 1927.

BUHEIRY, Marwan R., *U.S. Threats of Intervention Against Arab Oil: 1973-1979*, pamphlet, Institute for Palestine Studies, Beirut 1980.

BULLARD, Sir Reader, *The Camels Must Go*, Faber and Faber, London 1961.

BUSCH, B.C., *Britain and the Persian Gulf, 1894-1914*, Cambridge University Press, London 1968/ University of California, Berkeley and Los Angeles 1967.

CHURCHILL, Winston S., *The Second World War*, vol. VI *Triumph and Tragedy*, Cassell, London 1953/Houghton Mifflin, New York 1953.

COPELAND, Miles, *The Game of Nations: The Amorality of Power Politics*, Weidenfeld and Nicolson, London 1969/Simon and Schuster, New York 1970.

CRANE-EVELAND, Wilbur, *Ropes of Sand: America's Failure in the Middle East*, Norton, New York 1980.

DE GAURY, Gerald, *Faisal: King of Saudi Arabia*, Arthur Barker, London 1966/Praeger, New York 1967.

DICKSON, H.R.P., *Kuwait and Her Neighbours*, Allen and Unwin, London and Winchester, MA, 1956.

DOBSON, Christopher, and PAYNE, Ronald, *The Carlos Complex: A Study in Terror*, Coronet, London 1977/Putnam, New York 1977.

DOUGHTY, Charles M., *Travels in Arabia Deserta*, abridged by Edward Garnett, Penguin, London 1956/Doubleday, New York 1955.

EDDY, William, *F.D.R. Meets Ibn Saud*, American Friends of the Middle East, New York 1954.

Egyptian Gazette, Cairo.

EISENHOWER, Dwight D., *The White House Years*, vol. II *Waging Peace*, Heinemann, London 1965/Doubleday, New York 1965.

HABIB, John S., *Ibn Saud's Warriors of Islam: The Ikhwan of Najd and their role in the creation of the Saudi Kingdom, 1910-1930*, E. J. Brill, Leiden 1978/Humanities Press, New Jersey 1978.

HEIKAL, Muhammad, *Nasser: The Cairo Documents*, New English Library, London 1972/Doubleday, New York 1973.

———, *The Road to Ramadan*, Collins, London 1975/Quadrangle, New York 1975.

———, *Sphinx and Commissar: The Rise and Fall of Soviet Influence in the Arab World*, Collins, London 1978/Harper and Row, New York 1979.

HOLDEN, David, *Farewell to Arabia*, Faber and Faber, London 1966/Walker, New York 1966.

HOWARTH, David, *Desert King: A Life of Ibn Saud*, Collins, London 1964/McGraw-Hill, New York 1964.

INGRAMS, Harold, *The Yemen: Imams, Rulers and Revolutions*, John Murray, London 1963/Praeger, New York 1964.

International Currency Review, London.

King Feisal Speaks, Saudi Arabian Ministry of Information.

KISSINGER, Henry, *The White House Years*, Weidenfeld and Nicolson/Michael Joseph, London 1979/Little, Brown and Co., Boston 1979.

The Koran, tr. N. J. Dawood, 4th ed., Allen Lane, London 1974.

LAWRENCE, T.E., *Seven Pillars of Wisdom*, new edition, Cape, London 1973.

LILIENTHAL, Alfred M., *The Zionist Connection*, Dodd, Mead and Co., New York 1978.

MANSFIELD, Peter, *The Arabs*, revised edition, Penguin, London 1978.

The Middle East, London.

The Middle East Annual Review, Saffron Walden, Essex.

Middle East Contemporary Record, Israel.

Middle East Economic Survey, Nicosia, Cyprus.

Middle East Journal, Washington.

Middle East Intelligence Survey, Tel Aviv.

MILLER, Aaron David, *Search for Security: Saudi Arabian Oil and American Foreign Policy, 1939-1949*, University of North Carolina Press, Chapel Hill 1980.

MONROE, Elizabeth, *Philby of Arabia*, Faber and Faber, London 1973.

Multinational Corporations and U.S. Foreign Policy, published in full and in summary form by the U.S. Senate Foreign Relations Committee Subcommittee on Multinational Corporations, Washington 1976.

NEWTON, Lord (Thomas Wodehouse Legh), *Lord Lansdowne*, Macmillan, London 1929.

NIXON, Richard, *Memoirs*, Sidgwick and Jackson, London 1978/Grosset and Dunlap, New York 1978.

PHILBY, H. St John B., *Arabia*, Ernest Benn, London 1930/Charles Scribner's Sons, New York 1930.

———, *Arabian Days: An Autobiography*, Robert Hale, London 1948.

———, *Arabian Jubilee*, Robert Hale, London 1952/Day, New York 1953.

RAUNKIAER, Barclay, *Through Wahhabiland on Camelback*, Routledge and Kegan Paul, London 1969/Praeger, New York 1969.

RIHANI, Ameen, *Ibn Saud of Arabia: His People and His Land*, Constable, London 1928/Houghton Mifflin, New York 1928.

RUBIN, Barry, *The Great Powers in the Middle East, 1941-1947*, Frank Cass, London and New Jersey 1980.

SACHAR, Howard M., *Europe Leaves the Middle East*, Allen Lane, London 1974 (revised from U.S. edition, Knopf, New York 1972).

SADAT, Anwar el-, *In Search of Identity: An Autobiography*, Fontana, London 1978/Harper and Row, New York 1978.

SAMPSON, Anthony, *The Seven Sisters: The Great Oil Companies and the World They Made*, Hodder and Stoughton, London 1975/Viking Press, New York 1975.

SCHMIDT, Dana Adams, *Yemen: The Unknown War*, Bodley Head, London 1968/Holt, Rinehart and Winston, New York 1968.

SEALE, Patrick, *The Struggle for Syria: A study of post-war Arab politics, 1945-1958*, issued under the auspices of the Royal Institute of International Affairs, Oxford University Press, London and New York 1965.

SHEEHAN, Edward R. F., *The Arabs, Israelis and Kissinger: A secret history of American diplomacy in the Middle East*, Readers Digest Press, New York 1976.

SMILEY, David, with Peter Kemp, *Arabian Assignment*, Leo Cooper, London 1975.

SMITH, Colin, *Carlos: Portrait of a Terrorist*, Andre Deutsch, London 1976/Holt, Rinehart and Winston, New York 1977.

STEGNER, Wallace, *Discovery! The Search for Arabian Oil*, as abridged for *Aramco World Magazine*, Middle East Export Press, Beirut 1971.

STORRS, Sir Ronald, *Orientations*, 2nd ed., Love and Malcomson, London and Redhill 1945.

TROELLER, Gary, *The Birth of Saudi Arabia: Britain and the Rise of the House of Saud*, Frank Cass, London and Oregon, U.S.A., 1976.

VAN DER MEULEN, Daniël, *The Wells of Ibn Saud*, John Murray, London 1957/Praeger, New York 1957.

VON HORN, Major-General Carl, *Soldiering for Peace*, Cassell, London 1966/D. McKay, New York 1967.

WAHBA, Hafiz, *Arabian Days*, Arthur Barker, London 1964.

WINSTONE, H.V.F., *Captain Shakespear: A Portrait*, Cape, London 1976.

Notes

M.N.C. = *Multinational Corporations and U.S. Foreign Policy* (U.S. Senate Foreign Relations Committee Subcommittee on Multinational Corporations, Washington 1976). This is published in full and in summary form. Page references given here are to the summary.
M.E.E.S. = *Middle East Economic Survey* (Nicosia, Cyprus).

Chapter 1: Rebirth 1902
1. For varying accounts of the capture of the Musmak see e.g., H.C. Armstrong, *Lord of Arabia* (Penguin, London 1938), H. P. R. Dickson, *Kuwait and Her Neighbours* (Allen and Unwin, London 1956), David Howarth, *Desert King: A Life of Ibn Saud* (Collins, London 1964), H. St. J. B. Philby, *Arabian Jubilee* (Robert Hale, London 1952), and Hafiz Wahba, *Arabian Days* (Arthur Barker, London 1964). I am also indebted for some details in my account to individual members of the Saudi family.

Chapter 2: Out of the Desert
1. Charles M. Doughty's *Travels in Arabia Deserta* was first published in 1888. This quotation is from an abridgement by Edward Garnett (Penguin, London 1956), p.20.
2. For instance, the current ruling families of Kuwait, Abu Dhabi, and Sharjah, the former ruling family of Yemen, and many lesser sheikhly families, as well as the Rashids of Hail and the Saudis. Muhammad bin Rashid killed at least six of his own family before he was established as ruler, and the late King Feisal of Saudi Arabia was murdered by one of his nephews.

Chapter 3: The Consul and the Oilman 1903-8
1. Lord Lansdowne to the House of Commons, 5 May 1903, quoted in B.C. Busch, *Britain and the Persian Gulf, 1894-1914* (Cambridge University Press, London 1968), p.256.
2. India Office Library, London.
3. Quoted in Busch, op. cit., pp.109-10.
4. ibid., p.256.
5. Lord Lansdowne in a letter to Lord Cromer, 7 December 1903, quoted in Lord Newton (Thomas Wodehouse Legh), *Lord Lansdowne* (Macmillan, London 1929), p.287.
6. Quoted in Busch, op. cit., p.225.
7. The Maria Theresa thaler (of which the word 'dollar' is a corruption) continued to be minted in Europe until the 1960s especially for use in Arabia and adjoining parts of East Africa. The date on the coin remained fixed for eternity, however, at 1780 in deference to conservative Arab views about its legitimacy. According to Arabian legend the coin retained its popularity among Arabs not only because of its high silver content but also because of the portrait of the Empress Maria Theresa which adorned its face, showing an imperious lady with a notably large and ill-clad bosom.
8. Burmah's original involvement in the Anglo-Persian Oil Company kept it going for many years after the Burma fields were exhausted, but the company crashed in 1974 and its B.P.

shareholding – converted from the original stake – was taken over by the Bank of England. However, it may have left a lasting mark on the Arabic language in Arabia and the Gulf where the standard metal oil drum is universally known as a 'burmail'. Although some sceptics suggest this is just a local corruption of the English 'barrel', other, more romantic, etymologists believe it is a derivation from Burmah Oil.

Chapter 4: Shakespear Rides Out 1910-15

1. *Through Wahhabiland on Camelback* (Routledge, London 1969), p.149.
2. Quoted in H.V.F. Winstone, *Captain Shakespear: A Portrait* (Cape, London 1976), p.84.
3. IO to FO, 8 June 1911 (L), No.22208; FO 424/227, quoted in Gary Troeller, *The Birth of Saudi Arabia: Britain and the Rise of the House of Saud* (Frank Cass, London and Oregon, U.S.A., 1976), p.40.
4. Quoted in Winstone, op. cit., p.224.
5. op. cit., p.122.
6. Quoted in Troeller, op. cit., p.63.
7. Cable from Viceroy to Foreign Office, 20 December 1913, quoted in John S. Habib, *Ibn Saud's Warriors of Islam: The Ikhwan of Najd and their role in the creation of the Saudi Kingdom 1910-1930* (E.J. Brill, Leiden 1978), pp.12-13.
8. Quoted in Winstone, op.cit., p.159.
9. ibid., p.160.
10. See the account of the battle in Winstone, op. cit., pp.108-10.

Chapter 5: The Arab Revolt 1916

1. *Seven Pillars of Wisdom* (Cape, London 1973), from the Introduction, p.26.
2. One of the main proponents of this view was St John Philby, who eventually took over where Shakespear left off as Ibn Saud's greatest British devotee. Writing of Shakespear's death, for example, he said: '. . . it was a disaster to the Arab cause. It must certainly be reckoned in the small category of individual events which have changed the course of history. Had he survived to continue a work for which he was so eminently fitted it is extremely doubtful whether subsequent campaigns of Lawrence would ever have taken place in the west . . . It is probable that a less tragic outcome of Shakespear's mission would have resulted in [Ibn Saud] being actively supported by Great Britain with money and arms.' *Arabia* (Ernest Benn, London 1930), pp.233-4.
3. For an account of the negotiations see Sir Ronald Storrs, *Orientations*, 2nd ed. (Love and Malcomson, London and Redhill, 1945), pp.152-6. In a note on p.153 Storrs states: 'By October 1916 Lawrence had written: "The Coast towns are glutted with Gold and the Rupee is only 10-12 to the sovereign." Yet the gold dispatched was less than 10 per cent of the total cost to the British taxpayer of the Revolt in the Desert, which amounted to £11,000,000. In addition to the initial sums I took, Husain received from August 8th, 1916, £125,000 a month; in all less than one million sterling. The remaining ten millions represent military operations and supplies from Great Britain.'
4. Letter from Sir Henry McMahon to the Sherif Hussein of 24 October 1915, quoted in George Antonius, *The Arab Awakening: The story of the Arab national movement* (Hamish Hamilton, London 1938), Appendix A, p.419.
5. ibid., pp.266-7.
6. ibid., p.395.
7. Minute to Lord Curzon, 11 August 1919, PRO FO 371, line 4183.

Chapter 6: Sowing the Wind 1917-21

1. *Arabian Days* (Arthur Barker, London 1964), p.125.

2. Quoted in David Howarth, *Desert King: A Life of Ibn Saud* (Collins, London 1964), pp.98-9.
3. Quoted in Elizabeth Monroe, *Philby of Arabia* (Faber, London 1973), p.80.
4. Most writers refer the name to the Koran, The Imrans III 99: 'Remember the favours He has bestowed upon you: how He united your hearts when you were enemies, so that you are now brothers through His grace.' *The Koran*, tr. N. J. Dawood, 4th ed. (Allen Lane, London 1974), p.416.
5. *Hujar* – plural of *Hijrah*, which is the word used for the Prophet's emigration from Mecca to Medina. For the Ikhwan, the word implied that they, like the Prophet, were abandoning the unbelievers for a life of rectitude.
6. H. R. P. Dickson, *Kuwait and Her Neighbours* (Allen and Unwin, London 1956), p.248.
7. Hafiz Wahba, op. cit., p.129.
8. In Arabia, the debate about Lawrence revolves around whether or not he knew about the wilful deceit of his masters. Daniël van der Meulen, who was Dutch representative at Jeddah, remembers discussing him with some Jeddah notables:

   'I replied: "Lawrence believed he spoke the truth. His Government let him down and then he was so ashamed that he went away never to return to this country."

   'Then the answer came: "You underrate the *Ingliz*. He knew and he understood."' *The Wells of Ibn Saud* (John Murray, London 1957), p.82.
9. *Gertrude Bell, Letters*, edited and selected by Lady Bell, volume II (Ernest Benn, London 1927), p.660.

Chapter 7: Reaping the Whirlwind 1922-30

1. PRO, Mss, Vol. 13736, Doc. no. 3457, quoted in John S. Habib, *Ibn Saud's Warriors of Islam: The Ikhwan of Najd and their role in the creation of the Saudi Kingdom 1910-1930* (E. J. Brill, Leiden 1978), Appendix p.182.
2. *Gertrude Bell, Letters*, edited and selected by Lady Bell, volume II (Ernest Benn, London 1927), pp.660-1.
3. H.R.P. Dickson, *Kuwait and Her Neighbours* (Allen and Unwin, London 1956), p.274.
4. *Arabian Days* (Arthur Barker, London 1964), p.20.
5. Sir Reader Bullard, *The Camels Must Go* (Faber and Faber, London 1961), pp.138-9, 140.
6. *Arabian Days: An Autobiography* (Robert Hale, London 1948), p.245.
7. Fuad Hamza, *Al Bilad al Arabiyal al Saudiyah* (Mecca AH 1355/ AD 1937), pp.90-1, quoted in Willard A. Beling (ed.), *King Faisal and the Modernization of Saudi Arabia*, 'The Saudi Monarchy' by George Rentz (Croom Helm, London 1980), p.28.
8. Dickson, op. cit., p.320.

Chapter 8: The Coming of Mammon 1931-4

1. H. St John B. Philby, *Arabian Days: An Autobiography* (Robert Hale, London 1948), p.291.
2. Quoted in Ameen Rihani, *Ibn Saoud of Arabia: His People and His Land* (Constable, London 1928), p.198.
3. *Arabian Jubilee* (Robert Hale, London 1952), p.110.
4. ibid., p.110.
5. Hafiz Wahba, *Arabian Days* (Arthur Barker, London 1964), p.170.
6. See Wahba, op. cit., p.60.
7. Quoted in Mohammed Almana, *Arabia Unified: A Portrait of Ibn Saud* (Hutchinson, London 1980), p.174.
8. To his profiteering even the sober Van der Meulen attests. But it is said that when some bedouin complained to the King that he was 'eating' (i.e. corrupt) the King replied, 'Aye, he eats but, by God, he feeds.'

9. *Arabian Jubilee,* p.230.
10. Philby, *Arabian Days: An Autobiography,* p.291.
11. *The Wells of Ibn Saud* (John Murray, London 1957), p.120.

Chapter 9: The Land of Promise 1932-8

1. Princess Alice writing of Dhahran in 1938, in *For My Grandchildren* (Evans, London 1966), pp.240-1.
2. H. St John B. Philby, *Arabian Days: An Autobiography* (Robert Hale, London 1948), p.281.
3. Quoted in Mohammed Almana, *Arabia Unified: A Portrait of Ibn Saud* (Hutchinson, London 1980), p.129. Another account, given by Sheikh Mohammed bin Abdullah Alireza, who was later Saudi Ambassador in Paris, also leaves Philby quite out of the picture. He says that Crane, during a visit to Cairo in 1930, so impressed the Saudi Envoy, Sheikh Fazan Sabeq, with his judgement of horseflesh that he was given a horse as a present. In return, Crane offered to send a geologist to Saudi Arabia.
4. The Mahd al Dhahab mine near Medina did produce reasonable quantities of gold before its closure in 1951. It was poised to reopen again, with the vastly improved price of gold, at publication time.
5. *Seven Pillars of Wisdom* (Cape, London 1973), pp.72-3.
6. 'Although they might be said to have an acquired right to the task and the necessary political experience it was not the British who were to transform Arabia into this new Land of Promise, but the newcomers whose qualifications were technical rather than historical or political and for whom the lack of past contacts was an advantage rather than a drawback . . . I could not help feeling sad at this turn in the wheel of fortune and of fate. Sad, too, for the Englishmen in Arabia who saw the reins slipping from their worn hands.' Van der Meulen, *The Wells of Ibn Saud* (John Murray, London 1957), p.141.
7. Wallace Stegner, *Discovery! The Search for Arabian Oil,* as abridged for *Aramco World Magazine* (Middle East Export Press, Beirut 1971), p.95.
8. From a letter to Sir L. Oliphant dated 21 March 1938, Foreign Office Records, E1687/189/25.
9. ibid.
10. *For My Grandchildren,* p.234.
11. ibid., pp.236-7.
12. ibid., p.240.

Chapter 10: America Takes Over 1939-45

1. Executive Order 8926, 18 February 1943, President Franklin D. Roosevelt to Under-Secretary of State Edward Stettinus, Foreign Relations of the United States records 1943, vol. IV, p.859.
2. See Aaron David Miller, *Search for Security: Saudi Arabian Oil and American Foreign Policy, 1939-49* (University of North Carolina Press, Chapel Hill 1980), p.70.
3. Quoted in Miller, op. cit., pp.101-2.
4. Quoted in Barry Rubin, *The Great Powers in the Middle East, 1941-1947* (Frank Cass, New Jersey and London 1950), p.52.
5. Anwar Sadat belonged to one such German-financed anti-British faction.
6. *The Koran,* tr. N. J. Dawood, 4th ed. (Allen Lane, London 1974).
7. Memorandum by Hoskins dated Cairo, 31 August 1943, Foreign Relations of the United States records – Hoskins' report – 1943, vol. IV, pp.807-10, quoted in Elizabeth Monroe, *Philby of Arabia* (Faber and Faber, London 1973), p.224.
8. See David Howarth, *Desert King: A Life of Ibn Saud* (Collins, London 1964), p.208.
9. Quoted in William Eddy, *F.D.R. Meets Ibn Saud* (American Friends of the Middle East, New York 1954), p.34.

546 NOTES TO PAGES 137–97

10. See Eddy, op. cit., p.15.
11. Winston S. Churchill, *The Second World War*, vol. VI *Triumph and Tragedy* (Cassell, London 1953), pp.348-9.
12. Eddy, op. cit., p.15

Chapter 11: Mammon Triumphant 1946-53
1. Quoted in William Eddy, *F. D. R. Meets Ibn Saud* (American Friends of the Middle East, New York 1954), p.37.
2. Gerald de Gaury, *Faisal: King of Saudi Arabia* (Arthur Barker, London 1966), p.73.
3. See Willard A. Beling (ed.), *King Faisal and the Modernization of Saudi Arabia*, 'The Saudi-American Relationship and King Faisal' by Malcolm C. Peck (Croom Helm, London 1980), p.234.
4. Quoted in Howard M. Sachar, *Europe Leaves the Middle East* (Allen Lane, London 1974), p.513.
5. M.N.C., p.81.

Chapter 12: The End of an Era 1949-53
1. *The Sunday Times*, 23 October 1953.
2. Dr Hjalmar Schacht was acquitted by the Nazi War Tribunals. His expertise was such that the U.S. occupying authorities engaged him to reform the West German currency in 1949.
3. See articles by Arthur N. Young, 'Saudi Arabian Currency and Finance', in *The Middle East Journal*, Washington 1953, vol. 7, no. 3, p.363, and no. 4, pp.539-52.
4. *The Sunday Times*, 23 October 1953.

Chapter 13: The Years of Ozymandias 1954-8
1. See José Arnold, *Golden Pots and Swords and Pans* (Gollancz, London 1964), pp.143, 222-6.
2. Quoted in George Baroody (ed.), *Saudi Arabia and Arabian-American Oil Company, First Memorial* (privately published by the Arabian-American Oil Company, 1958), p.270.
3. Quoted in Baroody, op. cit., p.265.
4. See Arnold, op. cit., pp.198-9.
5. Quoted in Elizabeth Monroe, *Philby of Arabia* (Faber and Faber, London 1973), p.281.
6. Reported in *The Times*.
7. See Arnold, op. cit., p.206.
8. See Wilbur Crane-Eveland, *Ropes of Sand: America's Failure in the Middle East*, (Norton, New York 1980), pp.208-13.
9. Peter Mansfield, *The Arabs* (Penguin, London 1978), p.304.
10. *The White House Years*, vol. II *Waging Peace* (Heinemann, London 1965), pp.115-16.
11. Reported by Alistair Cook, *Manchester Guardian*, 30 January 1957.
12. 28 January 1957.
13. 31 January 1957.
14. op. cit., pp.101 and 102.
15. Reported in the *New York Times*, 7 February 1957.
16. *New York Times*, 26 June 1957.
17. 16 June 1957.
18. See Patrick Seale, *The Struggle for Syria: A study of post-war Arab politics, 1945-1958*, issued under the auspices of the Royal Institute of International Affairs (Oxford University Press, London 1965), p.291.
19. Quoted in Seale, op. cit., p.304.
20. Osgood Caruthers, *New York Times*, 29 March 1957.
21. op. cit., vol. II, p.263

Chapter 14: A House Divided 1958-62
1. Reported in *The Times*, 30 April 1958.
2. Reported in the *New York Times*, 18 August 1958.
3. Reported in the *New York Times*, 4 June 1958.
4. Reported in the Egyptian newspaper *Al Gomhouriya*, 25 May 1960.
5. Reported in the Beirut newspaper *Al Hayat*, 1 January 1961.
6. Quoted in *Middle East Contemporary Record*, Israel 1961.
7. ibid.
8. ibid.
9. Muhammad Heikal, *Nasser: The Cairo Documents* (New English Library, London 1972), p.182.

Chapter 15: The Fox and the Lion 1962-7
1. Quoted in David Holden, *Farewell to Arabia* (Faber and Faber, London 1966), p.94.
2. Reported in the *Daily Telegraph*, 3 June 1963.
3. Quoted in Dana Adams Schmidt, *Yemen: The Unknown War* (Bodley Head, London 1968), p.186.
4. *The American Approach to the Arab World* (Harper and Row, New York 1968), p.143.
5. See Schmidt, op. cit., p.194.
6. Major-General Carl von Horn, *Soldiering for Peace* (Cassell, London 1966), p.295
7. See Von Horn, op. cit., p.300.
8. ibid., pp.348-9
9. Reported by Richard Beeston, *Daily Telegraph*, 8 July 1963.
10. Reported in the *Egyptian Gazette*, Cairo, 24 November 1964.
11. David Smiley with Peter Kemp, *Arabian Assignment* (Leo Cooper, London 1975), p.192
12. Reported in the *Financial Times*, 22 December 1965.
13. See Anthony Sampson, *The Arms Bazaar* (Hodder and Stoughton, London 1977), p.159.
14. Reported in the *New York Times*, 25 June 1966.
15. Reported in *Le Monde*, 22 March 1967.
16. Reported in *The Times*, 24 April 1967.
17. B.B.C. Monitoring Reports ME/2485/A/7, 1967.

Chapter 16: Uneasy Interlude 1967-70
1. *King Feisal Speaks* (Saudi Arabian Ministry of Information), p.17.
2. For the full text of Feisal's ten-point reform programme see Gerald de Gaury, *Faisal: King of Saudi Arabia* (Arthur Barker, London 1966), pp.147-51. This quote, p.150.
3. ibid., p.149.
4. ibid., p.150.
5. See *Le Monde*, 28 September 1965.
6. Reported in the *New York Times*, 5 June 1966.
7. *The Road to Ramadan* (Collins, London 1975), p.78.

Chapter 17: Friendship Across the Red Sea 1970-2
1. *The White House Years* (Weidenfeld and Nicolson/Michael Joseph, London 1979), p.1285.
2. *Sphinx and Commissar: The Rise and Fall of Soviet Influence in the Arab World* (Collins, London 1978), p.219.
3. op. cit., p.1293.
4. *Arab Report and Record* 1971, 1-15 January, p.40.
5. ibid., 1-15 March 1971, p.147.
6. Anwar el-Sadat, *In Search of Identity: An Autobiography* (Fontana, London 1978), p.259.
7. *Sphinx and Commissar*, p.219.
8. See Kissinger, op. cit., p.1293.

9. *Arab Report and Record* 1971, 16-30 June, p.332.
10. ibid., p.333.
11. ibid., 1-15 December 1971, p.630.
12. op. cit., p.1289.
13. ibid., p.1289
14. ibid., p.1292.
15. ibid., p.1295
16. ibid., p.1295

Chapter 18: The Jewel in Hand 1970-2
1. M.E.E.S., 3 November 1972.
2. Abdul Amir Kubbah, *O.P.E.C. Past and Present* (Vienna, September 1974), p.54, quoted in Anthony Sampson, *The Seven Sisters: The Great Oil Companies and the World They Made* (Hodder and Stoughton, London 1979), p.215.
3. M.N.C., p.134.
4. Reported in the *Financial Times*, 14 February 1971.
5. See M.N.C., p.134.
6. M.N.C., p.47.
7. M.N.C., p.136.
8. Reported in the *Financial Times*, 3 July 1972.

Chapter 19: Oil Power Unleashed 1973
1. M.E.E.S., 18 May 1973, p.6.
2. *In Search of Identity: An Autobiography* (Fontana, London 1978), p.273.
3. ibid., p.278.
4. M.E.E.S., 20 April 1973, pp.1 and 2.
5. M.E.E.S., 18 May 1973, p.6.
6. M.E.E.S., 1 June 1973, p.1.
7. M.N.C., p.142.
8. M.E.E.S., 13 July 1973, pp.1 and 3.
9. *Arab Report and Record* 1973, 1-15 August, p.354.
10. M.E.E.S., 7 September 1980, supplement p.iii.
11. ibid., p.ii.
12. ibid., 31 August 1973, p.1.
13. *Arab Report and Record* 1973, 16-30 September, p.421.
14. ibid., p.421.
15. *The Road to Ramadan* (Collins, London 1975), pp.269-70.
16. ibid., p.271.
17. *Memoirs* (Sidgwick and Jackson, London 1978), p.927.
18. M.E.E.S., 19 October 1973, supplement p.iii.
19. Quoted in Nixon, op. cit., p.930.
20. M.E.E.S., 26 October 1973, p.4.

Chapter 20: Beyond the Dreams of Avarice 1973-4
1. M.E.E.S., 30 November 1973, p.10.
2. ibid., p.10.
3. Quoted in Edward R. F. Sheehan, *The Arabs, Israelis and Kissinger* (Readers Digest Press, New York 1976), p.73.
4. *Washington Post*, 9 February 1974, quoted in Alfred M. Lilienthal, *The Zionist Connection* (Dodd, Mead and Company, New York 1978), p.674.
5. *A World Restored: Castlereagh, Metternich and the Restoration of Peace, 1812-1822* (Harvard University, 1957), quoted in Lilienthal, op. cit., p.667.
6. M.E.E.S., 23 November 1973, p.9.

7. ibid., p.9.
8. ibid., p.11.
9. ibid., 21 December 1973, p.1.
10. ibid., 23 December 1973, p.2.
11. ibid., 4 January 1974, p.11.
12. ibid., 25 January 1974, p.x.
13. ibid., 22 March 1974, p.1.
14. ibid., 26 April 1974, p.20.

Chapter 21: The Arabian Janus 1974-5
1. M.E.E.S., 15 November 1974, supplement p.vii.
2. M.N.C., p.154.
3. M.E.E.S., 14 June 1974, p.ii.
4. *Memoirs* (Sidgwick and Jackson, London 1978), p.1012.
5. *Multinational Corporations and U.S. Foreign Policy* (M.N.C.), published in full and in summary form by the U.S. Foreign Relations Committee Subcommittee on Multinational Corporations, Washington 1976.
6. 17 September 1975.
7. Reported in the *Washington Post*, 16 September 1975.
8. Reported in the *Sunday Times*, 14 March 1976.
9. Reported in the *Washington Post*, 15 September 1975.
10. ibid., 15 September 1975.
11. *International Currency Review*, vol.12, no.1, 1980, p.8.
12. ibid., pp.10 and 8.
13. United Press International dispatch quoted in M.E.E.S., 8 November 1974, p.4.
14. M.E.E.S., 15 November 1974, supplement p.vii.
15. Interview with *Business Week*, 13 January 1975, reported in M.E.E.S. 10 January 1975, 'Kissinger on Oil', p.iii.
16. ibid., p.vi.
17. M.E.E.S., 17 January 1975, supplement p.i.
18. *Fiches du Monde Arabe*, no.900, 8 March 1978.
19. *Time*, 19 November 1973, quoted in Willard A. Beling (ed.), *King Faisal and the Modernization of Saudi Arabia* (Croom Helm, London 1980), p.62.
20. ibid., p.62.
21. M.E.E.S., 4 April 1975, p.i.
22. *Arab Report and Record* 1975, 16-31 March, p.199.
23. *Middle East Intelligence Survey*, 15 April 1975, pp.15-16.

Chapter 22: Mammon Rampant 1975-7
1. Quoted by Paul Balta, *Le Monde*, 27 March 1975.
2. *Financial Times*, 10 June 1974.
3. *khawaja*: a word of Turkish origin meaning 'master', customarily used in the Kingdom to describe a foreigner of fairer complexion.
4. International Monetary Fund f.o.b. statistics based on the returns of Saudi Arabia's trading partners.
5. Interview with Richard Johns published by *Ahlan Wasahlan*, the magazine of the Saudi Arabian Airlines Corporation, no.1, vol.1, 1977.
6. Quoted by Paul Balta, *Le Monde*, 27 March 1975.

Chapter 23: Lords of the Ascendant 1975-7
1. *Al Anwar* (Beirut newspaper), 24 May 1975.

2. In an interview with *U.S. News and World Report*, 19 May 1975.
3. *Washington Post*, 23 May 1975.
4. *Al Anwar*, 24 May 1975.
5. *Al Thawra* (Sana newspaper), 3 July 1975.
6. *Al Anwar*, 4 July 1975.
7. *Al Rai Al Amm* (Kuwaiti newspaper), 2 August 1975.
8. M.E.E.S., 4 April 1975.
9. See Christopher Dobson and Ronald Payne, *The Carlos Complex: A Study in Terror* (Coronet, London 1977) and Colin Smith, *Carlos: Portrait of a Terrorist* (Andre Deutsch, London 1976). A full transcript of Yamani's account of the kidnapping is in Dobson and Payne's book.
10. *Washington Post*, 17 September 1976.
11. M.E.E.S., 27 December 1976, 'O.P.E.C. Round-up', p.iii.
12. ibid.
13. ibid.
14. ibid., 'O.P.E.C. Round-up', p.iv.
15. ibid.
16. ibid.
17. M.E.E.S., 10 January 1977, supplement p.6.
18. Reported in *Le Monde*, 21 January 1977, *Arab Report and Record* 1977, 16-31 January, p.55.
19. M.E.E.S., 27 December 1976, 'O.P.E.C. Round-up', p.iv.
20. M.E.E.S., 31 January 1977, p.1.
21. *Toward Peace in the Middle East*, Washington, April 1977.
22. M.E.E.S., 30 May 1977, p.1.
23. ibid.
24. Reported by Richard Johns in the *Middle East Annual Review*, Saffron Walden, Essex, 1979, p.60.
25. ibid., p.60.

Chapter 24: The Wages of Wealth 1977-9
1. From a paper entitled 'Projecting Oil Supply and Demand: A Science, an Art or Just Politics' presented by the former U.S. Ambassador to Saudi Arabia at the Conference on World Energy Economics No. 4 held in London on 26-28 February 1979 and reproduced in M.E.E.S., 12 March 1979, supplement p.1.
2. In an interview with David Holden.
3. In an interview with David Holden.
4. Edward R. F. Sheehan, *New York Times Magazine*, 14 November 1976, p.121.
5. In an interview with David Holden.
6. In an interview with David Holden.
7. In an interview with David Holden.
8. In an interview with David Holden.
9. *Arab Report and Record* 1977, 1-15 March, p.176.
10. *Washington Post*, 16 May 1977.
11. *Arab Report and Record* 1977, 15-28 February, p.131.
12. Interview with *Newsweek*, 27 June 1977, *Arab Report and Record* 1977, 16-30 June, p.504.
13. *Arab Report and Record* 1977, 1-15 August, p.650.
14. *The Middle East*, London, September 1977, p.16.
15. *The Middle East*, August 1977, p.35.
16. M.E.E.S., 4 July 1977, p.4.
17. *Arab Report and Record* 1977, 16-30 June, p.517.
18. ibid., 1-31 October, p.881.
19. M.E.E.S.

20. In an interview with *Al Nahar* (Beirut newspaper), 27 October 1977, M.E.E.S., 31 October 1977, p.2.
21. *Arab Report and Record* 1977, 1-30 November, p.922.
22. *Events* (London journal), 16 December 1977, pp.11-13.
23. *Arab Report and Record* 1977, 1-31 December, p.1011.
24. Fern Racine Gold and Melvin Conant, *Access to Oil: The U.S. Relationship With Saudi Arabia and Iran*, U.S. Senate Committee on Energy, Washington, December 1977, p.84, quoted in Marwan R. Buheiry, *U.S. Threats of Intervention Against Arab Oil: 1973-1979* (Institute for Palestine Studies), Beirut 1980, p.44.
25. 'Saving Camp David (1), Improve the Framework', *Foreign Policy* (Washington journal), No.41, Winter 1980.
26. *International Herald Tribune,* 20 September 1978.
27. *Newsweek,* 2 October 1978.
28. *Arab Report and Record* 1978, 1-15 October, p.715.
29. *The Middle East,* December 1978, p.34.
30. ibid., p.35.
31. M.E.E.S., 13 November 1978, p.1.
32. *Arab Report and Record* 1978, 1-15 November, p.810.
33. ibid., p.810.
34. *Arab Report and Record,* 31 January 1979, 1-15 January, p.27.
35. ibid., 16-31 January, p.29.
36. *Newsweek,* 5 March 1979, p.22.
37. *Arab Report and Record,* 28 March 1979, p.1.
38. *Fiches du Monde Arabe,* no.1238, 4 April 1979.
39. *Observer,* 24 March 1979.
40. 9 April 1980.

Chapter 25: The Return of the Ikhwan 1979

1. Reported in *Arab News,* 30 November 1979.
2. Juhaiman bin Muhammad Utaibi, Pamphlet 1, *Rules of Allegiance and Obedience: The Misconduct of Rulers.*
3. *Arabian Days* (Arthur Barker, London 1964), p.138.
4. Pamphlet 3, *The Call of the Brethren.*
5. ibid.
6. Reported in the Saudi newspaper *Al Riyadh,* 4 December 1979.
7. Pamphlet 3, *The Call of the Brethren.*
8. *Mekka in the Latter Part of the Nineteenth Century,* tr. J. H. Monahan (E. J. Brill, Leiden, Luzac and Co., London, 1931), pp.195-6.
9. M. Th. Houtsma, A. J. Wensinck, E. Lévi-Provençal, H.A.R. Gibb, and W. Heffening (eds), *The Encyclopaedia of Islam,* vol. III, article on the Al Mahdi (E. J. Brill, Leiden, Luzac and Co., London, 1936), p.115.

Chapter 26: The Ruins of Diriya

1. I am indebted for this description of the Israeli leader to Nicholas von Hoffman, *Spectator,* 20 June 1981.
2. *The Wells of Ibn Saud* (John Murray, London 1957), p.239.

The House Of Saud

ABDUL AZIZ fathered forty-five recorded sons by at least twenty-two different mothers. There were as many daughters from an even wider range of women. The sheer number of his progeny and the reluctance of the House of Saud to discuss such family matters make the compilation of a complete and detailed record of its members no easy task. The following table of Abdul Aziz's male descendants, based on the one given in *Burke's Royal Families of the World,* Volume II, is as comprehensive as possible.

ABDUL AZIZ bin Abdul Rahman bin Feisal al Saud, *b* at Riyadh 24 Nov 1880, entered Riyadh 15 Jan 1902 and became Imam of the Wahhabis and Prince of Nejd, proclaimed King of the Hejaz 8 Jan 1926 and King of Nejd and its Dependencies Feb 1927, unified the dual monarchy and became King of Saudi Arabia 22 Sept 1932, *d* at Taif 9 Nov 1953, having had issue:

1 Turki, *b* at Kuwait 1900 (son of Wadhba bint Hazzam), *d* at Riyadh during the influenza epidemic 1919, leaving issue:

  1. Feisal, *b* (posthumously) at Riyadh 1920, and has issue:

    (1) Turki, *b* 1943 and has issue:

      Ghalib

    (2) Abdullah, *b* 1946

    (3) Abdul Aziz

    (4) Khaled

    (5) Muhammad

2 Saud (son of Wadhba bint Hazzam), succeeded his father (see below)

3 Khaled, *b* 1903, *d* an infant

4 Feisal (son of Tarfa bint Abdullah), succeeded his half-brother King Saud (see below)

5 Fahd, *b* at Riyadh 1905, *d* during the influenza epidemic 1919

6 Muhammad, *b* at Riyadh 1910 (son of Jauharah bint Musaid), and has issue:

  1. Fahd, and has issue:

    (1) Feisal

    (2) Saad

    (3) Abdullah

  2. Bandar, and has issue:

    (1) Khaled

    (2) Abdul Aziz

  3. Badr

  4. Saad

  5. Abdullah

  6. Abdul Aziz

7 Khaled, succeeded his half-brother King Feisal (see below)

8 Saad, *b* at Riyadh 1914, *d* during the influenza epidemic 1919

9 Nasir, *b* 1920 (son of Bazza), and has issue:
1. Saud, and has issue:
   (1) Khaled
   (2) Abdul Aziz
2. Khaled, and has issue:
   (1) Feisal
   (2) Fahd
3. Abdullah, *b* 1941, and has issue:
   (1) Feisal
   (2) Khaled
4. Fahd, and has issue:
   (1) Khaled
   (2) Nauf
5. Muhammad
6. Abdul Rahman, and has issue:
   (1) Feisal
   (2) Jawahir
7. Turki
8. Ahmed
9. Mansour, *dec*
10. Feisal
11. Mansour
12. Sultan
13. Abdul Aziz
14. Thamir
15. Nawwaf

10 Saad, *b* at Riyadh 1920 (son of Jauharah bint Saad al Sudairi), and has issue:
1. Fahd
2. Saud
3. Muhammad
4. Feisal
5. Fahd (2), and has issue:
   (1) Abdul Aziz
6. Khaled

11 Fahd, Crown Prince of Saudi Arabia, *b* at Riyadh 1921 (son of Hassa bint Ahmed al Sudairi), and has issue (order uncertain):
Feisal

Khaled
Saud
Saad
Sultan
Muhammad
Turki

12 Mansour, *b* at Riyadh 1922 (son of Shahida), *d* at Paris May 1951, leaving issue:
1. Talal (Mansour), *b* posthumously 1951

13 Abdullah, *b* at Riyadh 1923 (son of Bint Asi al Shuraim of the Shammar), and has issue:
1. Mutaib, *dec*
2. Mutaib
3. Khaled
4. Abdul Aziz
5. Feisal

14 Bandar, *b* at Riyadh 1923 (son of Bazza), and has issue:
1. Feisal, *b* 1943
2. Muhammad
3. Turki
4. Feisal (2)
5. Abdul Aziz
6. Abdullah
7. Mansour
8. Khaled

15 Musaid, *b* at Riyadh 1923 (son of Jauharah bint Saad al Sudairi), and has issue:
1. Feisal, shot and killed his uncle King Feisal, *b* 1948 (son of a Rashidi mother), *d* (executed) at Riyadh June 1975
2. Yusuf
3. Mashhur
4. Saud, and has issue:
   (1) Feisal
5. Bandar
6. Khaled, *d* 1965 (shot while taking part in a demonstration opposing the introduction of television in Saudi Arabia), leaving issue:
   (1) Abdul Rahman

(2) Abdul Aziz

16 Sultan, *b* at Riyadh 1924 (son of Hassa bint Ahmed al Sudairi), and has issue:
  1. Khaled, *b* 1949
  2. Feisal, *b* 1950
  3. Bandar
  4. Muhammad
  5. Turki
  6. Fahd
  7. Muhammad

17 Abdul Mohsen, *b* at Riyadh 1925 (son of Jauharah bint Saad al Sudairi), and has issue:
  1. Bandar
  2. Badr
  3. Fawwaz, and has issue:
    (1) Walid
    (2) Abir
  4. Saud

18 Mishaal, *b* at Riyadh 1926 (son of Shahida), and has issue:
  1. Feisal, *b* 1947
  2. Mishairi, *b* 1948
  3. Muhammad, *b* 1949
  4. Mansour
  5. Turki
  6. Abdul Aziz
  7. Khaled
  8. Bandar

19 Abdul Rahman, *b* at Riyadh 1926/7 (son of Hassa bint Ahmed al Sudairi), and has issue:
  1. Khaled
  2. Fahd
  3. Turki

20 Mitab, *b* at Riyadh 1928 (son of Shahida), and has issue:
  1. Mansour

21 Talal, *b* 1930 (son of Munaiyer), *d* at Taif 1931

22 Badr, *b* 1931 (son of Haya bint Saad al Sudairi), *d* 1931/2

23 Talal, *b* at Riyadh 1931 (son of Munaiyer), and has issue:
  1. Feisal, *b* 1949
  2. Turki
  3. Walid

  4. Khaled

24 Mishairi, *b* at Riyadh 1932 (son of Bushra), killed the British Vice-Consul Cyril Ousman, Nov 1951, and has issue:
  1. Muhammad
  2. Mansour
  3. Feisal
  4. Talal

25 Badr, *b* at Riyadh 1933 (son of Haya bint Saad al Sudairi), and has issue:
  1. Khaled
  2. Fahd, *dec*
  3. Talal
  4. Fahd

26 Naif, *b* at Riyadh 1933 (son of Hassa bint Ahmed al Sudairi), and has issue:
  1. Saud
  2. Muhammad

27 Nawwaf, *b* at Riyadh 1934 (son of Munaiyer), and has issue:
  1. Muhammad, *b* 1952
  2. Nawwaf

28 Turki, *b* at Riyadh 1934 (son of Hassa bint Ahmed al Sudairi), and has issue:
  1. Feisal
  2. Sultan
  3. Fahd
  4. Khaled

29 Fawwaz, *b* at Riyadh 1934 (son of Bazza)

30 Majid, *b* at Riyadh 1934 (son of Mudhi), *d* 1940

31 Abdul Illah, *b* at Riyadh 1935 (son of Haya bint Saad al Sudairi), and has issue:
  1. Abdul Aziz

32 Salman, *b* at Riyadh 1936 (son of Hassa bint Ahmed al Sudairi), and has issue:
  1. Fahd
  2. Sultan
  3. Ahmed
  4. Abdul Aziz

33 Majid, *b* at Riyadh 1937 (full

brother of Sattam), and has issue:
1. Abdul Aziz
2. Naif

34 Thamir, *b* at Riyadh 1937 (son of Bint al Shalan of the Ruwala), wished for a sex change operation, *d* at Miami, Florida, U.S.A., June 1959 (committed suicide by drenching his clothes with petrol and setting himself alight), leaving issue:
1. Feisal

35 Ahmed, *b* at Riyadh 1937 (7th and youngest son of Hassa bint Ahmed al Sudairi)

36 Abdul Majid, *b* at Riyadh 1940 (son of Haya bint Saad al Sudairi) and has issue:
1. Feisal

37 Mamduh, *b* at Riyadh 1940 (son of Bint al Shalan of the Ruwala), and has issue:
1. Muhammad
2. Nawwaf
3. Miqrin

38 Abdul Salam, *b* at Riyadh 1941, *dec*

39 Hidhlul, *b* at Riyadh 1941 (son of Saida al Yamaniya)

40 Mashhur, *b* at Riyadh 1942 (son of Bint al Shalan of the Ruwala)

41 Sattam, *b* at Riyadh 1943 (full brother of Majid)

42 Miqrin, *b* at Riyadh 1943 (son of Baraka al Yamaniya), and has issue:
1. Fahd
2. Abdul Aziz
3. Feisal

43 a son, *b* at Riyadh 1942 (son of Khadra al Yamaniya), *d* 1944

44 Hamud, *b* at Riyadh 1947 (son of Futaima al Yamaniya)

45 a son, *b* at Riyadh June 1952, *dec*

SAUD, King of Saudi Arabia and Imam and Protector of the Wahhabis 1953–64, *b* at Kuwait 12 Jan 1902 (son of Wadhba bint Hazzam), apptd Viceroy of Nejd 1926, proclaimed Crown Prince at Mecca 1933, apptd C-in-C of the Army Aug 1953, succeeded his father 9 Nov 1953, deposed in favour of his half-brother Crown Prince Feisal 2 Nov 1964, *d* at Athens 23 Feb 1969, had many wives, leaving issue:

1 Sultan, *d* an infant
2 Fahd, *b* 1923
3 Musaid, and has issue:
1. Khaled
2. Nawwaf
3. Fahd
4. Walid
5. Turki
6. Abdul Aziz
7. Feisal

4 Muhammad, and has issue:
1. Khaled
2. Sultan
3. Feisal
4. Mishaal

5 Abdullah, and has issue:
1. Mitab, *b* 1948
2. Nawwaf
3. Abdul Aziz
4. Khaled
5. Muhammad
6. Nahar
7. Naif
8. Feisal

6 Feisal, and has issue:
1. Mansour
2. Nawwaf
3. Naif
4. Abdul Rahman

7 Khaled, *b* 1935
8 Abdul Mohsen
9 Abdul Rahman, *b* 1946, and has issue:
1. Khaled
2. Feisal
3. Abdul Aziz

4. Mamdur
10 Mansour, *b* 1947
11 Abdul Illah, *b* 1949
12 Saad, *d* 1967/8, leaving issue:
  1. Fahd
  2. Khaled
  3. Nawwaf
  4. Mansour
13 Badr, and has issue:
  1. Salman
  2. Mishaal
  3. Abdul Aziz
  4. Turki
  5. Mansour
  6. Khaled
  7. Abdul Majid
  8. Talal
  9. Hosan
  10. Walid
14 Bandar, and has issue:
  1. Feisal
15 Majid, and has issue:
  1. Mansour
  2. Naif
16 Thamir, *d* 1967/8, leaving issue:
  1. Abdul Mohsen
  2. Feisal
  3. Khaled
  4. Turki
  5. Bandar
17 Sultan
18 Abdul Majid
19 Talal
20 Naif
21 Miqrin
22 Ahmed
23 Turki, *b* 1952/3, and has issue:
  1. Saud, *b* 1973/4
  2. Sara, *b* 1972/3
24 Mashhur, *b* 1953
25 Mishairi
26 Saif ad Daula
27 Hosan ad Din
28 Mishaal, and has issue:
  1. Feisal
  2. Khaled
  3. Abdul Aziz

  4. Nawwaf
29 Fawwaz, and has issue:
  1. Salman
30 Saif al Islam
31 Hasan
32 Mohsab
33 Saif an Nasr
34 Jalawi
35 Nasir
36 Ghalib
37 Yusuf
38 Hamud
39 Abdul Karim
40 Yazid
41 Sattam
42 Muhtasim
43 Saif ad Din
44 Muntasir
45 Nawwaf

FEISAL, King of Saudi Arabia and Imam and Protector of the Wahhabis 1964 – 75, *b* at Riyadh 1904 (or 1905?)(son of Tarfa bint Abdullah), apptd Viceroy of the Hejaz 1926 and Minister for Foreign Affairs 1930, proclaimed Crown Prince Nov 1953, succeeded his half-brother King Saud 2 Nov 1964, shot and killed in the Royal Palace at Riyadh by his nephew Feisal bin Musaid 25 March 1975, leaving issue by his three wives:

1 Abdullah, *b* at Riyadh 1921 (son of Sultana bint Ahmed bin Muhammad al Sudairi), and has issue:
  1. Khaled, and has issue:
    (1) Muhammad
  2. Muhammad, and has issue:
    (1) Turki
  3. Abdul Rahman
  4. Saud
  5. Talal, *b* 1948, killed in a shooting accident 25 Sept 1960
  6. Bandar
  7. Turki
  8. Sultan
2 Muhammad, *b* at Riyadh 1937

(son of Iffat bint Ahmed bin
Abdullah al Saud), and has issue:
   1. Amr
3 Khaled, *b* at Riyadh (son of
Haya bint Muhammad bin Abdul
Rahman), and has issue:
   1. Sultan
   2. Bandar
4 Saud, *b* at Riyadh (son of Iffat),
and has issue:
   1. Muhammad
   2. Khaled
5 Abdul Rahman, *b* at Riyadh (son
of Iffat), and has issue:
   1. Saud
   2. Sara

6 Saad, *b* at Riyadh (son of Haya)
7 Bandar, *b* at Riyadh (son of Iffat)
8 Turki, *b* at Riyadh (son of Iffat)

KHALED, King of Saudi Arabia and
Imam and Protector of the Wahhabis
1975 – , *b* at Riyadh 1912 (or 1913?),
proclaimed Crown Prince March 1965,
succeeded his half-brother King Feisal
25 March 1975, and has issue by Sitta:
1 Bandar, and has issue:
   1. Feisal
2 Abdullah, and has issue:
   1. Fahd
3 Feisal
4 Fahd

Index

Members of royal families are listed under their first name and non-royals under their patronymic name.